Modeling and Analysis of Bio-molecular Networks

Jinhu Lü • Pei Wang

Modeling and Analysis of Bio-molecular Networks

Springer

Jinhu Lü
School of Automation Science
and Electrical Engineering
State Key Laboratory of Software
Development Environment
Beijing Advanced Innovation Center for Big
Data and Brain Computing
Beihang University
Beijing, China

Pei Wang
School of Mathematics and Statistics
Henan University
Kaifeng
Henan, China

ISBN 978-981-15-9143-3 ISBN 978-981-15-9144-0 (eBook)
https://doi.org/10.1007/978-981-15-9144-0

This Springer imprint is published by the registered company Springer Nature Singapore Pte Ltd.
The registered company address is: 152 Beach Road, #21-01/04 Gateway East, Singapore 189721, Singapore

We dedicate this book to our families for their patience, love and support over the years.

Preface

Systems biology is an interdisciplinary scientific area, which aims at system-level understanding of biological systems. It applies mathematical modeling, as well as computational and statistical analysis techniques to explore complex biological systems. Biological systems can be described by different levels of bio-molecular networks, which include gene regulatory networks, protein–protein interaction networks, metabolic networks, signal transduction networks, gene co-expression networks, etc. Different networks contain different biological entities and interactive relationships. The investigations on bio-molecular networks can help us to understand the origin, development, and evolution of life phenomena at the systems level. They can also provide insights as well as real-world applications to medicine and life sciences.

This book aims to introduce the mathematical modeling, as well as dynamical and statistical analysis on several typical bio-molecular networks. We will discuss the topics on the reconstruction of bio-molecular networks, mathematical modeling methods for gene regulatory networks, dynamical analysis of several simple genetic circuits (also well known as network motifs) as well as statistical analysis on some large-scale bio-molecular networks. Finally, we also introduce the readers a focus research field that is full of opportunities and challenges—data-driven statistical approaches for omics data analysis. The detailed organization of the book is as follows.

In Chap. 1, we introduce the related backgrounds of this book, where we review the basic concepts and discuss the brief history of systems biology and the complex networks science. Moreover, we introduce some traditional metrics for complex networks and statistical analysis methods for practical problems. Finally, we introduce some software tools to perform dynamical and statistical analysis, as well as to visualize complex data.

In Chap. 2, we introduce some methods for reconstructing bio-molecular networks. Four approaches for bio-molecular networks reconstruction were discussed, namely: (1) construction of bio-molecular networks based on online databases; (2) generating bio-molecular networks through artificial algorithms that mimic the general features of real-world networks, such as those based on the duplication and divergence mechanisms; (3) inferring bio-molecular networks via various mathematical and statistical models based on behavioral data of bio-molecules; and (4) topology identification based on dynamical system theory. We summarize some online databases, various artificial algorithms, mathematical and statistical methods to reconstruct each type of bio-molecular networks. The reconstruction of bio-molecular networks facilitates the further mathematical modeling, dynamical and statistical analysis on them.

Chapter 3 deals with the mathematical modeling and dynamical analysis of several simple genetic circuits. The whole bio-molecular networks for model organisms are too complex, which have hindered our understanding of them. However, it was found that real-world bio-molecular networks consist of network motifs. Network motifs are those simple circuits that appeared far frequently than that in its randomized counterparts. The investigations on network motifs are the first step to systems-level exploration. Through mathematical modeling and dynamical analysis, we classify the structural advantages of several simple circuits. In Chaps. 4 and 5, we overview and discuss the modeling and dynamical analysis of several coupled and large-sized gene regulatory networks, which gradually promote our explorations of bio-molecular networks to systems level.

Based on statistical analysis and the duplication-divergence model, we clarify several evolutionary mechanisms of network motifs in undirected protein interaction networks in Chap. 6. In Chaps. 7 and 8, for large-scale bio-molecular networks, we introduce some methods on the identification of important nodes (also known as gene prioritization) in biological networks and clarify the statistical features of some functional genes in large-scale human protein–protein interaction networks. Chapter 9 deals with the topic of data-driven approaches for omics data analysis, and some classical methods in this area have been introduced. Especially, we summarized some typical statistical models with different penalization terms (for different purposes during data analysis).

The topics discussed in this book cover four aspects: bio-molecular network reconstruction; mathematical modeling and dynamical analysis of bio-molecular networks; statistical analysis of bio-molecular networks; data-driven mathematical modeling and statistical analysis of biological systems. Some of the methods introduced in this book are adapted from our published papers. This book has the following merits:

1. Introduces various approaches to reconstruct bio-molecular networks.
2. Covers the analysis of various networks ranging from simple circuits, middle-sized networks, to large-scale bio-molecular networks.
3. Introduces various dynamical and statistical analysis tools for gene regulatory networks and protein interaction networks.

The expected audience of this book are undergraduates, graduates, and researchers who are interested in systems biology, dynamical systems, and complex networks. Prerequisites needed to understand this book include complex network theory, multivariate statistical analysis theory, linear algebra, matrix theory, and theories of dynamical systems.

Due to the festinate time and limited knowledge of the authors, some mistakes and errors in the book are unavoidable. Thus, we welcome to all readers to bring to our attention any errors they may find in the book and also give suggestions for adding new material to improve this book. We hope that this book can help our audience during their exploration of the truth of nature.

Beijing, China
Kaifeng, China
May 2020

Jinhu Lü
Pei Wang

Acknowledgments

We acknowledge many people who have contributed in various ways to the completion of this book. First of all, I would like to thank many academic predecessors for their long-term academic guidance, support and help, including the following academicians: Prof. Lei Guo (CAS), Prof. Bohu Li (BAUU), Prof. Weimin Bao (BAUU), Prof. Jinpeng Huai (BAUU), Prof. Jiancheng Fang (BAUU), Prof. Zhiming Zheng (BAUU), Prof. Wei Li (BAUU), Prof. Jun Zhang (BIT), Prof. Jifeng He (ECNU), Prof. Jianhua Lu (TsinghuaU), Prof. Hao Ying (CAMS), Prof. Wei Wang (CASC), Prof. Jie Jiang (CASC), Prof. Zhongben Xu (XJTU), Prof. Xiaohong Guan (XJTU), Prof. Tieniu Tan (CAS) and Prof. Guangren Duan (HIT). Sorry, I can only mention a few of them due to the space limitation.

We also want to express our appreciation to Professor Jun-an Lu at the School of Mathematics and Statistics, Wuhan University, Professor Xinghuo Yu at the School of Electrical and Computer Engineering, RMIT University, and Professor Tianshou Zhou at the Sun Yat-sen University for discussion of some contents in this book. We thank Shuang Xu, Mingqiu Li, Peipei Huang, Can Wei, Qiong Xu, Lingling Wang, Chunfang Liu, Shunjie Chen, and Lixin Li for their help during the preparation of the book. Dr. Shuang Xu has contributed to some contents in Chaps. 2 and 7. Dr. Haibo Gu, Qiong Xu, and Lingling Wang have contributed to help us to apply for the rights and permissions to reuse some figures and contents of some published papers. We are also thankful for the support and encouragement from our colleagues and families.

This work was supported in part by the National Key Research and Development Program of China under Grants 2016YFB0800401, in part by the National Natural Science Foundation of China under 61532020 and 61773153. It was also sponsored by Natural Science Foundation of Henan under grant 202300410045, the Program for Science and Technology Innovation Talents in Universities of Henan Province under grant 20HASTIT025, and the Training Plan of Young Key Teachers in Colleges and Universities of Henan Province under grant 2018GGJS021.

Contents

Acronyms

ADMM	Alternating direction method of multipliers
AHB	Andronov–Hopf bifurcation
AI	Autoinducer
AIC	Akaike information criterion
ANOVA	Analysis of variance
APFL	Additional positive feedback loop
APL	Average path length
ASE	Average synchronization error
AUC	Area under curve
BIC	Bayesian information criterion
CFFL	Coherent feed-forward loop
CGS	Conserved gene network
CLE	Chemical Langevin equation
CME	Chemical master equation
CPNFC	Coupled positive and negative feedback circuit
CPNFGC	Coupled positive and negative feedback genetic circuit
CV	Coefficient of variance
DAG	Directed acyclic graph
DD	Duplication-divergence
DDC	Degree–degree correlation
DDE	Delay differential equation
DEG	Differentially expressed gene
DGS	Disease gene network
DNA	Deoxyribonucleic acid
EGS	Essential gene network
ER	Erdös–Rényi
FFL	Feed-forward loop
FPC	First principal component
GCN	Gene co-expression network
GO	Gene ontology
GPS	Global parameter sensitivities

GRIOS	Global relative input–output sensitivities
GRN	Gene regulatory network
GRPS	Global relative parameter sensitivities
GWAS	Genome-wide association study
HB	Hopf bifurcation
HCB	Homoclinic bifurcation
HK	Housekeeping
HKGS	Housekeeping gene network
HMRFM	Hidden Markov random field model
HPIN	Human protein interaction network
HTS	High-throughput sequencing
IFFL	Incoherent feed-forward loop
LASSO	Least absolute shrinkage and selection operator
LAR	Least angle regression
LCC	Largest connected component
MC	Monte-Carlo
MLE	Maximum likelihood estimation
MM	Michaelis–Menten
MV	Mean variance
ODE	Ordinary differential equation
OLS	Ordinary least square
PCA	Principal component analysis
PCC	Pearson correlation coefficient
PLE	Power-law exponent
PPI	Protein–protein interaction
QS	Quorum sensing
RNA	Ribonucleic acid
RNA-seq	RNA sequencing
ROC	Receiver operating characteristic
RRE	Reaction rate equation
SCC	Spearman correlation coefficient
SDDE	Stochastic delay differential equation
SDE	Stochastic differential equation
SDN	Structurally dominant node
SDP	Structurally dominant protein
SF	Scale-free
SN	Saddle node
SNIC	Saddle node on invariant circle
SNP	Single-nucleotide polymorphism
SSA	Stochastic simulation algorithm
SSS	Stable steady state
Std	Standard deviation
SW	Small-world
SWING	Sliding window inference for network generation
TE	Tissue-enriched

TEGS	Tissue-enriched gene subnetwork
TF	Transcription factor
TRN	Transcriptional regulatory network
TS	Tissue-specific
VGS	Viable gene subnetwork
WGCNA	Weighted gene co-expression network analysis
Y2H	*yeast* two-hybrid

Chapter 1
Introduction and Preliminaries

Abstract The main focus of systems biology is to understand in quantitative, predictable ways the regulation of complex cellular pathways and of intercellular communication, so as to shed light on complex biological functions, such as metabolism, cell signaling, cell cycle, apoptosis, differentiation, and transformation. In this chapter, we firstly introduce the background of systems biology, and then, we introduce some preliminaries of the book, including some statistical analysis methods, complex networks theory as well as some software tools.

1.1 Systems Biology

1.1.1 Overviews

Systems biology is the computational and mathematical modeling of complex biological systems, which applies computational, mathematical, statistical, and engineering approach to biomedical and biological scientific research. Particularly from the year 2000, the concept has been used widely in the biosciences in a variety of contexts. For example, the Human Genome Project is an example of applied systems thinking in biology, which has led to new, collaborative ways of working on problems in the biological field of genetics [1–3]. One of the outreaching aims of systems biology is to model and discover emergent properties, properties of cells, tissues and organisms functioning as a system whose theoretical description is only possible using techniques which fall under the remit of systems biology [4]. The definition of systems biology has been given by different people or organizations. One definition is described as: Systems biology is a biology-based interdisciplinary field that focuses on complex interactions within biological systems, using a holistic approach to biological and biomedical research, instead of the traditional reductionism [5]. Another alternative definition is described as: Systems biology is the science that discovers the principles underlying the emergence of the functional properties of living organisms from interactions among macromolecules (DNA, mRNA, proteins, etc.) [2].

J. Lü, P. Wang, *Modeling and Analysis of Bio-molecular Networks*,
https://doi.org/10.1007/978-981-15-9144-0_1

Systems biology can be understood from different ways. Firstly, systems biology is the study of the interactions between the components of certain biological system, and how these interactions give rise to the functions and behaviors of that system. For example, the interactions among the enzymes and metabolites in a metabolic pathway, and how the interactions give rise to the functions and behaviors of the metabolic pathway [2, 5].

Secondly, as a paradigm, systems biology is usually defined in antithesis to the so-called reductionist paradigm, although fully consistent with the scientific method. The distinction between the two paradigms is as follows: The reductionist approach has successfully identified most of the components and many of the interactions, but unfortunately, offers no convincing concepts or methods to understand how system properties emerge. The pluralism of causes and effects in biological networks is better addressed by observing multiple components simultaneously and by rigorous data integration with mathematical models [6]. Systems biology is about putting together rather than taking apart, integration rather than reduction. It requires to develop different ways of thinking about integration that are as rigorous as reductionist programmers [7].

Thirdly, systems biology uses a series of operational protocols for performing research, namely a framework composed of theory, analytical, or computational modeling to propose specific testable hypotheses about a biological system, experimental validation, and then using the newly acquired quantitative description of cells or cell processes to refine the computational model or theory [8]. Since the first objective is to model the interactions in a system, the experimental techniques that most suit systems biology are those that are system-wide and attempt to be as complete as possible. Therefore, high-throughput omics techniques have been used to collect quantitative data for the construction and validation of models [9].

Fourthly, systems biology applies the dynamical systems theory to investigate problems in molecular biology. Indeed, the focus on the dynamics of the studied systems is the main conceptual difference between systems biology and bioinformatics. Bioinformatics is also an interdisciplinary field but endeavors to develop methods and software tools for understanding biological data. Bioinformatics combines computer science, statistics, mathematics, and engineering to analyze and interpret biological data. Whereas, dynamical modeling and analysis of biological systems are important directions in systems biology research.

Finally, systems biology is a socio-scientific phenomenon defined by the strategy of pursuing integration of complex data about the interactions in biological systems from diverse experimental sources using interdisciplinary tools and personnel [10]. This variety of viewpoints is illustrative of the fact that systems biology refers to a cluster of peripherally overlapping concepts rather than a single well-delineated field.

1.1.2 Developments

Systems biology finds its roots in the quantitative modeling of enzyme kinetics, a discipline that flourished between 1900 and 1970. Initial investigations or foundations involve the mathematical modeling of population dynamics, the simulations developed to study neurophysiology, control theory, and cybernetics [5].

One of the theorists who can be seen as one of the precursors of systems biology is Ludwig von Bertalanffy with his general systems theory [11]. One of the first numerical simulations in cell biology was published in 1952 by the British neurophysiologists and Nobel Prize winners Alan Lloyd Hodgkin and Andrew Fielding Huxley, who constructed a mathematical model that explained the action potential propagating along the axon of a neuronal cell [12]. Their model described a cellular function emerging from the interaction between two different molecular components, a potassium and a sodium channel, and can therefore be seen as the beginning of computational systems biology [13]. Also in 1952, Alan Turing published the book named "The Chemical Basis of Morphogenesis," describing how non-uniformity could arise in an initially homogeneous biological system [14]. In 1960, Denis Noble developed the first computer model of the heart pacemaker [15].

The formal study of systems biology, as a distinct discipline, was launched by systems theorist Mihajlo Mesarovic in 1966 in an international symposium at the Case Institute of Technology in Cleveland, Ohio, entitled "Systems Theory and Biology" [16]. The 1960s and 1970s saw the development of several approaches to study complex molecular systems, such as the metabolic control analysis and the biochemical systems theory. The successes of molecular biology throughout the 1980s, coupled with a skepticism toward theoretical biology, that then promised more than it achieved, caused the quantitative modeling of biological processes to become a somewhat minor field.

However, the birth of functional genomics in the 1990s meant that large quantities of high-quality data became available, while the computing power exploded, making more realistic models possible. From 1992 to 1994, serial articles [17–21] on systems medicine, systems genetics, and systems biological engineering by B.J. Zeng were published in China, and Zeng gave a lecture on biosystems theory and systems approach research at the "First International Conference on Transgenic Animals", Beijing, 1996. In the year 1997, the group of Masaru Tomita published the first quantitative model of the metabolism of a whole (hypothetical) cell [22]. In the year 2000, Leroy Hood co-founded the Institute for Systems Biology (ISB) in Seattle, Washington, an independent, nonprofit organization that develops strategies and technologies for systems approaches to biology and medicine. Since the year 2000, systems biology emerged as a movement in its own right, spurred on by the completion of various genome projects, the large increase in data from the omics (e.g., transcriptomics, genomics, proteomics, methylomics, and phenomics) and the accompanying advances in high-throughput experiments (e.g., *yeast*-two-hybrid (Y2H) and mass spectrometry) and bioinformatics.

In the year 2002, the National Science Foundation of the USA put forward a grand challenge for systems biology in the twenty-first century to build a mathematical model of the whole cell [23]. And around the year 2002, the world's leading scientific journals—Science and Nature, published special issues on systems biology, successively. In 2003, the Massachusetts Institute of Technology began to launch CytoSolve, a method to model the whole cell by dynamically integrating multiple molecular pathway models. Since then, various research institutes dedicated to systems biology have been founded. For example, the National Institute of General Medical Sciences of the National Institutes of Health in the USA established a project grant that is currently supporting over ten systems biology centers. As of the summer of 2006, due to a shortage of researchers in systems biology, several doctoral training programs in systems biology have been established in many regions of the world. In the year 2011, The Cell journal published another special issue on systems biology [24–26], which greatly promoted the development of systems biology.

Several achievements of systems biology have been made during the last decades. For example, a new mechanism for cell size control during the cell cycle has been formulated, which is experimentally testable [2, 27, 28]. The investigations have also been able to get the experiments in gear with the theory and modeling, and to motivate the measurement of the dynamics of cell cycling quantitatively. While some researchers have discovered some new drug targets, understood why trypanosomes have glycosomes and why phosphatases are sometimes more important than kinases for signal transduction, figured out why phosphofructokinase is not the rate-limiting step, and found a way to determine experimentally rather than assume intuitively how much of regulation was actually transcriptional [2].

1.1.3 Implications and Applications

Systems biology has important meaning and applications in our life [2]. The most significant potential application of systems biology is to improve human health. Quite a few diseases are due to dysregulation of cellular systems, since systems biology aims at understanding cell regulation by experimentation and computer simulation, it could contribute to a better understanding of such disease processes. Dysregulation of cellular systems implies the involvement of a multitude of molecular factors, encoded by a multitude of gene functions. Accordingly, the mentioned diseases are observed as multifactorial diseases. Their impact on a patient is affected by many polymorphisms. Because other diseases have been decimated through successful therapies, most diseases that require treatment are multifactorial, involving pleiotropic dysregulation. Examples include cancer, neurological, and cardiovascular diseases. The same is true for the aging process. In order to improve our understanding of such complex diseases, it is essential to develop new strategies, based on the understanding of how the functional system is controlled simultaneously by many factors, i.e., via systems biology approaches.

Systems biology also has the potential applications to become part of the drug development process. Network-based drug design makes use of systems properties, for instance, by trying to identify the "weak points" in an otherwise robust parasite or tumor cell. Whole organism models could then be used to simulate the effects, and side-effects, of drug treatments. In the foreseeable future, this concept is likely to become a requirement of the Food and Drug Administration (FDA) for any new drug to be approved. Because they may well substitute for animal testing and for part of the clinical testing, computer simulations could reduce that testing period as well as eliminate false drug at a much earlier stage. In this manner, systems biology could do something in the drug development process. Systems biology should also open the possibility for development of drugs and drug combinations that are directed to specific genetic and physiological backgrounds. The concept of personalized medicine should become more realistic if the parameters of a generic model of the human body (or parts of it) are adapted to an individual and their values are obtained from personalized functional genomics. A better understanding of diseases based on systems biology could lead to alternative treatment strategies, for instance, based on special diets or on tailor-made nutraceuticals. Part of systems biology should be developed towards these venues, as its applications are likely to increase public health quite significantly and to reduce public health cost.

Furthermore, systems biology could play a major role in preventing diseases and developing an improved, healthier life style. While it is not difficult to the common human sense to foresee the consequences of massive calorie or alcohol intake, the consequences of many other human activities are more difficult to assess in advance. In principle, it may be possible in the future that everyone who so wishes runs its own personalized simulation to assess the consequences or intake of certain foods, physical exercise, traveling, sleeping, etc. Systems biology will certainly be developed and applied in various areas of biotechnology, such as the engineering of industrially important microorganisms as well as breeding of productive crops and animals. Today, 25% of all medicines are plant derived [29], and the spectrum of medicinal feedstocks and efficiency of production can still be greatly enhanced. Systems biology also holds great potential to foster sustainable development, by accelerating and rationalizing the production of plant derived biofuels and chemical feedstocks in preparation for the inevitable depletion of fossil carbon. Probably closest to actual application is the use of simulations in the design of physiologically adapting microorganisms for biotechnological processes, such as *yeasts* for bio-ethanol production.

1.2 Complex Networks

1.2.1 Overviews

A complex network is a graph (network) with non-trivial topological features—features that do not occur in simple networks such as lattices or random graphs but often occur in real systems [30]. The study of complex networks is a young and active area of scientific research inspired largely by the empirical study of real-world networks such as computer networks, biological networks, technological networks, and social networks.

Most social, biological, and technological networks display substantial non-trivial topological features, with patterns of connection between their elements that are neither purely regular nor purely random. Such features include a heavy tail in the degree distribution, a high clustering coefficient, assortativity or disassortativity (which will be defined in the following sections) among vertices, community structure, and hierarchical structure. In the case of directed networks, these features also include reciprocity, triad significance profile, and other features. In contrast, many of the mathematical models of networks that have been studied in the past, such as lattices and random graphs, do not show these features. The most complex structures can be realized by networks with a medium number of interactions. This corresponds to the fact that the maximum information content (entropy) is obtained for medium probabilities.

Two well-known and much studied classes of complex networks are scale-free (SF) networks [31] and small-world (SW) networks [32, 33], whose discovery and definition are canonical case-studies in the field. Both are characterized by specific structural features—power-law degree distributions for the former, short path lengths and high clustering for the latter (which will be discussed in the following sections). However, as the study of complex networks has continued to grow in importance and popularity, many other aspects of network structure have attracted attention as well. Recently, the study of complex networks has been expanded to networks of networks [34]. If those networks are interdependent, they become significantly more vulnerable to random failures and targeted attacks, and exhibit cascading failures and first-order percolation transitions [35].

The field of complex networks continues to develop quickly, and has brought together researchers from many areas, including mathematics, physics, biology, telecommunications, computer science, sociology, epidemiology, and others [36]. Ideas from network science and engineering have been applied to the analysis of metabolic and gene regulatory networks (GRNs); the modeling and design of scalable communication networks, such as the generation and visualization of complex wireless networks [37]; the development of vaccination strategies for the control of disease [38–42]; and a broad range of other practical issues [30, 43–45]. Research on networks has seen regular publications in some of the most visible scientific journals and vigorous funding in many countries, has been the topic of conferences in a variety of different fields, and has been the subject of numerous books [30, 43–45].

1.2.2 Mathematical Description

A network is defined as a set of N nodes (representing the entities in the system under study) connected to each other by M edges. This definition is equivalent to a graph in the mathematical or computer science literature. In these fields, nodes are often denoted by vertices, and edges by links. A graph is denoted by $G = (V, E)$, where V denotes the node set, with the number of nodes $|V| = N$, and E denotes the edge set with the number of edges $|E| = M$.

Networks are usually described by their adjacency matrix $A = (a_{ij})_{N\times N}$, where a_{ij} characterizes the connection from node i to node j (In some publication, from node j to i). In general, $a_{ij} \neq 0$ indicates the presence of an edge, while $a_{ij} = 0$ stands for the absence of an edge from node i to node j. Different kinds of networks can be distinguished according to the values of a_{ij}. Simple networks have symmetric and binary connections: either two nodes are connected or not, A is symmetric. Weighted symmetric networks allow for different values of the edge weight w_{ij}, accounting for the variability in the nature of the connections between nodes. For weighted networks, the elements in the adjacency matrices take positive values, $a_{ij} = w_{ij} \geq 0$, where $a_{ij} = 0$ indicates the absence of edges between nodes i and j, where A is still symmetric. The weight of a node or the strength of a node is given by the sum of the weights of its edges, $w_i = \sum_{j=1}^{N} a_{ij}$. In a weighted symmetric network, the degree k_i of a node i is naturally defined as the number of connections it has, without including the weight of the connections. Although less informative than the weight, k_i indicates the total number of nodes that interact with node i. If a_{ij}'s are positive integers, the network is a multi-graph and weights are interpreted as multiple edges. Directed networks are necessary to be considered if the connections between nodes are not symmetric. Typically, if nodes are defined as transcription factors (TFs) and connections as regulation relations (activation or repression) between them, the network is clearly directed.

1.2.3 Four Types of Networks

1.2.3.1 Regular Networks

Globally coupled networks, the K-nearest-neighbor coupled networks, and the star networks are all regular networks. Figure 1.1 shows some examples of the mentioned regular networks. The globally coupled networks are also called as complete graph, in which any two nodes are connected. For a complete graph with N nodes, if the network is undirected, then the total number of edges equals $M = N(N-1)/2$; while for a directed complete graph, $M = N(N-1)$. In a complete graph, all nodes have the same degree, the same clustering coefficients (equals 1), and the same shortest path length (equals 1).

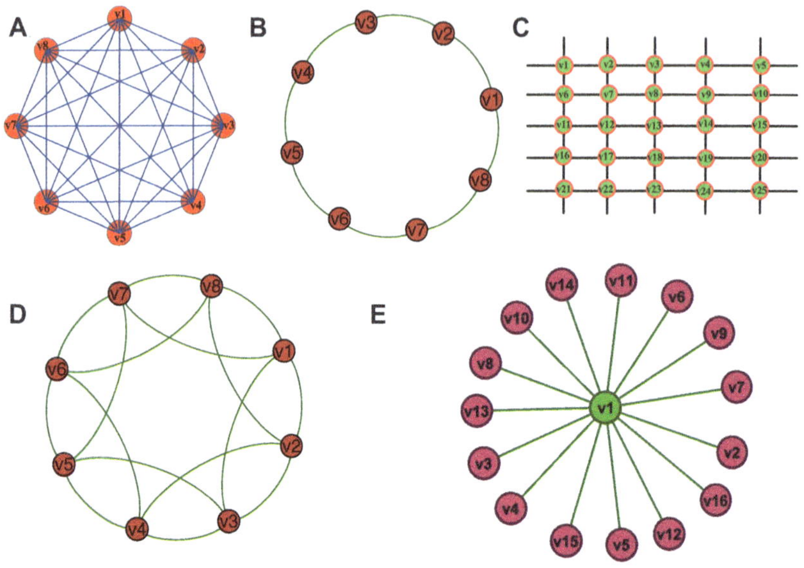

Fig. 1.1 Some regular networks. (**a**) A globally coupled network. (**b–d**) Three K-nearest-neighbor coupled networks. (**e**) A star network

In the K-nearest-neighbor coupled networks, each node connects with its $K(K \leq N-1, K \in Z^+)$ nearest neighbors, and thus all nodes have the same node degree and clustering coefficient. A K-nearest-neighbor coupled network has high clustering coefficient and long average path length (APL) for large N.

The star networks consist of one center node and $N-1$ leaf nodes, the center node connects with each of the leaf nodes. The center node is with degree $N-1$, and the leaf nodes are with degree 1. The star networks are sparse. As $N \to +\infty$, its APL tends to 2.

1.2.3.2 Erdös–Rényi (ER) Random Networks

Historically, random graph is the first generic model of networks. It was introduced by the Hungarian mathematicians P. Erdös and A. Rényi in 1959 [46, 47] and for many decades remained the paradigm of graph theory. In mathematics, random graph is the general term to refer to probability distributions over graphs. Random graphs may be described simply by a probability distribution, or by a random process which generates them. The theory of random graphs lies at the intersection between graph theory and probability theory.

From a mathematical perspective, random graphs are used to answer questions about the properties of typical graphs. Its practical applications are found in all areas in which complex networks need to be modeled—a large number of random graph models are thus known, mirroring the diverse types of complex networks encountered in different areas. Mathematically, random graph refers almost exclusively to the ER random graph model. In other contexts, any graph model may be referred to as a random graph. The model simply assumes that, for each pair of nodes, an edge is drawn with probability $0 < p < 1$. The degree distribution is given by

$$P(k) = C_N^k p^k (1-p)^{N-k} \approx \frac{z^k e^{-z}}{k!}, \tag{1.1}$$

where $z = Np$ represents the average degree.

A random graph is obtained by starting with a set of N isolated vertices and adding successive edges between them at random. The aim of the study in this field is to determine at what stage a particular property of the graph is likely to arise [48]. Different random graph models produce different probability distributions. Most commonly studied is the one proposed by Edgar Gilbert, denoted as $G(N, p)$, in which every possible edge occurs independently with probability $0 < p < 1$. The probability of obtaining any one particular random graph with m edges is $p^m(1-p)^{n-m}$ with the notation $n = C_N^2$ [49].

A closely related model, the ER model is denoted as $G(N, M)$, which assigns equal probability to all graphs with exactly M edges. With $0 \leq M \leq n$, $G(N, M)$ has C_n^M elements and every element occurs with probability $1/C_n^M$. The latter model can be viewed as a snapshot at a particular time (M) of the random graph process $\tilde{G}_N$, which is a stochastic process that starts with n vertices and no edges, and at each step adds one new edge chosen uniformly from the set of missing edges. If instead we start with an infinite set of vertices, and again let every possible edge occur independently with probability $0 < p < 1$, then we get an object G called an infinite random graph. Except in the trivial cases when p is 0 or 1, such a graph almost surely has the following property: Given any $N + m$ elements $a_1, \ldots, a_N, b_1, \ldots, b_m \in V$, there is a vertex c in V that is adjacent to each of $a_1, \ldots, a_N$ and is not adjacent to any of $b_1, \ldots, b_m$.

The second equality of Eq. (1.1) is exact only in the limit of $n \rightarrow \infty$. Equation (1.1) shows that as k becomes larger than z, the probability to observe a node with degree k decreases exponentially with k. Random graphs are not always connected, since a very small p results in very few edges. An important result about random graphs is that, if $p > 1/N$, the size of the largest connected component (LCC) scales as N, while the size of other connected components scales at most as $N^{-\alpha}$ with $\alpha < 1$. Therefore, the LCC is also called the giant component (GC).

For the clustering coefficient of an ER random graph, since p is uniform over the network, the average clustering coefficient c is simply given by p. Close to the critical point $p_c = 1/N$ (which is also the condition for the graph to be sparse), $c \propto 1/N$, which indicates that the clustering coefficient decreases strongly with the network size. Another important result about random graphs concerns the average shortest path length and the diameter. Both quantities were shown to scale as $ln(N)$ [48] if $p > 1/N$, implying that even for very large graphs, the average distance between nodes remains small.

Random graphs are fascinating mathematical objects exhibiting interesting properties. Because of these properties, they have been used as models of real-world complex networks for a long period of time. However, despite some success, several characteristics of real-world networks are not described in the framework of random graphs. In the next two sections, two features that are not present in random graphs, and two classes of models providing a possible explanation for these features, are discussed. Although these models might not be relevant for all situations, they take on a historical importance as the two models that triggered a huge amount of work and shaped the actual way of looking at complex networks.

1.2.3.3 Scale-Free Networks

A network is named SF if its degree distribution follows a particular mathematical function called as power law [50]. The power law implies that the degree distribution of these networks has no characteristic scale. In contrast, networks with a single well-defined scale are somewhat similar to a lattice in that every node has (roughly) the same degree. Examples of networks with a single scale include the ER random graph and hypercubes. In a network with a SF degree distribution, some vertices have a degree that are in orders of magnitude larger than the average—these vertices are often called "hubs," although it is a bit misleading as there is no inherent threshold above which a node can be viewed as a hub. If there were such a threshold, the network would not be SF.

SF networks began to attract people's attention in the late 1990s with the reporting of the discoveries of power-law degree distributions in real-world networks, such as the World Wide Web (WWW), the network of autonomous systems, some networks of Internet routers, protein–protein interaction (PPI) networks, email networks. There are many different ways to build a network with a power-law degree distribution. The preferential attachment proposed by Barabási and Albert (BA) for power-law degree distributions is a well-known one. The BA SF algorithm is described in Algorithm 1.

Algorithm 1 The SF network model proposed by Barabási and Albert [50]

1: Generate an initial connected network with m_0 nodes.
2: **repeat**
3: **Growth**: Add a new node to the network, and connect the new node with randomly chosen m existing nodes, where $m \leq m_0$.
4: **Preferential attachment**: At each time step, the newly added node will link an existing node i with probability:

$$\pi_i = \frac{k_i}{\Sigma_j k_j}, \tag{1.2}$$

where k_i denotes the degree of node i.
5: **until** Network reaches an expected size N.

The BA algorithm is a model of a growing network in which nodes can be continuously added. But instead of connecting them randomly with nodes already present in the network, the connections are drawn with a probability proportional to the degree of the nodes, leading to a preferential attachment. This simply states that nodes with a high degree receive even more new connections, exemplifying the "rich-get-richer" principle. Working out the degree distribution arising from the preferential attachment rule, a power law with exponent $\gamma = 3$ is obtained [50].

Interestingly, the topology of the BA SF networks displays important differences with the one obtained with random graphs. While all nodes are almost equivalent in random graphs, SF networks exhibit several hubs characterized by a much higher degree than the rest of the nodes. The APL [51, 52] for the BA SF network is

$$APL \propto \frac{logN}{log(logN)}.$$

Such graphs are called ultra small-world (SW) networks [51, 52]. Clustering coefficient (CC) for the SF network is [53]

$$CC = \frac{m^2(m+1)^2}{4(m-1)}\left[ln\left(\frac{m+1}{m}\right) - \frac{1}{m+1}\right]\frac{[ln(t)]^2}{t}. \tag{1.3}$$

Equation (1.3) indicates that the BA SF networks are just like the ER random networks, which have no clustering characteristic.

Another observation that could not be described in the random graph model is the shape of the degree distribution. Random graphs show a clear exponential decay, while several real-world networks display a much slower decay, often better described by a power law (also called Zipf's law [54]), $P(k) \sim k^{-\gamma}$. The degree distribution of the BA SF network follows [50]

$$P(k) = \frac{2m(m+1)}{k(k+1)(k+2)} \propto 2m^2k^{-3}, \tag{1.4}$$

which indicates that the BA algorithm can only generate networks with power-law exponent (PLE) $\gamma = 3$. The power-law behavior has important consequences on the topology of the network, since it implies with a significant probability the existence of high-degree nodes that act as hubs in the network. Furthermore, power laws are typical of SF systems in statistical physics, i.e., systems having the same statistical properties at any scale. Networks with a power-law degree distribution can be highly resistant to the random deletion of vertices—i.e., the vast majority of vertices remain connected together in a GC. Such networks can also be quite sensitive to target attacks aimed at fracturing the network quickly.

The BA model is especially intuitive to describe the evolution of the WWW: as new pages are added to the network, they will most likely connect to relevant and well-known pages that are already pointed by several other ones. However, the exact exponent of the BA model is far from universal (for instance, the WWW seems to have rather a $\gamma = 2.2$). To explain this discrepancy, and many others, variations of the BA model have been proposed, including a nonlinear preferential attachment [55], addition of edges between existing nodes [56], preferential attachment based on other quantities than the degree (typically including the effect of clustering coefficient, betweenness, etc.) [57–60], preferential attachment including weights [61–63], algorithm to generate SF networks with given expected degree sequences [64] and so forth.

Another important aspect is the validity of the preferential attachment rule. Though the BA model is rather intuitive for networks like the WWW, it does not always make sense for other kinds of networks, such as PPI networks or some social networks. Other models have been developed, aiming at providing a reasonable explanation for the origin of the degree distribution of these kinds of networks [65–68]. Finally, a general formalism to generate networks with power-law degree distributions with any exponent γ has been developed by Molloy and Reed [69, 70].

To conclude the short discussion about SF networks, the vast amount of models used to describe different features of real-world networks raise the question of how relevant a model is if it describes some properties of a network. In other words, how can one be sure that a model actually describes what happens in a given type of networks, only by showing that some properties of the network are well described in the model. The question becomes even more delicate when noise is present in the data, such that for instance the exponent of a power law is very difficult, if not impossible, to estimate exactly. Despite some attempts [71], it is likely that this question has received too little attention, being diluted in the flow of enthusiasm for the new science of complex networks.

1.2.3.4 Small-World Networks

A network is called a small-world (SW) network [32] by analogy with the SW phenomenon (popularly known as six degrees of separation). The SW hypothesis, which was first described by the Hungarian writer Frigyes Karinthy in 1929, and tested experimentally by Stanley Milgram (1967), is the idea that two arbitrary people will be connected averagely through another five people, i.e., the diameter of the corresponding graph of social connections is not much larger than six.

In the year 1998, Watts and Strogatz proposed the first SW network model, which smoothly interpolates between a random graph and a lattice via tuning a single parameter. The Watts–Strogatz (WS) SW network model includes two steps. Given the desired number of nodes N, the mean degree K (assumed to be an even integer), and a special parameter p, satisfying $0 \leq p \leq 1$ and $N \gg K \gg \ln(N) \gg 1$, in the first step, we construct a regular ring lattice, a graph with N nodes each connected to K neighbors, $K/2$ on each side. That is, if the nodes are labeled $v_0 \dots v_{N-1}$, there is an edge (v_i, v_j) if and only if $0 < |i-j| \; mod \; (N-1-K/2) \leq K/2$. In the second step, for every node $v_i (i = 0, \cdots, N-1)$, take every edge (v_i, v_j) with $i < j$, and rewire it with probability p. Rewiring is done by replacing (v_i, v_j) with (v_i, v_k) where k is chosen with uniform probability from all possible values that avoid self-loops ($k \neq i$) and link duplication (there is no edge $(v_i, v_{k'})$ with $k' = k$ at this point in the algorithm; in other words, no repeated edges are allowed). It is noted that the rewiring processes in the WS model may destroy the connectivity of the networks. The WS model demonstrates that with the addition of only a small number of long-range links, a regular graph can be transformed into a "small world," in which the average number of edges between any two vertices is very small (mathematically, it should grow as the logarithm of the size of the network), while the clustering coefficient stays large.

The WS SW model [32] aims at providing an explanation for two properties observed in several kinds of real-world networks. First, networks often have a large clustering coefficient, as expected if connections are drawn only locally. In the same time, the average distance between any two nodes is short, in the sense that it does not scale as the network size. In the framework of random graphs, these two features can only be obtained if p becomes close to one, i.e., the graph is very dense, which is not the case for most real networks. The model of WS starts from an ordered configuration of edges. In such a configuration, the clustering coefficient of the network is large, but the average distance between any two nodes scales as N. Then, each edge is rewired with probability p to a randomly selected node. For large p, the network becomes indeed completely random, exhibiting similar properties as random graph (Fig. 1.2). The transient behavior is the most interesting one. It has

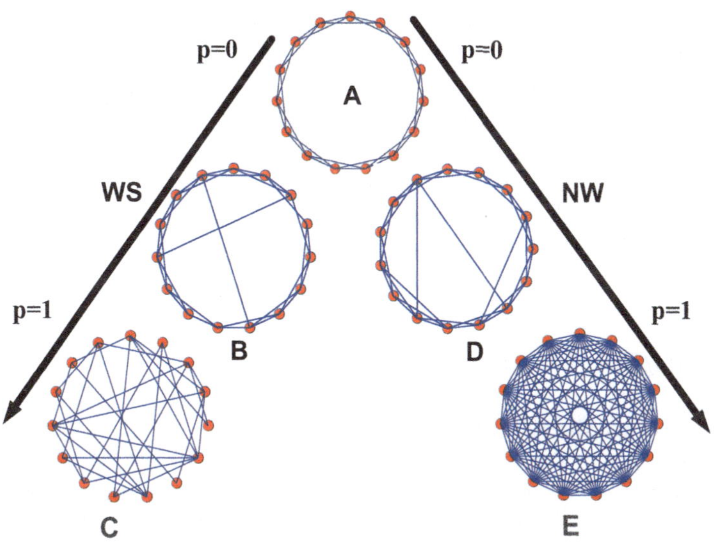

Fig. 1.2 The WS and NW SW networks with $N = 15$. (**a**) $p = 0$ corresponds to a regular network. (**b**) For $p << 1$ but $p > 0$, a SW network is generated by rewiring edges with probability p. (**c**) When p is large, the network is equivalent to a random graph. (**d**) For $p << 1$ but $p > 0$, a NW network is generated by adding edges with probability p between node pairs without edges. (**e**) When $p = 1$, the NW model generates a complete network

been shown that a small value of p is sufficient to decrease significantly the average distance between nodes, while the clustering coefficient remains almost similar to the one with $p = 0$. The word "small-world" can also be understood in terms of social networks. Most individuals on earth have the impression that their friends live in the same area and are part of the same social groups (local interactions resulting in a high clustering coefficient). However, when meeting an unknown person, it often happens that we share a common friend, or at least that some of our respective friends know each other. The model of WS shows that only a few long-range connections are necessary to explain this phenomenon in social networks. The presence of a large clustering coefficient together with a small average distance has been observed in a very large number of different networks that have been grouped under the "small-world" label. Although this is often an indication of a particular topology, it is not sufficient in itself to conclude that the graph is "small-world." For instance, dense random graphs ($p \approx 1$) meet both criteria of small average distance and large clustering coefficient. Therefore, to conclude that a network is really SW, it is also essential to check that the network is sparse.

Another SW model is the Newman–Watts (NW) SW model [72], which is based on the randomly adding edges to a regular network. Starting with a regular network of N vertices in which every vertex has degree K, we go through each pair of nodes without edges between them in turn, and with some probability p, we add an edge between them (Fig. 1.2). The randomly added edges are commonly referred to as shortcuts.

In the scientific literature of complex networks, there is some ambiguity associated with the term "small-world." In addition to referring to the size of the diameter of the network, it can also refer to the co-occurrence of a small diameter and a high clustering coefficient. The clustering coefficient is a metric that represents the density of triangles in the network. For instance, sparse random graphs have a vanishingly small clustering coefficient while real-world networks often have a coefficient significantly larger. Scientists point to this difference as suggesting that edges are correlated in real-world networks.

1.2.4 *Statistical Metrics of Networks*

1.2.4.1 Average Degree and Degree Distribution

Average degree of an undirected unweighted complex network with adjacency matrix $A = (a_{ij})_{N\times N}$ is defined as

$$< k >= \frac{1}{N}\sum_{i=1}^{N} k_i, \tag{1.5}$$

where k_i denotes the degree of node i.

The degree distribution is the probability distribution of these degrees over the whole network. The degree distribution $P(k)$ of a network is defined to be the fraction of nodes in the network with degree k. Thus if there are N nodes in total in a network and n_k of them have degree k, we have

$$P(k) = \frac{n_k}{N}.$$

The degree distribution is very important in studying both real and theoretical networks, such as the Internet, social networks, and the theoretical SF networks. The degree distributions for the ER networks follow the Poisson distributions. Most networks in the real world, however, have degree distributions very different from this. Most are highly right-skewed, meaning that a large majority of nodes have low degrees but a small number, known as hubs, have high degrees. Such networks are the so-called SF networks and have attracted particular attention for their structural and dynamical properties.

1.2.4.2 Average Path Length

APL is a concept in network topology that is defined as the average number of steps along the shortest paths for all possible pairs of nodes. It is a measure of the efficiency of information or mass transport in a network.

Consider an unweighted graph G with the set of vertices V. Let $d(v_1, v_2)$, where $v_1, v_2 \in V$, denotes the shortest distance between v_1 and v_2. Assume that $d(v_1, v_2) = 0$ if v_2 cannot be reached from v_1. Then, the APL is defined as

$$APL = \frac{1}{N(N-1)} \sum_{i \neq j} d(v_i, v_j), \tag{1.6}$$

where N is the number of vertices in G.

1.2.4.3 Diameter

Network's diameter D is defined as the maximum in the shortest path length between any two nodes in the network. Mathematically, if we define the eccentricity $\epsilon(v_i)$ of a vertex v_i as the greatest geodesic distance between v_i and any other vertex. Then the diameter D of a graph is the maximum eccentricity of any vertex in the graph. That is, D is the greatest distance between any pair of vertices or, alternatively,

$$D = \max_{v_i \in V} \epsilon(v_i). \tag{1.7}$$

To find the diameter of a graph, one needs to first find the shortest path between each pair of vertices. The greatest length of any of these paths is the diameter of the graph.

1.2.4.4 Assortativity and Disassortativity

Assortativity, or assortative mixing is a preference for the network's nodes to attach to others that are similar in some way. Though the specific measure of similarity may vary, network theorists often examine assortativity in terms of a node's degree [73].

The assortativity coefficient is the Pearson correlation coefficient (PCC) of degree between pairs of linked nodes [74]. The PCC [75] can act as an indicator of assortativity and disassortativity, which is defined as

$$PCC = \frac{M^{-1}\sum_i j_i k_i - [M^{-1}\sum_i \frac{1}{2}(j_i + k_i)]^2}{M^{-1}\sum_i \frac{1}{2}(j_i^2 + k_i^2) - [M^{-1}\sum_i \frac{1}{2}(j_i + k_i)]^2}. \tag{1.8}$$

Here, M is the total number of edges, k_i, j_i are the degrees of the nodes at the ends of the $i'th(i = 1, 2, \cdots, M)$ edge. $PCC > 0$ indicates a correlation between nodes of similar degrees, while $PCC < 0$ indicates relationships between nodes of different degrees. In general, PCC lies between -1 and 1. When $PCC = 1$, the network is said to have perfect assortative mixing patterns, when $PCC = 0$, the network is non-assortative, while for $PCC = -1$, the network is completely disassortative.

1.2.4.5 Small Worldness

A network with high clustering coefficient and short APL is deemed as SW network [32]. In the year 2008, Humphries and Gurney [76] proposed a measure to explore the small worldness of networks. The SW index is defined as

$$SW = \frac{C^*/C^*_{rand}}{APL/APL_{rand}}. \tag{1.9}$$

$SW > 1$ indicates that the test network is SW. Here, C^*, C^*_{rand} are the clustering coefficients [76, 77] for the test network and the average of that for randomized networks. APL_{rand} is the APL for randomized networks. For ER random networks with N nodes and average degree $\langle k \rangle$, the clustering coefficient C^*_{rand} can be approximated by [75]

$$C^*_{rand} \approx \frac{\langle k \rangle}{N}. \tag{1.10}$$

The APL_{rand} can be approximated by [75]

$$APL_{rand} \approx \frac{ln(N)}{ln(\langle k \rangle)}. \tag{1.11}$$

Generally, one can use the approximation of C^*_{rand} and APL_{rand} as defined in Eqs. (1.10) and (1.11) to obtain the SW value.

1.2.4.6 Hierarchical Modularity

Many real-world networks in nature and society share two generic properties: they are SF and display a high degree of clustering. These two features are the consequence of a hierarchical organization [78], implying that small groups of nodes organize in a hierarchical manner into increasingly large groups, while maintaining a SF topology.

In the year 2003, Ravasz and Barabási [78] introduced an index to investigate the hierarchical modularity of networks. If the average clustering coefficient [32] $C(k) \propto k^{-\theta}$ for nodes with the degree k ($k = k_{min}, \cdots, k_{max}$), and $\theta \approx 1$, then the network is hierarchical modularity. The scaling law quantifies the coexistence of a hierarchy of nodes with different degrees of clustering, while in the study of networks, modularity (networks) is a benefit function that measures the quality of a division of a network into groups or communities.

1.2.4.7 Modularity

Modularity [79] is a measure of the structure of networks or graphs. It was designed to measure the strength of division of a network into modules (also called groups, clusters, or communities). Networks with high modularity have dense connections between the nodes within modules but sparse connections between nodes in different modules. Optimization methods are often used for detecting community structure in networks. However, it has been shown that modularity suffers a resolution limit and, therefore, it is unable to detect small communities.

Biological networks, including animal brains, exhibit a high degree of modularity. Biology scholars have provided a list of features that should characterize a module. For instance, Rudy Raff [80] provided the following list of characteristics that developmental modules should possess: (1) discrete genetic specification; (2) hierarchical organization; (3) interactions with other modules; (4) a particular physical location within a developing organism; (5) the ability to undergo transformations on both developmental and evolutionary timescales.

Mathematically, a large number of quality functions of community structure have been proposed. The most popular quality function is the Girvan–Newman modularity [81], defined as follows:

$$Q(G, V) = \sum_{k=1}^{D} \left(\frac{E_k^i}{2M} - \left(\frac{m_k}{2M} \right)^2 \right). \tag{1.12}$$

Here, $G = (V, E)$ denotes the complex network with node set V and edge set E. E_k^i contains the internal edges of V_k, where $V_k \subset V$. $V = \{V_1, V_2, \ldots, V_D\}$, the size of partition is $D = |V|$. $m_k = \sum_{x \in V_k} \sum_{y \in V} a_{xy}$ is the total degree of cluster V_k. $a_{xy} = 1$ if $(x, y) \in E$. $M = |E|$ denotes the total number of undirected edges. For a network with high $Q(G, V)$, it indicates that the associated network is with community structure. For details of some other methods, one can refer to reference [79], we omit the detailed discussions. Figure 1.3 shows the modularity structures for a coauthorship network of scientists working on network theory and experiment.

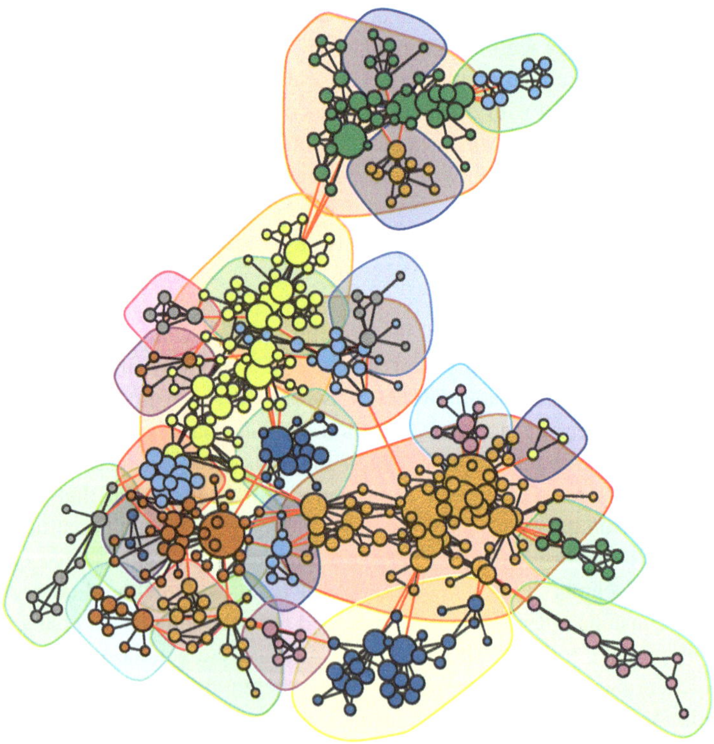

Fig. 1.3 Examples of modularity in a coauthorship network of scientists working on network theory and experiment [82]. This network has 26 communities. Edges within the same community are shown in black, while those across two communities are shown in red

1.2.4.8 Network Structure Entropy

The entropy of network ensembles characterizes the amount of information encoded in the network structure, and can be used to quantify network complexity, and the relevance of given structural properties observed in real network datasets with respect to a random hypothesis [83]. Suppose k_i is the degree of node $v_i (i = 1, 2, \cdots, N)$, the importance of node v_i can be evaluated by

$$I_i = \frac{k_i}{\sum_{j=1}^{N} k_j}, i = 1, 2, \cdots, N. \tag{1.13}$$

Without considering nodes with $k_i = 0$, one can define the structure entropy of the network as

$$E = -\sum_{i=1}^{N} I_i ln I_i. \tag{1.14}$$

For homogeneous network, since $I_i = 1/N$, it has the maximum structure entropy $E_{\max} = lnN$. For the most heterogeneous network—the star networks, denote $I_1 = 1/2, I_i = 1/[2(N-1)](i > 1)$, the corresponding minimum structure entropy is $E_{\min} = [ln(4(N-1))]/2$.

To eliminate the effect of network size N when comparing among different networks, we can use the following normalized structure entropy:

$$E^* = \frac{E - E_{\min}}{E_{\max} - E_{\min}} = \frac{-2\sum_{i=1}^{N} I_i ln I_i - ln[4(N-1)]}{2lnN - ln[4(N-1)]}. \tag{1.15}$$

Here, $0 \leq E^* \leq 1$.

1.2.5 Datasets for Real-World Complex Networks

With the development of network science, many network datasets have been collected. KONECT (http://konect.uni-koblenz.de/) contains over a hundred network datasets of various types, including directed, undirected, bipartite, weighted, unweighted, signed, and rating networks. The networks of KONECT are collected from diverse areas such as social networks, hyperlink networks, authorship networks, physical networks, interaction and communication networks. NetWiki (http://netwiki.amath.unc.edu/SharedData/SharedData) contains links to a large collection of network data. Stanford Large Network Dataset Collection (http://snap.stanford.edu/data/) contains a collection of large networks mainly from social and web-based domains. easyN (http://www.esyn.org/) allows the creation of gene interaction networks (either physical or genetic) but also the creation of Petri net models. The tool also allows the users to save their networks online and, if they want, publish them. Hereinafter, we describe some real-world complex networks, which will be provided as test datasets in the following chapters.

The following are 15 frequently used directed networks, which are described as follows:

1. Advogato [84]: This is a trust network of Advogato. Advogato is an online community platform for developers of free software launched in 1999. Nodes are users of Advogato and directed edges represent trust relationships. The original network is with 6541 nodes and 51,127 edges. We consider its LCC, which contains 5042 nodes and 49,631 edges.
2. Anybeat [85]: An online community from a public gathering place where one can interact with people from its neighborhood or across the world. Nodes are users and directed edges represent the relationships. This network encompasses 12,645 nodes and 67,053 edges.
3. HighSchool [86]: This directed network contains friendships between boys in a small highschool in Illinois. Each boy was asked once in the fall of 1957 and the spring of 1958. This dataset aggregates the results from both dates. A node represents a boy and an edge between two boys shows whether the left boy chose the right boy as a friend. This network encompasses 70 nodes and 366 edges.
4. JameMoody [87]: This directed network was created from a survey that took place in 1994/1995. Each student was asked to list his/her 5 best female and his/her 5 male friends. A node represents a student and an edge between two students shows whether the left student chose the right student as a friend. The JameMoody network encompasses 2539 nodes and 12,969 edges.
5. ResidenceHall [88]: This directed network contains friendship ratings between 217 residents living at a residence hall located on the Australian National University campus. A node represents a person and an edge represents the friendship between two persons. This network contains 217 nodes and 2672 edges.
6. OpenFlights [89]: This directed network contains flights between airports of the world. A directed edge represents a flight from one airport to another. This dataset is extracted from Openflights.org. This network has 2939 nodes and 30,501 edges.
7. USAirport [90]: This is a directed network of flights between US airports in 2010. Each edge represents a connection from one airport to another, and the weight of an edge shows the number of flights on that connection in the given direction, in 2010. This network contains 1574 nodes and 28,236 edges.
8. SpaBook [91]: It reflects word adjacency relationships of a Spanish book. Nodes in the network are words and an edge denotes that two words occurred one after another in the book. The network is directed, i.e., the edge (u, v) denotes that word u was followed by word v. Since a word can occur twice in a row, the network contains loops. This network encompasses 12,643 nodes and 57,453 edges.

9. Cora [92]: This is a cora citation network. The network is directed. Nodes represent scientific papers. An edge between two nodes indicates that the left paper cites the right paper. This network includes 23,166 nodes and 91,500 edges.
10. DBLP [93]: This is the citation network of DBLP, a database of scientific publications such as papers and books. Each node in the network is a publication, and each edge represents a citation of a publication by another publication. This network encompasses 12,591 nodes and 49,743 edges.
11. Cite-th [94]: This is the network of publications in the arXiv's High Energy Physics-Theory (hep-th) section. The directed links that connect the publications are citation relationships.
12. Cite-ph [94]: This is the collaboration graph of authors of scientific papers from the arXiv's High Energy Physics-Phenomenology (hep-ph) section. An edge between two authors represents a common publication.
13. UCsocial [95]: This directed network contains sent messages between the users of an online community of students from the University of California, Irvine. A node represents a user. A directed edge represents a sent message. This network includes 1899 nodes and 20,296 edges.
14. WikiVote [96]: This is a network of users from the English Wikipedia that voted for and against each other in admin elections. Nodes represent individual users, and edges represent votes. Edges can be positive ("for" vote) and negative ("against" vote). Each edge is annotated with the date of the vote. This network encompasses 7118 nodes and 103,675 edges.
15. RockLake [97]: This is a food web of Little Rock Lake, Wisconsin in the USA. Nodes in this network are autotrophs, herbivores, carnivores, and decomposers; links represent food sources. This network encompasses 183 nodes and 2494 edges.

The detailed information for 12 frequently used undirected networks are described as follows:

1. Hamster [98]: This network contains friendships between users of the website hamsterster.com. A node represents the user and an edge represents the friendship. This network encompasses 1858 nodes and 12,534 edges.
2. Vidal [99]: This network represents an initial version of a proteome-scale map of Human binary PPIs. PPIs extracted from PMID: 16189514. This network has 3133 nodes and 6726 edges.
3. *Yeast* [100]: This undirected network contains protein interactions for the *yeast*. Research showed that proteins with a high degree were more important for the survival of the *yeast* than others. A node represents a protein and an edge represents a metabolic interaction between two proteins. This network encompasses 1870 nodes and 2277 edges.
4. Router [101]: A network of autonomous systems of the Internet connected to each other. Nodes are autonomous systems, and edges denote communication. This network encompasses 5022 nodes and 6258 edges.

5. USAir [102]: The US air transportation network. Nodes are airports, edges represent airways. This network encompasses 332 nodes and 2126 edges.
6. Bible [103]: This undirected network contains nouns (places and names) of the King James Version of the Bible and information about their co-occurrences. A node represents one of the above noun types and an edge indicates that two nouns appeared together in the same Bible verse. This network encompasses 1773 nodes and 9131 edges.
7. David [82]: This is the undirected network of common noun and adjective adjacencies for the novel "David Copperfield" by English nineteenth century writer Charles Dickens. A node represents either a noun or an adjective. An edge connects two words that occur in adjacent positions. The network is not bipartite, i.e., there are edges connecting adjectives with adjectives, nouns with nouns, and adjectives with nouns. This network encompasses 112 nodes and 425 edges.
8. Email [104]: This is an email communication network at the University Rovira i Virgili in Tarragona in the south of Catalonia in Spain. Nodes are users and each edge represents that at least one email was sent. The direction of emails or the number of emails is not stored. This network encompasses 1133 nodes and 5451 edges.
9. Jazz [105]: A collaboration network between Jazz musicians. Each node is a Jazz musician and an edge denotes that two musicians have played together in a band. The data was collected in 2003. This network encompasses 198 nodes and 2742 edges.
10. NS [82]: A coauthorship network of scientists working on network theory and experiment, as compiled by Newman in May 2006. Node is a scientist and edge represents that two scientist wrote at least one joint work. The original network have 1589 nodes, and only the LCC is considered, which has 379 nodes and 914 edges.
11. PrettyGood [106]: This is the interaction network of users of the Pretty Good Privacy algorithm. The network contains only the LCC of the whole network, which contains 10,680 nodes and 24,316 edges.
12. PB [107]: A network contains front-page hyperlinks between blogs in the context of the 2004 US election. A node represents a blog and an edge represents a hyperlink between two blogs. The original network is directed, the current one is its undirected version, which has 1222 nodes and 16,714 edges.

Details for 5 binary networks are as follows:

1. AmeRev [108]: This bipartite network contains membership information of 136 people in 5 organizations dating back to the time before the American Revolution. The list includes well-known people such as the American activist Paul Revere. Left nodes represent persons and right nodes represent organizations. An edge between a person and an organization shows that the person was a member of the organization. This network encompasses 141 nodes and 160 edges.
2. Leadership [109]: This bipartite network contains person–company leadership information between companies and 20 corporate directors. Data was collected

in 1962. Left nodes represent persons and right nodes represent companies. An edge between a person and a company shows that the person had a leadership position in that company. This network encompasses 44 nodes and 99 edges.

3. SexEsc [110]: This is a bipartite network of sex buyers and their escorts. Nodes are buyers and escorts. An edge denotes sexual intercourse between a male sex-buyer and a female escort. Edges are weighted with the rating of the escort given by the buyer. Three ratings are possible: bad (−1), neutral (0), good (+1). The unweighted version is considered. This network encompasses 16,730 nodes and 35,051 edges.
4. WikiBooks [111]: This is a bipartite edit network of the French Wikibooks (https://fr.wikibooks.org/wiki/Accueil). It contains users and pages from the French Wikipedia, connected by edit events. Each edge represents an edit. This network encompasses 30,616 nodes and 67,613 edges.
5. WikiNews [111]: This is a bipartite edit network of the French Wikinews (https://fr.wikinews.org/wiki/Accueil). It contains users and pages from the French Wikipedia, connected by edit events. Each edge represents an edit. This network encompasses 26,447 nodes and 68,703 edges.

1.3 Central Dogma of Molecular Biology

Every cell contains DNA and genes in its nucleus [1]. DNA contains all the information required to build the cells and tissues of an organism [1, 112, 113]. The exact replication of this information in any species assures its genetic continuity from generation to generation, which is critical to the normal development of an individual. The information stored in DNA is arranged in hereditary units, now known as genes. Genes are the specific instructions for cells that make every organism unique. DNA replication occurs in cells preparing to divide, deoxyribonucleoside triphosphate monomers (dNTPs) are polymerized to yield two identical copies of each chromosomal DNA molecule. Each daughter cell receives one of the identical copies. In the process of transcription (Fig. 1.4), the information stored in DNA is copied into ribonucleic acid (RNA).

mRNA carries the instructions from DNA that specify the correct order of amino acids during protein synthesis. The remarkably accurate, stepwise assembly of

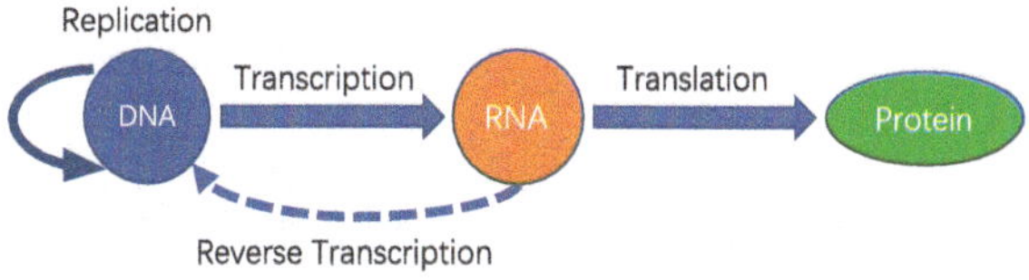

Fig. 1.4 The central dogma of molecular biology. Genetic information can transfer from DNA to RNA to protein. In viruses, RNA may modify DNA, called reverse transcription

amino acids into proteins occurs by translation of mRNA (Fig. 1.4). In this process, the information in mRNA is interpreted by a second type of RNA called transfer RNA (tRNA) with the aid of a third type of RNA, ribosomal RNA (rRNA), and its associated proteins. As the correct amino acids are brought into sequence by tRNAs, they are linked by peptide bonds to make proteins.

Discovery of the structure of DNA in 1953 and subsequent elucidation of how DNA directs synthesis of RNA, which then directs assembly of proteins—the so-called central dogma—were monumental achievements marking the early days of molecular biology. The simplified representation of the central dogma as $DNA \rightarrow RNA \rightarrow protein$ will be considered during the modeling of GRNs in this book. Proteins are largely responsible for regulating gene expression, the entire process whereby the information encoded in DNA is decoded into the proteins that characterize various cell types.

Gene expression is the process of how a gene works within a cell. Cells contain many genes, and not all of them are active. Within any given cell, some genes will be "on" or "off." When a gene is "on," it is making proteins or RNA products and affecting the functioning of the organism in some way. If a gene is "on," scientists consider that it is being expressed.

1.4 Bio-Molecular Networks

Biological systems can be described by complex networks and investigated through complex networks theory. There are different types of bio-molecular networks [114], such as GRNs, TRNs, PPI networks, signal transduction networks, metabolic networks [114], and the integration of them [115]. Different networks describe different levels of the life phenomena [116], as described in Fig. 1.5a. The TRNs describe the regulatory relationships among transcriptional regulators and DNAs;

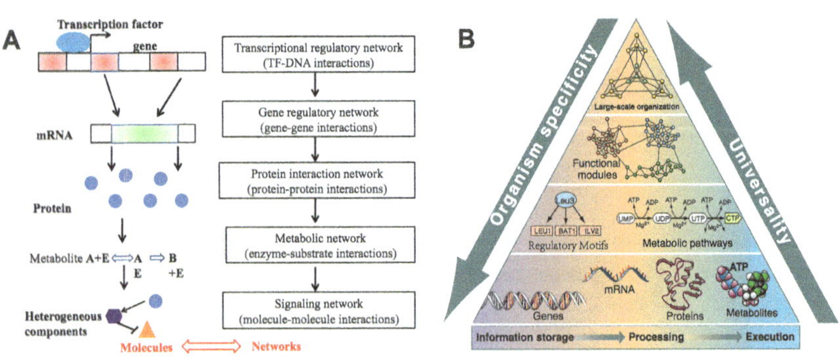

Fig. 1.5 Different levels of bio-molecular networks and life's complexity pyramid. (**a**) Different levels of bio-molecular networks (Copyright ©(2009) Wiley. Used with permission from ref. [114]). (**b**) Life's complexity pyramid. From ref. [116], reprinted with permission from AAAS

the GRNs mimic the interactions among different genes; nodes in the PPI networks denote proteins, and edges represent the interactions among them. The metabolic networks show the interactions among enzymes and substrates, while the signaling networks describe the molecule–molecule interactions. Figure 1.5b describes the life's complexity pyramid, which is composed of various molecular components of cells, genes, RNAs, proteins, and metabolites. The bottom of the pyramid shows the traditional representation of the cell's functional organization (level 1). There is a remarkable integration of various layers at both the regulatory and the structural levels. Insights into the logic of cellular organization can be gained when one views the cell as an individual complex network in which the components are connected by functional links. At the lowest level, these components form genetic regulatory motifs or metabolic pathways (level 2), which in turn are the building blocks of functional modules (level 3). Finally, these modules are nested, generating a SF hierarchical architecture (level 4).[1]

Figure 1.6 presents four typical real-world biological networks, which correspond to a GRN in hepatocellular carcinoma [117], the cell signal transduction network [118], a *yeast* PPI network [119], an integration of transcriptomics and metabolomics networks [115] for *Arabidopsis*. In this book, we mainly consider the GRNs and PPI networks.

1.5 Several Statistical Methods

1.5.1 Descriptive Statistics

Denote $\mathbf{X} = (\mathbf{X}_1, \mathbf{X}_2, \cdots, \mathbf{X}_p)^T$ as a m-dimensional stochastic vector, $X_{(1)}, X_{(2)}, \cdots, X_{(n)}$ are n observations of $\mathbf{X}$. The observations consist of the following rectangular array, called sample material matrix or sample data matrix, denoted as X:

$$X = \begin{bmatrix} x_{11} & x_{12} & \cdots & x_{1p} \\ x_{21} & x_{22} & \cdots & x_{2p} \\ \vdots & \vdots & \ddots & \vdots \\ x_{n1} & x_{n2} & \cdots & x_{np} \end{bmatrix} = \begin{bmatrix} X_{(1)}^T \\ X_{(2)}^T \\ \vdots \\ X_{(n)}^T \end{bmatrix} = [X_1, X_2, \cdots, X_p]. \tag{1.16}$$

Here $X_{(i)} = (x_{i1}, x_{i2}, \cdots, x_{ip})^T$ represents the i'th $(i = 1, 2, \cdots, n)$ observation of $\mathbf{X}$. $X_j = (x_{1j}, x_{2j}, \cdots, x_{nj})^T$ denotes the observation vector for the j'th variable $\mathbf{X}_j (j = 1, 2, \cdots, p)$.

[1]From Oltvai, Z.N., Barabasi, A.L.: Life's complexity pyramid. Science 298, 763–764 (2002). Reprinted with permission from AAAS.

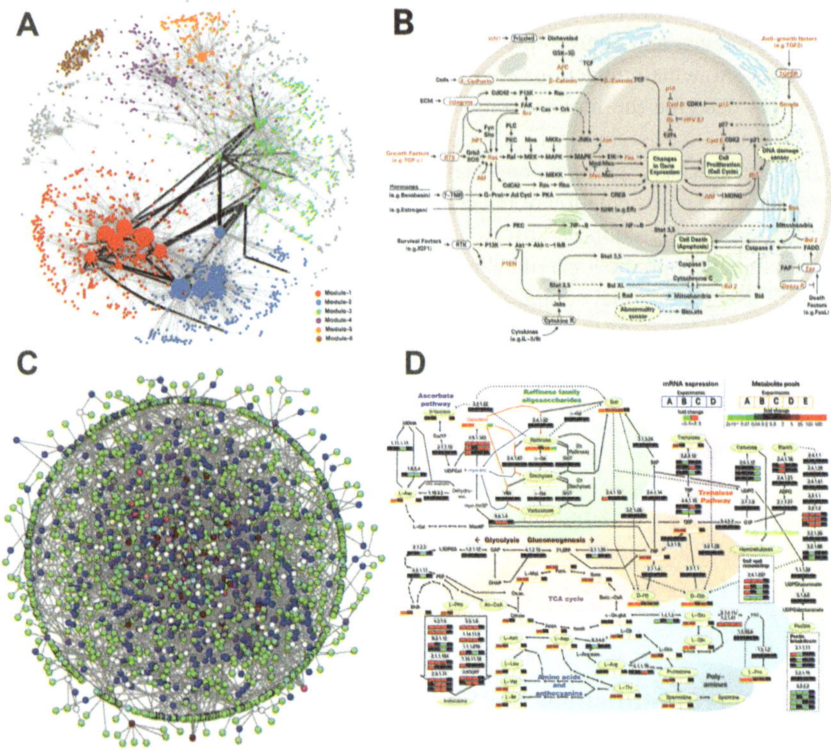

Fig. 1.6 Four real-world biological networks. (**a**) A GRN in hepatocellular carcinoma [117]. (**b**) A cell signal transduction network for human; Reprinted from ref. [118], with permission from Elsevier. (**c**) A *yeast* PPI network; ©[2014] IEEE. Reprinted, with permission, from ref. [119]. (**d**) An integration of transcriptomics and metabolomics networks for *Arabidopsis* [115] (Copyright (2004) National Academy of Sciences, USA)

A large dataset is bulky, and it always brings a serious obstacle to any attempt to visually extract pertinent information. Much of the information contained in the data can be accessed by calculating summary numbers, known as descriptive statistics [120].

The **sample mean** is a simple descriptive statistics, which can be computed from the n measurements on each of the p variables. The sample mean for each of the p variables can be defined as

$$\bar{x}_k = \frac{1}{n}\sum_{j=1}^{n} x_{jk}, k = 1, 2, \cdots, p. \tag{1.17}$$

or the sample mean vector can be calculated as follows:

$$\bar{X} = (\bar{x}_1, \bar{x}_2, \cdots, \bar{x}_p)^T = \frac{1}{n} X^T 1_n, \tag{1.18}$$

where $1_n = (1, 1, \cdots, 1)^T$ is a n-dimensional constant vector.

A measure of spread is provided by the sample variance, for the p variables, we have

$$s_k^2 = \frac{1}{n-1} \sum_{j=1}^{n} (x_{jk} - \bar{x}_k)^2, k = 1, 2, \cdots, p. \tag{1.19}$$

It is noted that the sample variance with a divisor of $n - 1$ rather than n is always used to derive the unbiased estimation of the population variance. The square root of the sample variance s_k is known as the sample standard deviation. The sample covariance is defined as

$$s_{ik} = \frac{1}{n-1} \sum_{j=1}^{n} (x_{ji} - \bar{x}_i)(x_{jk} - \bar{x}_k), i, k = 1, 2, \cdots, p. \tag{1.20}$$

The covariance reduces to the sample variance when $i = k$, that is $s_k^2 = s_{kk}$. Moreover, $s_{ik} = s_{ki}$ for all i, k. The sum of squares of the deviations from the mean and the sum of cross-product deviations are often of interest themselves. These quantities are

$$a_{kk} = \sum_{j=1}^{n} (x_{jk} - \bar{x}_k)^2, k = 1, 2, \cdots, p. \tag{1.21}$$

and

$$a_{ik} = \sum_{j=1}^{n} (x_{ji} - \bar{x}_i)(x_{jk} - \bar{x}_k), i, k = 1, 2, \cdots, p. \tag{1.22}$$

Obviously, $s_{ik} = a_{ik}/(n - 1)$.

Based on the sample covariance, the sample correlation coefficient is defined as

$$r_{ik} = \frac{s_{ik}}{\sqrt{s_{ii}}\sqrt{s_{kk}}} = \frac{\sum_{j=1}^{n} (x_{ji} - \bar{x}_i)(x_{jk} - \bar{x}_k)}{\sqrt{\sum_{j=1}^{n} (x_{ji} - \bar{x}_i)^2}\sqrt{\sum_{j=1}^{n} (x_{jk} - \bar{x}_k)^2}}, i, k = 1, 2, \cdots, p. \tag{1.23}$$

Note that $r_{ik} = r_{ki}$ for all i and k.

Some other descriptive statistics include the median, the mode, and so on. The median is the number separating the higher half of a data sample, a population, or a probability distribution, from the lower half. The median of a finite list of numbers can be found by arranging all the observations from the lowest value to the highest value and picking the middle one. If there is an even number of observations, then there is no single middle value; the median is then usually defined to be the mean of the two middle values, which corresponds to interpret the median as the fully trimmed mid-range. The mode is the value that appears most often in a set of data.

The descriptive statistics for stochastic vector **X** can be written as vector or matrix form. The **sample means vector** is

$$\bar{X} = \begin{bmatrix} \bar{x}_1 \\ \bar{x}_2 \\ \vdots \\ \bar{x}_p \end{bmatrix}. \tag{1.24}$$

The **sample cross-product matrix** is

$$A = \begin{bmatrix} a_{11} & a_{12} & \cdots & a_{1p} \\ a_{21} & a_{22} & \cdots & a_{2p} \\ \vdots & \vdots & \ddots & \vdots \\ a_{p1} & a_{p2} & \cdots & a_{pp} \end{bmatrix} = \sum_{i=1}^{n} (X_{(i)} - \bar{X})(X_{(i)} - \bar{X})^T \tag{1.25}$$

$$= X^T \left[I_n - \frac{1}{n} 1_n 1_n^T \right] X = \tilde{X}^T \tilde{X}.$$

Here, $\tilde{X} = \left[I_n - \frac{1}{n} 1_n 1_n^T \right] X$ corresponds to the centralized matrix of X.

The **sample covariance matrix** is

$$S = \begin{bmatrix} s_{11} & s_{12} & \cdots & s_{1p} \\ s_{21} & s_{22} & \cdots & s_{2p} \\ \vdots & \vdots & \ddots & \vdots \\ s_{p1} & s_{p2} & \cdots & s_{pp} \end{bmatrix} = \frac{1}{n-1} A. \tag{1.26}$$

The **sample correlation matrix** is

$$R = \begin{bmatrix} 1 & r_{12} & \cdots & r_{1p} \\ r_{21} & 1 & \cdots & r_{2p} \\ \vdots & \vdots & \ddots & \vdots \\ r_{p1} & r_{p2} & \cdots & 1 \end{bmatrix}. \tag{1.27}$$

One can easily find that

$$S = diag\{\sqrt{s_{11}}, \sqrt{s_{22}}, \cdots, \sqrt{s_{pp}}\} \times R \times diag\{\sqrt{s_{11}}, \sqrt{s_{22}}, \cdots, \sqrt{s_{pp}}\}.$$

In multivariate statistical analysis, if the p dimensional stochastic vector $\mathbf{X} \sim N_p(\mu, \Sigma)$, and the observational matrix X as defined in Eq. (1.16) is a sample of $\mathbf{X}$, then, $\bar{X}$ and $S = A/(n-1)$ are the unbiased maximum likelihood estimation (MLE) of the parameters vector μ and $\Sigma (n > p)$, respectively.

1.5.2 Cluster Analysis

Cluster analysis [120] is the task of grouping a set of objects in such a way that objects in the same group are more similar to each other than to those in other groups, which is also well known as unsupervised learning in machine learning and with wide real-world applications.

1.5.2.1 Hierarchical Clustering

Hierarchical clustering, also known as connectivity based clustering, is based on the core idea of objects being more related to nearby objects than to objects farther away. These algorithms connect "objects" to form clusters based on their distance. A cluster can be described largely by the maximum distance needed to connect parts of the cluster. At different distances, different clusters will form, which can be represented using a dendrogram. These algorithms do not provide a single partitioning of the dataset, but instead provide an extensive hierarchy of clusters that merge with each other at certain distances. In a dendrogram, the y-axis marks the distance at which the clusters merge, while the objects are placed along the x-axis such that the clusters do not mix.

Connectivity based clustering is a whole family of methods that differ by the way distances are computed. Two classes of distance functions are needed to perform the hierarchical clustering analysis: the distance between two observations and the distance between two clusters. The first class of distances is used to measure the closeness among samples. Classical distances include the Minkowski distance, the Lance–Williams distance, and so on. Suppose $\mathbf{X} = (\mathbf{X_1}, \mathbf{X_2}, \cdots, \mathbf{X_p})^T$ is a stochastic vector, there are totally n observations $X_{(i)}, i = 1, 2, \cdots, n$. The Minkowski distance between observations $X_{(i)}$ and $X_{(j)}$ is defined as

$$d_{ij}(q) = \left[\sum_{k=1}^{p} |x_{ik} - x_{jk}|^q \right]^{1/q}, i, j = 1, 2, \cdots, n. \tag{1.28}$$

Here, q is a parameter. If $q = 1$, the Minkowski distance reduces to the classical absolute distance, also known as the city-block distance:

$$d_{ij}(1) = \sum_{k=1}^{p} |x_{ik} - x_{jk}|, i, j = 1, 2, \cdots, n. \tag{1.29}$$

For $q = 2$, the Minkowski distance is just the well-known Euclidean distance:

$$d_{ij}(2) = \sqrt{\sum_{k=1}^{p} |x_{ik} - x_{jk}|^2}, i, j = 1, 2, \cdots, n. \tag{1.30}$$

For $q \to \infty$, the Minkowski distance corresponds to the Chebyshev distance:

$$d_{ij}(\infty) = \max_{k=1,2,\ldots,p} |x_{ik} - x_{jk}|, i, j = 1, 2, \cdots, n. \tag{1.31}$$

The Minkowski distance is with dimensions. To derive dimensionless distances, Lance and William introduced the Lance distance, defined as follows:

$$d_{ij}(L) = \sum_{k=1}^{p} \frac{|x_{ik} - x_{jk}|}{x_{ik} + x_{jk}}, i, j = 1, 2, \cdots, n. \tag{1.32}$$

Obviously, the Lance distance requires $x_{ij} > 0, i = 1, 2, \cdots, n, j = 1, 2, \cdots, p$. The Lance distance is insensitive to outliers, and it is dimensionless, which is appropriate for noisy data. However, due to relative computational complexity of the Lance distance, in practical applications, the absolute distance and the Euclidean distance are frequently used in hierarchical clustering analysis.

Apart from the usual choice of the above distance functions defined for samples, users also need to decide what is the linkage criterion to use, since a cluster consists of multiple objects, there are multiple candidates to compute the distance. Popular choices are known as the single-linkage clustering, the complete-linkage clustering, the average linkage clustering, and the WARD method.

Suppose $G_p, G_q, G_k, p \neq q \neq k$ are three disjoint clusters, the single-linkage distance between clusters G_p and G_q is defined as

$$D_{pq} = \min_{i \in G_p, j \in G_q} d_{ij}. \tag{1.33}$$

Suppose G_p, G_q are merged as a new cluster $G_r, r \neq k$ at a step, then the single-linkage distance between cluster G_r and an old cluster G_k for the next step can be derived as

$$D_{rk} = \min\{D_{pk}, D_{qk}\}, k \neq p, q. \tag{1.34}$$

The complete-linkage method defines the distance between cluster G_p and cluster G_q as

$$D_{pq} = \max_{i \in G_p, j \in G_q} d_{ij}. \tag{1.35}$$

And the corresponding recursive formula among newly formed cluster G_r and old cluster G_k is as follows:

$$D_{rk} = \max\{D_{pk}, D_{qk}\}, k \neq p, q. \tag{1.36}$$

For the average linkage method, the squared distance between cluster G_p and cluster G_q is defined as

$$D_{pq}^2 = \frac{1}{n_p n_q} \sum_{i \in G_p, j \in G_q} d_{ij}^2. \tag{1.37}$$

And the recursive formula of square distance equation among newly formed cluster G_r and old cluster G_k is as follows:

$$D_{rk}^2 = \frac{n_p}{n_r} D_{pk}^2 + \frac{n_q}{n_r} D_{qk}^2, k \neq p, q. \tag{1.38}$$

Here, n_p, n_q, n_r are the numbers of observations in clusters G_p, G_q, and G_r, respectively.

The WARD method is based on minimizing the loss of information from joining two groups. This method is usually implemented with loss of information taken to be an increase in an error sum of squares criterion, ESS. First, for a given cluster k, let ESS_k be the sum of the squared deviation of every item in the cluster from the cluster mean. If there are currently k clusters, define ESS as

$$ESS = ESS_1 + ESS_2 + \cdots + ESS_k. \tag{1.39}$$

At each step in the analysis, the union of every possible pair of clusters is considered, and the two clusters whose combination results in the smallest increase in ESS are joined. Initially, each cluster consists of a single item, and if there are n items, $ESS_k = 0, k = 1, 2, \cdots, n$, so $ESS = 0$. At the other extreme, when all the clusters are combined in a single group of N items, the value of ESS is given by

$$ESS = \sum_{j=1}^{n} (X_{(j)} - \bar{X})'(X_{(j)} - \bar{X}), \tag{1.40}$$

where $X_{(j)}$ is the j'th observation, $\bar{X}$ is the mean of all the items. For the WARD method, the recursive formula of square distance equation among newly formed cluster G_r and old cluster G_k can be deduced as

$$D_{rk}^2 = \frac{n_p + n_k}{n_r + n_k} D_{pk}^2 + \frac{n_q + n_k}{n_r + n_k} D_{qk}^2 - \frac{n_k}{n_r + n_k} D_{pq}^2, k \neq p, q. \tag{1.41}$$

Here, $n_r = n_p + n_q, n_p, n_q, n_k$ represent the numbers of observations in clusters G_r, G_p, G_q, and G_k, respectively.

The results of the hierarchical clustering methods can be displayed as dendrograms, or tree diagrams. The branches in the tree represent clusters. The branches come together at nodes whose positions along a distance axis indicate the level at which the fusions occur. The hierarchical clustering methods will not produce a unique partitioning of the dataset, but a hierarchy from which the user still needs to choose appropriate clusters. They are not very robust towards outliers, which will either show up as additional clusters or even cause other clusters to merge. In the general case, the complexity is $\mathcal{O}(n^3)$ for agglomerative clustering and $\mathcal{O}(2^{n-1})$ for divisive clustering, which makes them too slow for large datasets. For some special cases, optimal efficient methods (of complexity $\mathcal{O}(n^2)$) are known: single-linkage and complete-linkage clustering. In the data mining community, these methods are recognized as a theoretical foundation of cluster analysis, but often considered obsolete. They did however provide inspiration for many later methods, such as density based clustering and k-means clustering.

1.5.2.2 k-Means Clustering

The term "k-means" was first used by James MacQueen in 1967 [121], though the idea goes back to Hugo Steinhaus in 1957 [122]. The standard algorithm was firstly proposed by Stuart Lloyd in 1957 as a technique for pulse-code modulation, though it was not published outside of the Bell Labs until 1982 [123]. In 1965, E. W. Forgy published essentially the same method, which is why it is sometimes referred to as Lloyd–Forgy [124]. A more efficient version was proposed and published in Fortran by Hartigan and Wong in 1979 [125].

The k-means clustering is a method of vector quantization, originally from signal processing, that is popular for clustering analysis in data mining. The k-means clustering aims to partition n observations into k clusters in which each observation belongs to the cluster with the nearest mean, serving as a prototype of the cluster.

The flow of the k-means clustering algorithm is shown in Algorithm 2.

Algorithm 2 The k-means clustering algorithm

1: Partition the items into k initial clusters.
2: **repeat**
3: Proceed through the list of items, assigning an item to the cluster whose centroid is the nearest. Here, distance is usually computed using Euclidean distance with either standardized or unstandardized observations. Recalculate the centroid for the cluster receiving the new item and for the cluster losing the item.
4: **until** No more reassignments take place.

Rather than starting with a partition of all items into k preliminary groups in step 1, we would specify k initial centroids and then proceed to the following steps. The final assignment of items to clusters will be, to some extent, dependent upon the initial partition or the initial selection of seed points. Experience suggests that most major changes occur with the first reallocation step.

1.5.3 Principal Component Analysis

Principal component analysis (PCA) was invented in 1901 by Pearson [126], as an analogue of the principal axis theorem in mechanics. It was later independently developed by Harold Hotelling in the 1930s [127].

PCA is a statistical procedure that uses an orthogonal transformation to convert a set of observations of possibly correlated variables into a set of values of linearly uncorrelated variables called principal components. The number of principal components is less than or equal to the number of original variables. This transformation is defined in such a way that the first principal component has the largest possible variance (that is, accounts for as much of the variability in the data as possible), and each succeeding component in turn has the highest variance possible under the constraint that it is orthogonal to the preceding components. The resulting vectors are an uncorrelated orthogonal basis set. The principal components are orthogonal because they are the eigenvectors of the covariance matrix. The PCA is sensitive to the relative scaling of the original variables.

The PCA is mostly used as a tool in exploratory data analysis and for making predictive models. The PCA can be done by eigenvalue decomposition of a data covariance (or correlation) matrix or singular value decomposition of a data matrix, usually after mean centering (and normalizing or using Z-scores) the data matrix for each attribute. The results of a PCA are usually discussed in terms of component scores, sometimes called factor scores, and loadings (the weight by which each standardized original variable should be multiplied to get the component score). Usually, if components in the principal eigenvectors of the covariance matrix have the same sign, then the first principal component can be used to rank observations; for its applications in biological systems, one can refer to Chap. 7.

1.6 Software for Network Visualization and Analysis

1.6.1 Pajek

Pajek [128] is a noncommercial software for the Windows system, which is designed for analysis and visualization of large networks having thousands or even millions of vertices. Pajek is implemented in Delphi (Pascal), and the software is developed since the year 1996.

Pajek is a powerful software, which can be used to find clusters (components, neighborhoods of "important" vertices, cores, etc.) in a network, extract vertices that belong to the same clusters and show them separately, possibly with the parts of the context, shrink vertices in clusters and show relations among clusters. Besides ordinary (directed, undirected, mixed) networks, Pajek supports also multirelational networks, 2-mode networks (bipartite graphs—networks between two disjoint sets of vertices), and temporal networks (dynamic graphs—networks changing over time). For details of the Pajek, one can refer to Pajek textbook written by de Nooy et al. [128].

1.6.2 Gephi

Gephi is an open-source network analysis and visualization software package written in Java on the NetBeans platform [129], initially developed by students of the University of Technology of Compiègne in France.

Gephi has been used in a number of research projects in academia, journalism, and elsewhere, for instance, in visualizing the global connectivity of New York Times content [130] and examining Twitter network traffic during social unrest [131, 132] along with more traditional network analysis topics. It is a tool for data analysts and scientists keen to explore and understand graphs. Like Photoshop but for graph data, the user interacts with the representation and manipulates the structures, shapes, and colors to reveal hidden patterns. The goal is to help data analysts to make hypothesis, intuitively discover patterns, isolate structure singularities or faults during data sourcing. It is a complementary tool to traditional statistics, as visual thinking with interactive interfaces is now recognized to facilitate reasoning.

Gephi can be applied in (1) exploratory data analysis: intuition-oriented analysis by networks manipulations in real time; (2) link analysis: revealing the underlying structures of associations between objects; (3) social network analysis: easy creation of social data connectors to map community organizations and SW networks; (4) biological network analysis: representing patterns of biological data; (5) poster creation: scientific work promotion with high-quality printable maps.

The statistics and metrics framework offer the most common metrics for social network analysis and SF networks. Gephi can compute the following statistics and metrics for a network, including betweenness centrality, closeness, diameter, clustering coefficient, PageRank, modularity, the shortest path length. Furthermore, one can use ranking or partition data to make meaningful representation of the network. Customized colors, sizes, or labels help to bring sense to the network representation. The vectorial preview module lets one put the final touch and care about esthetics before exporting in SVG or PDF. Gephi supports customizable PDF, SVG, and PNG export.

1.6.3 Cytoscape

Cytoscape [133] is an open-source software project for integrating bio-molecular interaction networks with high-throughput expression data and other molecular states into a unified conceptual framework. Although applicable to any system of molecular components and interactions, Cytoscape is most powerful when used in conjunction with large databases of protein–protein, protein–DNA, and genetic interactions that are increasingly available for humans and model organisms. Cytoscape's software core provides basic functionality to layout and query the network, to visually integrate the network with expression profiles, phenotypes, and other molecular states, and to link the network to databases of functional annotations. The core is extensible through a straightforward plug-in architecture, allowing rapid development of additional computational analyses and features.

1.6.4 MATLAB Packages and Others

MATLAB [134] is a powerful software for scientific computing and graph drawing. Researchers have developed several MATLAB packages to perform network visualization and analysis in MATLAB, such as the Complex Networks Package for MATLAB developed by Lev Muchnik; A MATLAB toolbox for the construction of artificial complex networks developed by Gregorio Alanis-Lobato; The MatlabBGL package (http://www.mathworks.com/matlabcentral/fileexchange/loadFile.do?objectId=10922&objectType=file) and SBEToolbox (Systems Biology & Evolution Toolbox).

The Complex Networks Package for MATLAB [135] (http://www.levmuchnik.net/Content/Networks/ComplexNetworksPackage.html) comes to provide a comprehensive, efficient, and expandable framework for network research and education in MATLAB. It can help characterizing empirical networks of dozens of millions of nodes, generating artificial networks, running robustness experiments, testing the resilience of networks to different attacks, simulating arbitrarily complex contagion in the context of epidemiology, marketing, or social media, and generating nice network layouts and even movies representing processes on networks or network evolution.

MATLAB toolbox for the construction of artificial complex networks: Network representation of complex systems, which allows for the analysis of their constituents as important parts of a whole, has led to the development of network science and to the widespread application of its tools. In order to better understand the processes that lead to the presence of characteristics common to several real-world systems, like the network of interactions between proteins or genes within the cell, several network formation models have been proposed.

CNM (http://cn.mathworks.com/matlabcentral/fileexchange/45734-cnm) is a fast, easy-to-use, and well-documented MATLAB toolbox for the construction of artificial complex networks based on such models. It offers the possibility to test the generality of a hypothesis in different configurations, which can lead to important discoveries in the fields where the use of networks is becoming crucial.

MatlabBGL is a MATLAB package for working with graphs, which was written by David Gleich. It uses the Boost Graph Library to efficiently implement the graph algorithms. MatlabBGL is designed to work with large sparse graphs with hundreds of thousands of nodes.

Brain Connectivity Toolbox (http://www.brain-connectivity-toolbox.net) was developed by Mika Rubinov (University of Cambridge), Olaf Sporns, and a growing number of contributors worldwide, which was an open-source MATLAB toolbox for brain network analysis [136–139].

LaNet-vi [140] is a large network visualization tool. It provides images of large-scale networks on a two-dimensional layout. The algorithm is based on the k-core decomposition. A complete description of the algorithm and the visualization layout can be found in the article: k-core decomposition: a tool for the visualization of large-scale networks.

Social Network Visualizer (SocNetV) [141] is a cross-platform, user-friendly application for the analysis and visualization of Social Networks in the form of mathematical graphs, where vertices depict actors/agents and edges represent their relations. With SocNetV, one can construct social networks with a few clicks on a virtual canvas or load field data from various social network file formats such as GraphML, GraphViz, Adjacency, Pajek, UCINET, etc. Furthermore, one can create random networks using various random models (SF, ER, WS SW, ring lattice, d-regular, etc.) or recreate famous social network analysis datasets, i.e., Padgett's Florentine families. A simple web crawler is also included to automatically create networks from links found in a given initial URL. The crawler scans the given web page for links and visualizes the network of all webpages/sites linked from it. SocNetV enables one to edit the social network data through point-and-click, analyze their social and mathematical properties, produce reports for these properties, and apply visualization layouts for relevant presentation of each network. It also supports multirelational loading and editing. One can load a network consisting of multiple relations or create a network on our own and add multiple relations to it. SocNetV easily computes basic graph-theoretic properties, such as density, diameter, geodesics and distances (geodesic lengths), connectedness, eccentricity, etc. But it also computes advanced structural measures for social network analysis, such as centrality and prestige indices (i.e., closeness centrality, betweenness centrality, information centrality, power centrality, proximity, and PageRank prestige), triad census, clique census, clustering coefficient, etc. The application supports various layout algorithms based either on prominence indices (i.e., circular, level and nodal sizes by centrality score) or on force-directed models (i.e., Eades, Fruchterman–Reingold, etc.) for meaningful visualizations of the social networks. There is also comprehensive documentation, both online and inside the application, which explains each feature and algorithm of SocNetV in detail.

Network Workbench (NWB) [142] is a large-scale network analysis, modeling and visualization toolkit for biomedical, social science, and physics research. The NWB will support network science research across scientific boundaries. Users of the NWB will have online access to major network datasets or can upload their own networks. They will be able to perform network analysis with the most effective algorithms available. In addition, they will be able to generate, run, and validate network models to advance their understanding of the structure and dynamics of particular networks. NWB will provide advanced visualization tools to interactively explore and understand specific networks, as well as their interaction with other types of networks.

A major computer science challenge is the development of an algorithm integration framework that supports the easy integration and dissemination of existing and new algorithms and can deal with the multitude of network data formats in existence. Another challenge is the design and implementation of an easy-to-use menu-based, online portal interface for interactive algorithm selection, data manipulation, user and session management. The NWB will be evaluated in diverse research projects and educational settings in biology, social and behavioral science, and physics research. It will be well documented and available as open source for easy duplication and usage at other sites. The NWB will provide members of the scientific research community at large (biologists, physicists, computer scientists, social and behavioral scientists, engineers, etc.) with the means to carry out network analysis, modeling, and visualization projects in their own fields. This will result in a direct transfer of knowledge and results from the fields of specialist network research to a wider scientific community. Researchers will have access to validated algorithms that in the past have been obtained through time-consuming personal developments of ad hoc computer programs. The NWB is expected to enhance and encourage the empirical analysis and model validation of networks, generating an eventual acceleration in the development of network science research. Online instructional material will support the use of the NWB in educational settings. The NWB will provide a unique tool for network science researchers in many disciplines. In effect, NWB can deploy the knowledge accumulated in network theory and practice across sciences with just one web click to any interested researcher, practitioner, or student. The NWB shared resources environment will speed up and ease network science applications and education in biology, social and behavioral science, and large infrastructure analysis, thereby accelerating the rate of scientific discovery.

Except the software or packages mentioned above, there are many other related software tools. For the other related software, one can refer to [143].

1.7 Software for Statistical and Dynamical Analysis

1.7.1 SAS

The **SAS** software is one of three widely used commercial statistical software. A SAS program consists of two parts: The Data part and the Procedure part, which initiated by the words "DATA" and "PROC," respectively. The DATA part allows the user to input and edit data for analysis. The PROC part can perform various statistical analysis through built-in SAS procedures. The SAS procedures can be analogy to built-in functions in MATLAB.

To perform descriptive statistical analysis, the MEAN procedure, the UNIVARIATE procedure, the PLOT and GPLOT procedures are often used. Here, one can derive simple statistics via the MEAN procedure, while detailed statistics as well as normality test can be obtained by using the UNIVARIATE procedure. The PLOT and GPLOT procedures can draw graphs based on the input data or the statistical results.

The hypothesis test can be performed by using the IML (interactive matrix language) procedure, while the analysis on variance can be explored by the ANOVA (Analysis of variance) procedure and the GLM (General linear model) procedure.

For clustering analysis, the built-in procedures in SAS are the CLUSTER and the TREE procedures. The CLUSTER procedure can help to analyze the data and give the detailed results on clustering processes, while the TREE procedure can be used to generate dendrograms, which give graphical details of the clustering processes.

The PCA and the factor analysis can be performed by the PRINCOMP procedure and the FACTOR procedure. For details of the SAS software and the detailed syntax of the mentioned procedures, one can refer to the SAS user guide.

1.7.2 SPSS

SPSS (Statistical product and service solutions) is known to be the first powerful statistical software. SPSS has its advantages in user-friendly interface and clear output files. Data in Excel and DBF format can be freely read by the software.

SPSS is a widely used program for statistical analysis in social science. It is also used by market researchers, health researchers, survey companies, government, education researchers, marketing organizations, data miners, and others. The original SPSS manual has been described as one of "sociology's most influential books" for allowing ordinary researchers to do their own statistical analysis. In addition to statistical analysis, data management and data documentation are features of the base software.

Statistics included in the base software are as follows. Descriptive statistics: Cross tabulation, frequencies, descriptives, explore, descriptive ratio statistics; Bivariate statistics: Means, t-test, ANOVA, correlation (bivariate, partial, distances), nonparametric tests; Prediction for numerical outcomes: Linear regression; Prediction for identifying groups: Factor analysis, cluster analysis (two-step, k-means, hierarchical), discriminant. Most of the features of SPSS statistics are accessible via pull-down menus or can be programmed with a proprietary 4GL command syntax language. Command syntax programming has the benefits of reproducibility, simplifying repetitive tasks, and handling complex data manipulations and analyses. Additionally, some complex applications can only be programmed in syntax and are not accessible through the menu structure. The pull-down menu interface also generates command syntax: this can be displayed in the output, although the default settings have to be changed to make the syntax visible to the user. They can also be pasted into a syntax file using the "paste" button present in each menu. Programs can be run interactively or unattended, using the supplied production job facility.

1.7.3 MATLAB

MATLAB [134] is a multi-paradigm numerical computing environment and fourth-generation programming language. It is a proprietary programming language developed by MathWorks. MATLAB allows matrix manipulations, plotting of functions and data, implementation of algorithms, creation of user interfaces, and interfacing with programs written in other languages, including C, C++, Java, Fortran, and Python.

Although MATLAB is intended primarily for numerical computing, an optional toolbox uses the MuPAD symbolic engine, allowing access to symbolic computing capabilities. An additional package, Simulink, adds graphical multi-domain simulation and model-based design for dynamic and embedded systems.

The features of MATLAB are very similar to C++. There is a special MATLAB editor in which we are supposed to type our programs. It is also very simple to execute a MATLAB program. The following points give us a better picture of the features of MATLAB.

The computations which are done in matrix and vector calculus are very efficient and accurate. For the usage of graphics, it is easy to create them. Graphics are mostly used by scientific and engineering domains. MATLAB is an object-oriented programming language. There are many tools which will help us create a better performing program and will help us get better graphics. Programmer can file I/O functions. One can even develop an application using MATLAB programming language. To make it easy for the user, there is a graphical interface, which is built in.

MATLAB programming language has found itself very useful in many versatile fields. With features like vector computation, numerical matrix, and also the ability to manipulate the algorithm, the software is used for many applications. The first and foremost application is to produce solutions to some of the most complex system of equations. Many students are dependent on MATLAB software to solve high level computation. Simulation is another major application of this software. Some of the other trivial applications are imaging, analysis, visualization, and exploration.

Users can easily develop some specific toolboxes to solve specific questions arising from various branches of science. For example, various complex network toolboxes and the systems biology toolboxes have been developed in MATLAB, as discussed in the above section.

1.7.4 R

R is a programming language and software environment for statistical computing and graphics supported by the R Foundation for statistical computing. The R language is widely used among statisticians and data miners for developing statistical software and data analysis. Polls, surveys of data miners, and studies of scholarly literature databases show that R's popularity has increased substantially in recent years.

R was created by Ross Ihaka and Robert Gentleman at the University of Auckland, and is currently developed by the R Development Core Team. R is named partly after the first names of the first two R authors. The capabilities of R are extended through user-created packages, which allow specialized statistical techniques, graphical devices (ggplot2), import/export capabilities, reporting tools (knitr, Sweave), etc. These packages are developed primarily in R, and sometimes in Java, C, C++, and Fortran.

A core set of packages is included with the installation of R, with more than 15,000 additional packages (as of April 2020) available at the Comprehensive R Archive Network (CRAN), Bioconductor, Omegahat, GitHub, and other repositories. The "Task Views" page on the CRAN website lists a wide range of tasks (in fields such as Finance, Genetics, High Performance Computing, Machine Learning, Medical Imaging, Social Sciences, and Spatial Statistics) to which R has been applied and for which packages are available. R has also been identified by the FDA as suitable for interpreting data from clinical research. Other R package resources include Crantastic, a community site for rating and reviewing all CRAN packages, and R-Forge, a central platform for the collaborative development of R packages, R-related software, and projects. R-Forge also hosts many unpublished beta packages, and development versions of CRAN packages.

The Bioconductor project provides R packages for the analysis of genomic data, such as Affymetrix and cDNA microarray object-oriented data-handling and analysis tools, and has started to provide tools for analysis of data from next-generation high-throughput sequencing methods. The general consensus is that R compares well with other popular statistical packages, such as SAS, SPSS, and STATA.

1.7.5 Some Other Software

1.7.5.1 Small Software for Clustering Analysis

Cluster and **TreeView** [144, 145] are programs that provide a computational and graphical environment for analyzing data from DNA microarray experiments, or other genomic datasets. Cluster organizes and analyzes the data in a number of different ways. TreeView allows the organized data to be visualized and browsed.

Cluster was originally written by Michael Eisen while at Stanford University. In the year 2004, Michiel de Hoon modified the k-means clustering algorithm in Cluster [145], and extended the algorithm for self-organizing maps to include two-dimensional rectangular grids. The Euclidean distance and the city-block distance were added as new distance measures between gene expression data. The proprietary numerical recipes routines, which were used in the original version of Cluster/TreeView, have been replaced by open-source software. The latest version of Cluster—Cluster 3.0, is available for Windows, Mac OS X, Linux, and Unix. It provides several clustering algorithms. Hierarchical clustering methods organize genes in a tree structure, based on their similarity. Four variants of hierarchical clustering are available in Cluster, including the single linkage, the complete linkage, the average linkage, and the WARD method. In k-means clustering, genes are organized into k clusters, where the number of clusters k needs to be chosen in advance. Self-organizing maps create clusters of genes on a two-dimensional rectangular grid, where neighboring clusters are similar. Finally, in the PCA, clusters are organized based on the principal component axes of the distance matrix.

TreeView is a program that allows interactive graphical analysis of the results from Cluster. TreeView reads in matching *.cdt and *.gtr, *.atr, *.kgg, or *.kag files produced by Cluster. The Java program Java TreeView is recommended, which is based on the original TreeView. Java TreeView was written by Alok Saldanha at Stanford University; it can be downloaded from http://jtreeview.sourceforge.net/. Java TreeView runs on the Windows, Mac OS X, Linux, and Unix, and can show both hierarchical and k-means results.

Some other software for clustering analysis include Pycluster and Algorithm:: Cluster for PERL and so on [145].

1.7.5.2 Venn Diagrams

A Venn diagram [146, 147] (also known as a set diagram or logic diagram) is a diagram that shows all possible logical relations among a finite collection of different sets. It is thus a special case of Euler diagrams, which does not necessarily show all relations. Venn diagrams were conceived around 1880 by John Venn [146, 147]. They are used to teach elementary set theory, as well as illustrate simple set relationships in probability, logic, statistics, linguistics, and computer science.

Venn (http://bioinformatics.psb.ugent.be/webtools/Venn/), **VENNY** (http://bioinfogp.cnb.csic.es/tools/venny/index.html), and VENNTURE [148] (http://www.grc.nia.nih.gov) are online software that can draw Venn diagrams, which can help us to explore the relations among datasets. VENNY can draw Venn diagrams for as many as 4 datasets. Venn can calculate the intersections of list of elements. It will generate a textual output indicating which elements are in each intersection or are unique to a certain list. If the number of lists is lower than 6, it will also produce a graphical output in the form of a Venn diagram. You have the choice between symmetric (default) or non-symmetric Venn diagrams. Currently, one is able to calculate the intersections of at maximum 30 lists. The graphical output is produced in SVG and PNG format. VENNTURE can facilitate the visualization of up to six datasets (6-way) in a user-friendly manner. VENNTURE includes versatile output features, where grouped data points can be easily exported into a spreadsheet. Up to now, it is generally difficult to visualize the Venn diagrams for more than seven sets. For the visualization of seven sets Venn diagram, one can refer to http://moebio.com/research/sevensets/. The 7-way Venn Diagram is designed by Moebio Labs, which is inspired by Newton's theories on light and color spectrum, and is based on 128 color combinations from mixing 7 colors.

1.7.5.3 Software for Bifurcation and Dynamical Analysis

Oscill8 (http://oscill8.sourceforge.net) is a suite of tools for analyzing large systems of ordinary differential equations (ODEs), particularly with respect to understanding how the high dimensional parameter space controls the dynamics of the system. The suite includes the following user features:

1. Time course/integration (using CVODE);
2. Multiple time course generation (incrementing parameters, etc.);
3. One-parameter bifurcation diagrams (batch mode, several parameters);
4. Two-parameter bifurcation diagrams (batch mode, several parameters); Bifurcation searches (subsequent restart available);
5. Organization of output, including: notes for each run, analysis on subsets of parameters, graphical output is stored with model and state.

The goal of Oscill8 is to allow a user to concentrate less on the details of bifurcation analysis, and more on the results obtained from that analysis. When dealing with models containing a large number of parameters, one needs access to large amounts of graphical data revealing the bifurcation behavior of the model, and one needs to be able to get this data rapidly.

XPPAUT [149] (http://www.math.pitt.edu/~bard/xpp/xpp.html) is also known as XPP, which is a tool for simulating, animating, and analyzing dynamical systems. The program evolved from a DOS program that was originally written by John Rinzel and Bard Ermentrout, which was firstly used to illustrate the dynamics of a simple model for an excitable membrane. The XPPAUT can solve differential equations, difference equations, delay equations, functional equations, boundary value problems, and stochastic equations. XPP contains the code for the popular bifurcation program—AUTO. Thus, one can switch back and forth between XPP and AUTO, using the values of one program in the other and vice versa. The code brings together a number of useful algorithms and is extremely portable. All the graphics and interface are written completely in Xlib, which explains the somewhat idiosyncratic and primitive widgets interface. Now, XPP is available for the iPad/iPhone.

MATCONT (https://sourceforge.net/projects/matcont/) [150] is a MATLAB package, which can be used to perform bifurcation analysis and basic dynamical analysis. MATCONT provides means for continuing equilibria and limit cycles (periodic orbits) of ODEs systems, and their bifurcations (including branch points). It also provides access to all standard ODE solvers supplied by MATLAB, as well as to two new stiff solvers, ode78 and ode87. MATCONT computes Poincare maps, as well as phase response curves for limit cycles and their derivatives as a byproduct of the continuation of limit cycles. These curves are fundamental for the study of the behavior of oscillators and their synchronization in networks. For equilibria, the software supports the computation of critical normal form coefficients for all codimension 1 and 2 bifurcations. For limit cycles, it supports the computation of critical coefficients of periodic normal forms for codimension 1 bifurcations. Finally, the continuation of homoclinic orbits (both to hyperbolic saddles and to saddle-nodes) is supported by MATCONT, together with detection of a large number of codimension 2 bifurcations along the homoclinic curves. Most curves are computed with the same prediction-correction continuation algorithm based on the Moore–Penrose matrix pseudo-inverse. The continuation of bifurcation points of equilibria and limit cycles is based on bordering methods and minimally extended systems. Besides sophisticated numerical methods, MATCONT provides data storage and a modern graphical user interface.

1.8 Organization of the Book

Surrounding the topic of modeling and analysis of bio-molecular networks, various sophisticated mathematical and statistical approaches have been introduced in this book. This book first overviews approaches to reconstruct various bio-molecular networks (Chap. 2), and then we discuss the modeling and dynamical analysis of simple genetic circuits (Chap. 3), coupled genetic circuits (Chap. 4), middle-sized and large-scale biological networks (Chap. 5); Relationships among the structures, dynamics, and functions of the considered networks have been clarified. Subsequently, for large-scale bio-molecular networks, we introduce some statistical methods to explore important bioinformatics, including evolutionary mechanisms of network motifs (Chap. 6), the identification of important bio-molecules for the purpose of network medicine and genetic engineering (Chap. 7), the graphical features of functional genes in bio-molecular networks (Chap. 8). Finally, some state-of-the-art statistical methods have been discussed to analyze omics data that

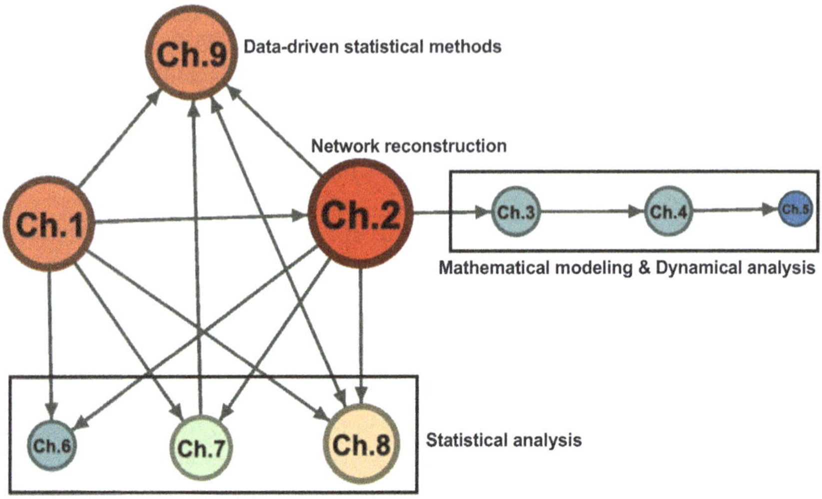

Fig. 1.7 Organization of the book. The arrowheads indicate the directions in which the chapters relate to each other. Chapter 1 describes the foundations of the book; Chap. 2 deals with network reconstruction; Chaps. 3, 4, and 5 introduce the mathematical modeling and dynamical analysis of bio-molecular networks, ranging from simple circuits to large-scale ones; Chaps. 6, 7, and 8 describe the statistical analysis of bio-molecular networks; Chap. 9 introduces the data-driven statistical approaches for omics data analysis in biological systems

are generated from high-throughput sequencing (Chap. 9). Figure 1.7 gives an overview on which chapters are closely related to each other.

References

1. Watson, J. D., Baker, T. A., Bell, S. P., Gann, A., Levine, M., Losick, R.: Molecular biology of the gene (5th Edition). Cold Spring Harbor Labor. Press, New York (2004)
2. Alberghina, L., Westerhoff, H. V. (eds.): Systems biology: definitions and perspectives. Springer-Verlag, Berlin (2005)
3. Ahmed, Z.: Physical biology: from atoms to medicine. Imperial College Press, London (2008)
4. Bu, Z., Callaway, D. J.: Proteins MOVE! Protein dynamics and long-range allostery in cell signaling. Adv. Protein. Chem. Struct. Biol. **83**, 163–221 (2011)
5. Systems biology: https://en.wikipedia.org/wiki/Systems_biology
6. Sauer, U., Heinemann, M., Zamboni, N.: Genetics: getting closer to the whole picture. Science **316**, 550–551 (2007)
7. Noble, D. : The music of life: Biology beyond the genome. Oxford Univ. Press, Oxford (2006)
8. Kholodenko, B.N., Sauro, H.M.: Mechanistic and modular approaches to modeling and inference of cellular regulatory networks. In: Alberghina, L., Westerhoff, H.V. (eds.) Systems biology: definitions and perspectives. 357–451. Springer-Verlag, Berlin (2005)
9. Chiara, R., Gerolamo, L., Statistical tools for gene expression analysis and systems biology and related web resources. In: Stephen, K. (eds.) Bioinformatics for Systems Biology (2nd ed.), 181–205. Humana Press, New York (2009)

10. Voit, E.: A first course in systems biology. Garland Science, New York, (2012)
11. Baitaluk, M.: System biology of gene regulation. Biomed. Informat. **569**, 55–87 (2009)
12. Bertalanffy, L.V.: General system theory: foundations, development, applications. George Braziller, New York (1968)
13. Hodgkin, A.L., Huxley, A.F.: A quantitative description of membrane current and its application to conduction and excitation in nerve. J. Physiol. **117**, 500–544 (1952)
14. Le Novére, N.: The long journey to a systems biology of neuronal function. BMC Syst. Biol. **1**, 1 (2007)
15. Noble, D.,: Cardiac action and pacemaker potentials based on the Hodgkin-Huxley equations. Nature **188**, 495–497 (1960)
16. Mesarovic, M.D.: Systems theory and biology. Springer-Verlag, Berlin (1968)
17. Zeng, B.J.: On the holographic model of human body. The 1st national conference of comparative studies traditional Chinese Medicine and West Medicine, Medicine and Philosophy, April (1992)
18. Zeng, B.J.: On the concept of system biological engineering. Commun. Transgenic Animals **6** (1994)
19. Zeng, B.J.: Transgenic animal expression system-transgenic egg plan. Commun. Transgenic Animals **1** (1994)
20. Zeng, B.J.: From positive to synthetic medical science. Commun. Transgenic Animals **11** (1995)
21. Zeng, B.J.: The structure theory of self-organization systems. Commun. Transgenic Animals **8–10** (1996)
22. Tomita, M., Hashimoto, K., Takahashi, K., et. al.: E-CELL: Software environment for whole cell simulation. Genome. Inform. **8**, 147–155 (1997)
23. Kling, J.: Working the systems. Science **311**, 1305-1306 (2006)
24. Macilwain, C.: Systems biology: evolving into the mainstream. Cell **144**, 839–841 (2011)
25. Arkin, A.P., Schaffer, D.V.: Network news: innovations in 21st century systems biology. Cell **144**, 844–849 (2011)
26. Nurse, P., Hayles, J.: The cell in an era of systems biology. Cell **144**, 850–854 (2011)
27. Novák, B., Chen, K.C., Tyson, J.J., Systems biology of the yeast cell cycle engine. In: Alberghina, L., Westerhoff, H.V. (eds.) Systems biology: definitions and perspectives, 305–324. Springer-Verlag, Berlin (2005)
28. Alberghina , L., Rossi, R.L., Porro, D., Vanoni, M.: A modular systems biology analysis of cell cycle entrance into S-phase. In: Alberghina, L., Westerhoff, H.V. (eds.) Systems biology: definitions and perspectives, 305–324. Springer-Verlag, Berlin (2005)
29. Mathan, S., Smith, A., A., Kumaran, J., Prakash, S.: Anticancer and antimicrobial activity of *Aspergillus protuberus* SP1 Isolated from marine sediments of South Indian coast, Chin. J. Nat. Med. **9(4)**, 0286–0292 (2011)
30. Newman, M.E.J.: Networks: an introduction. OUP Oxford, New York (2010)
31. Barabási, A.L., Bonabeau, E.: Scale-free networks. Sci. Amer. **288**, 50–59 (2003)
32. Strogatz, S.H., Watts, D.J.: Collective dynamics of 'small-world' networks. Nature **393**, 440–442 (1998)
33. Stanley, H.E., Amaral, L.A.N., Scala, A., Barthelemy, M.: Classes of small-world networks. Proc. Natl. Acad. Sci. USA. **97**, 11149–11152 (2000)
34. Buldyrev, S.V., Parshani, R., Paul, G., Stanley, H.E., Havlin, S.: Catastrophic cascade of failures in interdependent networks. Nature **464**, 1025–1028 (2010)
35. Parshani, R., Buldyrev, S.V., Havlin, S.: Interdependent networks: reducing the coupling strength leads to a change from a first to second order percolation transition. Phys. Rev. Lett. **105**, 048701 (2010)
36. Majdandzic, A., Podobnik, B., Buldyrev, S.V., Kenett, D.Y., Havlin, S., Stanley, H.E.: Spontaneous recovery in dynamical networks. Nat. Phys. **10**, 34–38 (2014)
37. Motter, A.E., Albert, R.: Networks in motion. Phys. Today **65**, 43–48 (2012)
38. Zhang, Z.K., Liu, C., Zhan, X.X., et al.: Dynamics of information diffusion and its applications on complex networks. Phys. Rep. **651**, 1–34 (2016)

39. Pastor-Satorras R., Castellano C., Van Mieghem P., et al.: Epidemic processes in complex networks. Rev. Mod. Phys. **87**, 925–946 (2015)
40. Wang, Z., Moreno, Y., Boccaletti, S., et al.: Vaccination and epidemics in networked populations-an introduction. Chaos, Solitons & Fractals **103**, 177–183 (2017)
41. Gallos, L.K., Liljeros, F., Argyrakis, P., et al.: Improving immunization strategies. Phys. Rev. E **75**, 045104 (2007)
42. Levine, M.M., Sztein, M.B.: Vaccine development strategies for improving immunization: the role of modern immunology. Nat. Immun. **5**(5), 460–464 (2004)
43. Wang, X., Li, X., Chen, G.: Complex network: theory & application. Qinghua University Press, 2006 (In chinese)
44. Wang, X., Li, X., Chen, G.: Network science: an introduction, Higher Education Press, 2012 (In chinese)
45. Lü, J., Tan, S.: Games and evolutionary dynamics on complex networks, Higher Education Press, 2019 (In chinese)
46. Erdös, P., Rényi, A.: On random graphs. Publ. Math. Debrecen **6**, 290–297 (1959)
47. Gfeller D.: Simplifying complex networks: from a clustering to a coarse graining strategy, Ph.D thesis, Univ. of Lausanne (2007)
48. Bollobás, B.: Random graphs. Academic Press, London (1985)
49. B. Bollobás, B., Chung, F.R.: Probabilistic combinatorics and its applications. In Proc. Symp. Appl. Math., Amer. Math. Soc. **44**, (1991)
50. Barabási, A.L., Albert, R.: Emergence of scaling in random networks. Science **286**, 509–512 (1999)
51. Bollobás, B., Riordan O.M.: Mathematical results on scale-free random graphs. In: Bornholdt, S., Schuster, H.G. (eds.) Handbook of Graphs and Networks: From the Genome to the Internet, 1–34. Wiley-VCH, Berlin (2003)
52. Reuven, C., Shlomo, H.: Scale-free networks are ultrasmall. Phys. Rev. Lett. **90**, 058701 (2002)
53. Fronczak, A., Fronczak, P., Holyst, J.A.: Mean-field theory for clustering coefficients in Barabási-Albert networks. Phys. Rev. E **68**, 046126 (2003)
54. Newman, M.E.J.: Power laws, Pareto distributions and Zipf's law. Contemporary Phys. **46**, 323–351 (2005)
55. Krapivsky, P.L., Redner, S., Leyvraz, F.: Connectivity of growing random networks. Phys. Rev. Lett. **85**, 4629 (2000)
56. Zhou, S., Mondragón, R.J.: Accurately modeling the Internet topology. Phys. Rev. E **70**, 066108 (2004)
57. Catanzaro, M., Caldarelli, G., Pietronero, L.: Social network growth with assortative mixing. Physica A **338**, 119–124 (2004)
58. Guimera, R., Amaral, L.A.N.: Modeling the world-wide airport network. Eur. Phys. J. B **38**, 381–385 (2004)
59. Fortunato, S., Flammini, A., Menczer, F.: Scale-free network growth by ranking. Phys. Rev. Lett. **96**, 218701 (2006)
60. Klemm, K., Eguiluz, V.M.: Growing scale-free networks with small-world behavior. Phys. Rev. E **65**, 057102 (2002)
61. Barrat, A., Barthelemy, M., Pastor-Satorras, R., Vespignani, A.: The architecture of complex weighted networks. Proc. Natl. Acad. Sci. USA. **101**, 3747–3752 (2004)
62. Barrat, A., Barthélemy, M., Vespignani, A.: Weighted evolving networks: coupling topology and weight dynamics. Phys. Rev. Lett. **92**, 228701 (2004)
63. Yook, S.H., Jeong, H., Barabási, A.L., Tu, Y.: Weighted evolving networks. Phys. Rev. Lett. **86**, 5835 (2001)
64. Chung, F., Lu, L.: Connected components in random graphs with given expected degree sequences. Ann. Combinat. **6**, 125–145 (2002)
65. Boguná, M., Pastor-Satorras, R.: Class of correlated random networks with hidden variables. Phys. Rev. E **68**, 036112 (2003)

66. Caldarelli, G., Capocci, A., De Los Rios, P., Munoz, M.A.: Scale-free networks from varying vertex intrinsic fitness. Phys. Rev. Lett. **89**, 258702 (2002)
67. Vázquez, A., Flammini, A., Maritan, A., Vespignani, A.: Modeling of protein interaction networks. Complexus **1**, 38–44 (2003)
68. Wagner, A.: The yeast protein interaction network evolves rapidly and contains few redundant duplicate genes. Mol. Biol. Evol. **18**, 1283–1292 (2001)
69. Molloy, M., Reed, B.: The size of the giant component of a random graph with a given degree sequence. Comb. Probab. Comput. **7**, 295–305 (1998)
70. Molloy, M., Reed, B.: A critical point for random graphs with a given degree sequence. Random. Struct. Algor. **6**, 161–180 (1995)
71. Garlaschelli, D., Loffredo, M.I.: Maximum likelihood: extracting unbiased information from complex networks. Phys. Rev. E **78**, 015101 (2008)
72. Newman, M.E.J., Watts, D.J.: Renormalization group analysis of the small-world network model. Phys. Lett. A **263**, 341–346 (1999)
73. Newman, M.E.J.: Mixing patterns in networks. Phys. Rev. E **67**, 026126 (2003)
74. Newman, M.E.J.: Assortative mixing in networks. Phys. Rev. Lett. **89**, 208701 (2002)
75. Newman, M.E.J.: A measure of betweenness centrality based on random walks. Soc. Netw. **27**, 39–54 (2005)
76. Humphries, M.D., Gurney, K.: Network 'small-worldness': a quantitative method for determining canonical network equivalence. PLoS One **3**, e0002051 (2008)
77. Newman, M.E.J.: The structure and function of complex networks. SIAM Rev. **45**, 167–256 (2003)
78. Ravasz, E., Barabási, A.L.: Hierarchical organization in complex networks. Phys. Rev. E **67**, 026112 (2003)
79. Newman, M.E.J.: Modularity and community structure in networks. Proc. Natl. Acad. Sci. USA. **103**, 8577–8582 (2006)
80. Raff, R.A.: The shape of life. Chicago Univ. Press, Chicago (1996)
81. Newman, M.E.J., Girvan, M.: Finding and evaluating community structure in networks. Phys. Rev. E **69**(2), 026113 (2003)
82. Newman, M.E.J.: Finding community structure in networks using the eigenvectors of matrices. Phys. Rev. E **74**(3), 036104 (2006)
83. Anand, K., Bianconi, G.: Entropy measures for networks: toward an information theory of complex topologies. Phys. Rev. E **80**, 045102 (2009)
84. Massa, P., Salvetti, M., Tomasoni, D.: Bowling alone and trust decline in social network sites in 2009 Eighth IEEE Int. Conf. Dependable, Autonomic and Secure Comput. 658–663 (2009)
85. Fire, M., Puzis, R., Elovici, Y.: Link prediction in highly fractional data sets, ed. Subrahmanian V. (Springer New York, New York, NY), 283–300 (2013)
86. Coleman, J.S.: Introduction to mathematical sociology. London Free Press Glencoe, (1964)
87. Moody, J.: Peer influence groups: identifying dense clusters in large networks. Soc. Netw. **23**(4), 261–283 (2001)
88. Freeman, L.C., Webster, C.M., Kirke, D.M.: Exploring social structure using dynamic three-dimensional color images. Soc. Netw. **20**(2),109–118 (1998)
89. Opsahl, T., Agneessens, F., Skvoretz, J.: Node centrality in weighted networks: generalizing degree and shortest paths. Soc. Netw. **3**(32), 245–251 (2010)
90. Opsahl, T.: Why anchorage is not (that) important: binary ties and sample selection (2011) (accessed on 2016.08.06)
91. Kunegis, J.: Spanish book network dataset, KONECT, (2016)(accessed on 2016.08.06)
92. Subelj, L., Bajec, M.: Model of complex networks based on citation dynamics. Proc. WWW Workshop on Large Scale Netw. Anal. 527–530 (2013)
93. Ley, M.: The DBLP computer science bibliography: Evolution, research issues, perspectives. Proc. Int. Symp. String Processing and Information Retrieval. 1–10 (2002)
94. Leskovec, J., Kleinberg, J., Faloutsos, C.: Graph evolution: Densification and shrinking diameters. ACM Trans. Knowledge Discovery from Data **1**(1),1–40 (2007)
95. Opsahl, T., Panzarasa. P.: Clustering in weighted networks. Soc. Netw. **31**(2),155–163 (2009)

96. Leskovec, J., Huttenlocher, D., Kleinberg, J.: Governance in social media: A case study of the Wikipedia promotion process in Proc. Int. Conf. on Weblogs and Social Media (2010)
97. Martinez, N.D., Magnuson, J.J., Kratz, T., Sierszen, M.: Artifacts or attributes? Effects of resolution on the Little Rock Lake food web. Ecological Monographs **61**, 367–392 (1991)
98. Kunegis, J., Hamsterster friendships network dataset, KONECT, (2016)
99. Rual, J.F., Venkatesan, K., Hao, T., et al.: Towards a proteome-scale map of the human protein- protein interaction network. Nature **437**(7062), 1173–1178 (2005)
100. Stumpf, M.P.H., Wiuf, C., May, R.M.: Subnets of scale-free networks are not scale-free: Sampling properties of networks. Proc. Natl. Acad. Sci. USA.**102**(12), 4221–4224 (2005)
101. Spring, N., Mahajan, R., Wetherall, D., Anderson, T.: Measuring ISP topologies with rocketfuel. IEEE/ACM Trans. Networking **12**(1), 2–16 (2004)
102. Batagelj, V., Mrvar, A.: Pajek datasets. (2006) (accessed on 2016.08.06)
103. Harrison, C.: Bible cross-references (http://chrisharrison.net/projects/bibleviz/index.html(accessed on 2014.08.22)) (2014)
104. Guimerá, R., Danon, L., Díaz-Guilera, A., Giralt, F., Arenas, A.: Self-similar community structure in a network of human interactions. Phys. Rev. E **68**(6), 065103 (2003)
105. Gleiser, P.M., Danon, L.: Community structure in jazz. Adv. Complex Syst. **6**(4), 565–573 (2003)
106. Boguna, M., Pastor-Satorras, R., Díaz-Guilera, A., Arenas, A.: Models of social networks based on social distance attachment. Phys. Rev. E **70**(5), 056122 (2004)
107. Adamic, L.A., Glance, N.: The political blogosphere and the 2004 U.S. election: Divided they blog. Proc. 3rd Int. Workshop on Link Discovery, LinkKDD'05. (ACM, New York, NY, USA), 36–43 (2005)
108. Kunegis, J.: American revolution network dataset, KONECT (2016)
109. Barnes, R., Burkett, T.: Structural redundancy and multiplicity in corporate networks. Int. Network for Social Netw. Anal. **30**(2), (2010)
110. Rocha, L.E.C., Liljeros, F., Holme, P.: Information dynamics shape the sexual networks of internet-mediated prostitution. Proc. Natl. Acad. Sci. USA. **107**(13), 5706–5711 (2010)
111. Wikimedia Foundation (2010) Wikimedia downloads (http://dumps.wikimedia.org/).
112. Lodish, H., Berk, A., Kaiser, C.A., et al.: Molecular cell biology. 8th Edition, Freeman & Co., New York (2016)
113. Carlberg, C., Molnár, F.: Overview: what is gene expression? In: Mechanisms of Gene Regulation. Springer, Dordrecht. (2014)
114. Chen, L., Wang, R.S., Zhang, X.S.: Biomolecular networks: methods and applications in systems biology. John Wiley & Sons, Hoboken (2009)
115. Lange, B.M.: Counting the cost of a cold-blooded life: Metabolomics of cold acclimation. Proc. Natl. Acad. Sci. USA. **101**, 14996–14997 (2004)
116. Oltvai, Z.N., Barabási, A.L.: Life's complexity pyramid. Science **298**, 763–764 (2002)
117. Gu, Z., Zhang, C., Wang, J.: Gene regulation is governed by a core network in hepatocellular carcinoma. BMC Syst. Biol. **6**, 32 (2012)
118. Hanahan, D., Weinberg, R.A.: The hallmarks of cancer. Cell **100**, 57–70 (2000)
119. Wang, P., Lü, J., Yu, X., Liu, Z.: Duplication and divergence effect on network motifs in undirected bio-molecular networks. IEEE Trans. Biomed. Circuits Syst. **9**, 312–320 (2015)
120. Johnson, R.A., Wichern, D.W.: Applied multivariate statistical analysis 6th Edition. Pearson Education, Upper Saddle River (2007)
121. MacQueen, J.B.: Some Methods for classification and Analysis of Multivariate Observations. Proc. 5th Berkeley Symp. Math. Statistics Prob. 281–297 (1967)
122. Steinhaus, H.: Sur la division des corps matériels en parties. Bull. Acad. Polon. Sci. (in French) **4**, 801–804(1957)
123. Lloyd, S.P. Least squares quantization in PCM. IEEE T. Inform. Theory **28**, 129–137 (1982)
124. Forgy, E.W.: Cluster analysis of multivariate data: efficiency versus interpretability of classifications. Biometrics **21**, 768–769 (1965)
125. Hartigan, J.A., Wong, M.A.: Algorithm AS 136: A k-means clustering algorithm. J. Roy. Stat. Soc. C-APP. **28**, 100–108 (1979)

126. Person, K.: On lines and planes of closest fit to system of points in space. Philiosophical Mag. **2**, 559–572 (1901)
127. Hotelling, H.: Analysis of a complex of statistical variables into principal components. J. Educ. Psychol. **24**, 417–441(1933).
128. De Nooy, W., Mrvar, A., Batagelj, V.: Exploratory social network analysis with Pajek. Cambridge Univ. Press, Cambridge (2011)
129. Bastian, M., Heymann, S., Jacomy, M.: Gephi: an open source software for exploring and manipulating networks. International AAAI Conference on Weblogs and Social Media. **8**, 361–362 (2009)
130. Leetaru, K.: Culturomics 2.0: Forecasting large-scale human behavior using global news media tone in time and space. First Monday **16** (2011)
131. Aouragh, M.: Collateral damage: Oslo attacks and proliferating islamophobia. Jadaliyya (2011)
132. Panisson: The Egyptian revolution on Twitter - featured on the PBS news hour. YouTube (2011)
133. Shannon, P., Markiel, A., Ozier, O., et al.: Cytoscape: a software environment for integrated models of biomolecular interaction networks. Genome Res. **13**, 2498–2504 (2003)
134. MATLAB programming language. Altius Directory. Retrieved 17 Dec. 2010
135. Muchnik, L.: Complex networks package for MatLab (Version 1.6). http://www.levmuchnik.net/Content/Networks/ComplexNetworksPackage.html
136. Sporns, O., Graph theory methods for the analysis of neural connectivity patterns. In: Kötter R, editor. Neuroscience databases. A practical guide. Boston: Klüwer, 171–186 (2002)
137. Sporns, O., Kötter, R.: Motifs in brain networks, PLoS Biol. **2**(11), e369 (2004)
138. Sporns, O., Tononi, G.: Classes of network connectivity and dynamics. Complexity **7**, 28–38 (2002)
139. Sporns, O., Zwi, J.: The small world of the cerebral cortex. Neuroinformat. **2**, 145–162 (2004)
140. http://lanet-vi.soic.indiana.edu
141. http://socnetv.sourceforge.net
142. NWB Team: Network Workbench Tool. Indiana University, Northeastern University, and University of Michigan, http://nwb.slis.indiana.edu
143. http://www.caida.org/projects/internetatlas/viz/viztools.html
144. Eisen, M.B., Spellman, P.T., Brown, P.O., Botstein, D.: Cluster analysis and display of genome-wide expression patterns. Proc. Natl. Acad. Sci. USA. **95**, 14863–14868 (1998)
145. de Hoon, M. J., Imoto, S., Nolan, J., Miyano, S.: Open source clustering software. Bioinformat. **20**, 1453–1454 (2004)
146. Venn, J.: On the diagrammatic and mechanical representation of propositions and reasonings. Philosophical Magazine and J. Sci. **10**, 1–18 (1880)
147. Venn, J.: On the employment of geometrical diagrams for the sensible representations of logical propositions. Proc. Cambridge Philosophical Society **4**, 47–59 (1880)
148. Martin, B., Chadwick, W., Yi, T., Park, S.S. et. al.: VENNTURE–a novel Venn diagram investigational tool for multiple pharmacological dataset analysis. PLoS One **7**, e36911 (2012)
149. Ermentrout, B.: Simulating, analyzing, and animating dynamical systems: a guide to XPPAUT for researchers and students. SIAM, Philadelphia (2002)
150. Dhooge, A., Govaerts, W., Kuznetsov, Y.A.: MATCONT: a MATLAB package for numerical bifurcation analysis of ODEs. ACM Trans. Math. Softw. **29**, 141–164 (2003)

Part I
Modeling and Dynamical Analysis of Bio-molecular Networks

This part deals with the reconstruction, mathematical modeling, and dynamical analysis of biological networks. Chapter 2 discusses some state-of-the-art methods on the reconstruction of bio-molecular networks. Chapter 3 introduces some works on the mathematical modeling and dynamical analysis of several simple network motifs. Chapters 4 and 5 will discuss how to perform mathematical modeling and dynamical analysis on several coupled genetic circuits and large-scale bio-molecular networks, respectively.

Chapter 2
Reconstruction of Bio-molecular Networks

Abstract Network reconstruction is the first step for subsequent network analysis, which is an inverse problem and open issue. In this chapter, we will introduce how to construct bio-molecular networks. Generally speaking, bio-molecular networks can be constructed from four approaches: (1) constructing bio-molecular networks from timely updated online databases or published papers; (2) generating artificial bio-molecular networks based on artificial computer algorithms; (3) inferring bio-molecular networks from behavioral data of biological entities via sophisticated mathematical or statistical methods; (4) topological identification of complex systems via complex dynamical network theory. Reconstruction of bio-molecular networks facilitates our further mathematical modeling, dynamical analysis, and statistical analysis on the related life systems.

2.1 Backgrounds

Network reconstruction is a typical inverse problem in the area of systems biology and complex networks science. In fact, network reconstruction is the first step for network analysis, it is an interesting and increasingly important scientific topic. Our aim of network reconstruction is to infer the relationships among entities in a system, based on experimental technologies, existing data, mathematical and statistical models or general evolution mechanisms of the system. Approaches of network reconstruction can be roughly classified into the following four cases.

Case 1: Construction of bio-molecular networks based on online databases. Various databases have collected timely updated both experimentally and statistically inferred interaction data among molecules [1–9]. Both physical interaction data and functional interaction data have been collected in the existing databases. Some of the collected data are from existing references that were predicted from mathematical or statistical models, but mostly are from high-throughput technologies. Various high-throughput technologies [10] are ceaselessly developed to experimentally determine the relationships among various biological molecules. Some typical

J. Lü, P. Wang, *Modeling and Analysis of Bio-molecular Networks*,
https://doi.org/10.1007/978-981-15-9144-0_2

methods or platforms to detect PPIs include the standard high-throughput *yeast*-two-hybrid (Y2H) assays [11], phage display technology, surface plasmon resonance, fluorescence resonance energy transfer, protein chip mass spectrometry technology, co-immunoprecipitation, GST pull-down technology, and CrY2H-seq [12], whereas ChIP-Seq (Chromatin Immunoprecipitation sequencing) and CLIP-Seq (cross-linking-immunoprecipitation and high-throughput sequencing (HTS)) can be used to detect the interaction between proteins and DNAs or RNAs.

Case 2: Evolutionary mechanisms of the system were known, how to generate networks that follow these mechanisms, and the generated networks can well mimic the statistical features of the system [13–34]. Under this circumstance, various artificial models have been developed, such as the famous WS or NW SW model [13, 15], the BA SF model [14, 15], models based on the duplication and divergence mechanisms for generating artificial bio-molecular networks [17–34].

Case 3: Behavioral data of each entity in the system was known, how to statistically infer the structure among the entities [35–37]. For this case, one can use various similarity measures, distance measures, mutual dependence measures or optimization techniques [37] to evaluate the relationships among entities in the system.

Case 4: Structure or network of the system was partly known, and we further know the dynamic equations of each node, our task is to infer the possible unknown structures (edges) of the system. For this case, various methods have been developed in the area of complex networks, which are based on networked control and synchronization, and are also known as topology identification [38–48].

In the following sections of this chapter, we will introduce the basic information and ideas, as well as necessary mathematical and statistical models for this topic.

2.2 Reconstruction of Bio-molecular Networks Based on Online Databases

2.2.1 Regulatory Networks

A gene is a locus or region of DNA that encodes a functional RNA or protein product, and is the molecular unit of heredity. The activity of genes is regulated by TFs, proteins that typically bind to DNA. Most TFs bind to multiple binding sites in a genome. As a result, all cells have complex GRNs. For instance, the human genome encodes on the order of 1400 DNA-binding TFs that regulate the expression of more than 20,000 human genes [49]. Technologies to detect GRNs include ChIP-chip, ChIP-seq, CliP-seq, and others. Mapping of the human regulatory network is still in its infancy, making this network perhaps the most incomplete among all biological networks. Data generated by experimental techniques, such as chromatin immunoprecipitation (ChIP) followed by microarrays (ChIP-chip) and ChIP followed by sequencing (ChIP-seq), have started to be collected in databases such as the Universal Protein Binding Microarray Resource for Oligonucleotide Binding Evaluation (UniPROBE) and JASPAR. Literature-curated and predicted protein–DNA interactions have been compiled in various databases, such as TRANSFAC and the B-cell interactome (BCI). Human posttranslational modifications can be found in databases such as Phospho.ELM, PhosphoSite, phosphorylation site database (PHOSIDA), NetPhorest, and the CBS prediction database.

Reconstruction of regulatory networks is known as reverse engineering of regulatory networks. The most convenience approach is based on continuously cumulated online databases and literature, such as the databases as listed in Table 2.1. Different databases encompass different amounts of datasets and are with different level of reliability. Reconstruction of regulatory networks should integrate the datasets from many different databases and need to filter out false positive data. Another

Table 2.1 Databases for regulatory networks

Database	Website	Description
UniPROBE	uniprobe.org	Universal Protein Binding Microarray Resource for oligonucleotide binding evaluation
JASPAR	jaspar.genereg.net	A curated, non-redundant set of profiles, derived from published collections of experimentally defined TF binding sites for eukaryotes. The prime difference to similar resources (TRANSFAC, etc.) consist of the open data access, non-redundancy, and quality
TRANSFAC	www.biobase-international.com/gene-regulation	Provides data on eukaryotic TFs, their experimentally-proven binding sites, consensus binding sequences (positional weight matrices), and regulated genes
PHOSIDA	141.61.102.18/phosida/index.aspx	Phosida allows the retrieval of phosphorylation, acetylation, and N-glycosylation data of any protein of interest. It lists posttranslational modification sites associated with particular projects and proteomes or, alternatively, displays posttranslational modifications found for any protein or protein group of interest. In addition, structural and evolutionary information on each modified protein and posttranslational modification site is integrated

(continued)

Table 2.1 (continued)

Database	Website	Description
CBS	www.cbs.dtu.dk/index.shtml	The Center for Biological Sequence Analysis at the Technical University of Denmark was formed in 1993, and conducts basic research in the field of bioinformatics and systems biology
TRRUST	https://www.grnpedia.org/trrust/	A manually curated database of human and mouse transcriptional regulatory networks. Current version of TRRUST contains 8444 and 6552 TF-target regulatory relationships of 800 human TFs and 828 mouse TFs, respectively
LncMAP	http://bio-bigdata.hrbmu.edu.cn/LncMAP/index.jsp	LncMAP systematic dissection of lncRNA mediated transcriptional regulations perturbations in 20 types of cancer
PlantCircNet	http://bis.zju.edu.cn/plantcircnet/index.php	It is the first database that provides plant circRNA-miRNA-gene regulatory networks, as well as circRNA information and circRNA expression profiles
DisNor	https://disnor.uniroma2.it/#about-disnor	A disease-focused resource that uses the causal interaction information annotated in SIGNOR and the PPI data in mentha to generate and explore PPI networks linking disease genes

extensively used approach is based on expression data, the objective is to reveal the underlying network of regulatory interactions from the measured datasets of expression. Many methods have been developed to reconstruct GRNs, such as the singular value decomposition method [37, 50] and the model-based optimization method [37, 51, 52]. We will introduce some of them in the following sections.

2.2.2 Protein–Protein Interaction Networks

Many PPIs in a cell form PPI networks, where proteins are nodes and their interactions are edges. PPI networks are the most intensely analyzed networks in biology. There are dozens of PPI detection methods to identify such interactions. The Y2H system is a commonly used experimental technique for the study of binary interactions [53, 54].

Recent studies have indicated conservation of molecular networks through deep evolutionary time [55]. Moreover, it has been discovered that proteins with high degrees of connectedness are more likely to be essential for survival than proteins with lower degrees [56]. This suggests that the overall composition of the network (not simply interactions between protein pairs) is important for the overall functioning of an organism.

In the past 5 years, significant efforts towards obtaining comprehensive protein interaction maps have been made. High-throughput Y2H maps for humans have been generated by several groups [57–61], yielding more than 7000 binary interactions. The immunoprecipitation and high-throughput mass spectrometry technique, which identifies co-complexes, is now being applied to humans as well [62]. There have also been major efforts to curate the interactions that have been validated individually in the literature into databases [63–72] such as the Münich Information Center for Protein Sequence (MIPS) protein interaction database, the Bio-molecular Interaction Network Database (BIND), the Database of Interacting Proteins (DIP), the Molecular Interaction database (MINT), and the protein Interaction database (IntAct). More recent PPI curation efforts, including the Biological General Repository for Interaction Datasets (BioGRID) and the Human Protein Reference Database (HPRD), have attempted larger-scale curation of data. Additionally, the STRING database contains known and predicted PPIs. The databases and related websites can be referred to Table 2.2. Despite these extensive curation efforts, the existing maps are considered incomplete [58], and the literature-based datasets, although richer in interactions, are prone to investigative biases [72] as they contain more interactions for the more explored disease proteins [65].

Table 2.2 Databases for PPI networks

Database	Website	Description
OPHID	ophid.utoronto.ca/ophidv2.204/	OPHID is designed to be a resource for the laboratory scientist to explore known and predicted PPIs
HPRD	www.hprd.org	HPRD represents a centralized platform to visually depict and integrate information pertaining to domain architecture, posttranslational modifications, interaction networks and disease association for each protein in the human proteome
BioGrid	thebiogrid.org	BioGrid curated sets of physical and genetic interactions
MIPS	mips.helmholtz-muenchen.de/proj/ppi/	The MIPS Mammalian PPI Database is a collection of manually curated high-quality PPI data collected from the scientific literature by expert curators
BOND	bond.unleashedinformatics.com/Action?	Bio-molecular object network databank
DIP	dip.doe-mbi.ucla.edu/dip/Main.cgi	The DIP database catalogs experimentally determined interactions between proteins
MINT	mint.bio.uniroma2.it/mint/	MINT is a database designed to store functional interactions between biological molecules (proteins, RNA, DNA)

(continued)

Table 2.2 (continued)

Database	Website	Description
Intact	www.ebi.ac.uk/intact/	IntAct provides a freely available, open-source database system and analysis tools for molecular interaction data. All interactions are derived from literature curation or direct user submissions and are freely available
STRING	string-db.org	A database of known and predicted PPIs, derived from four sources: Genomic Context, High-throughput experiments, (Conserved) co-expression and previous knowledge
APID	bioinfow.dep.usal.es/ apid/index.htm	APID is an interactive bioinformatic web-tool that has been developed to allow exploration and analysis of main currently known information about PPIs integrated and unified in a common and comparative platform
Predictome	predictome.bu.edu/	Predicted functional associations and interactions
InWeb_InBioMap	https://www.intomics.com/ inbio/map/#home	It provides a scored human PPI network with severalfold more interactions (>500,000) and better functional biological relevance than comparable resources

Except the PPI datasets as curated in various databases from high-throughput technologies, many efforts have also contributed to an attempt to computationally predict PPIs, such as the gene fusion methods, the phylogenetic profile methods, the sequence-based methods, the structure-based methods, the domain-based methods, and the coevolution-based methods. For details, one can refer to the reviews in reference [37].

2.2.3 *Signal Transduction Networks*

Signals are transduced within cells or between cells and thus form complex signaling networks. For instance, the MAPK/ERK pathway is transduced from the cell surface to the cell nucleus by a series of PPIs, phosphorylation reactions, and other events. Signaling networks typically integrate PPI networks, GRNs, and metabolic networks. Table 2.3 shows some signal transduction databases depositing experimentally determined signaling molecules and signaling pathways.

The experimental methods can only generate specific linear signaling pathways. The functions and mechanisms of complex signaling networks and their internal interactions are still unclear [37]. Therefore, many computational methods to

Table 2.3 Databases for signal transduction

Database	Website	Description
SPAD	www.grt.kyushu-u.ac.jp/spad/index/html	Signaling pathway database
KEGG	www.genome.jp/kegg	KEGG is a database resource for understanding high-level functions and utilities of the biological system, such as the cell, the organism, and the ecosystem, from molecular-level information, especially large-scale molecular datasets generated by genome sequencing and other high-throughput experimental technologies
DOQCS	doqcs.ncbs.res.in	DOQCS is a repository of models for signaling pathways. It includes reaction schemes, concentrations, rate constants, as well as annotations on the models. The database provides a range of search, navigation, and comparison functions
BBID	bbid.grc.nia.nih.gov	BBID is a WWW accessible relational database of archived images from research articles that describe regulatory pathways of higher eukaryotes
BioCarta	www.biocarta.com	A large database of pathways in organisms. Pathways are given in graphical representations supported by explaining text
NetPath	www.netpath.org	A manually curated resource of signal transduction pathways in humans
MiST	genomics.ornl.gov/mist	Microbial Signal Transduction database
TRANSPATH	www.biobase-international.com	A database system about gene regulatory networks that combines encyclopedic information on signal transduction with tools for visualization and analysis
BioModels	www.ebi.ac.uk/biomodels	A repository of computational models of biological processes. Models described from literature are manually curated and enriched with cross-references
SignaLink	signalink.org/	A signaling pathway resource with multi-layered regulatory networks
The Cell Collective	thecellcollective.org/	The Cell Collective is a web-based platform that enables laboratory scientists from across the globe to collaboratively build large-scale models of various biological processes, and simulate/analyze them in real time
SMPDB	http://smpdb.ca/	SMPDB (The Small Molecule Pathway Database) is an interactive, visual database containing more than 30,000 small molecule pathways found in humans only
WikiPathways	https://www.wikipathways.org/index.php/WikiPathways	It is a database of biological pathways maintained by and for the scientific community
Reactome	www.reactome.org	A free, open-source, curated, and peer reviewed pathway database

capture the details of signaling pathways by exploiting high-throughput genomic and proteomic data have been developed, such as the differential equation models, Petri net models, and so on. For details, one can refer to reference [37] and others therein.

2.2.4 Metabolic Networks

The chemical compounds of a living cell are connected by biochemical reactions which convert one compound into another. The reactions are catalyzed by enzymes. Thus, all compounds in a cell are parts of an intricate biochemical network of reactions, which is called metabolic network. It is possible to use network analyses to infer how selection acts on metabolic pathways.

The metabolic network maps are probably the most comprehensive of all biological networks. Databases such as the Kyoto Encyclopedia of Genes and Genomes (KEGG) and the Biochemical Genetic and Genomics knowledgebase (BIGG) contain the metabolic network of a wide range of species. Recently, Duarte et al. [35] published a comprehensive literature-based genome-scale metabolic reconstruction of human metabolism, with 2766 metabolites and 3311 metabolic and transport reactions. An independent manual construction by Ma et al. [36] contains nearly 3000 metabolic reactions, organized into about 70 human-specific metabolic pathways. Table 2.4 lists some main metabolic network databases. For details of reviews on the analysis and modeling of the metabolic networks, one can refer to reference [37].

2.3 Artificial Algorithms for Generating Bio-molecular Networks

Except the existing data for various real-world networks, one can also simulate real-world networks by artificial ones. In complex networks theory, the famous ER random model, the WS or NW SW models, the BA SF model are all artificial algorithms. The generated artificial networks follow the universal principals of real-world networks. For example, the BA SF model can generate networks that follow SF degree distribution with PLE $\gamma = 3$. There are also more biologically motivated models which have specifically been developed for PPIs and GRNs. Mostly, these models are based on two fundamental processes: occasional copying of individual genes/proteins (duplication) and subsequent mutations (divergence). These so-called duplication-divergence (DD) models start off with an initial network and update the network each time step by a duplication and divergence events. Network models derived from simple rate laws offer an intermediate level analysis, going beyond simple statistical analysis, but falling short of a fully quantitative

Table 2.4 Databases for metabolic networks

Database	Website	Description
KEGG	www.kegg.jp	A bioinformatics database containing information on genes, proteins, reactions, and pathways
BIGG	bigg.ucsd.edu	A knowledge base of biochemically, genetically, and genomically structured genome-scale metabolic network reconstructions
BioCyc	www.biocyc.com	A collection of both manually curated and computationally-derived biochemical pathways and genome databases
EcoCyc	www.ecocyc.com	Literature-based curation of the *E. coli* genome, and of *E. coli* transcriptional regulation, transporters, and metabolic pathways
MetaCyc	metacyc.com	A highly curated metabolic database that contains metabolic pathways, enzymes, metabolites, and reactions
AraCyc	www.arabidopsis.org/biocyc/index.jsp	A database containing biochemical pathways of *Arabidopsis*
GRAMENE	www.gramene.org/pathway	GRAMENE hosts ten plant pathways databases on the Pathway Tools platform
MRAD	capb.dbi.udel.edu/whisler	Metabolic reaction analysis database
PathCase	nashua.cwru.edu/PathwaysWeb	Pathways Database System
BRENDA	www.brenda-enzymes.info	A comprehensive enzyme database that allows for an enzyme to be searched by name, EC number, or organism
SMPDB	http://smpdb.ca/	SMPDB (Small Molecule Pathway Database) is an interactive, visual database containing more than 30,000 small molecule pathways found in humans only

description. In this section, we will briefly introduce some algorithms for the generation of artificial gene networks and PPI networks.

2.3.1 Algorithms for Artificial Regulatory Networks

Gene networks can be constructed through computer artificial algorithms. It is reported that gene duplication is the driving force for creating new genes in the genome, at least 50% of prokaryotic genes and over 90% of eukaryotic genes are products of gene duplication [25]. Moreover, it is also reported that gene duplication has a key role in network evolution, more than one-third of known regulatory interactions were inherited from the ancestral TF or target gene after duplication, and roughly one-half of the interactions were gained during divergence after duplication [25]. Therefore, based on the duplication and divergence mechanisms, many works have been performed to investigate the artificial construction of GRNs.

Duplication results in the creation of a daughter node that has inherited all the connectivity of the parent node. In the year 2002, Bhan et. al. [26] proposed some network growth models based on gene duplication events, where two models are considered. The first model includes gene duplication plus random removal of edges from the daughter node. The second model includes gene duplication plus preferential rewiring [14]. The detailed steps for the network growth models proposed by Bhan et al. [26] are described in Algorithm 3. In the mixed model with preferential rewiring, both duplication and rewiring are treated as random processes, each occurring with a probability of one-half. This parameter could also be varied, but Bhan et al. [26] found that this condition was sufficient to create satisfactory models. The schematic representation of network growth models is shown in Fig. 2.1a. The authors declared that the network growth models can yield networks with the same combination of global graphical properties that they inferred from the *yeast* expression data, where the graphical properties include the cluster coefficient, the APL, and the scaling exponent of degree distribution.

Algorithm 3 Algorithm for network growth model proposed by Bhan et. al.[26]

1: Initialization: Start with a small initial seed network. Two different seed networks were considered: a random network seed and a network seed with a high clustering coefficient. The influence of the seed reveals those graph parameters that are influenced by initial conditions and those that are due to the dynamics of the growth process.
2: **repeat**
3: Network growth from duplication: Starting with the seed graph, the network is grown in a probabilistic manner. A node from the entire network is chosen at random to be duplicated.
4: Divergence: After duplication, the newly duplicated edges are then removed at random from the new daughter node. On average, half of the edges will be removed.
5: **until** Network size reaches the desired one.

The basic unit of gene regulation consists of a TF, its DNA-binding site, and the target gene or transcription unit it regulates. This basic unit can be elaborated to form a complex network in two ways: some genes may be regulated by more than one TF, and some TFs may control more than one gene [25]. In 2004, Teichmann et al. [25] investigated the evolution properties of GRN growth by duplication. The authors considered a growth model with duplication and divergence. When duplication of a TF occurs, the new TF may initially recognize the same binding site and hence, regulate the same target gene as the duplicated TF. During the subsequent divergence steps, the duplicated TF may continue regulate the target genes as its ancestor but respond to a different signal, or it may recognize a new binding site upstream of some other target genes. The schematic representation

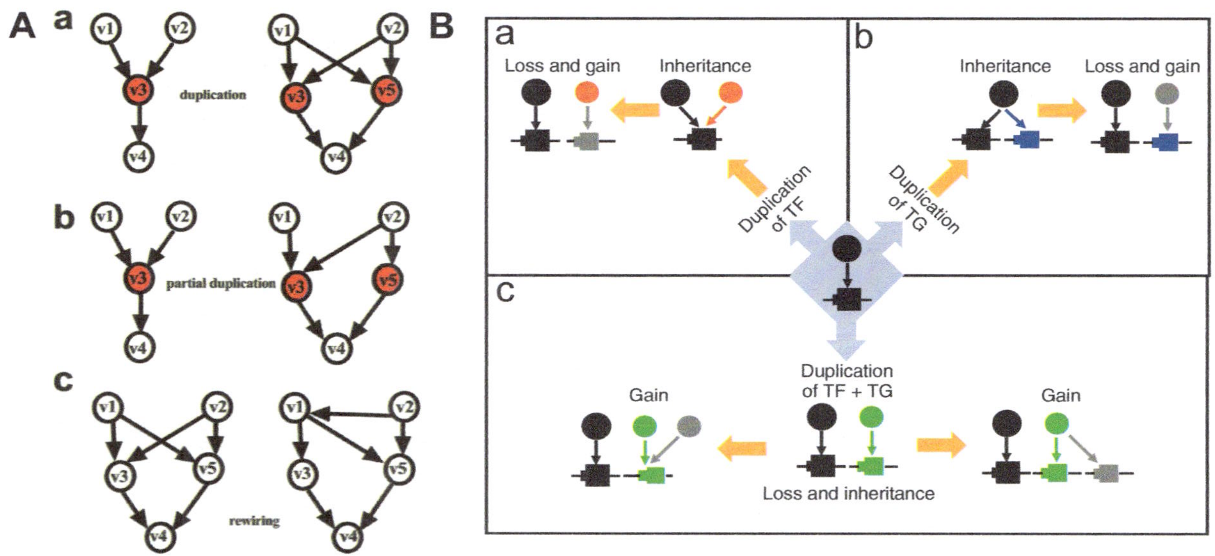

Fig. 2.1 Some gene network models. (**A**) Schematic representation of network growth through gene duplication; (**a**) Shows pure gene duplication where a new node $v5$ is created by duplicating the connectivity of the parent $v3$. (**b**) The partial duplication model where node $v3$ is duplicated to $v5$ but not all the original connections are retained. (**c**) Shows a rewiring process where edge $v2 \rightarrow v3$ is rewired to become $v2 \rightarrow v1$. (**B**) Duplication growth models and consequences for network evolution. The basic unit of gene regulation is shown in the center: the TF, the target gene (TG), and its binding site. The three panels describe the possible duplication events of this basic unit and the subsequent divergence resulting in new regulatory interactions. Duplication events are represented by light blue arrows and divergence events by orange arrows. Divergence may also result in the loss of the duplicated gene, but this case is not considered here. (**a**) Duplication of the TF leads to both TFs regulating the same gene. Divergence can result in the duplicated TF regulating the original target gene by competing for the same binding site used by the ancestral TF or regulating a different gene. (**b**) Duplication of a target gene results in both genes being regulated by the same TF. Divergence can lead to the duplicated gene remaining under the control of the same TF or coming under the control of a different TF. (**c**) Duplication of TF and its target genes gives rise to new regulatory interactions. Divergence can result in homologous TFs regulating homologous genes. Subsequent divergence of the TF or the TG can result in additional interactions. Reprinted by permission from Springer, ref. [25]

of network growth models is shown in Fig. 2.1b. The authors declared that gene duplication is the driving force for creating new genes in genome and it has a key role in network evolution. Furthermore, the authors suggested that evolution has been incremental, rather than making entire regulatory circuits or motifs by duplication with inheritance of interactions.

In the year 2006, Foster et al. [16] proposed and studied a class of growth algorithms for directed graphs that can be provided as candidate models for the evolution of GRNs. The algorithms involve partial duplication of nodes and their links, together with the innovation of new links, allowing for the possibility that input and output links from a newly created node may have different probabilities of survival. They found some counterintuitive trends as the parameters are varied, including the broadening of the in-degree distribution when the probability for retaining input links is decreased. They also found that both the scaling of TFs with genome size and the measured degree distributions for genes in *yeast* can be reproduced by the growth algorithm if and only if a special seed is used to initiate the process. For details, one can refer to reference [16] and Algorithm 4.

Algorithm 4 The partial duplication model for directed GRNs proposed by Foster et al. [16]

1: Start from an initial seed network, and define a time step to be the time between duplication events.
2: **repeat**
3: At each time step a gene g is chosen at random from the network and duplicated, forming a gene g' that has all the same input and output links as g. One of these identical nodes—say, g'—is then assumed to subsequently mutate.
4: Each input link inherited from g is independently tested and kept with probability c_i, and each output link with probability c_o. If g' should lose all of its inputs and outputs, it is considered to have lost all its function and thus is removed from the network entirely. There is no physical symmetry requiring $c_i = c_o$.
5: **until** Network size reaches the desired one.

In the year 2007, Enemark and Sneppen [30] proposed some gene duplication models for directed networks with limits on growth. The detailed steps for the directed gene duplication model with limits on growth are described in Algorithm 5. The duplication and kill moves in the evolving networks and some snapshots of networks generated with the given parameters are shown in Fig. 2.2.

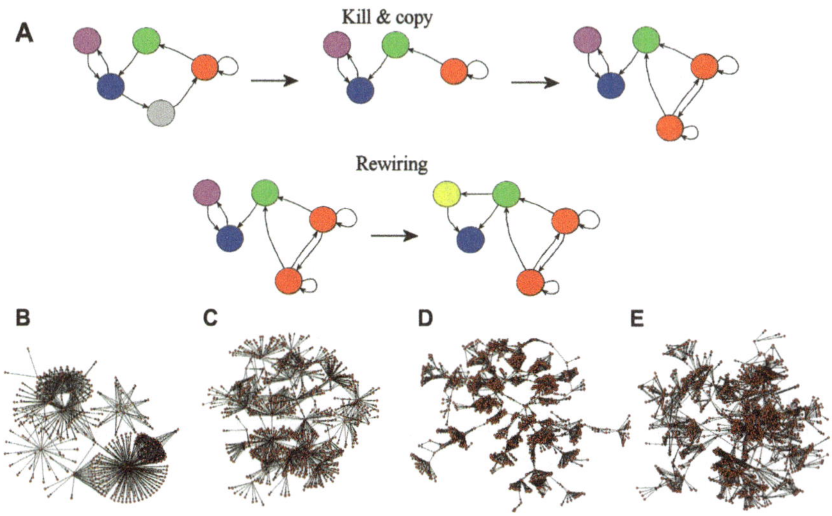

Fig. 2.2 The duplication and kill moves in the evolving networks and some snapshots of networks generated with the given parameters. (**a**) The two basic moves in evolving networks. The upper case refers to the removal and duplication move, where the gray node is "removed" and subsequently the red node duplicated along with its upstream region. The lower case illustrates a rewiring move in which the upstream region of the purple/yellow node is mutated. This results in a change in connections. A shape mutation in the purple node could similarly change its out links (not shown here). (**b**–**e**). Snapshots of networks generated with $N = 1000, 2000, 3000, 3000$, respectively. The difference between (**d**) and (**e**) illustrates that two steady-state samples of the system can be very different. Unconnected proteins are not shown. All of the networks are generated with parameters $\alpha = 0.72$, $\beta = 0.27$, and $\epsilon = 0.01$ while the size of the "shape" space is set by $s = 2.3 \times 10^5$ and the number of potential operator sites by $\nu = 100$. Reproduced from ref. [30], with permission of IOP Publishing

Algorithm 5 The directed gene duplication model with limits on growth proposed by Enemark and Sneppen [30]

1: Initially each node is assigned random shape and upstream numbers.
2: **repeat**
3: At each evolutionary step, one evolves the network by either duplicating or mutating a random node (protein). That is, at each time step, one preforms one of the following steps:

Case 1: With probability α, one duplicates a node and its upstream region, by making a complete copy of both the integers representing the upstream and the ones representing the shape. Subsequently one removes a random node and all its upstream sites.

Case 2: With probability β, one changes the shape number of a node.

Case 3: With probability $\epsilon = 1 - \alpha - \beta$, one selects ν random sites among all the $N\nu$ upstream sites in the system. Each of these chosen sites is assigned a new random number.

4: **until** Network size reaches the desired one.

Except the mentioned works, there are many other works on the constructions and analysis of artificial GRNs, such as references [28, 29, 31, 66, 67]. One can refer to these works for details.

2.3.2 Algorithms for Artificial PPI Networks

Experimentally constructions of PPI networks through high-throughput technologies are often costly. If one can design some artificial computer algorithms to mimic the statistical properties of PPI networks, then the artificial networks can be freely generated for further analysis [17–24]. For example, it is possible to explore the evolutionary mechanisms of PPI networks based on artificial algorithms [33].

In 1999, Barabási and Albert proposed the well-known BA algorithm to generate artificial complex networks with power-law degree distributions [14]. The basic idea of the BA algorithm is the "rich-gets-richer" rule. As a result, the degree–degree correlation coefficients of the generated networks are positive. However, for bio-molecular networks, the degree–degree correlation coefficients are often negative, that is, they are disassortative [22–24, 32, 33]. Therefore, the possible growth strategy for bio-molecular networks will be different from the BA strategy.

The PPI networks are disassortative, sparse, SF, SW and with modularity structures [24, 34]. Based on these properties, many DD models have been proposed [24, 34]. For example, in 2002 and 2003, Solé et al. [34] and Vázquez et al. [17] proposed two models to generate artificial PPI networks. Details for the firstly proposed models were shown in Algorithms 6, 7, and Fig. 2.3, respectively. The Solé model and the Vázquez model can both generate artificial PPI networks with similar properties as the real-world *yeast* PPI network. The two algorithms have a few differences in the divergence process, and the Vázquez model further considered the dimerization process for newly duplicated node.

In 2007, based on a random duplication model and an anti-preference duplication model, Zhao et al. [24] investigated the effect of duplication strategies on the disassortativity. In 2010, Xu et al. investigated several models with different ways of divergences [22], and they clarified how the divergence mechanisms influence the disassortativity of bio-molecular networks. Also in 2010, Wan et al. [23] proposed a simple DD model, which considered the anti-preference duplication, edge deletion, dimerization, and edge addition processes. They found the DD model under proper parameters can well mimic the real-world PPI networks.

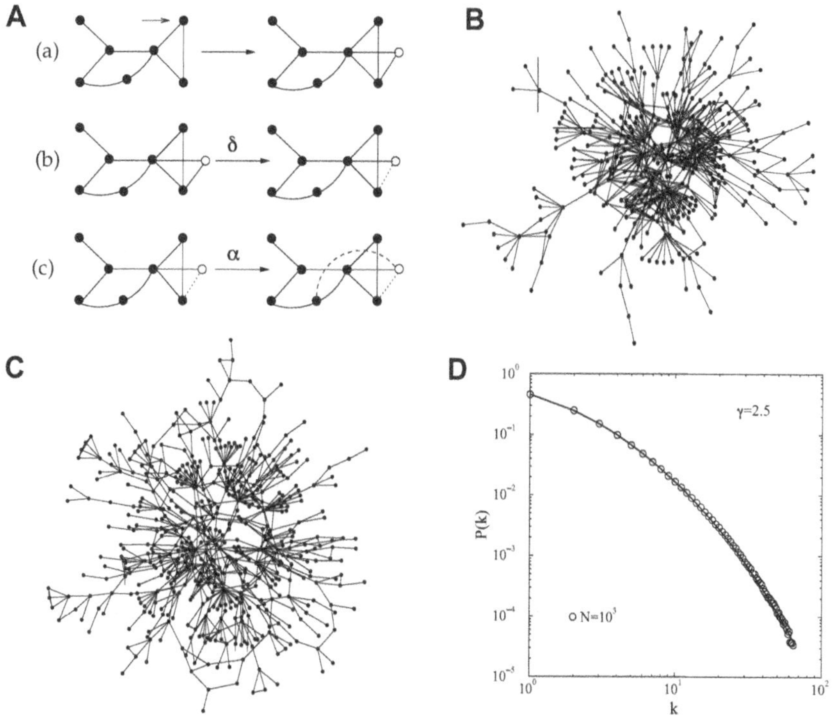

Fig. 2.3 The Solé model for PPI networks. (**A**) Growing network by duplication of nodes. First (**a**) duplication occurs after randomly selecting a node (arrow). The links from the newly created node (white) now can experience deletion (**b**) and new links can be created (**c**); these events occur with probabilities δ and α, respectively. (**B**) An example of a small proteome interaction map (giant component) generated by the model with $N = 103, \delta = 0.58$, and $\alpha = 0.16$. (**C**) Real *yeast* proteome map obtained from the MIPS database. One can observe the close similitude between the real map and the output of the model. (**D**) Degree distribution $P(k)$ for the model, averaged over 10^4 networks of size $N = 10^3$. The distribution shows a characteristic power-law behavior, with exponent $\gamma = 2.5 \pm 0.1$. Reprinted from ref. [34], with permission from World Scientific Publishing

Algorithm 6 Algorithm for the Solé model of PPI network [34]

1: Initialization with a given small network.
2: **repeat**
3: At each time step (duplication), one preforms the following steps (Fig. 2.3):

Rule 1: one node in the graph is randomly chosen and duplicated.
Rule 2: The links emerging from the new generated node are removed with probability δ.
Rule 3: New links (not previously present) can be created between the new node and all the rest of the nodes with probability α.

4: **until** Network size reaches the desired one.

Algorithm 7 Algorithm for the Vázquez model of PPI network [17]

1: Starts with an initial network of two connected nodes.
2: **repeat**
3: At each time step, a node is added to the network according to the following rules:

Rule 1: Duplication: a node v is selected at random. A new node v' with a link to all the neighbors of v is created.
Rule 2: Self-interaction: with probability p a link between v and v' is established (self-interacting proteins).
Rule 3: Divergence: for each of the nodes w linked to v and v' one of the two links is chosen randomly and removed with probability q.

4: **until** Network size reaches the desired one.

Generally, duplication can create new proteins, divergence in the newly created proteins or interactions can lead to the emergence of novelty [68]. Artificially, duplication and divergence can be reflected by various ways. Different ways of duplication and divergence result to different evolution models. For example, a pre-duplicated node can be selected randomly or by the anti-preference strategy [24]. Divergence can be reflected by node deletion, edge deletion, addition, or rewiring [22]. By referring to the existing work [23], procedures of a currently widely used DD algorithm are described in Algorithm 8:

Algorithm 8 Algorithms for PPI network by considering different duplication and divergence strategies [23, 33]

1: Generate an initial connected network with n_0 nodes.
2: **repeat**
3: Duplication. Two duplication approaches can be separately considered (corresponding to two different models):
(i) Anti-preference strategy: At each time step, node i with degree k_i is chosen to duplicate with probability:

$$p_i = \frac{1/k_i}{\Sigma_j(1/k_j)}. \tag{2.1}$$

(ii) Random strategy: randomly chose a node to replicate.
4: Divergence. Four approaches of divergence are simultaneously considered.
(a) Edge deletion: For each node l linked to the newly selected target node i and its replica i', randomly choose one of the two links (i, l) or (i', l) and remove it with probability α.
(b) Dimerization: the target node i and its replica i' can be dimerized with probability β_0, that is, an edge between them will be added with probability β_0. According to [23], $\beta_0 = \beta k_i'$, where k_i' denotes the degree of the replica i', β is a constant, β_0 equals one if it is bigger than one.
(c) Edge addition: randomly choose a non-target node j, add a link between node j and i' with probability γ.
(d) After all the above processes, remove isolated nodes.
5: **until** Network size reaches the desired one.

It is noted that the duplication and divergence processes correspond to real-world biological processes. For example, during the evolution of bio-molecular networks, a gene encoding an existing protein undergoes nucleotide substitutions, which leads to the creation of new links or deletion of existing links, this process can be mimicked by the edge deletion and addition processes in the DD model [23]. The dimerization process mimics the probability that a duplicated node is self-interacting protein, or the links between the duplicated node and its replica are conserved during divergence [23, 69].

In the following chapters, when considers the artificial PPI networks, we mainly consider Algorithm 8. It is noted that, some other duplication or divergence strategies can also be similarly considered. Existing works reported that biological networks are disassortative, the potential duplication strategy is unlikely the preferential attachment strategy as that in the BA SF model. In the following chapters, we usually suppose $n_0 = 2$, that is, all networks will be evolved from two ancestors. Parameters α, β, γ can be tuned. Available data and works [24, 34] indicate that α is far higher than γ, β is larger than γ, and γ is empirically very small. α, β, γ can be selected according to the following characteristics of real-world PPI networks. Firstly, researchers have reported that PPI networks are sparse. One result reported by Newman in 2003 is 2.12 [70], while Schwikowski et al. [71] and Yu et al. [72] showed that the average degrees of the *yeast* PPI networks are around 3. Secondly, PPI networks are SF, with PLE $r = 2.5$ [24, 34]. Thirdly, PPI networks are SW, with shorter APL and larger clustering coefficient than its random counterparts [70]. Moreover, biological networks are disassortativity. Highly connected nodes tend to be neighbors of nodes with low degrees [24, 70].

Under the anti-preference duplication strategy, Wan et al. [23] investigated the effect of each parameter on degree distributions, modularity, and disassortativity in PPI networks. They showed that the proposed model can not only reproduce SF connectivity and SW pattern but also exhibit hierarchical modularity and disassortativity (Fig. 2.4).The generated networks follow SF distributions with the PLEs close to the number 2.5, in accordance with the characteristics of real PPIs [34]–[24]. In addition, one can also observe that the power-law exponents basically keep unchanging with increasing α, remain constant with increasing β_0, and decay slightly with increasing γ in the left panels of Fig. 2.4. These results demonstrated that the SF feature in the proposed model is robust to the model parameters as general biological networks do. The modularity of a network can be measured by the average cluster coefficient of all k-degree nodes $C(k)$, defined as

$$C(k) = \frac{\sum_i C_i \delta(k_i - k)}{\sum_i \delta(k_i - k)}. \tag{2.2}$$

Here, C_i denotes the clustering coefficient of node i defined as $C_i = 2e_i/(k_i(k_i - 1))$, which measures the local cohesiveness of the network in the neighborhood of the node, where k_i is the degree of node i and e_i is the number of links connecting k_i neighbors of node i to one another [73]. If $C(k) \sim k^{-\theta_1}$ with θ_1 approximating to 1, it suggests that the network has a hierarchical structure. For the generated networks

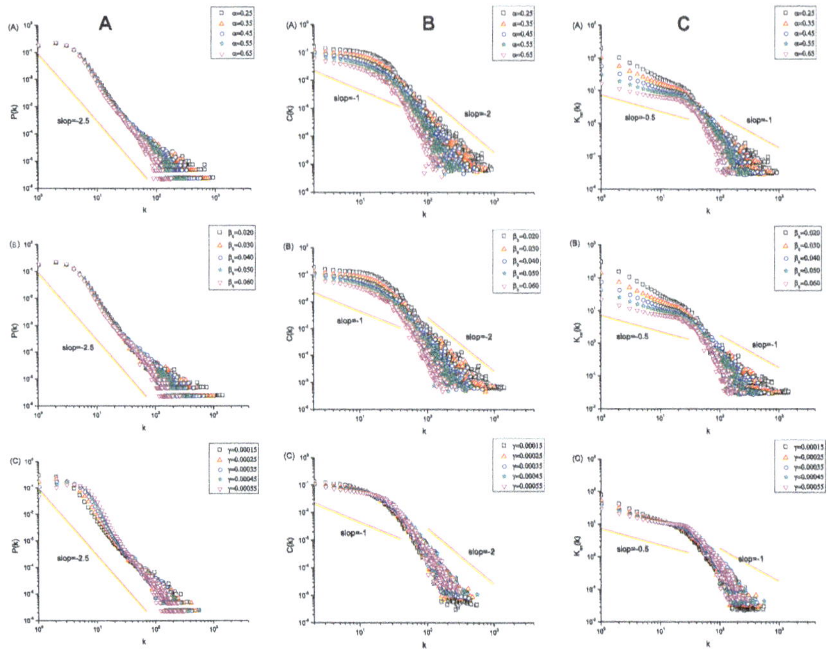

Fig. 2.4 Effect of parameters on degree distributions, modularity, and disassortativity in PPI networks that generated by Algorithm 8 under the anti-preference duplication strategy. (**a**) Degree distributions; (**b**) The effect of parameters on modularity index $C(k)$ [73]. (**c**) The distributions of the average nearest-neighbor degree of all k-degree nodes $K_{nn}(k)$ as a function k [74, 75]. Nominal parameters when considering α are taken as $\beta_0 = 0.035$, $\gamma = 0.00025$; When considering β_0, nominal parameters for α and γ are taken as $\alpha = 0.45$ and $\gamma = 0.00025$; as to the effect of γ, α, and β_0 are taken as $\alpha = 0.45$, $\beta_0 = 0.04$. Reprinted from ref. [23], with permission from AIP Publishing

under certain parameters, Fig. 2.4 reveals that θ_1 decays with increasing α and grows with increasing β, respectively. Moreover, one can obtain $C(k) \sim k^{-1}$ for small k and $C(k) \sim k^{-2}$ for large k. The average nearest-neighbor degree of all k-degree nodes $K_{nn}(k)$ can be used to measure the degree–degree correlation [74, 75]. $K_{nn}(k)$ is defined as

$$K_{nn}(k) = \frac{\sum_i \delta(k_i - k) k_{nn,i}}{\sum_i \delta(k_i - k)}. \tag{2.3}$$

$k_{nn,i}$ denotes the average nearest-neighbor degree of node i and its neighbors, written as $k_{nn,i} = \sum_{j \in O(i)} k_j / k_i$; $O(i)$ corresponds to the set of neighbors of node i, and the assortative behavior or disassortative behavior is represented if $K_{nn}(k)$ is an increasing or a decreasing function of k, respectively. Simulation results indicate that the PLE θ_2 of the distribution of $K_{nn}(k)$ mainly depends on the deletion probability α and the addition probability γ. More concretely, θ_2 decays with increasing α and decays slightly with increasing γ, respectively. Moreover, they also indicate $K_{nn}(k) \sim k^{-0.5}$ for small k and $K_{nn}(k) \sim k^{-1}$ for large k at around $\alpha = 0.45$, $\beta_0 = 0.040$, and $\gamma = 0.00025$.

Except the above features, one can also discuss whether the generated artificial networks preserve network structure entropy, as defined in Chap. 1. After comparing the features of the model with those of real-world PPI networks, the authors declared that the proposed model can provide relevant insights into the mechanism underlying the evolution of PPI networks. The artificially constructed bio-molecular networks facilitate the massive investigation of the evolving properties of PPI networks, which will be discussed in the following chapters.

2.4 Statistical Reconstruction of Bio-molecular Networks

To construct bio-molecular networks, except for the above mentioned interaction data from online databases and artificial algorithms, based on biological data, researchers have also developed many mathematical and statistical methods to construct them. With the repaid development of high-throughput technologies, we have accumulated massive omics data. How to explore bioinformatics from omics data is a challenge problem, whereas network analysis is a powerful tool to realize such target, which has been widely used in genome-wide association study (GWAS). Inferring networks based on biological data is the first step to perform subsequent network analysis. Various methods have been developed to realize the reconstruction of bio-molecule networks, including the correlation based method [76, 77], the optimization method [37, 78], the Bayesian method [79], the Granger causality method [80–85], and the method based on penalized regression [86].

The general formulation of network reconstruction based on data can be stated as follows. Suppose

$$X = \begin{bmatrix} x_{11} & x_{12} & \cdots & x_{1p} \\ x_{21} & x_{22} & \cdots & x_{2p} \\ \vdots & \vdots & \ddots & \vdots \\ x_{n1} & x_{n2} & \cdots & x_{np} \end{bmatrix} = \begin{bmatrix} X_{(1)}^T \\ X_{(2)}^T \\ \vdots \\ X_{(n)}^T \end{bmatrix} = [X_1, X_2, \cdots, X_p] \tag{2.4}$$

is the sample observation matrix or sample data matrix for the p-dimensional random vector $\mathbf{X} = (\mathbf{X}_1, \mathbf{X}_2, \cdots, \mathbf{X}_p)^T$. Here, $x_{ij}(i = 1, 2, \cdots, n; j = 1, 2, \cdots, p)$ denotes the i'th observation for the j'th variable $\mathbf{X}_j$; $X_{(i)}^T = (x_{i1}, x_{i2}, \cdots, x_{ip})(i = 1, 2, \cdots, n)$ represents the i'th observation for $\mathbf{X}$; $X_i = (x_{1i}, x_{2i}, \cdots, x_{ni})^T(i = 1, 2, \cdots, p)$ denotes the observation vector of $\mathbf{X}_i$ that consists of n observations. Generally, we require that the n samples $X_{(i)}(i = 1, 2, \cdots, n)$ are independent and identically distributed (i.i.d.). It is noted that, in omics data, each random variable

$\mathbf{X}_i$ corresponds to one molecular, such as a gene, a protein; n represents the amount of observations, where the observations typically include treatments and controls.

As to genome data or transcriptome data, the constructed networks are generally called gene co-expression network (GCN). If expression profiles of X_i for gene i and X_j for gene j were with high similarity, then the two genes are connected in the GCN, and the two genes are possibly with similar biological functions. After we constructed the GCN from omics data, it can be used to perform clustering analysis for genes, to predict functions of genes without Gene Ontology (GO) or KEGG annotations, as well as to identify crucial genes that closely related to the phenotype differences between treatments and controls. We will discuss the detailed applications of GCN in Chap. 9. In this section, we will introduce the main idea, basic mathematical and statistical methods of network reconstruction.

2.4.1 Association Methods

The basic ideas of the association methods are as follows. Firstly, one computes pairwise association coefficients or detects whether two genes were dependent, and then networks can be constructed based on certain pre-determined hard threshold values on similarity measures or hypothesis test P values. Subsequently, we introduce various correlation coefficients and the mean variance method.

2.4.1.1 Various Similarity Measures

For gene i and j, we denote the similarity measure between them as $r_{ij}(i, j = 1, 2, \cdots, p)$. For the p genes, we denote $R = (r_{ij})_{p\times p}$ as the similarity matrix among the p genes. The similarity measure r_{ij} can have many different definitions. The Pearson correlation coefficient (PCC) [87], the Spearman correlation coefficient (SCC) [88], and the Kendall correlation coefficient (KCC) [89] are three typical measures to measure the correlation between two random variables or two observation vectors. It is noted that the PCC, SCC, and KCC are equivalent when the considered two variables are jointly normal. Hereinafter, we introduce some of the widely used similarity measures.

1. **Pearson correlation coefficient:**

$$r_{ij}^P = \frac{\sum_{k=1}^n (x_{ki} - \overline{x}_i)(y_{kj} - \overline{y}_j)}{\sqrt{\sum_{k=1}^n (x_{ki} - \overline{x}_i)^2}\sqrt{\sum_{k=1}^n (y_{kj} - \overline{y}_j)^2}}$$
$$= \frac{1}{n-1}\sum_{k=1}^n \left(\frac{x_{ki} - \overline{x}_i}{s_i}\right)\left(\frac{y_{kj} - \overline{y}_j}{s_j}\right), \tag{2.5}$$

where $\overline{x}_i$ and $\overline{y}_j$ denote the sample means of genes i and j, respectively;

$$s_i = \sqrt{\frac{1}{n-1}\sum_{k=1}^{n}(x_{ki} - \overline{x}_i)^2}$$

and

$$s_j = \sqrt{\frac{1}{n-1}\sum_{k=1}^{n}(y_{kj} - \overline{y}_j)^2}$$

represent the standard deviation of X_i and Y_j, respectively. Obviously, $(x_{ki} - \overline{x}_i)/s_i$ and $(y_{kj} - \overline{y}_j)/s_j (i, j = 1, 2, \cdots, n)$ are z-scores of observations for X_i and Y_j, and the PCC is just computed as the average products of such standardized score pairs.

To test whether the linear correlation between X_i and X_j is statistical significant, one can use the following static:

$$t = r_{ij}^P \sqrt{\frac{n-2}{1-(r_{ij}^P)^2}}. \tag{2.6}$$

The t static follows the student t distribution with degree-of-freedom $n - 1$, and one can obtain the corresponding P value according to $P = P\{|t| > t_0\}$. Here, t_0 is the value computed from Eq. (2.6). If $P < \alpha$ (a predefined significance level), then we can accept the assertion that there is significant linear correlation between X_i and X_j.

Obviously, $-1 \leq r_{ij}^P \leq 1$. The PCC is sensitive to data distribution (appropriate for normal distribution), which only shows good performance in linear cases and it is inappropriate for nonlinear cases and small samples [90] (See Fig. 2.5). Moreover, the PCC is sensitive to data, a little change on one data point may greatly alter the PCC [90] (See Fig. 2.5).

2. **Spearman correlation coefficient:** The SCC appropriates for ranked data, which is defined as

$$\begin{aligned} r_{ij}^S &= \frac{\sum_{k=1}^{n}(R_{ki} - \overline{R}_i)(Q_{kj} - \overline{Q}_j)}{\sqrt{\sum_{k=1}^{n}(R_{ki} - \overline{R}_i)^2}\sqrt{\sum_{k=1}^{n}(Q_{kj} - \overline{Q}_j)^2}} \\ &= \frac{1}{n-1}\sum_{k=1}^{n}\left(\frac{R_{ki} - \overline{R}_i}{sR_i}\right)\left(\frac{Q_{kj} - \overline{Q}_j}{sQ_j}\right). \end{aligned} \tag{2.7}$$

Here, R_{ki}, Q_{kj} represent the ranks of x_{ki}, y_{kj} in X_i and Y_j; $\overline{R}_i, \overline{Q}_j$ denote the average ranks. $sR_i = \sqrt{\frac{1}{n-1}\sum_{k=1}^{n}(R_{ki} - \overline{R}_i)^2}$ and $sQ_j =$

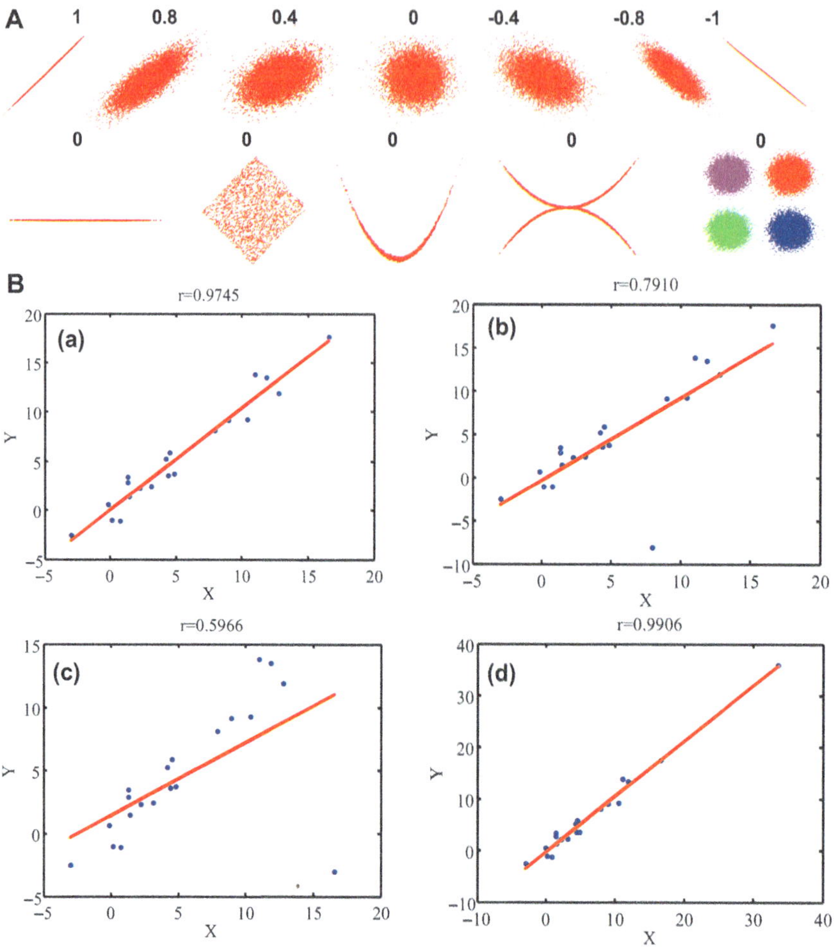

Fig. 2.5 The PCC is inappropriate for nonlinear cases and sensitive to data. (**A**) The PCC only shows good performance in linear cases and it is inappropriate for nonlinear cases. (**B**) The PCC is sensitive to data. (**a**) A dataset with no outliers; (**b**) the same data except for one outlier in the middle of the X values; (**c**) the same data except for one outlier at the high end of the X values (a point with leverage and influence); and (**d**) the same data except for one point that is an outlier with respect to the X (and Y) distribution, but not with respect to the regression line (a point with leverage, but little influence). The numbers represent the PCC values. Noted that only one point in (**b–d**) is different from (**a**), but they have different effect on r. Inspired and redrew from ref. [90]

$\sqrt{\frac{1}{n-1}\sum_{k=1}^{n}(Q_{kj}-\overline{Q}_j)^2}$ represent the standard deviation of the rank vector R_i and Q_j, respectively. $-1 \leq r_{ij}^S \leq 1$. The SCC can be not only used to evaluate linear correlation, but also used to evaluate nonlinear relationships. More importantly, the SCC is insensitive to the actual observation value, therefore, it is robust to outliers in the data.

3. **Kendall τ_a correlation coefficient** (KCC_a): The Kendall tau rank correlation coefficient [89–94] (or simply the Kendall tau coefficient, Kendall's τ or tau test(s)) is developed by Maurice Kendall in 1938, which is a non-parametric statistic used to measure the degree of correspondence between two rankings and assessing the significance of this correspondence. In other words, it measures the strength of association of the cross tabulations.

   If the agreement between the two rankings is perfect (i.e., the two rankings are the same) the coefficient has value 1. If the disagreement between the two rankings is perfect (i.e., one ranking is the reverse of the other) the coefficient has value -1. For all other arrangements, the value lies between -1 and 1, and increasing values imply increasing agreement between the rankings. If the rankings are completely independent, the coefficient has value 0 on average.

   Suppose we have two rankings R_1, R_2 for the n observations of variables $\mathbf{X}, \mathbf{Y}$. For any two observations i, j, $(R_1(i), R_2(i))$ and $(R_1(j), R_2(j))$ consist of a pair $\{(R_1(i), R_2(i)), (R_1(j), R_2(j))\}$. The pair is said to be concordant if either $R_1(i) > R_1(j)$ and $R_2(i) > R_2(j)$ or $R_1(i) < R_1(j)$ and $R_2(i) < R_2(j)$. The pair is said to be discordant if $R_1(i) > R_1(j)$ and $R_2(i) < R_2(j)$, or $R_1(i) > R_1(j)$ and $R_2(i) < R_2(j)$. If $R_1(i) = R_1(j)$ or $R_2(i) = R_2(j)$, the pair is said to be tied. Suppose the number of concordant pairs is N_C, the number of discordant pairs is N_D, the Kendall τ_a is defined as

$$\tau_a = \frac{2(N_C - N_D)}{n(n-1)}. \tag{2.8}$$

   τ_a measures the strength of association of the cross tabulations when both variables are measured at the ordinal level but makes no adjustment for ties.
4. **Kendall τ_b correlation coefficient** (KCC_b): Kendall's τ_b can quantitatively measure the correlation between two rankings with tied values [89–94]. The Kendall's τ_b is defined as

$$\tau_b = \frac{N_C - N_D}{\sqrt{(n_0 - n_1)(n_0 - n_2)}}, \tag{2.9}$$

   where $n_0 = n(n-1)/2$, $n_1 = \sum_{i=1}^{k_1} t_i(t_i - 1)/2$, $n_2 = \sum_{j=1}^{k_2} h_j(h_j - 1)/2$. Here, k_1, k_2 denote the number of tied groups in R_1, R_2, respectively; t_i denotes the number of tied values in the $i'th$ group of ties in R_1; h_j denotes the number of tied values in the $j'th$ group of ties in R_2.

   τ_b tests the strength of association of the cross tabulations when both variables are measured at the ordinal level. It makes adjustments for ties and is most suitable for square tables. Under the definition of Eq. (2.9), the τ_b ranges from -1 (100% negative association, or perfect inversion) to 1 (100% positive association, or perfect agreement). The more both ranking results go in the same direction, the higher τ_b. A value of zero indicates the absence of association.

5. **Kendall τ_c correlation coefficient** (KCC_c): The Kendall τ_c is defined as

$$\tau_c = \frac{N_C - N_D}{\frac{m-1}{2m}n^2}. \tag{2.10}$$

Here, m represents the minimum value of the row number and the column number of the rectangular tables.

Kendall τ_c tests the strength of association of the cross tabulations when both variables are measured at the ordinal level. It makes adjustments for ties and is most suitable for rectangular tables. Values range from -1 (100% negative association, or perfect inversion) to 1 (100% positive association, or perfect agreement). A value of zero indicates the absence of association.

6. **Included angle cosine** (IAC): The IAC is just the PCC for normalized data, which defined as

$$r_{ij}^C = \frac{< X_i^*, X_j^* >}{||X_i^*||||X_j^*||} = \frac{\sum_{k=1}^n x_{ki}^* y_{kj}^*}{\sqrt{\sum_{k=1}^n (x_{ki}^*)^2}\sqrt{\sum_{k=1}^n (y_{kj}^*)^2}}. \tag{2.11}$$

Here X_i^* is normalized $X_i (i = 1, 2, \cdots, p)$. $-1 \le r_{ij}^C \le 1$.

7. **Gray correlation coefficient (GCC):** The gray correlation coefficient is appropriate for small sample data [95]. Assume that $X_0 = (x_0(1), x_0(2), \cdots, x_0(n))'$ was a reference sequence, $X_i = (x_i(1), x_i(2), \cdots, x_i(n))', i = 1, 2, \cdots, p$ are observation sequences for p genes. Define the GCC between x_0 and x_i at time point or condition k as

$$\xi_i(k) = \frac{min_i min_k |x_0(k) - x_i(k)| + \rho max_i max_k |x_0(k) - x_i(k)|}{|x_0(k) - x_i(k)| + \rho max_i max_k |x_0(k) - x_i(k)|}. \tag{2.12}$$

Here, ρ is resolution ratio, usually taken as $\rho = 0.5$. Then the overall GCC between x_0 and x_i is defined as

$$r_{0i} = \frac{1}{n}\sum_{k=1}^{n} \xi_i(k). \tag{2.13}$$

Obviously, the gray correlation coefficient is always non-negative and locates in [0, 1]. For omics data, each observation for each gene X_i can successively be treated as reference sequence. Finally, we can obtain a correlation matrix R. Since the computation of different rows or columns of R relies on different reference sequences, the obtained R is unnecessary symmetry. To symmetrize matrix R, we can revise it as

$$R^* = \frac{R + R^T}{2}. \tag{2.14}$$

The GCC is appropriate for data with a few sample points, and it can be used to evaluate nonlinear correlations.

8. **Various distance measures:** Except the abovementioned various correlation measure, we can also use various distance metrics to measure the similarity between two variables. The distance measures include the Minkowski distance, the Lance distance, Mahalanobis distance, and the skew space distance, which have been introduced in Chap. 1, we will give no details here. Shorter distance indicates higher similarity, and longer distance indicates lower similarity. It is noted that, the distance measures have no fixed boundary, which increased the difficulty for comparison.

After we obtained the symmetry correlation matrix $R = (r_{ij})_{p\times p}$ or distance matrix $D = (d_{ij})_{p\times p}$ for the p genes, based on predefined hard thresholds r_0 or d_0, one can construct the corresponding co-expression network with adjacency matrix $A = (a_{ij})_{p\times p}$. If $r_{ij} \geq r_0$ or $d_{ij} \leq d_0$, we deem there exists a co-expression relationship between genes i and j; and a_{ij} can be taken as r_{ij}, $1/d_{ij}$, or 1; otherwise, $a_{ij} = 0$. If a_{ij} are taken as r_{ij}, $1/d_{ij}$, then the constructed network is weighted and undirected, while if a_{ij} is taken as either 1 or 0, the constructed network is unweighted and undirected.

2.4.1.2 The Mean Variance Method

Similarity between two variables can also be evaluated from various independence tests. As a fundamental testing problem, testing whether two random variables are independent or not has received much attention in the literature. When two random variables are both categorical, the classic Pearson's Chi-square test is able to test their statistical independence. Hoeffding proposed a test of independence based on the difference between the joint distribution function and the product of marginals of two random variables [96]. The Hoeffding's test statistic is

$$H = n\int\int\left[\hat{F}_{XY}(x,y) - \hat{F}_X(x)\hat{F}_Y(y)\right]^2 d\hat{F}_{XY}(x,y), \tag{2.15}$$

where $\hat{F}_{XY}(x, y)$ denotes the empirical joint distribution function of random variables X and Y, $\hat{F}_X$ and $\hat{F}_Y$ denote the empirical marginal distributions of X and Y, respectively. This is also the well-known Cramér–von Mises criterion between the joint distribution function and the product of marginals. Rosenblatt [97] considered a measure of dependence based on the difference between the joint density function and the product of marginal densities. Székely et al. [98] and Székely and Rizzo [99] defined a distance covariance (DC) between two random vectors $X \in R^p$ and $Y \in R^q$ by

$$V^2(X,Y) = \int_{R^{p+q}} |\phi_{XY}(t,s) - \phi_X(t)\phi_Y(s)|^2\, \omega(t,s)dtds, \tag{2.16}$$

where $\phi_{XY}(t,s)$, $\phi_X(t)$, $\phi_Y(s)$ denote the joint characteristic function, the marginal characteristic functions of X and Y, respectively, and $\omega(t,s)$ is a positive weight function. $V^2(X,Y) = 0$ if and only if X and Y are independent. They further proposed a test of independence based on the statistic $nV_n^2(X,Y)/S_2$, where $V_n^2(X,Y)$ is the estimator for $V^2(X,Y)$ by using the corresponding empirical characteristic functions and

$$S_2 = n^{-4} \sum_{k,l=1}^{n} |X_k - X_l|_p \sum_{k,l=1}^{n} |Y_k - Y_l|_q,$$

in which $\{(X_i, Y_i), i = 1, 2, \cdots, n\}$ is a random observation of (X, Y). Under the existence of moments, it was proved that $nV_n^2(X,Y)/S_2$ converges in distribution to a quadratic form. Without the explicit null distribution, one needs the permutation test to find p-value in practice, which is computationally inefficient when the sample size or the number of tests is very large.

The Hoeffiding's test static (2.15) and the DC index (2.16) need to know probability density functions or characteristic functions, which all rely on large datasets. Moreover, the two indexes are all appropriate to be applied to evaluate the independence between two continuous variables. However, they are both not appropriate to evaluate the independence between categorical data and continuous data.

Recently, Cui et al. [100, 101] proposed a new test based on mean variance (MV) index to test the independence between a categorical random variable Y and a continuous one X. The MV index can be considered as the weighted average of Cramér–von Mises distances between the conditional distribution functions of X given each class of Y and the unconditional distribution function of X. The MV index is zero if and only if X and Y are independent. The MV index is defined as

$$MV(X|Y) = \frac{1}{n} \sum_{k=1}^{C} \sum_{i=1}^{n} \hat{p}_k \left[\hat{F}_k(X_i) - \hat{F}(X_i) \right]^2, \tag{2.17}$$

where $\hat{F}(x) = n^{-1} \sum_{i=1}^{n} I\{X_i \le x_i\}$ is the empirical unconditional distribution function of X, $\hat{F}_k(x) = \sum_{i=1}^{n} I\{X_i \le x_i, y_i = y_k\} / \sum_{i=1}^{n} I\{y_i = y_k\}$ is the empirical conditional distribution function of X given $Y = y_k$, and $\hat{p}_k = n^{-1} \sum_{i=1}^{n} I\{y_i = y_k\}$ denotes the sample proportion of the k'th ($k = 1, 2, \cdots, C$) class, where $I(.)$ represents the indicator function.

Cui et al. [101] further proposed an algorithm for the MV test of independence. The testing hypothesis is as follows:

$$H_0 : X \textit{ and } Y \textit{ are statistically independent.}$$

versus

$$H_1: X \textit{ and } Y \textit{ are not statistically independent}.$$

The previous hypothesis is equivalent to the following:

$$H_0: F_k(x) = F(x) \; for \; any \; x \; and \; k = 1, 2, \cdots, C.$$

versus

$$H_1: F_k(x) \neq F(x) \; for \; some \; x \; and \; k = 1, 2, \cdots, C.$$

To test H_0, Cui et al. [101] proposed the following new test statistic based on the sample-level MV index:

$$T_n = nMV(X|Y) = \sum_{k=1}^{C} \sum_{i=1}^{n} \hat{p}_k \left[\hat{F}_k(X_i) - \hat{F}(X_i) \right]^2. \tag{2.18}$$

The larger value of T_n provides a stronger evidence against the null hypothesis H_0. The new test was called as the MV test of independence. Suppose X is continuous and Y is categorical with a fixed number C of classes. Under H_0, Cui et al. [101] proved that

$$T_n = nMV(X|Y) \xrightarrow{d} \sum_{j=1}^{+\infty} \frac{\chi_j^2(C-1)}{\pi^2 j^2}, \tag{2.19}$$

where $\chi_j^2(C-1)(j = 1, 2, \cdots)$ are i.i.d. χ^2 random variables with $C - 1$ degrees of freedom, and $\xrightarrow{d}$ denotes the convergence in distribution. Figure 2.6 shows the asymptotic null distributions of $C - 1$ degrees of freedom with $C = 2, 3, \cdots, 10$. Theoretical and empirical studies both show that

$$E\left[\sum_{j=1}^{+\infty} \frac{\chi_j^2(C-1)}{\pi^2 j^2}\right] = \frac{C-1}{6}, \; SD\left[\sum_{j=1}^{+\infty} \frac{\chi_j^2(C-1)}{\pi^2 j^2}\right] = \sqrt{\frac{C-1}{45}}. \tag{2.20}$$

Empirical studies show that the MV test based on the asymptotic null distribution performs well when the sample size is relatively moderate or large. However, if the sample size is very small, it is better to use the permutation test to compute the P-value for the MV test. The algorithm was listed as below.

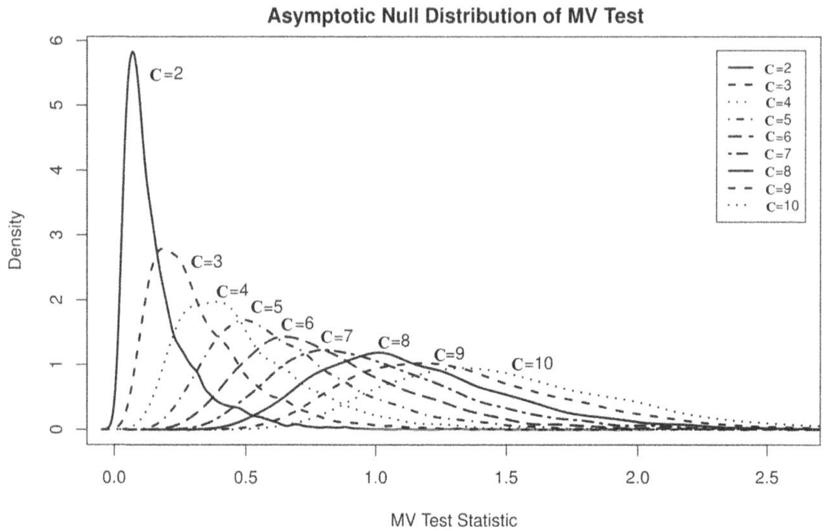

Fig. 2.6 The asymptotic null distributions of the MV test statistic with $C-1$ degrees of freedom. C denotes the number of classes of the categorical random variable. Reprinted from ref. [101], with permission from Elsevier

---

Algorithm 9 The permutation test to compute the P-value for the MV test [101]

1: Compute the MV test statistic T_{n0} for the given sample $\{(X_i, Y_i), i = 1, 2, \cdots, n\}$ by $T_0 = nMV(X|Y) = \sum_{k=1}^{C} \sum_{i=1}^{n} \hat{p}_k \left[\hat{F}_k(X_i) - \hat{F}(X_i)\right]^2$.

2: Generate a permutation response sample $\{Y_i^*, i = 1, 2, \cdots, n\}$ from the original response, and compute the corresponding MV test statistic $T_n^* = nMV(X|Y^*)$.

3: Repeat Step 2 K times and obtain the K values of the permutation MV test statistic,$\{T_{n1}^*, T_{n2}^*, \cdots, T_{nK}^*\}$.

4: The P-value is estimated by $\sum_{k=1}^{K} I(T_{nk}^* \geq T_0)/K$.

The MV test between random variables X and Y enjoys several appealing merits. First, an explicit form of the asymptotic null distribution is derived under the independence between X and Y. It provides an efficient way to compute critical values and P-value. Second, no assumption on the distributions of two random variables is required and the new test statistic is invariant under one-to-one transformations of the continuous random variable. It is essentially a rank test and distribution-free, so it is resistant to heavy-tailed distributions and extreme values in practice. Monte-Carlo (MC) simulations demonstrate its excellent finite-sample performance. The MV test can be used in high dimensional omics data to detect independence between genes and response, as well as to detect the significant genes associated with tumor types [101].

2.4.2 Information Theoretic Approaches

Generally, information theoretic methods use a generalization of pairwise correlation coefficient, termed mutual information (MI), to compare gene expression profiles [102–104]. Similar to correlation coefficient, MI is a measure that detects statistical dependence between two variables. MI coefficient for two variables X and Y is defined as follows:

$$I(X,Y) = \sum_{i,j} p(x_i, y_j) log \frac{p(x_i, y_j)}{p(x_i)p(y_j)}, \tag{2.21}$$

where $p(x_i)$ and $p(y_j)$ are marginal probabilities of $X = x_i$ and $Y = y_j$ for genes X and Y, respectively. Also, $p(x_i, y_j)$ is the joint probability of expression levels related to these two genes. If the gene pair X and Y have low or zero MI, then the two genes are not correlated; but if MI is greater than a predefined threshold, then the two genes are considered to be correlated [104].

Maximal information coefficient (MIC) is another information theory based measure for capturing dependencies, introduced by Reshef et al. [105]. MIC has two important features: the ability to find different types of association (e.g., linear and nonlinear) and the assignment of similar scores to equally noisy data. It has been shown recently that replacement of MI with MIC improves the performance of MI-based reconstruction algorithm [106].

MIC is based on the idea that if a relationship exists between two variables, then a grid can be drawn on the scatterplot of the two variables that partitions the data to encapsulate that relationship. Thus, to calculate the MIC of a set of two-variable data, one can explore all grids up to a maximal grid resolution, dependent on the sample size (Fig. 2.7a), computing for every pair of integers (x, y) the largest possible MI achievable by any x-by-y grid applied to the data. one then normalizes these MI values to ensure a fair comparison between grids of different dimensions, and to obtain modified values between 0 and 1. One defines the characteristic matrix $M = (m_{x,y})$, where $m_{x,y}$ is the highest normalized MI achieved by any x-by-y grid, and the statistic MIC to be the maximum value in M (Fig. 2.7b, c). Every entry of M falls between 0 and 1, and so MIC does as well. MIC is also symmetric. The MIC has good performance on detecting associations not well modeled by a function, as noise level varies (Fig. 2.7d). For details of the algorithm and its comparison with other association measures, one can refer to reference [105].

Except MI and MIC, various algorithms have been proposed based on information theory, such as RelNet [107], ARACNE [108], MRNET [109], CLR [110], and so on [111]. For details of these algorithms, one can refer to the mentioned works.

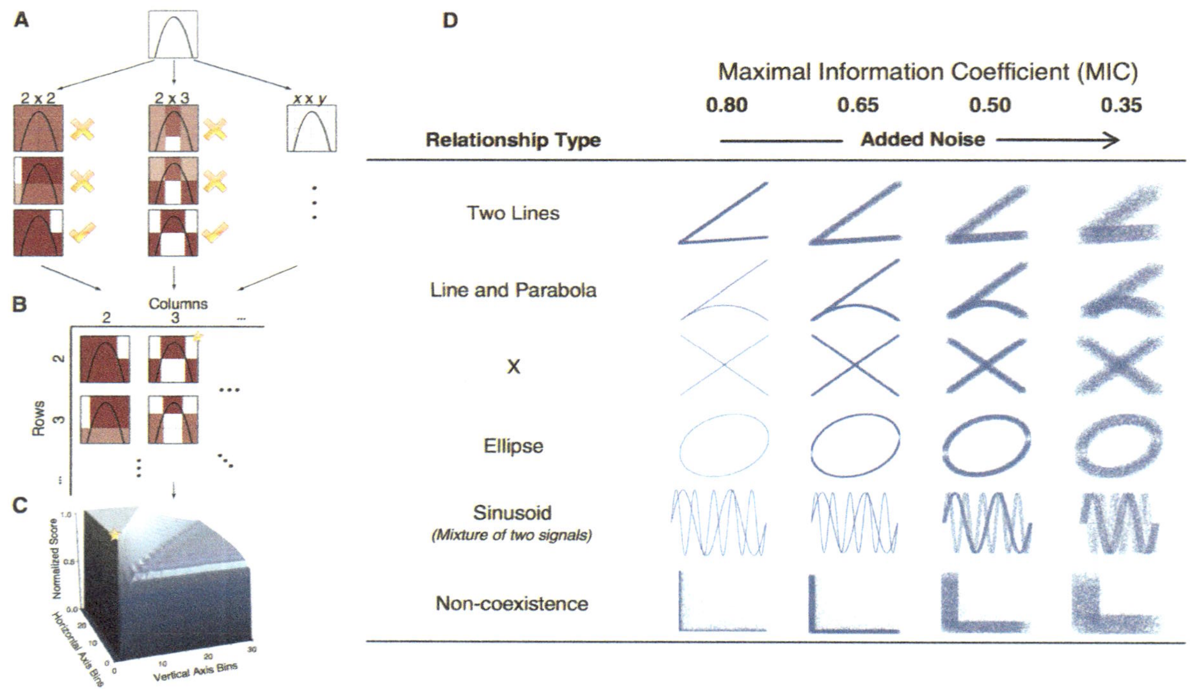

Fig. 2.7 The flowchart of MIC and its performance. (**a**) For each pair (x, y), the MIC algorithm finds the x-by-y grid with the highest induced MI. (**b**) The algorithm normalizes the MI scores and compiles a matrix that stores, for each resolution, the best grid at that resolution and its normalized score. (**c**) The normalized scores form the characteristic matrix, which can be visualized as a surface; MIC corresponds to the highest point on this surface. In this example, there are many grids that achieve the highest score. The star in (**b**) marks a sample grid achieving this score, and the star in (**c**) marks that grid's corresponding location on the surface. (**d**) Performance of MIC on associations not well modeled by a function, as noise level varies. Reprinted from ref. [105], with permission from AAAS

2.4.3 Partial Correlation/Gaussian Graphical Models

Both correlation methods and the information theoretical methods only consider pairwise relationships between genes. However, in a real biological pathway, a gene may interact with a group of genes but not possess a strong marginal relationship with any individual member of the group [112]. Such higher-level interactions can be potentially missing in the networks constructed by pairwise measures. In this sense, Gaussian graphical models (GGM) offer a more realistic way to represent complex gene networks due to its interpretation in terms of conditional correlations. Assuming a multivariate normal distribution for the expression vectors for a set of genes U, the GGM uses Σ^{-1}, the inverse of the gene covariance matrix (or precision matrix), as a measure for gene association patterns. This approach is closely related with the concept of partial correlations, noting that the partial correlation between genes i and j can be expressed as

$$\rho_{ij} = corr(i, j|U/\{i, j\}) = \begin{cases} -\frac{u_{ij}}{\sqrt{u_{ii}}\sqrt{u_{jj}}}, & i \neq j, \\ 1, & i = j, \end{cases} \tag{2.22}$$

where u_{ij} is the element in the precision matrix. Therefore, genes i and j being conditionally independent is equivalent to the corresponding partial correlation and element in the precision matrix being zero. And non-zero entries in the precision matrix correspond to the presence of direct interaction between two genes having controlled for the effect of the other genes.

The major difficulty of estimating the precision matrix arises from the high dimensional nature of gene expression data [112]. Various regularized estimation methods have been proposed to address this "curse of dimensionality." Edwards [113] proposed a backward selection scheme to remove weak edges in the estimated Σ^{-1}. Schäfer and Strimmer [114] chose to estimate Σ^{-1} directly using the Moore-Penrose pseudo-inverse [115] and using the bagged average of all bootstrap estimates. Since gene networks are believed to be inherently sparse, Li and Gui [116] introduced in-built sparsity in their estimated Σ^{-1} by a threshold gradient descent algorithm. Noting that regressing the expression vector X_i for gene i on the other expression vectors X_j,

$$X_i = \sum_{j \neq i} \beta_{ij} X_j + \varepsilon_i, \tag{2.23}$$

where the coefficients $\beta_{ij} = \rho_{ij}\sqrt{u_{jj}/u_{ii}}$, sparsity can be more naturally incorporated in a penalized regression setting [117–119]. A rich wealth of the literature exists on the problem of estimating sparse precision matrix in high dimensional GGMs [120].

The above partial correlation based approaches have attractive theoretical properties and their asymptotic behaviors have been extensively studied. However, the kind of biological inference they are capable of achieving is still limited. In the current literature, partial correlation is usually calculated conditioned on either all of the available genes or a more or less arbitrary subset of them that may contain noisy (biologically unrelated) genes [112].

2.4.4 Granger Causality Methods

The measures introduced in the above two sections can only detect associations between two variables, they cannot infer causation relationships [80, 82] between two variables. In this subsection, we will introduce some approaches for causal network reconstruction, paying special attention to methods based on the Granger causality methods [80, 84]. The Granger causality methods are generally used to explore time series data, it can also be applied to omics data if one treats samples in the omics data as time series.

2.4.4.1 Granger Causality

The major approach to causality analysis is to examine if the prediction of one time series could be improved by incorporating information from the other, as proposed by Granger [81]. Specifically, given two time series X_t and Y_t which are jointly stationary, consider the autoregressive prediction of the current value of X_t based on its past measurements, described by

$$X_t = \sum_{i=1}^{\infty} a_{1i} X_{t-i} + \varepsilon_{1t}, \tag{2.24}$$

and the prediction using information of past measurements of both processes X_t and Y_t, given by

$$X_t = \sum_{i=1}^{\infty} a_{2i} X_{t-i} + \sum_{i=1}^{\infty} c_{2i} Y_{t-i} + \varepsilon_{2t}, \tag{2.25}$$

where $\varepsilon_{it} (i = 1, 2)$ represents the prediction error. According to the definition of Granger causality [81], if $var(\varepsilon_{2t}) < var(\varepsilon_{1t})$, then Y_t influences X_t. The causal influence can be quantified by

$$F_{Y \to X} = ln \left(\frac{var(\varepsilon_{1t})}{var(\varepsilon_{2t})} \right). \tag{2.26}$$

Obviously, $F_{Y \to X} = 0$ indicates that there is no causal connection from Y_t to X_t, and $F_{Y \to X} > 0$ suggests that there is. The causal connection from X_t to Y_t can be defined similarly.

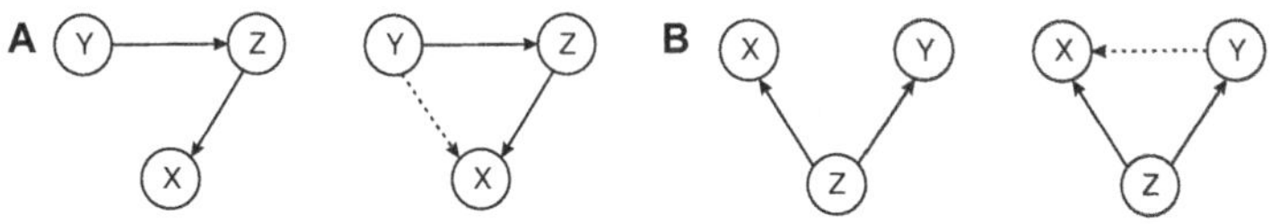

Fig. 2.8 The pairwise Granger causality might infer false connections. (**a**) Left: the true topology; right: a false link from Y to X is given. (**b**) Left: the true topology; right: a false link from Y to X is inferred. Reprinted from ref. [85], with permission from APS

2.4.4.2 Partial Granger Causality

For a system having numerous nodes, various possibilities for causal connections among nodes arise. From the above pairwise Granger causality analysis for more than two time series, some false connections might be given due to the influence of observable or hidden variables in the network [85]. For example, consider three variables X, Y, and Z, whose connection pattern is shown in Fig. 2.8a. However, the false link from Y to X is likely to be incorrectly inferred by a pairwise Granger causality test due to the mediation of Z. Another possible causal connection is shown in Fig. 2.8b, where X and Y are simultaneously driven by Z. If the driving signal Z is powerful enough, then X, Y, and Z might evolve into generalized synchronization and it is very likely that one will get a false causal link between X and Y, such as the dashed line from Y to X.

In 1984, Geweke [121] introduced conditional Granger causality, which has the ability to resolve whether the interaction between two time series is direct or mediated by another recorded time series and whether the causal influence is simply due to different time delays in their respective driving input. Critically, conditional Granger causality is effective only when all relevant variables in a network are observable. This is practically impossible, since both environmental inputs and unmeasured hidden variables can obscure accurate causal connections. In 2008, Guo et al. introduced partial Granger causality to detect causal connections, which is said to be capable of eliminating the influence of exogenous inputs and latent variables [122].

According to Guo et al. [122], partial Granger causality can be explained in the following way. Given two processes X_t and Z_t, the joint autoregressive representation for X_t and Z_t can be written as

$$X_t = \sum_{i=1}^{\infty} a_{2i} X_{t-i} + \sum_{i=1}^{\infty} c_{2i} Z_{t-i} + u_{1t}, \quad Z_t = \sum_{i=1}^{\infty} b_{1i} X_{t-i} + \sum_{i=1}^{\infty} d_{2i} Z_{t-i} + u_{2t}. \tag{2.27}$$

The noise covariance matrix for the model can be represented by

$$\begin{bmatrix} var(u_{1t}) & cov(u_{1t}, u_{2t}) \\ cov(u_{2t}, u_{1t}) & var(u_{2t}) \end{bmatrix}, \tag{2.28}$$

with var and cov representing variance and covariance, respectively. Based on partial correlation in statistics, the value of

$$var(u_{1t}) - cov(u_{1t}, u_{2t})var(u_{2t})^{-1}cov(u_{2t}, u_{1t}) \tag{2.29}$$

measures the accuracy of the autoregressive prediction of X_t based on its previous values conditioned on Z_t by eliminating the influence of all other variables present in the network, such as common exogenous input and hidden variables.

Wu et al. [85] further extended the concept of partial Granger causality to the vector autoregressive representation for a system involving three time series X_t, Y_t, and Z_t, which can be written as follows:

$$\begin{aligned} X_t &= \textstyle\sum_{i=1}^{\infty} a_{2i} X_{t-i} + \sum_{i=1}^{\infty} b_{2i} Y_{t-i} + \sum_{i=1}^{\infty} c_{2i} Z_{t-i} + u_{3t}, \\ Y_t &= \textstyle\sum_{i=1}^{\infty} d_{2i} X_{t-i} + \sum_{i=1}^{\infty} e_{2i} Y_{t-i} + \sum_{i=1}^{\infty} f_{2i} Z_{t-i} + u_{4t}, \\ Z_t &= \textstyle\sum_{i=1}^{\infty} g_{2i} X_{t-i} + \sum_{i=1}^{\infty} h_{2i} Y_{t-i} + \sum_{i=1}^{\infty} k_{2i} Z_{t-i} + u_{5t}. \end{aligned} \tag{2.30}$$

The noise covariance matrix for the above model can be represented by

$$\begin{bmatrix} var(u_{3t}) & cov(u_{3t}, u_{4t}) & cov(u_{3t}, u_{5t}) \\ cov(u_{4t}, u_{3t}) & var(u_{4t}) & cov(u_{4t}, u_{5t}) \\ cov(u_{5t}, u_{3t}) & cov(u_{5t}, u_{4t}) & var(u_{5t}) \end{bmatrix}. \tag{2.31}$$

Similarly, the value of

$$var(u_{3t}) - cov(u_{3t}, u_{5t})var(u_{5t})^{-1}cov(u_{5t}, u_{3t}) \tag{2.32}$$

represents the accuracy of predicting the present value of X_t based on the previous information of both X_t and Y_t conditioned on Z_t by eliminating the effect of other variables in the system. According to Guo et al. [122], the partial Granger causality from Y_t to X_t conditioned on Z_t by eliminating the effect of the common exogenous inputs and hidden variables present in the network can be expressed as

$$F_{Y\to X} = ln\left(\frac{var(u_{1t}) - cov(u_{1t}, u_{2t})var(u_{2t})^{-1}cov(u_{2t}, u_{1t})}{var(u_{3t}) - cov(u_{3t}, u_{5t})var(u_{5t})^{-1}cov(u_{5t}, u_{3t})}\right). \tag{2.33}$$

$F_{Y\to X} = 0$ indicates that there is no direct causal influence from Y_t to X_t, and $F_{Y\to X} > 0$ suggests that there is.

For the applications of the partial Granger causality method, one can refer to Wu et al. [85] for details.

2.4.4.3 Windowed Granger Causality

High-throughput technologies can provide a wealth of time series data to better interrogate the complex regulatory dynamics inherent to organisms, but many network inference strategies do not effectively use temporal information [80]. In the year 2018, Finkle et al. [80] addressed this limitation by introducing Sliding Window Inference for Network Generation (SWING). SWING embeds existing multivariate methods, both linear and nonlinear, into a Granger causal framework that concurrently considers multiple time delays to infer causal regulators for each node. SWING also uses sliding windows to create many sensitive, but noisy, inference models that are aggregated into a more stable and accurate network. The following contents are from reference [80].

GRN SWING addresses the challenge of inferring GRNs from gene expression data. GRNs are directed graphs with N nodes, where each node represents a gene. An edge from gene g_i to gene g_j indicates that g_i regulates the expression of g_j. For simplicity, we use the notations in Finkle et al. [80] to describe their methods.

Time Series Data Suppose we have the time series measurement of expression for gene, i, with T time points, is defined as $G_i = [g_i^1, g_i^2, \cdots, g_i^T]^T$. Thus, a time series experiment is defined as $\mathbf{T} = [G_1, \cdots, G_N]$. $\mathbf{T}$ is a $T \times N$ matrix which provides an ordered sequence of values for each observed gene (columns) at each time point (rows).

$$\mathbf{T} = \begin{bmatrix} g_1^1 & \cdots & g_N^1 \\ \vdots & \ddots & \vdots \\ g_1^T & \cdots & g_N^T \end{bmatrix}. \tag{2.34}$$

For simplicity, the following describes the case where there are no replicates. However, if there are multiple time series, D, of the same length for each gene, such as experiments with multiple biological replicates or experimental perturbations, they are stacked into a $(T \bullet D) \times N$ matrix such that $\mathbf{T} = [\mathbf{T}_1, \cdots, \mathbf{T}_D]^T$.

SWING Window Creation SWING employs a fixed-length sliding window to divide time series observations into ensembles of training data with the same measured features within each time series. Given a time series dataset $\mathbf{T}$, SWING creates Q consecutive windows. Q is defined as

$$Q = (T - \omega + 1)/s, \tag{2.35}$$

where ω is the window width, such that $\omega \leq T$, and s is the step size between windows. Both ω and s are specified by the user. Each window W_q, where $q \in \{1, \cdots, Q\}$, is a subset of rows from the time series data $\mathbf{T}$, such that:

$$W_q = \begin{bmatrix} g_1^{s(q-1)+1} & \cdots & g_N^{s(q-1)+1} \\ g_1^{s(q-1)+2} & \cdots & g_N^{s(q-1)+2} \\ \vdots & \ddots & \vdots \\ g_1^{s(q-1)+\omega} & \cdots & g_N^{s(q-1)+\omega} \end{bmatrix}. \tag{2.36}$$

If $\omega = T$, then there is only one window and SWING performs network inference equivalent to the base method. Additional parameters for window creation are described in the subsection of SWING parameter selection.

Edge Inference Once the temporal windows are delimited, one can apply multivariate Granger causality to generate training sets for inference algorithms. Traditional Granger causality models assess pairwise predictions with a set delay between the variables. Previous methods expanded the Granger models to be multivariate, but do not simultaneously compare multiple delays between explanatory and response variables. Finkle et al. [80] described the formulation of a Granger model that is both multivariate and includes multiple delays.

SWING utilizes a general statistical framework where weights between explanatory variables and a response variable are calculated using supervised learning algorithms. For each window, W_q, Finkle et al. [80] sequentially defined a response vector for each gene, j, as $y_j = W_{q,j}$, which is the jth column of window W_q. The explanatory data is created based on two user-specified parameters. The maximum lag, k_{max}, and the minimum lag, k_{min}, define the number of time points that can exist between the explanatory variables and the response. They are used to define the user-allowed set of delays, $L = \{k_{\min}, k_{min} + 1, \cdots, k_{\max}\}$.

$|L|$ is the cardinality of the set L, and is used to calculate the maximum number of explanatory variables. For most windows, the number of user-allowed delays is $|L| = k_{\max} - k_{\min} + 1$, but there will be fewer when $q \leq k_{\max}$. The explanatory data matrix for each response vector is constructed by concatenating data from the delayed windows, and is defined as $X = [W_{q-k_{\min}}, \ldots, W_{q-k_{\max}}]$ if $q > k_{\max}$, and $X = [W_{q-k_{\min}}, \ldots, W_1]$ if $k_{\min} < q \leq k_{\max}$.

To maintain consistency between SWING and existing methods, if $k_{\min} = 0$, the response variable is excluded from the explanatory data, prohibiting self-edges within the same window. X has an augmented number of explanatory variables, corresponding to an explanatory variable for each gene at each delay. The number of columns in X is $N \bullet |L|$ if $k_{\min} > 0$, or $N \bullet |L| - 1$ if $k_{\min} = 0$. Finkle et al. [80] did not include any self-edges, regardless of delay, during their testing, because the in silico and in vitro data was collected in a way that does not account for self-edges.

Model Aggregation SWING aggregates the results from several weak, but sensitive, windowed models to generate a ranked list of edges. Each window generates an $N \times (N \bullet |L|)$ adjacency matrix, A, of edge scores where $A^k_{i,j}$ is the inferred score for gene i as the upstream regulator of gene j with delay k.

The time series data are naturally left censored, as one cannot know measurements before the experiment occurs. As such, depending on the user-specified k_{min} and k_{max}, some windows, particularly the earlier ones, will not infer interactions for larger values of k (e.g., $g_i^{q-2} \rightarrow g_j^q$ cannot be inferred if $q < 2$). Therefore, each window W_q infers at most $|L|$ scores, for each gene pair.

In order to combine scores across multiple windows and different delays into a single score $g_i \rightarrow g_j$, SWING does two aggregations. Confidence values from windowed subsets are aggregated into a single network by taking the mean rank of the edge at each delay k, and then taking the mean rank of the edge across all delays. Additionally, community networks estimated from multiple classifiers are built by computing the mean rank of edges outputted from RandomForest (RF)[123], the least absolute shrinkage and selection operator (LASSO)[124], and partial least-squares regression (PLSR) [125]. We use the edge rank because scores between window models and methods may not have equivalent distributions. The median of edge ranks may also be used, but in preliminary testing it did not significantly change the results.

SWING Graph Generation A directed SWING graph shows causal relationships between N nodes in a system and can be represented by the adjacency matrix A in which each element $A_{i,j}$ is the confidence that an edge exists between parent node g_i and child node g_j. Given Q user-defined windows, for each window, W_q, there are at most $N^2|L| - N$ possible edges that exist in the inferred model. Therefore, the adjacency matrix for each window is

$$A_q = \begin{bmatrix} A^{k_{\text{min}}}_{1,1} & \cdots & A^{k_{\text{max}}}_{1,N} \\ \vdots & \ddots & \vdots \\ A^{k_{\text{min}}}_{N,1} & \cdots & A^{k_{\text{max}}}_{N,N} \end{bmatrix}, \tag{2.37}$$

where $A^k_{i,j}$ is the confidence of the interaction whereby the parent node g_i is said to be Granger causal of the child node g_j with a delay of k time points. Self-edges within the same window are prohibited, and therefore values $A^0_{i,i}$ are set to be 0. In this way, a network model with N targets and at most $N \bullet |L|$ regulators is created for each window.

For each window, SWING estimates the confidence of each edge and generates a ranked list of edges based on method-specific criteria. Specifically, RF uses the importance score calculated with the mean squared error [123]; LASSO uses a stability selection metric [124], and PLSR uses the variable importance in projection (VIP) score [125]. The rank of an edge in each windowed model can be used as the confidence metric to compare across methods. One computes a consensus model (SWING-Community) by calculating the mean rank across methods for each possible edge:

$$\bar{A}_{i,j} = \frac{R_{i,j}^{SWING-RF} + R_{i,j}^{SWING-LASSO} + R_{i,j}^{SWING-PLSR}}{3}, \quad (2.38)$$

where $R_{i,j}$ are the ranks of the edge for each of the tested methods, and $\bar{A}_{i,j}$ is the average rank of the edge $g_i \rightarrow g_j$ used as the confidence metric in the consensus network.

SWING Parameter Selection SWING is a generalized framework that can be used with any multivariate machine learning inference method. In developing and testing SWING, Finkle et al. [80] implemented three different existing methods: RF, LASSO, and PLSR. Each algorithm requires different tuning parameters. When using RF, they selected the number of trees, the maximum depth of the tree, and the number of trees based on guidelines from the GENIE3 manuscript [123]. For LASSO, the authors utilized two methods to select the regularization parameters [124]: for in silico studies, they selected the regularization parameters based on the cross-validation score; for in vitro datasets with comparatively less data, they selected the regularization parameters based on sensitivity analysis for a single random subnetwork and evaluated all subnetworks with the subsequent parameters. For PLSR, the authors selected the number of principal components to use based on the elbow criterion [125].

In addition to the base method's specific parameters, SWING has user-selected parameters that require knowledge of the system and data. For optimal performance, it is suggested that the window size be selected such that $T/2 < \omega < T$, where T is the number of time points in the time series. If $\omega < T/2$, increased noise can lead to inference of more false positive edges. In general, the step size can be set to $s = 1$, unless the user has an abundance of time points and wishes to train on only a subset of the data.

The allowed delay range is specified by the user in setting k_{max} and k_{min}. They recommended the user set these values based on the range of dynamics expected in the system, or by prior delay analysis such as cross-correlation. Since $k_{\max}$ and $k_{\min}$ are integer values, they also depend on the sampling interval of the experimental data. Specifying $k_{\min} = 0$ allows the SWING to infer edges with no delay, as many existing methods do. If, however, the user specifies null SWING parameters—specifically, $\omega = T$, $k_{\max} = 0$, $k_{\min} = 0$, and $s = 1$—there is only a single window with no delays between the explanatory and response variables. This condition corresponds to running the base methods independent of the SWING.

The flowchart of the SWING is shown in Fig. 2.9. Finkle et al. [80] demonstrated that SWING elucidates network structure with greater accuracy in both in silico and experimentally validated in vitro systems. They estimated the apparent time delays present in each system and demonstrated that SWING infers time-delayed, gene-gene interactions that are distinct from baseline methods. By providing a temporal framework to infer the underlying directed network topology, the SWING generates testable hypotheses for gene-gene influences.

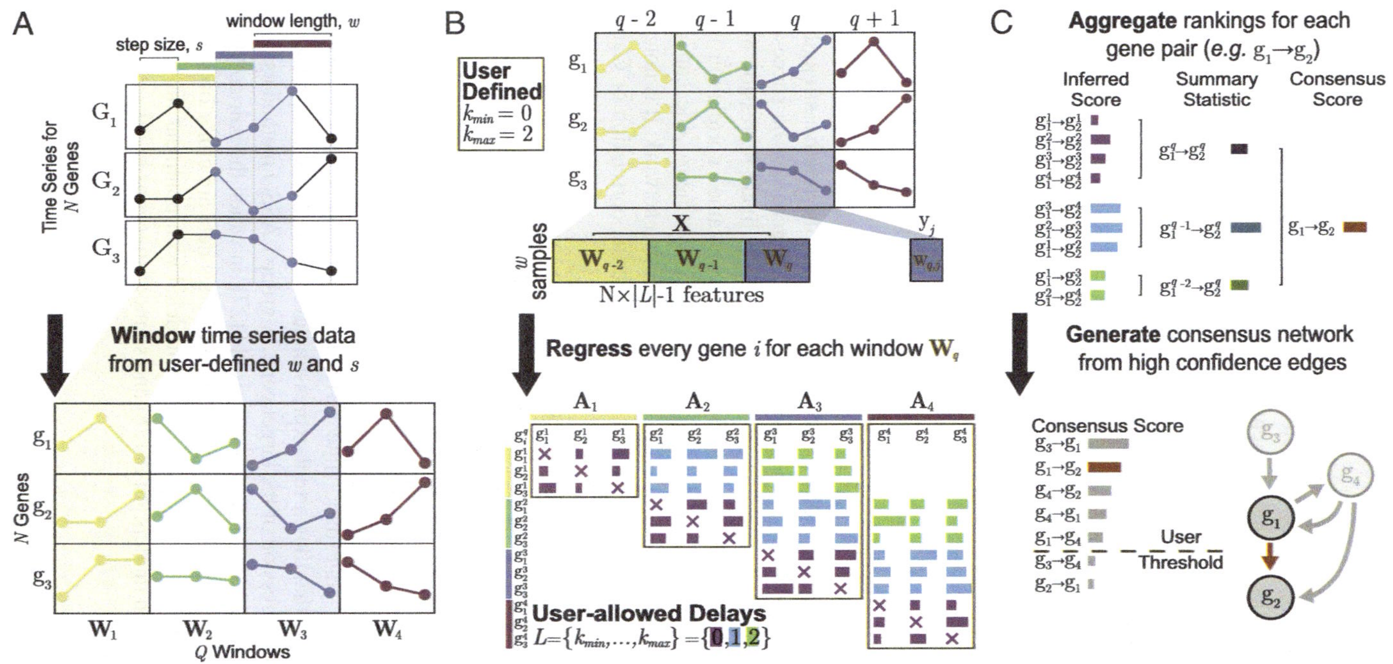

Fig. 2.9 Overview of the SWING framework. (**a**) Time series data are divided into windows with a user-specified width, ω. (**b**) For each window, inference is performed by iteratively selecting response and explanatory genes. The subset of available explanatory genes is defined by the minimum and maximum user-allowed time delays. (**c**) Edges from each window model are aggregated into a single network representation of the biological interactions between measured variables. Reprinted from ref. [80]

2.4.5 Statistical Regression Methods

Interring GRNs can be treated as a feature selection problem [126]. The expression level of a target gene can be predicted by its direct regulatory TFs. It is often assumed that the links among the genes are sparse. For example, LASSO (also known as compressed sensing) [117] is a widely used regression method for network inference [86], and it is a shrinkage and selection method for linear regression. Specifically, it minimizes the usual sum of squared errors, with a bound on the L_1 norm of the coefficients, which is used to regularize the model. Given the expression data of genes, in a steady-state experiment, the feature selection problem can be formulated as

$$\min ||Y - X\beta||_2^2 + \lambda||\beta||_1, \tag{2.39}$$

where λ is an adjustable parameter for controlling the sparsity of the coefficient vector β. Y represents the response, and X denotes the covariates that may have relationships with Y (Suppose X and Y are both normalized data). In real-world applications of omics data, Y can be the expression profile of a gene, and X represents the other genes; the components of β reflect the relative strength of a corresponding gene with the responsive gene Y. Each gene can be successively served as the response Y. For a dataset with p genes, p linear regression models (2.39) with $p - 1$ covariates should be solved. The obtained correlation matrix R that consist of the estimated βs are unnecessary symmetry, one can define $R^* = (R + T^T)/2$ and symmetrize the final co-expression network. It is noted that the regression method is just similar to the above partial correlation or Gaussian graphical models.

A direct use of LASSO [117] to infer networks has two shortcomings [126]: (a) it is known to be an unstable procedure in terms of the selected features, and (b) it does not provide confidence scores for the selected features. As a result, a stability selection procedure is often integrated into LASSO [124]. One first resamples the data into several sub-data based on bootstrapping, then applies the LASSO to solve these sub-data regression problems, and aggregates the final score for each feature to select more confident features.

Beside the steady-state data, time series data can also be tackled using the LASSO in the sense that the current expression level of the TFs is the predictors of the change in the expression of the target gene. A group LASSO method using both steady-state and time series data was proposed [127], in which the pair coefficients of a single TF across both steady-state and time series data are either both zero or both non-zero. Gene regulation can also be modeled with nonlinear models, e.g., polynomial regression models [111, 128] and sigmoid functions [129].

2.4.6 Bayesian Methods

From marginal dependencies in co-expression measure to conditional dependencies in partial correlation based approaches, the methods all attempt to capture gene relationships using probabilistic dependencies of different kinds [112]. However, they all lead to the reconstruction of undirected graphs and hence unable to represent causal relationships between genes. Except the Granger methods, Bayesian networks (BNs) for GRNs, pioneered by Friedman et al. [79], are directed acyclic graphs (DAGs) that characterize the joint distribution of nodes (genes) as a series of local probability distributions. Denoting gene i as $X_i (i = 1, 2, \cdots, p)$, the joint distribution of all nodes is given by

$$P(X_1, \cdots, X_p) = \Pi_{i=1}^{p} P(X_i | P_\alpha^G(X_i)), \tag{2.40}$$

where $P_\alpha^G(X_i)$ are all the parent nodes of X_i in the DAG G. The joint distribution can be factorized this way because of the Markov assumption of BNs: given its parents, each node is independent of its non-descendants. In this sense, each directed edge can be interpreted as a causal link. A BN implies both a set of conditional dependencies and conditional independences. Two different DAGs can encode the same set of conditional independences [130], and the goal of BN inference algorithms is to infer these equivalent classes of DAGs.

The first problem of reconstructing a BN based on expression data D involves finding the best DAG G that describes D, and each G is evaluated using a Bayesian score which is the posterior probability of a graph G given the data [79]:

$$S(G : D) = log P(G|D) = log P(D|G) + log P(G) + C. \tag{2.41}$$

Here, C is a constant independent of G and

$$P(D|G) = \int P(D|G, \Theta) P(\Theta|G) d\Theta \tag{2.42}$$

is the marginal likelihood, which averages the probability of the data over all possible parameter assignments Θ to G. The particular choice of priors P(G) and $P(\Theta|G)$ for each G determines the exact Bayesian score [79]. The computation of the posterior probability is two-fold: (1) learning the graph G given observed data; (2) learning the local conditional probabilities given G.

The second problem amounts to parameter estimation, which can be accomplished via a number of algorithms such as sum-product, maximum likelihood estimation (MLE), maximum a posteriori (MAP), and expectation maximization (EM) depending on the form of the conditional probabilities (discrete, continuous, or mixture distribution [79] and whether any node has missing information. Prior information concerning the distributions of parameters and graphs is also incorporated in the final computation of the scoring function. It is important that the scoring function chosen should be decomposable to the local scores from each node for computational efficiency. The function should also contain features that guard against over-fitting. Popular schemes to achieve this goal include using the Bayesian information criterion (BIC) and Bayesian Dirichlet equivalent [131, 132]. A comparison of different scoring schemes can be found in Ref.[133, 134].

The first problem, however, is a lot more challenging, since theoretically it requires us to consider all possible topologies of DAGs, which is super-exponential in search space dimension [112]. Furthermore, the high dimensional nature of expression data lead to many DAGs that score equally well. A number of heuristic algorithms have been developed to walk through the space of possible DAGs, including greedy hill-climbing, simulated annealing and genetic algorithms [134]. Often the algorithms explore the neighborhood of a topology by adding, deleting, or reversing the direction of an edge to make incremental changes at a time. To further reduce the search space, biological assumptions and priors can be employed to limit the number of parents a child node is allowed to have, and co-expression clustering can be applied to arrive at a set of most likely parent/child nodes. Rather than choosing a single optimal G, a number of DAGs scoring comparably can be compared for the selection of consistent topological features. Summaries of how to infer BNs can be found in Ref.[135].

The BN has a number of advantages as a modeling framework. The probabilistic set up offers a natural way to incorporate latent variables, prior knowledge, and the possibility that gene expression levels are stochastic with noise. Some missing data can also be handled. However, in order to infer all this additional information, more parameters need to be estimated and hence more high-quality data is required. For this reason, the application of BNs has been centered around the *yeast* data [136], and the success in higher organisms and larger networks are still limited. Conceptually, feedback loops, which are common features in many pathways, cannot be modeled under this framework since all BNs are acyclic. Although the linkages can be potentially causal, they are still qualitative and do not indicate whether a regulation is activation or repression [112].

2.4.7 Variational Bayesian Methods

Variational Bayesian methods (VBM) are a family of techniques for approximating intractable integrals arising in Bayesian inference and machine learning [137–140]. They are typically used in complex statistical models consisting of observed variables (usually termed "data") as well as unknown parameters and latent variables, with various sorts of relationships among the three types of random variables, as might be described by a graphical model. As typical in Bayesian inference, the parameters and latent variables are grouped together as "unobserved variables." VBMs are primarily used for two purposes: (1) To provide an analytical approximation to the posterior probability of the unobserved variables, in order to do statistical inference over these variables; (2) To derive a lower bound for the marginal likelihood (sometimes called the "evidence") of the observed data (i.e., the marginal probability of the data given the model, with marginalization performed over unobserved variables). This is typically used for performing model selection, the general idea being that a higher marginal likelihood for a given model indicates a better fit of the data by that model and hence a greater probability that the model in question was the one that generated the data.

Recently, we built a new framework to cope with the reconstruction of the weighted complex networks [140]. The key idea was to employ a series of linear regression problems to model the relationship between network nodes, and then an efficient VBM rather than Markov chain Monte-Carlo (MCMC) sampling [141–143] is employed to approximate the posterior distributions of the unknown coefficients. The matrix form of the linear regression model is

$$Y = XD(\alpha)\omega + \varepsilon. \tag{2.43}$$

Here, ε denotes the noise item. The response vector $y \in R^M$ and the design matrix $X \in R^{M\times N}$ are observable, while the binary coefficient vector α and the continuous coefficient vector ω are needed to be estimated. Here $D(\alpha)$ denotes a diagonal matrix whose main diagonal is the vector α. M represents that there are M time points or M observations for the observational variables. The VBM is applied to estimate α and ω. For details of the algorithm, one can refer to our recent work [140]. Numerical experiments conducted on both synthetic and real data demonstrated that the new method outperforms LASSO with regard to both reconstruction accuracy and running speed (Fig. 2.10).

2.5 Topological Identification via Dynamical Networks

In the area of complex networks science, topological identification via dynamical networks is a hot research focus. The general formulation of the problem is as follows. For many real-world complex networks, their real topological structures are always difficult to be fully known by us, therefore, a critical issue is that whether we can estimate the network topology via dynamics of nodes. Of course, we must suppose that the dynamics of each node in the network can be measured or with explicit formula. In recent years, some important progresses have been made in this area [38–48].

Materassi and Innocenti [47] considered the problem of reconstructing the tree-like topological structure of a network of linear dynamical systems. A distance function was defined in order to evaluate the "closeness" of two processes and some useful mathematical properties are derived. Theoretical results to guarantee the correctness of the identification procedure for networked linear systems characterized by a tree topology were provided as well. They also suggested the approximation of a complex connected network with a tree in order to detect the most meaningful interconnections. The application of the techniques to the analysis

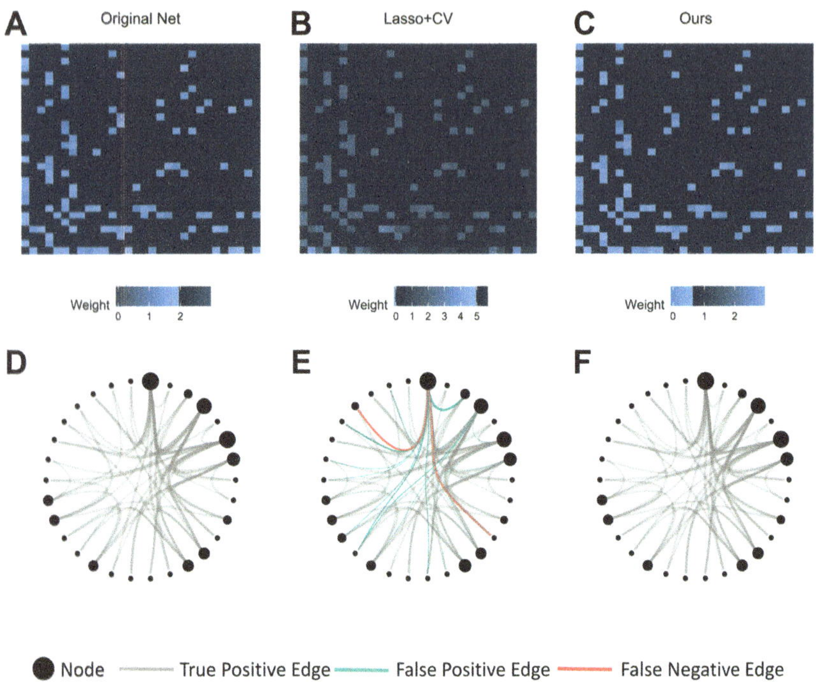

Fig. 2.10 Performance of the proposed VBM in weighted complex network reconstruction. (**a–c**). The heatmap of networks, the three panels correspond to the original network, the network reconstructed by Lasso and the network reconstructed by the proposed VBM. (**d–e**) The visualization of the original network as well as the reconstructed ones by LASSO with CV and our method, respectively. The size of each node and edge is proportional to its degree and weight, respectively. It is shown that our method correctly identifies all the edges, while LASSO with CV fails to detect 2 existing edges and identifies 36 fake edges. The error of connection strength of LASSO and our method is 0.297 and 0.006, respectively. The running time of LASSO and our method is 0.851 and 0.016 second, respectively. The original network is a BA network of 30 nodes; the number of data points is $M = 30$ and the scale of noise is $\sigma = 0.05$. Reprinted from ref. [140], with permission from Elsevier

of an actual complex network, i.e., to high frequency time series of the stock market, is extensively illustrated. One of the disadvantages of the method proposed by Materassi and Innocenti [47] is that they only considered linear dynamical systems, and it mainly used to infer tree-like topological structure.

In the year 2007, Zhou and Lu [44] performed topology identification of weighted complex dynamical networks. Using the Lyapunov theory, they proposed an adaptive feedback control method to identify the exact topology of a rather general weighted complex dynamical network model. By receiving the network nodes evolution, the topology of such kind of network with identical or different nodes, or even with switching topology can be monitored. In 2009, Zhou et al. [43] proposed a criterion for identifying the uncertain topology of a neurobiological network by using an adaptive feedback control method. Unlike similar approaches which monitor all the states of all the nodes to reconstruct network topology, they presented a different mechanism. By receiving the membrane potentials of only a fraction of the neurons, an estimated model is designed to identify the unknown weight couplings in the original neural network. Simulated examples are shown to illustrate the effectiveness of the proposed approach. In addition to the application in neurobiology, this technology was expected to be implemented on many other fields in which the dynamics of each agents can be monitored and received, such as remote control and diagnostics, disease transmission, management, administration of Internet cafe, and so on.

Recently, Liu et al. [48] inferred the structures and the dynamics of the complex networked systems based on time series data. They developed a framework to reconstruct the structures of networks with binary-state dynamics, in which the knowledge of the original dynamical processes is unknown. Within the reconstruction framework, the transition probabilities of binary dynamical processes were described by Sigmoid function in logistic regression, they then applied the mean-field approximation to enable MLE, which gives rise to that the network structure can be inferred by solving the linear system of equations. Meanwhile, the original dynamical processes can be simulated by estimating the parameters in the Sigmoid function. They validated their framework by a variety of binary dynamical processes on synthetic and empirical networks, indicating that their method can not only reveal the network structures but also estimate the dynamical processes.

2.6 Discussions and Conclusions

Network reconstruction or network inferring is an inverse question in systems science. It is more difficult than direct problems in that the inferred structure was often not unique. Different assumptions and methods can result to different solutions. Network reconstruction is a rapidly changing scientific topic. In this chapter, paying special attention to the reconstruction of bio-molecular networks, we introduced some approaches to realize structure inference. Except diverse databases for various types of biological interactions, we introduced many theoretical methods that are based on mathematical and statistical tools, including various association methods, the mean variance index, the information theoretical methods, the Granger causality analysis, the regression methods, the Bayesian methods, the VBM and dynamical network approaches. Moreover, we also introduced some concrete algorithms to generate artificial bio-molecular networks, these artificial networks can be provided as a powerful tool to explore the evolution of some properties in bio-molecular networks. Different methods have different advantages, and may be appropriate for certain types of questions (Table 2.5). When encountering a network reconstruction problem, one should keep in mind which type of networks (undirected or directed; weighted or unweighted; association or causality) she/he will be reconstructed, and what type of data (time series or different samples; $p > n$ or $p < n$; reliable data or

Table 2.5 Summaries of approaches for bio-molecular network reconstruction. By comprehensively considering references [145, 147]

Methods	Type	Summary
PCC	Undirected	Only linear relationships are detected; Need sufficient observations; Makes assumptions based on distribution; Very sensitive to data; Symmetry similarity matrix. High number of false positive relationships; Confusion between indirect and direct relationships
SCC,KCC	Undirected	Applicable for nonlinear data; un-sensitive to data values; Need for sufficient observations; No assumptions based on distribution; Confusion between indirect and direct relationships; Symmetry similarity matrix
GCC	Undirected	Applicable for nonlinear data; un-sensitive to data values robust to noise; Appropriate for problems with a few observations; No assumptions based on distribution; Confusion between indirect and direct relationships
Various distance measures	Undirected	Applicable for nonlinear data; Sensitive to data values; No assumptions based on distribution; Symmetry similarity matrix; Confusion between indirect and direct relationships
Mean variance index	Association	Applicable to data with categorical response; Model free; Has probability meaning
Information theoretic methods	Undirected	Relies on data distribution; Relies on sufficient observations
Granger causality	Directed	Can detect causality; Applicable for detecting hidden variable and GRNs
SWING	Directed	Elucidates network structure with greater accuracy in both in silico and in vitro systems; Estimated the apparent time delays present in the system; Provides a temporal framework to infer the underlying directed network topology; Generates testable hypotheses for gene-gene influences
Graphical Gaussian models	Association	No loops or feedbacks as in the BNs; More complex and computationally costly than the PCC-based methods; Eliminates the effect of other genes when similarity is calculated.
Bayesian network	Directed	Cope with uncertainties; Two or more transition function for each variable is allowed; The use of positive feedback and probabilities can make the model work more effectively; Difficult to apply for large-scale network; Cannot cope with instantaneous interactions between variables; Computationally costly
Variational Bayesian method	Weighted directed	Computationally costly; Directionality and weights of relationships and loops can be depicted

(continued)

Table 2.5 (continued)

Methods	Type	Summary
Linear regression methods	Undirected	Need sufficient observations; Only linear relationships are detected
Dynamical network approaches	Various types	Relies on dynamical equations; Computationally costly; Suited for steady-state or time series expression profiles; Only applicable to smaller network
Boolean network	Undirected	Capable to analyze large regulatory networks; Easier to interpret due to its simplicity; Phenomena of biological realistic complex can represent by simplistic Boolean formalism; Large set of algorithms are provided, which are available in already supervised learning in the binary domain; Deterministic in nature; Unable to handle incomplete regulatory network data only involves two representative states for gene expression levels; Computationally costly; Most of the BNs can only use with a small number of genes

noisy data) she/he has, and then carefully choose the appropriate approaches. Some other methods include conditional correlation analysis [77], deep learning method [144], and so on. For details and more recent progresses of these methods and their applications, readers can refer to the related review papers [102, 112, 126, 145–147]. Finally, we note that network reconstruction is still an open problem and a research area full of challenges.

References

1. Wang, P., Chen, Y., Lü, J., Wang, Q., Yu, X.: Graphical features of functional genes in human protein interaction network. IEEE T. Biomed. Circuits Syst. **10(3)**, 707–720 (2016)
2. Brown, K.R., Jurisica, I.: Online predicted human interaction database. Bioinformat. **21**, 2076–2082 (2005)
3. Peri, S., Navarro, J.D., Amanchy, R., et al.: Development of human protein reference database as an initial platform for approaching systems biology in humans. Genome Res. **13**, 2363–2371 (2003)
4. Stark, C., Breitkreutz, B.J., Reguly, T., et al.: BioGRID: a general repository for interaction datasets. Nucl. Acids Res. **34**, D535–D539 (2006)
5. Güldener, U., Münsterkötter, M., Oesterheld, M., et al.: MPact: the MIPS protein interaction resource on yeast. Nucl. Acids Res. **34**, D436–D441 (2006)
6. Bader, G.D., Hogue, C.W.: BIND–a data specification for storing and describing biomolecular interactions, molecular complexes and pathways. Bioinformat. **16**, 465–477 (2000)
7. Xenarios, I., Rice, D.W., Salwinski, L., et al.: DIP: the database of interacting proteins. Nucl. Acids Res. **28**, 289–291 (2000)
8. Zanzoni, A., Montecchi-Palazzi, L., Quondam, M., et al.: MINT: a molecular interaction database. FEBS Lett. **513**, 135–140 (2002)

9. Aranda, B., Achuthan, P., Alam-Faruque, Y., et al.: The IntAct molecular interaction database in 2010. Nucl. Acids Res. **38**, D525–D531 (2010)
10. Fan, L.: Bioinformatics. Hangzhou, Zhejiang Univ. Press, (2017) (in Chinese)
11. Roberts, G.G., Parrish, J.R., Mangiola, B.A., et al.: High-throughput yeast two-hybrid screening. Meth. Mol. Biol. (Clifton, N.J.) **812**, 39–61 (2012)
12. Trigg, S.A., Garza, R.M., Macwilliams, A., et al.: CrY2H-seq: a massively multiplexed assay for deep-coverage interactome mapping. Nat. Meth. **14**(8), 819–825 (2017)
13. Watts, D.J., Strogatz, S.H.: Collective dynamics of 'small-world' networks. Nature **393** (6684), 440–442 (1998)
14. Barabási, A.L., Albert, R.: Emergence of scaling in random networks. Science **286**, 509–512 (1999)
15. Dorogovtsev, S.N., Mendes, J.F.F.: Evolution of networks. Adv. Phys. **51**, 1079–1187 (2002)
16. Foster, D.V., Kauffman, S.A., Socolar, J.E.S.: Network growth models and genetic regulatory networks. Phys. Rev. E **73**, 031912 (2006)
17. Vázquez, A., Flammini, A, Maritan, A., Vespignani, A.: Modeling of protein interaction networks. Complexus **1**, 38–44 (2003)
18. Rutjes, T.: Duplication-Divergence and proteome evolution networks. Traineeship report, Eindhoven Univ. Technol. (2007)
19. Ispolatov, I., Krapivsky, P.L., Yuryev, A.: Duplication-divergence model of protein interaction network. Phys. Rev. E **71**, 061911 (2005)
20. Berg, J., LäNassig, M., Wagner, A.: Structure and evolution of protein interaction networks: a statistical model for link dynamics and gene duplications. BMC Evol. Biol. **4**, 51 (2004)
21. Pastor-Satorras, R., Smith, E., Solé, R.V.: Evolving protein interaction networks through gene duplication. J. Theor. Biol. **222**, 199–210 (2003)
22. Xu, C., Liu, Z., Wang, R.: How divergence mechanisms influence disassortative mixing property in biology. Physica A **389**, 643–650 (2010)
23. Wan, X., Cai, S., Zhou, J., Liu, Z.: Emergence of modularity and disassortativity in protein-protein interaction networks. Chaos **20**, 045113 (2010)
24. Zhao, D., Liu, Z., Wang, J.: Duplication: a mechanism producing disassortative mixing networks in biology. Chin. Phys. Lett. **24**, 2766–2768 (2007)
25. Teichmann, S.A., Babu, M.M.: Gene regulatory network growth by duplication. Nat. Genet. **36**, 492–496 (2004)
26. Bhan, A., Galas, D.J., Dewey, T.G.: A duplication growth model of gene expression networks. Bioinformat. **18**, 1486–1493 (2002)
27. Chung, F., Lu, L., Dewey, T.G., Galas, D.J.: Duplication models for biological networks. J. Comput. Biol. **10**(5), 677–687 (2003).
28. Solé, R.V., Valverde, S., Rodriguez-Caso, C.: Convergent evolutionary paths in biological and technological networks. Evolution: Edu. Outreach **4**, 415–426 (2011)
29. Lynch, M.: The evolution of genetic networks by non-adaptive processes. Nat. Rev. Genet. **8**, 803–813 (2007)
30. Enemark, J., Sneppen, K.: Gene duplication models for directed networks with limits on growth. J. Stat. Mech-Theory E **11**, P11007 (2007)
31. Leier, A., Kuo, P.D., Banzhaf, W.: Analysis of preferential network motif generation in an artificial regulatory network model created by duplication and divergence. Adv. Complex Syst. **10**(02), 155–172 (2007)
32. Wang, P., Yu, X., Lü, J.: Identification and evolution of structurally dominant nodes in protein-protein interaction networks. IEEE Trans. Biomed. Circ. Syst. **8**(1), 87–97 (2014)
33. Wang, P., Lü, J., Yu, X., Liu, Z.: Duplication and divergence effect on network motifs in undirected bio-molecular networks. IEEE T. Biomed. Circ. Syst. **9**, 312–320 (2015)
34. Solé, R.V., Pastor-Satorras, R., Smith, E., Kepler, T.B.: A model of large-scale proteome evolution. Adv. Complex Syst. **5**, 43–54 (2002)
35. Duarte, N.C., Becker, S. A., Jamshidi N., et al.: Global reconstruction of the human metabolic network based on genomic and bibliomic data. Proc. Natl. Acad. Sci. USA. **104(6)**, 1777–1782 (2007)

36. Ma, H.,Sorokin, A., Mazein, A., et al.: The Edinburgh human metabolic network reconstruction and its functional analysis. Mol. Syst. Biol. **3**(1), 135 (2007)
37. Chen, L., Wang, R.S., Zhang, X.S.: Biomolecular networks: methods and applications in systems biology. John Wiley & Sons., Hoboken (2009)
38. Mei, G., Wu, X., Wang, Y., Hu, M., Lu, J., Chen, G.: Compressive-sensing-based structure identification for multilayer networks. IEEE Trans. Cyber. **48**(2), 754–764 (2018)
39. Wang, Y., Wu, X., Lü, J., Lu, J., DSouza, R.: Topology identification in two-layer complex dynamical networks. IEEE Trans. Network Sci. Eng. **7**(1), 538–548 (2020)
40. Wang, X., Lü, J., Wu, X.: Recovering network structures with time-varying nodal parameters. IEEE Trans. Syst., Man Cyber.: Syst. https://doi.org/10.1109/TSMC.2018.2822780 (2018)
41. Liu, J., Mei, G., Wu, X., Lü, J.: Robust reconstruction of continuously time-varying topologies of weighted networks. IEEE Trans. Circuits Syst. I **65**(9), 2970–2982 (2018)
42. Wu, X., Zhao, X., Lü, J., Tang, L., Lu, J.: Identifying topologies of complex dynamical networks with stochastic perturbations. IEEE Trans. Control Netw. Syst. **3**(4), 379–389 (2016)
43. Zhou, J., Yu, W., Li, X., Small, M., Lu, J.: Identifying the topology of a coupled FitzHugh-Nagumo neurobiological network via a pinning mechanism. IEEE Trans. Neural Netw. **20**(10), 1679–1684 (2009)
44. Zhou, J., Lu, J.: Topology identification of weighted complex dynamical networks. Physica A **386**(1), 481–491 (2007)
45. Chen, J., Lu, J., Zhou, J.: Topology identification of complex networks from noisy time series using ROC curve analysis. Nonlinear Dyn. **75**(4), 761–768 (2014)
46. Yu, D., Righero, M., Kocarev, L.: Estimating topology of networks. Phys. Rev. Lett. **97**(18),188701 (2006)
47. Materassi, D.W., Innocenti, G.W.: Topological identification in networks of dynamical systems. IEEE Trans. Automat. Contr. **55**(8), 1860–1871 (2010)
48. Liu, Q., Ma, C., Xiang, B., Chen, H., Zhang, H.: Inferring network structure and estimating dynamical process from binary-state data via logistic regression. IEEE Trans. Syst., Man Cyber.: Syst. doi: 10.1109/TSMC.2019.2945363 (2019)
49. International Human Genome Sequencing Consortium: Finishing the euchromatic sequence of the human genome. Nature **431**, 931–945 (2004)
50. Yeung, M.K.S., Tegner, J., Collins, J.: Reverse engineering gene networks using singular value decomposition and robust regression. Proc. Natl. Acad. Sci. USA. **99**, 6163–6168 (2002)
51. Das, D., Banerjee, N., Zhang, M.Q.: Interacting models of cooperative gene regulation. Proc. Natl. Acad. Sci. USA. **101**: 16234–16239 (2004)
52. Thomas, R., Paredes, C.J., Mehrotra, S., Hatzimanikatis, V., Papoutsakis, E.T.: A model-based optimization framework for the inference of regulatory interactions using time-course DNA microarray expression data. BMC Bioinformat. **8**(1), 228 (2007)
53. Mashaghi, A.,Ramezanpour, A., Karimipour, V.: Investigation of a protein complex network. Eur. Phys. J. **41(1)**, 113–121 (2004)
54. Terentiev, A.A., Moldogazieva, N.T., Shaitan, K.V.: Dynamic proteomics in modeling of the living cell protein-protein interactions. Biochem. Biokhimiia **74**(13),1586–607 (2009)
55. Sharan, R., Suthram, S., Kelley, R.M., et al.: Conserved patterns of protein interaction in multiple species. Proc. Natl. Acad. Sci. USA. **102 (6)**, 1974–1979 (2005)
56. Jeong, H., Mason, S.P., Barabási A.L., et al.: Lethality and centrality in protein networks. Nature **411 (6833)**: 41–42 (2001)
57. Barabási, A.L., Gulbahce, N., Loscalzo, J.: Network medicine: a network-based approach to human disease. Nat. Rev. **12**, 56–68 (2011)
58. Venkatesan, K., Rual, J., Vázquez, A., Stelzl, U., et al.: An empirical framework for binary interactome mapping. Nat. Meth. **6**, 83–90 (2009)
59. Rual, J.F., Venkatesan, K., Hao, T., et al.: Towards a proteome-scale map of the human protein-protein interaction network. Nature **437**, 1173–1178 (2005)
60. Stelzl, U., Worm, U., Lalowski, M., et al.: A human protein-protein interaction network: a resource for annotating the proteome. Cell **122**, 957–968 (2005)

61. Dreze, M.: High-quality binary interactome mapping. Meth. Enzymol. **470**, 281–315 (2010)
62. Ewing, R.M., Chu, P., Elisma, F., et al.: Large-scale mapping of human protein-protein interactions by mass spectrometry. Mol. Syst. Biol. **3**, 89 (2007)
63. Cusick, M.E., Yu, H., Smolyar, A., et al.: Literature-curated protein interaction datasets. Nat. Meth. **6**, 39–46 (2009)
64. Uetz, P., Giot, L., Cagney, G., et al.: A comprehensive analysis of protein-protein interactions in *Saccharomyces cerevisiae*. Nature **403**, 623–627 (2000)
65. Xu, J., Li, Y.: Discovering disease-genes by topological features in human protein-protein interaction network. Bioinformat. **22**, 2800–2805 (2006)
66. Kuo, P.D., BanZhaf, W., Leier, A.: Network topology and the evolution of dynamics in an artificial genetic regulatory network model created by whole genome duplication and divergence. Biosyst. **85**(3), 177–200 (2006).
67. Adler, M., Anjum, M., Berg, O.G., Andersson, D.I., & Sandegren, L.: High fitness costs and instability of gene duplications reduce rates of evolution of new genes by duplication-divergence mechanisms. Mol. Biol. Evol. **31**(6), 1526–1535 (2014)
68. Patthy, L.: Protein evolution. Blackwell, Oxford (1999)
69. Wagner, A.: The *yeast* protein interaction network evolves rapidly and contains few redundant duplicate genes. Mol. Biol. Evol. **18**, 1283–1292 (2001)
70. Newman, M.E.J.: The structure and function of complex networks. SIAM Rev. **45**, 167–256 (2003)
71. Schwikowski, B., Uetz, P., Fields, S.: A network of protein-protein interaction in *yeast*. Nat. Biotech. **18**, 1257–1261 (2000)
72. Yu, H., Braun, P., Yildirim, M.A., et al.: High-quality binary protein interaction map of the *yeast* interactome network. Science **322**, 104–110 (2008)
73. Ravasz, E., Barabási, A.L.: Hierarchical organization in complex networks. Phys. Rev. E **67**: 026112 (2003)
74. Newman, M.E.J.: Assortative mixing in networks. Phys. Rev. Lett. **89**(20), 208701 (2002)
75. Costa, L. da F., Rodrigues, F. A., Travieso, G., Boas, P.R.V.: Characterization of complex networks: a survey of measurements. Adv. Phys. **56**, 167–242 (2007)
76. Remondini, D., Neretti, N., Franceschi, C., et al.: Networks from gene expression time series: characterization of correlation patterns. Int. J. Bifur. & Chaos **17(7)**: 2477–2483 (2007)
77. Rice, J., Tu, Y., Stolovitzky, G.: Reconstructing biological networks using conditional correlation analysis. Bioinformat. **21(6)**, 765–773 (2005)
78. Wang, Y., Joshi, T., Zhang, X., Xu, D., Chen, L.: Inferring gene regulatory networks from multiple microarray datasets. Bioinformat. **22(19)**, 2413–2420 (2006)
79. Friedman,N., Linial, M., Nachman, I., Peér, D.: Using Bayesian networks to analyze expression data. J. Comput. Biol. **7(3/4)**, 601–620 (2000)
80. Finkle, J.D., Wu, J.J., Bagheri, N.: Windowed Granger causal inference strategy improves discovery of gene regulatory networks. Proc. Natl. Acad. Sci. USA. **115**(9), 201710936 (2018)
81. Granger, C.W.J.: Investigating causal relations by econometric models and cross-spectral methods. Econometrica **37**, 424–438 (1969)
82. Sun, J., Taylor, D., Bollt, E.M.: Causal network inference by optimal causation entropy. SIAM J. Appl. Dyn. Syst. **14**(1), 73–106 (2015)
83. Runge,J.: Causal network reconstruction from time series: from theoretical assumptions to practical estimation, Chaos **28**, 075310 (2018)
84. Hu, S., Wang, H., Zhang, J., Kong, W., Cao, Y., Kozma, R.: Comparison analysis: Granger causality and new causality and their applications to motor imagery. IEEE Trans. Neur. Netw. Learn. Syst. **27**(7), 1429–1444 (2016)
85. Wu, X., Wang, W., Zheng, W.X.: Inferring topologies of complex networks with hidden variables. Phys. Rev. E **86**(4), 046106 (2012)
86. Han, X., Shen, Z., Wang,W., Di, Z.: Robust reconstruction of complex networks from sparse data. Phys. Rev. Lett. **114**, 028701 (2015)

87. Galton, F., Regression towards mediocrity in hereditary stature. J. Anthrop. instit. Great Britain and Ireland **15**, 246–263 (1886)
88. Spearman, C., The proof and measurement of association between two things. American J. Psych. **15**, 72–101 (1904)
89. Abdi, H.: The Kendall rank correlation coefficient. In Neil Salkind (Ed.): Encyclopedia of Measurement and Statistics. Thousand Oaks (CA): Sage (2007)
90. Samuels, M.L., Witmer, J.A., Schaffner, A.A.: Statistics for the life sciences. Pearson Edu. Limited (2016)
91. Kruskal, W.H.: Ordinal measures of association. J. Amer. Stat. Assoc. **53**(284), 814–861 (1958)
92. Kendall, M.G.: Rank correlation methods (4th ed.). Griffin & Company Limited (1976)
93. Kendall, M.: A new measure of rank correlation. Biometrika **30**, 81–89 (1938)
94. Agresti, A.: Analysis of ordinal categorical data (2nd ed.). New York, John Wiley & Sons (2010)
95. Wang, P., Yang, C., Chen, H., et al.: Exploring transcriptional factors reveals crucial members and regulatory networks involved in different abiotic stresses in *Brassica napus* L.. BMC Plant Biol. **18**: 202 (2018)
96. Hoeffding, W.: A non-parametric test of independence. Ann. Math. Stat. **19**, 546–557 (1948)
97. Rosenblatt, M.: A quadratic measure of deviation of two-dimensional density estimates and a test of independence. Ann. Statist. **3**, 1–14 (1975)
98. Székely, G.J., Rizzo, M.L., Bakirov, N.K.,: Measuring and testing dependence by correlation of distances. Ann. Statist. **35**, 2769–2794 (2007)
99. Székely, G.J., Rizzo, M.L.: Brownian distance covariance. Ann. Appl. Stat. **3**, 1236–1265 (2009)
100. Cui, H., Li, R., Zhong, W.: Model-free feature screening for ultrahigh dimensional discriminant analysis. J. Amer. Stat. Assoc. **110**(510): 630–641 (2015)
101. Cui, H., Zhong, W.: A distribution-free test of independence based on mean variance index. Computat. Stat. Data Anal. **139**, 117–133 (2019)
102. Emamjomeh, A., Saboori, R.E., Zahiri, J., et al.: Gene co-expression network reconstruction: a review on computational methods for inferring functional information from plant-based expression data. Plant Biotech. Rep. **11**(2), 71–86 (2017)
103. Steuer, R., Kurths, J., Daub, C.O., et al.: The mutual information: detecting and evaluating dependencies between variables. Bioinformat. **18**(suppl 2): S231-S240 (2002)
104. Bansal, M., Belcastro, V., Ambesi-Impiombato, A., Di Bernardo, D.: How to infer gene networks from expression profiles. Mol. Syst. Biol. **3**(1):1–10 (2007)
105. Reshef, D.N., Reshef, Y.A., Finucane, H.K., et al.: Detecting novel associations in large data sets. Science **334**(6062), 1518–1524 (2011)
106. Khosravi, P., Gazestani, V., et al.: Comparative analysis of coexpression networks reveals molecular changes during the cancer progression. World Congress on Medical Physics and Biomedical Engineering, Toronto, Springer 1481–1487 (2015)
107. Butte, A.J., Kohane, I.S.: Mutual information relevance networks: functional genomic clustering using pairwise entropy measurements. Pac. Symp. Biocomput. **5**, 418–429 (2000)
108. Margolin, A.A., Nemenman, I., Basso, K., et al.: ARACNE: an algorithm for the reconstruction of gene regulatory networks in a mammalian cellular context. BMC Bioinformat. **7**(Suppl 1): S7 (2006)
109. Meyer, P.E., Kontos, K., Lafitte, F., Bontempi, G.: Information-theoretic inference of large transcriptional regulatory networks. EURASIP J. Bioinformat. Syst. Biol. **2007**,1–9 (2007)
110. Faith, J.J., Hayete, B., Thaden, J.T., et al.: Large-scale mapping and validation of *Escherichia coli* transcriptional regulation from a compendium of expression profiles. PLoS Biol. **5**(1), e8 (2007)
111. Lin, S., Peter, L., Steve, H.: Comparison of co-expression measures: mutual information, correlation, and model based indices. BMC Bioinformat. **13**(1), 328–328 (2012)
112. Wang ,Y.X.R., Huang, H.: Review on statistical methods for gene network reconstruction using expression data. J. Theor. Biol. **362**, 53–61 (2014)

113. Edwards, D.I.: Introduction to graphical modelling (2nd ed.) Springer, New York, USA (2000)
114. Schäfer, J., Strimmer,K.: An empirical Bayes approach to inferring large-scale gene association networks. Bioinformat. **21**(6),754–764 (2005)
115. Penrose, R.: A generalized inverse for matrices. Math. Proc. Cambridge Philos. Soc. **51**,406–413 (1955)
116. Li, H., Gui, J.: Gradient directed regularization for sparse Gaussian concentration graphs, with applications to inference of genetic networks. Biostat. **7**, 302–317 (2006)
117. Tibshirani, R.: Regression shrinkage and selection via the LASSO. J. R. Stat. Soc. Ser. B Stat. Methodol. **58**, 267–288 (1996)
118. Meinshausen, N., Bühlmann, P.: High-dimensional graphs and variable selection with the LASSO. Ann. Stat. **34**,1049–1579 (2006)
119. Peng, J., Wang, P., Zhou, N., Zhu, J.: Partial correlation estimation by joint sparse regression models. J. Amer. Stat. Assoc. **104**,736–746 (2009)
120. Zhou, S., Rütimann,P., Xu, M., Bühlmann, P.: High-dimensional covariance estimation based on Gaussian graphical models. J. Mach. Learn. Res. **12**, 2975–3026 (2011)
121. Geweke, J.F.: Measures of conditional linear dependence and feedback between time series. J. Amer. Stat. Assoc. **79**(388), 907–915 (1984)
122. Guo, S., Seth, A.K., Kendrick, K.M., et al.: Partial granger causality-eliminating exogenous inputs and latent variables. J. Neurosci. Meth. **172**(1), 79–93 (2008)
123. Irrthum, A., Wehenkel, L., Geurts, P., et al.: Inferring regulatory networks from expression data using tree-based methods. PLoS One **5**(9), e12776 (2010)
124. Haury, A.C., Mordelet, F., Vera-Licona, P., Vert, J.P.: Tigress: trustful inference of gene regulation using stability selection. BMC Syst. Biol. **6**(1),145 (2012)
125. Ciaccio, M.F., Chen, V.C., Jones, R.B., Bagheri, N.: The DIONESUS algorithm provides scalable and accurate reconstruction of dynamic phosphoproteomic networks to reveal new drug targets. Integr. Biol. **7**(7), 776–791 (2015)
126. Wang, W., Lai, Y.C., Grebogi, C.: Data based identification and prediction of nonlinear and complex dynamical systems. Phys. Rep. **644**, 1–76 (2017)
127. Yuan, M., Lin, Y.: Model selection and estimation in regression with grouped variables. J. R. Stat. Soc. Ser. B Stat. Methodol. **68**, 49–67 (2006)
128. Li, H., Zhan, M.: Unraveling transcriptional regulatory programs by integrative analysis of microarray and transcription factor binding data. Bioinformat. **24**, 1874–1880 (2008)
129. Maetschke, S.R., Madhamshettiwar, P.B., Davis, M.J., Ragan, M.A.: Supervised, semi-supervised and unsupervised inference of gene regulatory networks. Brief. Bioinformat. **15**, 195–211 (2014)
130. Pearl, J., Verma, T.: A theory of inferred causation. in: KR1991, 441–452 (1991)
131. Cooper, G.F., Herskovits, E.: A Bayesian method for the induction of probabilistic networks from data. Mach. Learn. **9**, 309–347 (1992)
132. Yoo, C., Thorsson, V., Cooper, G.: Discovery of causal relationships in a gene regulation pathway from a mixture of experimental and observational DNA microarray data. Pacific Sym. Biocomput. 498–509 (2002)
133. Hartemink, A.J., Gifford, D.K., Jaakkola, T.S., Young, R.A.: Using graphical models and genomic expression data to statistically validate models of genetic regulatory networks. Pacific Sym. Biocomput. 422–433 (2001)
134. Yu, J., Smith, V.A., Wang, P.P., Hartemink, A.J., Jarvis, E.D.: Using Bayesian network inference algorithms to recover molecular genetic regulatory networks. Int. Conf. Syst. Biol. Stockholm, Sweden. Karolinska Institute (2002)
135. Needham, C.J., Bradford, J.R., Bulpitt, A.J., Westhead, D.R.: A primer on learning in Bayesian networks for computational biology. PLoS Comput. Biol. **3**, e129 (2007)
136. Jansen, R., Yu, H., Greenbaum, D., et al.: A Bayesian networks approach for predicting protein-protein interactions from genomic data. Science **302**(5644), 449–453 (2003)
137. Blei, D.M., Kucukelbir, A., McAuliffe, J.D.: Variational inference: a review for statisticians. J. Amer. Stat. Assoc. **112** (518), 859–877 (2017)

138. Jordan, M.I., Ghahramani, Z., Jaakkola, T.S., Saul, L.K.: An introduction to variational methods for graphical models. Mach. Learn. **37** (2), 183–233 (1999)
139. Wainwright, M.J., Jordan, M.I.: Graphical models, exponential families, and variational inference. Found. Trends Mach. Learn. **1** (1–2), 1–305 (2008)
140. Xu, S., Zhang, C., Wang, P., Zhang, J.: Variational Bayesian weighted complex network reconstruction. Inform. Sci. **521**, 291–306 (2020).
141. Haario, H., Laine, M., Mira, A., Saksman, E.: DRAM: Efficient adaptive MCMC. Stat. Comput. **16**, 339–354 (2006)
142. Haario, H., Saksman, E., Tamminen, J.: An adaptive Metropolis algorithm. Bernoulli **7**, 223–242 (2001)
143. Kirimasthong, K., Manorat, A., Chaijaruwanich, J., et. al.: Inference of gene regulatory network by Bayesian network using Metropolis-Hastings algorithm. ADMA 2007, Lect. Notes Comput. Sci. **4632**, 276–286 (2007)
144. Yuan, Y., Bar-Joseph, Z.: Deep learning for inferring gene relationships from single-cell expression data. Proc. Natl. Acad. Sci. USA. **116** (52), 27151–27158 (2019)
145. Chai, L.E., Loh, S.K., Low, S.T., et al.: A review on the computational approaches for gene regulatory network construction. Comput. Biol. Med. **48**, 55–65 (2014)
146. Björn, U., Obayashi, T., Mutwil, M., et al.: Coexpression tools for plant biology: opportunities for hypothesis generation and caveats. Plant Cell Environ. **32**(12),1633–1651 (2009)
147. López-Kleine, L., Luis, L., López, C.: Biostatistical approaches for the reconstruction of gene co-expression networks based on transcriptomic data. Brief. Funct. Genom. **12**(5), 457–467 (2013)

Chapter 3
Modeling and Analysis of Simple Genetic Circuits

Abstract Complex bio-molecular networks often consist of simple circuits, which are called as network motifs, the thorough investigations on network motifs are the first step to understand the complex biological system. The feed-forward loops, the single gene auto-activated circuit, the single gene auto-repressed circuit, the coupled positive and negative feedback genetic circuits are all typical simple circuits, which have been extensively investigated from the perspective of both mathematical modeling and experiments. In this chapter, we firstly review some mathematical models for simple biological networks. Then, based on mathematical modeling and dynamical analysis, we investigate the relations among the structures, functions, and dynamics of several simple circuits. Finally, we introduce some works on the large-scale exploration of simple bio-molecular networks with specific biological functions.

3.1 Backgrounds

Biological networks can be quantitatively investigated through mathematical models [1, 2]. Since GRNs have an important role in every process of life, including cell differentiation, metabolism, the cell cycle, and signal transduction, we mainly consider the mathematical modeling of genetic circuits. By understanding the dynamics of genetic networks, we can shed light on the mechanisms of diseases that occur when these cellular processes are dysregulated. Accurate prediction of the behavior of regulatory networks will also speed up biotechnological projects, as such predictions are quicker and cheaper than lab experiments. Computational methods, both for supporting the development of network models and for the analysis of their functionality, have already been proved to be valuable research tools [3]. Another important application area of related investigations is the synthetic biology. During the last decades, there have been three important experimental studies involving the design of synthetic genetic networks, including (a) a single autorepressive promoter utilized to demonstrate the interplay between negative feedback and internal noise; (b) two repressive promoters used to construct a

J. Lü, P. Wang, *Modeling and Analysis of Bio-molecular Networks*,
https://doi.org/10.1007/978-981-15-9144-0_3

genetic toggle switch; (c) three repressive promoters employed to exhibit sustained oscillations.

Different sizes of genetic networks can be described by different kinds of models [4]. Single genes can be modeled in molecular detail with stochastic simulations [5] based on the chemical master equation (CME); a differential equation (reaction rate equation: RRE) representation of gene dynamics is more practical when turning to circuits of genes [6]; Approximating gene dynamics by switch-like ON/OFF (Boolean dynamics) behavior allows modeling of mid-sized genetic circuits and still faithfully represents the overall dynamics of the biological systems [7, 8]. Large genetic networks are currently out of reach for predictive simulations (Fig. 3.1).

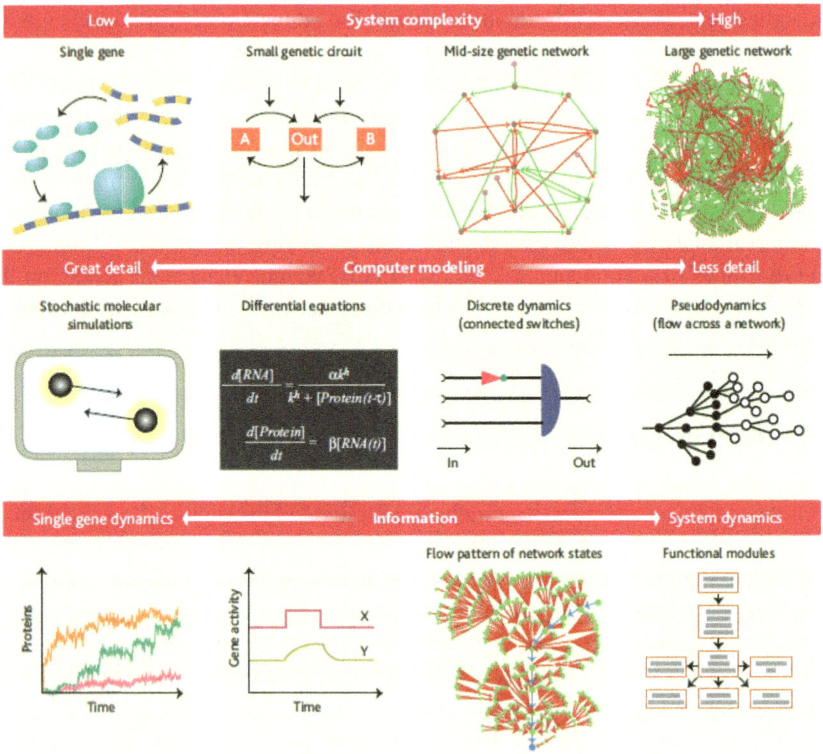

Fig. 3.1 The different levels of description in models of genetic networks. Single genes can be modeled in molecular detail with stochastic simulations (left column), a differential equation representation of gene dynamics is more practical when turning to circuits of genes (center left column). Approximating gene dynamics by switch-like ON/OFF behavior allows modeling of mid-sized genetic circuits (center right column) and still faithfully represents the overall dynamics of the biological system. Large genetic networks are currently out of reach for predictive simulations. However, more simplified dynamics, such as percolating flows across a network structure, can teach us about the functional structure of a large network (right column). Reprinted from ref. [4], with permission from AAAS

However, more simplified dynamics, such as percolating flows across a network structure, can teach us about the functional structure of a large network [4, 9].

In this chapter, we will firstly introduce the related mathematical models for simple genetic circuits, and then we introduce some works on several typical simple genetic circuits.

3.2 Mathematical Modeling Techniques of Biological Networks

In this section, we mainly review three categories of frequently used mathematical models, that is, the CME, the chemical Langevin equation (CLE), and the RRE.

3.2.1 The Chemical Master Equation

A biological system is a chemical reaction system. A well-known model that is able to exactly describe the chemical system is the CME. Suppose that there are N chemical species, $S_1, S_2, \cdots, S_N$, taking part in M different chemical reactions. In the CME formulation, we have a state vector $X(t) \in R^N$, whose $i'th$ component $X_i(t)$ denotes the number of molecules of S_i presents at time t. Hence, each $X_i(t)$ is a non-negative integer. For each $1 \le j \le M$, we have a stoichiometric vector $\nu_j \in R^N$, and a propensity function $a_j(X(t))$, such that the $j'th$ reaction takes place over the infinitesimal interval $[t, t+dt)$ with probability $a_j(X(t))dt$ and causes the change $X(t) \mapsto X(t) + \nu_j$ to the state vector.

Denote $P(x,t)$ as the probability that $X(t) = x$, the CME is an ODE system described as follows:

$$\frac{dP(x,t)}{dt} = \sum_{j=1}^{M}(a_j(x-\nu_j)P(x-\nu_j,t) - a_j(x)P(x,t)). \tag{3.1}$$

Generally speaking, the CME cannot be solved analytically, especially for systems with more than two species. However, one can compute its moments from CME by the generation function approach or through the following method: For the first-order moment, since $E(x_i) = \sum_x x_i P(x,t)$,

$$\sum_x x_i \frac{dP(x,t)}{dt} = \sum_x x_i \sum_{j=1}^{M}(a_j(x-\nu_j)P(x-\nu_j,t) - a_j(x)P(x,t))$$

$$= \sum_{x} \sum_{j=1}^{M} ((x_i + \nu_{ij}) a_j(x) P(x,t) - x_i a_j(x) P(x,t))$$

$$= \sum_{j=1}^{M} \nu_{ij} \sum_{x} a_j(x) P(x,t) = \sum_{j=1}^{M} \nu_{ij} E(a_j(x)). \tag{3.2}$$

That is

$$\frac{dE(x_i)}{dt} = \sum_{j=1}^{M} \nu_j E(a_j(x)). \tag{3.3}$$

Similarly,

$$\sum_{x} x_i x_k \frac{dP(x,t)}{dt} = \sum_{x} x_i x_k \sum_{j=1}^{M} (a_j(x-\nu_j) P(x-\nu_j, t) - a_j(x) P(x,t))$$

$$= \sum_{x} \sum_{j=1}^{M} ((x_i + \nu_{ij})(x_k + \nu_{kj}) a_j(x) P(x,t) - x_i x_k a_j(x) P(x,t))$$

$$= \sum_{j=1}^{M} \sum_{x} (x_i \nu_{kj} + x_k \nu_{ij} + \nu_{ij} \nu_{kj}) a_j(x) P(x,t)$$

$$= \sum_{j=1}^{M} (\nu_{kj} E(x_i a_j(x)) + \nu_{ij} E(x_k a_j(x)))$$

$$+ \sum_{j=1}^{M} \nu_{ij} \nu_{kj} E(a_j(x)). \tag{3.4}$$

That is

$$\frac{dE(x_i x_k)}{dt} = \sum_{j=1}^{M} (\nu_{kj} E(x_i a_j(x)) + \nu_{ij} E(x_k a_j(x))) + \sum_{j=1}^{M} \nu_{ij} \nu_{kj} E(a_j(x)). \tag{3.5}$$

Equations (3.3) and (3.5) can help us to investigate the average behaviors of the chemical systems.

3.2.2 Stochastic Simulation Algorithms

Except the investigations on the moments of the CME, stochastic simulations can be performed to simulate the chemical system. There are several stochastic simulation algorithms based on the MC algorithm, which include the direct Gillespie stochastic simulation algorithm [10], the τ-leap algorithm [11], the exact hybrid stochastic simulation algorithm [12], the rejection algorithm [15, 16], and the Gillespie algorithm with time delay [13]. For simplicity, we mainly review the direct Gillespie stochastic simulation algorithm, the rejection algorithm, and the algorithm with time delays, for details of the other algorithms, one can refer to the related references therein [14].

In the year 1977, Gillespie proposed the direct Gillespie algorithm [10] to simulate biochemical reactions. Suppose that the fixed volume Ω contains a spatially uniform mixture of N chemical species, which can interact through M specified chemical reaction channels. The N chemical species are denoted by $S_i, i = 1, 2, \cdots, N$, and X_i denotes molecular number of S_i. The $\mu'th$ reaction channel is denoted as R_μ, and $a_\mu(\mu = 1, 2, \cdots, M)$ denotes the propensity of the reaction channel R_μ. State-change vector $\nu_j = (\nu_{1j}, \nu_{2j}, \cdots, \nu_{Nj})^T$ denotes the dynamic of the reaction channel R_j, where ν_{ij} represents the changes of species S_i in the R_j reaction channel. t denotes the time, and t_{stop} denotes the maximum time that are considered in the reaction systems. There are two key points in the direct Gillespie algorithm, namely which is the next reaction and when the next reaction will happen. Based on the idea of the MC simulation, Gillespie has resolved the two key points and given the following exact simulation algorithm, which is shown in Algorithm 10.

Algorithm 10 The direct Gillespie algorithm [10]

1: Input values for initial state $X(0) = (X_1(0), \cdots, X_N(0))^T$ for the N species, set time $t = 0$ and reaction counter $i = 1$.
2: When time $t < t_{stop}$, compute the propensities for the M reaction channels a_μ, $\mu = 1, \cdots, M$ and compute the total propensity $a_0 = \Sigma_\mu a_\mu$; if $a_0 = 0$, stop; else go to steps 3–6.
3: Generate uniform random numbers $u_1, u_2 \in [0, 1]$.
4: Compute the time interval until the next reaction $\Delta t_i = -lnu_1/\Sigma_\mu a_\mu$.
5: Find the channel of the next reaction j, namely take j to be the integer for which $\Sigma_{\upsilon=1}^{j-1} a_\upsilon < u_2 a_0 \leq \Sigma_{\upsilon=1}^{j} a_\upsilon$.
6: Update X as $X + \nu_j$ according to the $j'th$ reaction channel, update time $t = t + \Delta t_i$ and increase counter $i = i + 1$, go to step 2.

The direct Gillespie algorithm is very time-consuming, and it is almost infeasible even with a few species. Therefore, based on the direct Gillespie algorithm, there have been some accelerated algorithms, such as the τ-leap method [11], the next reaction method, the accurate hybrid stochastic simulation method proposed in Ref. [12].

To cope with chemical reactions with time delays, Bratsun and coauthors [13], Barrio and coauthors [15], Cai [16], and Chen et al. [17] have proposed some stochastic simulation algorithms to deal with chemical reactions with time delays. Among these algorithms, the direct exact stochastic simulation algorithm for chemical reactions with delays [16] was proved to be equivalent to the rejection method proposed by Barrio et al. [15]. For simplicity, we mainly overview the rejection method.

Suppose some of the reaction channels or all the reaction channels incur a time delay, and we use $\mathscr{R}_D$ to denote reaction channels with time delays. A reaction $R_i \in \mathscr{R}_D$ will finish with a delay of τ_i after it is initiated. Consequently, the product of reaction R_i will be available after a delay of τ_i, and thus the population of the product will change after a time delay of τ_i. The delayed reactions can further be classified into nonconsuming reactions and consuming reactions, where for nonconsuming reactions, the reactants of an unfinished reaction can participate in a new reaction, and when the nonconsuming reaction occurs, the population of the reactants does not change. While for consuming ones, the reactants of an unfinished reaction cannot participate in a new reaction, and when a consuming reaction occurs, the population of the reactants changes immediately. Following the notations in [16], denote the set of nonconsuming reactions as R_{D1} and the set of consuming reactions as R_{D2}. The rejection algorithm is described in Algorithm 11 [15, 16]:

Algorithm 11 The rejection algorithm [15, 16]

1: Input values for the initial state $X(0) = (X_1(0), \cdots, X_N(0))^T$ for the N species, set time $t = 0$ and reaction counter $COUNT = 1$.
2: When time $t < t_{stop}$, compute the propensities for the M reaction channels $a_\mu (\mu = 1, \cdots, M)$ and compute the total propensity $a_0 = \Sigma_\mu a_\mu$; if $a_0 = 0$, stop; else go to steps 3–4.
3: Generate a uniform random number $u_1 \in [0, 1]$. Based on u_1, generate $\Delta t_i = -ln u_1 / \Sigma_\mu a_\mu$.
4: If there are delayed reaction(s) to finish in the interval $[t, t + \Delta t_i)$, discard Δt_i, update time t by t_d, where t_d is the time when the first delayed reaction finishes, update the state vector X, update $COUNT$ by $COUNT + 1$ and repeat steps 2–4. If there is no delayed reaction in the interval $[t, t + \Delta t_i)$, proceed to steps 5–7.
5: Generate a uniform random number $u_2 \in [0, 1]$. Find the channel of the next reaction j, namely take j to be the integer for which $\Sigma_{v=1}^{j-1} a_v < u_2 a_0 \leq \Sigma_{v=1}^{j} a_v$.
6: If $R_j \notin \mathscr{R}_D$, update X according to the $j'th$ reaction channel, update $COUNT$ by $COUNT + 1$. If $R_j \in \mathscr{R}_{D2}$, update X by $X + \widetilde{v}_j$, where $\widetilde{v}_j = (\widetilde{v}_{1j}, \widetilde{v}_{2j}, \cdots, \widetilde{v}_{Nj})^T$, $\widetilde{v}_{ij} = v_{ij}$ if $v_{ij} \leq 0$, and $\widetilde{v_{ij}} = 0$ if $v_{ij} > 0$. update $COUNT$ by $COUNT + 1$. If $R_j \in \mathscr{R}_{D1}$, marking the reaction as $\mathscr{R}_{D1}$, change the state vector at $t = t + \Delta t_i + \tau_j$.
7: Set $t = t + \Delta t_i$, go to step 2.

Algorithm 11 is effective for some delayed chemical reactions, but it is invalid for some other delayed chemical reactions. For example, the above algorithm is not suitable for the undeveloped delayed chemical reactions that transformed from ODEs. Hereinafter, based on the notations in this section, we propose our stochastic simulation algorithm to cope with undeveloped chemical reactions with time delays [18]. For simplicity, we call this algorithm as the undeveloped delayed stochastic simulation algorithm (UDSSA). The UDSSA is also based on the direct Gillespie algorithm. Suppose we have established the following delayed differential equation model for a biochemical system:

$$
\begin{aligned}
\dot{x_i} = {} & f_i(t; x_1(t), x_2(t), \cdots, x_N(t); \\
& x_1(t-\tau_1), x_2(t-\tau_2), \cdots, x_N(t-\tau_N); \lambda_i) \\
& -g_i(t; x_1(t), x_2(t), \cdots, x_N(t); \\
& x_1(t-\tau_1), x_2(t-\tau_2), \cdots, x_N(t-\tau_N); \theta_i). \\
& i = 1, 2, \cdots, N,
\end{aligned}
\tag{3.6}
$$

where $x_i(i = 1, 2, \cdots, N)$ denotes the concentration of the $i'th$ species. τ_i denotes the time delay for the $i'th$ species. λ_i and θ_i are two sets of parameter vectors for the $i'th$ equation of system (3.6). $f_i(t; x_1(t), x_2(t), \cdots, x_N(t); x_1(t-\tau_1), x_2(t-\tau_2), \cdots, x_N(t-\tau_N); \lambda_i)$ and $g_i(t; x_1(t), x_2(t), \cdots, x_N(t); x_1(t-\tau_1), x_2(t-\tau_2), \cdots, x_N(t-\tau_N); \theta_i)$ are non-negative functions of $t, x_1(t), \cdots, x_N(t), x_1(t-\tau_1), \cdots, x_N(t-\tau_N)$, which represent the comprehensive generation rate and consumption rate respectively.

For systems without time delays, Gonze, Goldbeter, and coauthors [19] have demonstrated that the deterministic ODE models for biochemical systems can be rewritten as birth-death biochemical reactions. And the birth-death biochemical reactions were called as undeveloped stochastic models, the stochastic models can be simulated by the Gillespie's direct method. They have shown the effectiveness of the stochastic models in investigating the dynamics of the circadian system.

Following the method of Gonze et al. [19], we rewrite system (3.6) as stochastic birth-death reactions, which are listed in Table 3.1.

Here

$$
\begin{aligned}
a_i = {} & \Omega f_i(t; X_1(t)/\Omega, \cdots, X_N(t)/\Omega; \\
& X_1(t-\tau_1)/\Omega, \cdots, X_N(t-\tau_N)/\Omega; \lambda_i); \\
b_i = {} & \Omega g_i(t; X_1(t)/\Omega, \cdots, X_N(t)/\Omega; \\
& X_1(t-\tau_1)/\Omega, \cdots, X_N(t-\tau_N)/\Omega; \theta_i); \\
& i = 1, 2, \cdots, N.
\end{aligned}
\tag{3.7}
$$

Table 3.1 Undeveloped stochastic model directly from system (3.6)

Reaction	Propensity function	Increment of the molecular numbers
$\emptyset \xrightarrow{a_i} X_i$	a_i	$(0, 0, \cdots, 0, \underbrace{1}_{i'th}, 0, \cdots, 0, 0)^T$
$X_i \xrightarrow{b_i} \emptyset$	b_i	$(0, 0, \cdots\cdots, 0, \underbrace{-1}_{i'th}, 0, \cdots, 0, 0)^T$

Ω denotes system volume. $X_i(t)$ represents the molecular number for the $i'th$ species at time t. It is noted that, if $f_i(.)$ and $g_i(.)$ were the sum of several separate terms, then the reactions in (3.6) should be divided into several separate birth and death reactions for the species X_i.

Let us give our stochastic simulation algorithm for biochemical reactions as presented in Table 3.1 and Eq. (3.7). Most parts of the stochastic simulation processes are similar to the direct Gillespie algorithm, except that when one computes the propensity functions with time delays. The main idea of treating the delayed propensity functions is that, when we compute the propensity functions with time delays at time t, we use the history values of X_i at time $t - \tau_i$ to replace the terms $X_i(t - \tau_i)(i = 1, 2, \cdots, N)$. We note that, since the reaction time steps are randomly generated, therefore, the values of $X_i(t-\tau_i)(i = 1, 2, \cdots, N)$ cannot be always exactly derived, one can only guarantee to call the closest history values at time points t_d, where $|t_d - (t - \tau_i)|$ has the minimum value. Detailed procedures of the UDSSA algorithm are described in Algorithm 12.

Algorithm 12 The UDSSA [18]

1: Input values for the initial history state $X(t_0) = (X_1(t_0), \cdots, X_N(t_0))^T$ for the N species, where $t_0 \leq 0$. Input time delays $\tau_1, \cdots, \tau_N$. Input the stop time t_{stop} and set time $t = 0$ and reaction counter $COUNT = 1$.
2: When time $t < t_{stop}$, compute the propensities $a_\mu, b_\mu, \mu = 1, \cdots, N$ for the $2N$ reaction channels, where the closest history values of $X_i(t - \tau_i)$ is used to compute a_μ, b_μ. Then compute the total propensity $a_0 = \Sigma_\mu(a_\mu + b_\mu)$; if $a_0 = 0$, stop; else go to steps 3–6.
3: Generate uniform random numbers $u_1, u_2 \in [0, 1]$.
4: Compute the time interval until the next reaction $\Delta t_i = -lnu_1/\Sigma_\mu a_\mu$.
5: Find the channel of the next reaction j, namely take j to be the integer for which $\Sigma_{\upsilon=1}^{j-1} a_\upsilon < u_2 a_0 \leq \Sigma_{\upsilon=1}^{j} a_\upsilon$.
6: Update X according to the $j'th$ reaction channel, update time $t = t + \Delta t_i$, and increase counter $COUNT = COUNT + 1$, go to step 2.

Obviously, this method is not an exact algorithm, however, we can numerically show the power of the UDSSA to simulate stochastic models directly from the deterministic delayed models.

In the following, we consider the two-component toggle switch system [20], where there are two genes X, Y in the circuit. The product of one gene represses the expression of the other gene. And the two genes consist a positive feedback loop. We use D_x, D_y to denote free promoter binding sites for gene X and Y. For simplicity, we also use X, Y to denote proteins that produced by genes X, Y. X_2, Y_2 represent dimers, which can combine with the promoter sites of target genes and regulate the expression of the other gene. We further suppose the total concentration of promoter binding sites for the two genes are constants, and denoted by $[D_xT], [D_yT]$ respectively. Basic biochemical reactions are listed in Table 3.2.

Table 3.2 Chemical reactions in the genetic toggle switch system

Fast reactions	Dissociation constant
$X + X \rightleftharpoons X_2$	K_1
$Y + Y \rightleftharpoons Y_2$	K_2
$D_y + X_2 \rightleftharpoons D_y X_2$	K_3
$D_x + Y_2 \rightleftharpoons D_x Y_2$	K_4
Slow reactions	Reaction rates
$D_x \rightarrow D_x + X$	r_1
$D_y \rightarrow D_y + Y$	r_2
$X \rightarrow \phi$	r_3
$Y \rightarrow \phi$	r_4
Conservation laws	
$[D_x] + [D_x Y_2] = [D_x T]$	$[D_y] + [D_y X_2] = [D_y T]$

Table 3.3 Undeveloped stochastic model directly from system (3.8)

Reaction	Propensity function	Increment of the molecular numbers
$\emptyset \xrightarrow{a_1} X$	$a_1 = \frac{\Omega^3 \alpha_1}{\Omega^2 + Y^2(t-\tau_2)}$	$(1, 0)^T$
$X \xrightarrow{b_1} \emptyset$	$b_1 = \beta_1 X$	$(-1, 0)^T$
$\emptyset \xrightarrow{a_1} Y$	$a_2 = \frac{\Omega^3 \alpha_2}{\Omega^2 + X^2(t-\tau_1)}$	$(0, 1)^T$
$Y \xrightarrow{b_1} \emptyset$	$b_2 = \beta_2 Y$	$(0, -1)^T$

Similar to the case of the single gene circuit, one can deduce the delayed differential equation model for the toggle switch system as follows:

$$\begin{cases} \dot{x} = \frac{\alpha_1}{1+y^2(t-\tau_2)} - \beta_1 x, \\ \dot{y} = \frac{\alpha_2}{1+x^2(t-\tau_1)} - \beta_2 y, \end{cases} \tag{3.8}$$

where τ_1, τ_2 are time delays. α_i and $\beta_i (i = 1, 2)$ are dimensionless maximal transcription rates and degradation rates respectively. The corresponding undeveloped stochastic models are shown in Table 3.3.

Hereinafter, we take $\alpha_1 = 4, \alpha_2 = 5, \beta_1 = 0.25, \beta_2 = 0.5, \tau_1 = 5, \tau_2 = 8$. In stochastic simulations, we set system volume $\Omega = 10$. From deterministic simulation result of Fig. 3.2a, we see that the system has two stable steady states, since under two sets of arbitrarily chosen initial conditions, the system converges to two sets of different steady states. Figure 3.2b shows the stochastic simulation results by using the proposed UDSSA. From the results of the UDSSA, one can also derive two sets of steady states under different sets of initial molecular numbers, and the molecular numbers are just about tenfold of the deterministic ones, which demonstrates the effectiveness of the UDSSA to reflect system dynamics.

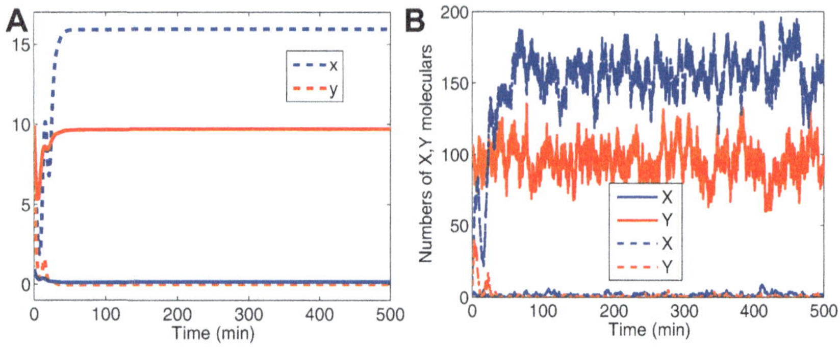

Fig. 3.2 Deterministic (**a**) and stochastic simulation (**b**) results for the undeveloped delayed toggle switch system. Where $\tau_1 = 5, \tau_2 = 8, \Omega = 10$ for panel (**b**). ©[2013] IEEE. Reprinted, with permission, from ref. [18]

Intrinsic noise in the toggle switch system can induce bistable switch behaviors [18, 21]. For the cases in Fig. 3.2, although the system is bistable, there are no bistable switch in the stochastic simulations. To verify whether the UDSSA can simulate bistable switch behaviors, we randomly choose another set of parameters, that is, $\alpha_1 = 1, \beta_1 = 0.5, \alpha_2 = 0.5, \beta_2 = 0.25, \tau_1 = 1, \tau_2 = 2, \Omega = 10$. Then, the system is also bistable, and stochastic simulation results show that intrinsic noise can induce bistable switch behavior, and the molecular number distribution for X shows bimodal distribution, which is a typical feature of bistability (see Fig. 3.3). Therefore, the example of the two genes system demonstrates the effectiveness of the UDSSA in investigating intrinsic noise-induced behaviors. The differences between Figs. 3.2 and 3.3 are mainly because that the two steady states in Fig. 3.2 have greater difference than those in Fig. 3.3, high potential is needed to overcome the potential well to realize switch under stochastic environment, and the intrinsic noise is not strong enough to overcome such potential well; thus, no bistable switch behaviors were observed in Fig. 3.2.

3.2.3 The Chemical Langevin Equation

The CLE uses a real-valued random variable $Y(t) \in R^N$ to describe the state of the system at time t. The $i'th$ component $y_i(t)$ represents the amount of species i at time t. In moving from the CME to the CLE, we typically make a dramatic reduction in the number of components, but at the cost of that each component is a real-valued random variable, rather than a non-negative integer. The CLE takes the form of a stochastic differential equation (SDE):

$$dY(t) = \sum_{j=1}^{M} \nu_j a_j(Y(t))dt + \sum_{j=1}^{M} \nu_j \sqrt{a_j(Y(t))} dW_j(t). \tag{3.9}$$

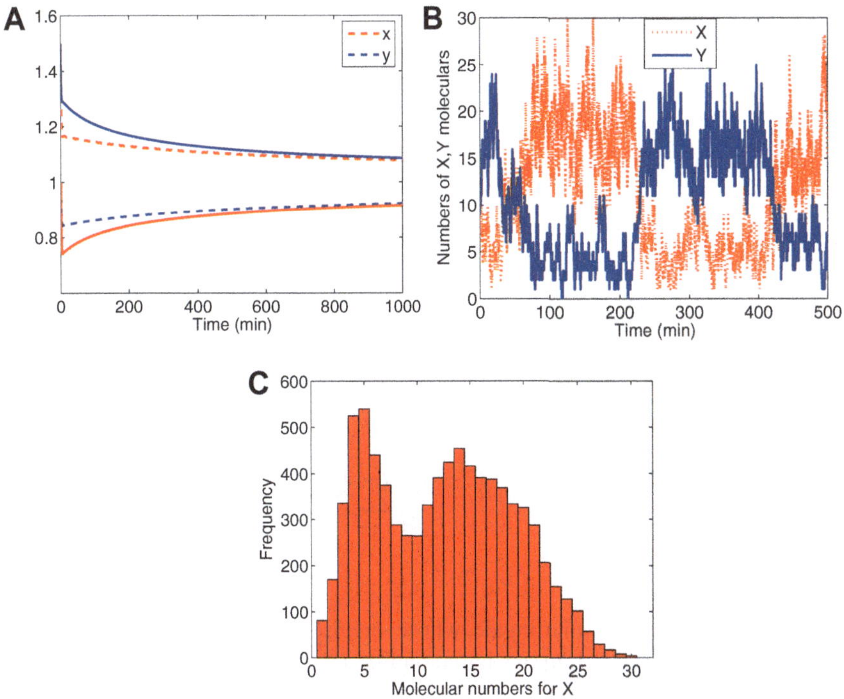

Fig. 3.3 Deterministic (**a**) and stochastic simulation (**b**, **c**) results for the toggle switch system. Panel (**c**) shows the distribution of molecular numbers for X. Here, we have set $\alpha_1 = 1, \beta_1 = 0.5, \alpha_2 = 0.5, \beta_2 = 0.25, \tau_1 = 1, \tau_2 = 2, \Omega = 10$. ©[2013] IEEE. Reprinted, with permission, from ref. [18]

Here, $W_j(t)$ is the standard Brownian motion or Wiener process defined in the interval $[0, T]$, which satisfy the following conditions: (1) $W_j(0) = 0$; (2) For $0 \leq s < t \leq T$, $W_j(t) - W_j(s) \sim \sqrt{t-s}N(0,1)$; (3) For $0 \leq s < t < u < v \leq T$, $W_j(t) - W_j(s)$ and $W_j(v) - W_j(u)$ are independent with each other. $dW_j = W_j(t + dt) - W_j(t)$ follows the Gaussian distribution with zero mean and $< (dW_j)^2 >= dt$. Moreover, dW_j is independent of t. $\nu_j, a_j(.)$ is the same as those defined in the CME as discussed in the above section.

Then, component wise equation of Eq. (3.9) can be described as

$$dy_i(t) = \sum_{j=1}^{M} \nu_{ij} a_j(Y(t))dt + \sum_{j=1}^{M} \nu_{ij} \sqrt{a_j(Y(t))} dW_j(t), 1 \leq i \leq N. \quad (3.10)$$

Before we compute the moments of the CLE, let us introduce the Itô's Lemma. For the general SDE system with N components and M independent Brownian motions:

$$dy_i(t) = b_i(Y(t))dt + \sum_{j=1}^{M} \sigma_{ij}(Y(t)dW_j(t), 1 \leq i \leq N. \tag{3.11}$$

We let

$$a(Y(t)) = \sigma(Y(t))\sigma(Y(t))^T \in R^{N\times N}.$$

Then for any function $f : R^N \to R$ that is twice continuously differentiable, the Itô's Lemma tells us:

$$df(Y(t)) = \sum_{i=1}^{N} \frac{\partial f(Y(t))}{\partial y_i} b_i(Y(t))dt + \frac{1}{2}\sum_{i=1}^{N}\sum_{j=1}^{N} \frac{\partial^2 f(Y(t))}{\partial y_i \partial y_j} a_{ij}(Y(t))dt + mart., \tag{3.12}$$

where $mart.$ denotes a martingale whose precise form is not relevant to our considerations. From this Lemma, based on Eqs. (3.9) and (3.10), we have

$$\frac{dE(y_i)}{dt} = \sum_{j=1}^{M} \nu_{ij} E(a_j(Y(t))), 1 \leq i \leq N. \tag{3.13}$$

When $f(Y(t)) = y_i y_k$, from the Lemma, also based on Eqs. (3.9) and (3.10), we have

$$\begin{aligned}\frac{dE(y_i y_k)}{dt} &= \sum_{j=1}^{M} \nu_{kj} E(y_i a_j(Y(t))) + \sum_{j=1}^{M} \nu_{ij} E(y_k a_j(Y(t))) \\ &+ \sum_{j=1}^{M} \nu_{kj}\nu_{ij} E(a_j(Y(t))), 1 \leq i, k \leq N.\end{aligned} \tag{3.14}$$

Except the analysis on the moments of the CLE, one can numerically solve the SDE. Some frequently used numerical regimes for the SDEs include the Euler–Maruyama method, the Milstein method [22, 23], and so on.

3.2.4 Numerical Regimes for Stochastic Differential Equations

Firstly, we briefly overview the Euler–Maruyama method, which is a strong approximation regime with order of strong convergence 0.5. Suppose the general form of the SDE is as follows:

$$dX_t = a(t, X_t)dt + b(t, X_t)dW_t, \tag{3.15}$$

with initial condition $X_0 = x_0$, where W_t denotes the Wiener process, and suppose that we wish to solve this SDE on some time interval $[0, T]$. Then the Euler–Maruyama approximation to the true solution X is the Markov chain Y defined as follows:

1. Partition the interval $[0, T]$ into N equal subintervals of width $\Delta t > 0$:
$$0 = \tau_0 < \tau_1 < \cdots < \tau_N = T$$
and $\Delta t = T/N$.
2. Set $Y_0 = x_0$.
3. Recursively define Y_n for $1 \leq n \leq N$ by
$$Y_{n+1} = Y_n + a(t_n, Y_n)\Delta t + b(t_n, Y_n)\Delta W_n. \tag{3.16}$$
Here, $\Delta W_n = W_{\tau_{n+1}} - W_{\tau_n}$. The random variables ΔW_n are i.i.d. normal random variables with expected value zero and variance Δt.

Secondly, we overview the Milstein method, which is firstly proposed by Grigori N. Milstein in the year 1974 [24]. For the SDE as shown in Eq. (3.15), the Milstein approximation to the true solution X is also the Markov chain Y, but defined as follows:

1. Partition the interval $[0, T]$ into N equal subintervals of width $\Delta t > 0$:
$$0 = \tau_0 < \tau_1 < \cdots < \tau_N = T$$
and $\Delta t = T/N, \tau_n = n\Delta t$.
2. Set $Y_0 = x_0$.
3. Recursively define Y_n for $1 \leq n \leq N$ by
$$\begin{aligned} Y_{n+1} &= Y_n + a(t_n, Y_n)\Delta t + b(t_n, Y_n)\Delta W_n \\ &\quad + \frac{1}{2}b(t_n, Y_n)b'(t_n, Y_n)((\Delta W_n)^2 - \Delta t), \end{aligned} \tag{3.17}$$
where b' denotes the derivative of $b(t, x)$ with respect to x, and $\Delta W_n = W_{\tau_{n+1}} - W_{\tau_n}$. The random variables $\Delta W_n (n = 1, 2, \cdots, N)$ are also independent and

identically distributed normal random variables with expected value zero and variance Δt. Note that when $b' = 0$, i.e., the diffusion term does not depend on X_t, then the Milstein Method is equivalent to the Euler–Maruyama method.

The Milstein scheme has both weak and strong order of convergence Δt, which is superior to the Euler–Maruyama method that has the same weak order of convergence Δt, but inferior strong order of convergence $\sqrt{\Delta t}$ [25]. Except the Euler–Maruyama method and the Milstein scheme, there are also some other methods, such as the Runge–Kutta method, for details, one can refer to reference [22, 23].

3.2.5 The Reaction Rate Equation

The RRE describes the concentrations of the reaction species. Let $Z(t) \in R^N$ denotes the species concentrations at time t. $a_j(Z(t))$ still denotes the propensity of the jth reaction. The RRE has the form:

$$\frac{dZ(t)}{dt} = \sum_{j=1}^{M} \nu_j a_j(Z(t)). \tag{3.18}$$

What we should pay attention to is that, here $a_j(Z(t))$ is different from the propensity function in Eqs. (3.1) and (3.9). Denote k_i as the stochastic reaction rate for the $i'th$ reaction, c_i the deterministic reaction rate for the $i'th$ reaction; Ω is the volume of the reactants, we suppose it is constant during the time. The relationships between c_i and k_i [26] are summarized in Table 3.4.

The RREs are ODEs. For biological systems, frequently used equations to describe biological systems are the Michaelis–Menten (MM) equations and the Hill equations.

The MM kinetics is one of the best-known models of enzyme kinetics [27]. It is named after German biochemist Leonor Michaelis and Canadian physician Maud

Table 3.4 Relationships between the stochastic reaction rate c_i and the deterministic reaction rate k_i [26]

Reactions	Rate relations	Propensity functions
$\phi \xrightarrow{k_i} \cdots$	$c_i = k_i$	$a_i(x) = k_i$
$X_m \xrightarrow{k_i} \cdots$	$c_i = k_i$	$a_i(x) = k_i x_m$
$X_m + X_n \xrightarrow{k_i} \cdots, (m \neq n)$	$c_i = k_i/\Omega$	$a_i(x) = k_i x_m x_n$
$X_m + X_m \xrightarrow{k_i} \cdots$	$c_i = 2k_i/\Omega$	$a_i(x) = k_i x_m^2$
$X_m + X_n + X_p \xrightarrow{k_i} \cdots, (m \neq n \neq p)$	$c_i = k_i/\Omega^2$	$a_i(x) = k_i x_m x_n x_p$
$X_m + X_n + X_n \xrightarrow{k_i} \cdots, (m \neq n)$	$c_i = 2k_i/\Omega^2$	$a_i(x) = k_i x_m x_n^2$

Menten. The model takes the form of an equation describing the rate of enzymatic reactions, by relating reaction rate v to $[S]$, the concentration of a substrate S. Its formula is given by

$$v = \frac{d[P]}{dt} = \frac{V_{max}[S]}{K_M + [S]}. \tag{3.19}$$

Here, V_{max} represents the maximum rate achieved by the system, at maximum (saturating) substrate concentrations. The Michaelis constant K_M is the substrate concentration at which the reaction rate is half of V_{max}. Biochemical reactions involving a single substrate are often assumed to follow the MM kinetics, without regard to the model's underlying assumptions.

The Hill equation was originally formulated by Archibald Hill in 1910 to describe the sigmoidal O_2 binding curve of hemoglobin [28]. In biochemistry and pharmacology, the binding of a ligand to a macromolecule is often enhanced if there are already other ligands present on the same macromolecule (this is known as cooperative binding). The Hill coefficient provides a way to quantify this effect. It describes the fraction of the macromolecule saturated by ligand as a function of the ligand concentration. It is used in determining the degree of cooperativeness of the ligand binding to the enzyme or receptor. A coefficient of 1 indicates completely independent binding, regardless of how many additional ligands are already bound. Numbers greater than one indicate positive cooperativity, while numbers less than one indicate negative cooperativity. The Hill coefficient of oxygen binding to hemoglobin is 2.3–3.0.

The Hill equation has the following form:

$$\theta = \frac{[L]^n}{K_d + [L]^n} = \frac{[L]^n}{(K_A)^n + [L]^n} = \frac{1}{\left(\frac{K_A}{[L]}\right)^n + 1}. \tag{3.20}$$

Here, θ represents the fraction of the ligand-binding sites on the receptor protein which are occupied by the ligand. $[L]$ denotes the free (unbound) ligand concentration; K_d denotes apparent dissociation constant derived from the law of mass action (equilibrium constant for dissociation); K_A denotes the ligand concentration producing half occupation (ligand concentration occupying half of the binding sites). This is also the microscopic dissociation constant. n is the Hill coefficient, describing the cooperativity (or possibly other biochemical properties, depending on the context in which the Hill equation is being used).

3.2.6 Numerical Regimes for Ordinary Differential Equations

It is generally difficult to derive the analytical solution of many ODEs, such as many of the ODEs with the MM forms or the Hill equation forms. However, numerical solutions based on computer software are frequently used to investigate the dynamics of complex ODEs. The Euler method and the Runge–Kutta method are two basic numerical regimes for the ODEs [29]. Hereinafter, we overview the two numerical regimes.

Suppose we have the following ODE:

$$\frac{dx}{dt} = f(t, x), \tag{3.21}$$

with initial condition $x(0) = x_0$, $t \in [0, T]$. The Euler method is as follows:

1. Partition the interval $[0, T]$ into N equal subintervals of width $\Delta t > 0$:

$$0 = t_0 < t_1 < \cdots < t_N = T$$

   and $\Delta t = T/N$, $t_n = n\Delta t$.
2. Set $x(0) = x_0$.
3. Recursively define

$$x_{n+1} = x_n + hf(t_n, x_n).$$

The x_n represents the numerical approximation of $x(t_n)$, $h = \Delta t$.

The Euler method is a first-order method, which means that the local error (error per step) is proportional to the square of the step size, and the global error (error at a given time) is proportional to the step size. The Euler method often serves as the basis to construct more complex methods. The Runge–Kutta method is more complex and accurate, which was developed around 1900 by the German mathematicians C. Runge and M.W. Kutta. The numerical regime of the classical Runge–Kutta method is as follows:

$$x_{n+1} = x_n + \frac{h}{6}(k_1 + 2k_2 + 2k_3 + k_4), \tag{3.22}$$

where

$$k_1 = f(t_n, x_n),$$

$$k_2 = f(t_n + \frac{h}{2}, x_n + \frac{h}{2}k_1),$$

$$k_3 = f(t_n + \frac{h}{2}, x_n + \frac{h}{2}k_2),$$

$$k_4 = f(t_n + h, x_n + hk_3),$$

$$t_{n+1} = t_n + h.$$

Here, $h = \Delta t$. The Runge–Kutta method is a fourth-order method, meaning that the local truncation error is with the order of $O(h^5)$, while the total accumulated error has order $O(h^4)$.

3.3 Network Motifs and Motif Detection

In the year 2002, Uri Alon et al. [1, 2, 30–33] proposed the concept of network motif, which is defined as subgraph that appears in a network significantly more frequent than those in its randomized counterparts. One can easily understand this concept from Fig. 3.4. By testing whether the appearance of a simple circuit is significantly more frequent in a real-world network than those in its randomized counterparts, one can determine whether the simple circuit could be a network motif.

It was reported that network motifs are building blocks of complex biological networks [30]. Two-node motifs include the double negative feedback loop, double positive feedback loop, and that with auto-activation or repression loops [20, 34, 35]. Three-node motifs include the FFLs, the repressilator, and so on [34, 36], with some of them are shown in Fig. 3.5. Figure 3.5 shows some representative 3 and 4-node motifs [30]. Functions of some motifs have been extensively investigated. For example, for the FFLs, researchers have theoretically and experimentally found its functional and structural advantages [30–49], which will be discussed in the following sections.

To detect network motifs, Milo et al. [30] scanned all possible i-node subgraphs in a network and its randomized counterparts, and defined network motifs as subgraphs for which the probability of occurrences in the real network are greater than those in its randomized counterparts. Since 2002, many motif detection algorithms and software [50–55] have been developed. For example, gSpan [52], Mfinder [50], FANMOD [54], Mavisto [55], and mDraw (http://www.weizmann.ac.il/mcb/UriAlon). In the following sections, we use mDraw to detect network motifs. For each network, we generate 100 randomized networks. The number of a subgraph in the real-world network is denoted as N_{real}. The average number in random networks is denoted as N_{rand}, with standard deviation denoted by SD. The Z_{score} measures the significance of the subgraph [30], which is defined as

$$Z_{\text{score}} = \frac{N_{\text{real}} - N_{\text{rand}}}{SD}.$$

Another index U is defined as the number of times a subgraph appears in the investigated network with distinct sets of nodes. In the subsequent sections, subgraphs with $Z_{\text{score}} \geq 2$, $U \geq 4$, and $N_{\text{real}} - N_{\text{rand}} \geq 0.1 N_{\text{rand}}$ are identified as motifs.

Among network motifs, it is founded that the FFLs are typical ones in various real-world systems [30] and received extensive investigations [1, 2, 30, 32]. In the following sections, we will overview some works on such network motifs.

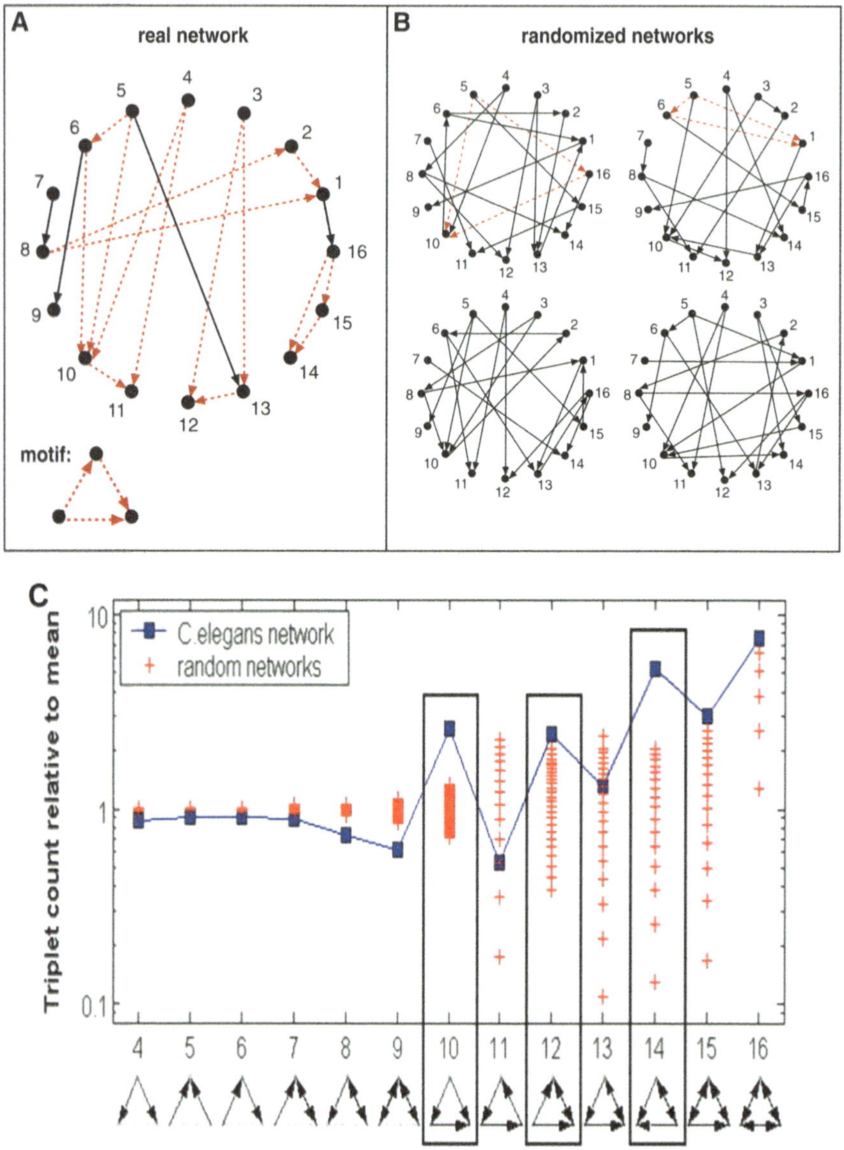

Fig. 3.4 Network motifs. (**a**, **b**) Schematic view of network motif detection. Network motifs are patterns that recur much more frequently (**a**) in the real network than (**b**) in an ensemble of randomized networks. Each node in the randomized networks has the same number of incoming and outgoing edges as does the corresponding node in the real network. Red dashed lines indicate edges that participate in the feed-forward loops (FFLs), which occurs five times in the real network. Reprinted from ref. [30], with permission from AAAS. (**c**) Network motifs in the *C. elegans* nervous system. Reprinted by permission from Springer, ref. [31]

Network	Nodes	Edges	N_{real}	$N_{rand} \pm$ SD	Z score	N_{real}	$N_{rand} \pm$ SD	Z score	N_{real}	$N_{rand} \pm$ SD	Z score
Gene regulation (transcription)					**Feed-forward loop**			**Bi-fan**			
E. coli	424	519	40	7 ± 3	10	203	47 ± 12	13			
S. cerevisiae	685	1,052	70	11 ± 4	14	1812	300 ± 40	41			
Neurons					**Feed-forward loop**			**Bi-fan**			**Bi-parallel**
C. elegans	252	509	125	90 ± 10	3.7	127	55 ± 13	5.3	227	35 ± 10	20
Food webs					**Three chain**			**Bi-parallel**			
Little Rock	92	984	3219	3120 ± 50	2.1	7295	2220 ± 210	25			
Ythan	83	391	1182	1020 ± 20	7.2	1357	230 ± 50	23			
St. Martin	42	205	469	450 ± 10	NS	382	130 ± 20	12			
Chesapeake	31	67	80	82 ± 4	NS	26	5 ± 2	8			
Coachella	29	243	279	235 ± 12	3.6	181	80 ± 20	5			
Skipwith	25	189	184	150 ± 7	5.5	397	80 ± 25	13			
B. Brook	25	104	181	130 ± 7	7.4	267	30 ± 7	32			
Electronic circuits (forward logic chips)					**Feed-forward loop**			**Bi-fan**			**Bi-parallel**
s15850	10,383	14,240	424	2 ± 2	285	1040	1 ± 1	1200	480	2 ± 1	335
s38584	20,717	34,204	413	10 ± 3	120	1739	6 ± 2	800	711	9 ± 2	320
s38417	23,843	33,661	612	3 ± 2	400	2404	1 ± 1	2550	531	2 ± 2	340
s9234	5,844	8,197	211	2 ± 1	140	754	1 ± 1	1050	209	1 ± 1	200
s13207	8,651	11,831	403	2 ± 1	225	4445	1 ± 1	4950	264	2 ± 1	200
Electronic circuits (digital fractional multipliers)					**Three-node feedback loop**			**Bi-fan**			**Four-node feedback loop**
s208	122	189	10	1 ± 1	9	4	1 ± 1	3.8	5	1 ± 1	5
s420	252	399	20	1 ± 1	18	10	1 ± 1	10	11	1 ± 1	11
s838	512	819	40	1 ± 1	38	22	1 ± 1	20	23	1 ± 1	25
World Wide Web					**Feedback with two mutual dyads**			**Fully connected triad**			**Uplinked mutual dyad**
nd.edu§	325,729	1.46e6	1.1e5	2e3 ± 1e2	800	6.8e6	5e4±4e2	15,000	1.2e6	1e4 ± 2e2	5000

Fig. 3.5 Network motifs in biological and technological networks. Reprinted from ref. [30], with permission from AAAS.

3.4 The Feed-Forward Genetic Circuits

The FFLs are typical network motifs in many real-world biological networks. The structures, functions, as well as noise characteristics of the FFLs have received increasing attention. This section aims to further investigate the global relative parameter sensitivities (GRPS) [47] and global relative input–output sensitivities (GRIOS) [48] of the FFLs in genetic networks modeled by the Hill kinetics, some simple novel approaches will be introduced. Our results indicate that: (1) For the coherent FFLs (CFFLs), the most abundant type 1 configuration (C1) is the most globally sensitive to system parameters, while for the incoherent FFLs (IFFLs), the most abundant type 1 configuration (I1) is the least globally sensitive to system parameters; (2) The less noisy of a FFL configuration, the more globally sensitive of this circuit to its parameters; (3) The most abundant FFL configurations are often either the least sensitive (robust) to system parameters variation (such as IFFLs) or the least noisy (such as CFFLs), (4) The results from the GRIOS analysis indicate that the most frequently appeared C1 and I1 configurations are quite insensitive under various levels of inputs and also rather robust to system parameters. It follows that the functions of the above circuits are robust against different inputs. It is also the reason why C1 and I1 are ubiquitous in various real-world genetic networks, especially in sensory related TRNs. Therefore, the above results can well explain the reason why the FFLs are network motifs and could be selected by nature from long time evolution. Furthermore, the proposed GRPS and GRIOS approaches shed some light on the potential real-world applications, such as the synthetic genetic circuits, predicting the effect of interventions in medicine and biotechnology, and so on.

3.4.1 Related Works and Motivations

Nowadays, complex networks are everywhere [56, 57]. As we know now, real-world complex networks consist of some simple building blocks, called network motifs. Network motifs were uncovered to be one of the shared global statistical features from biochemistry, neurobiology, ecology to engineering [30]. Moreover, the FFLs are typical 3-node network motifs from GRNs to neurons, from prokaryote *E. coli* [32] to eukaryote *S. cerevisiae* [33], and so on.

Over the last several decades, FFLs have been intensively investigated in various fields [37–62]. Figure 3.6 shows the eight basic types of FFLs. Over the last few years, the structures and functions of the above eight FFLs have received increasing attention from various disciplines [37–62]. It is well known that the FFLs can be classified into two fundamental categories: CFFLs and ICFFLs, as shown in Fig. 3.6. From [37], the ICFFLs can act as the sign-sensitive accelerators, however, CFFLs can act as the sign-sensitive delays. Very recently, these functions have been experimentally verified [38–40]. Following this line, Ghosh et al. [41] further investigated the noise characteristics of these FFLs. They found that the C1 FFL is the least noisy among the four CFFLs, while the I4 FFL is the least noisy among the four ICFFLs (see Figs. 2–4 in [41]). These results can well explain the real reason why some types of FFLs (e.g., the C1) are selected in the real-world organisms to some extent. Also, Kittisopikul et al. [43] recently investigated the noise effects in the FFLs. Different from the results in [41], they found that the FFL architectures can be classified into two different categories from the status whether their ON (stimulated) or OFF (unstimulated) steady states exhibit noise. Furthermore, Goentoro et al. [44, 63, 64] discovered a novel phenomenon in the I1 FFL, called fold-change detection. It is very similar to the famous Weber's law, a fundamental feature of many sensory systems (e. g., vision, smell, hearing, taste,

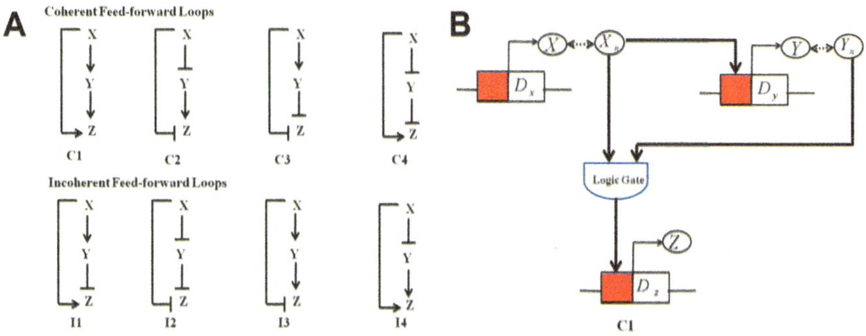

Fig. 3.6 The FFLs in the GRNs. (**a**) Schematic diagrams of eight configurations of the FFLs. The CFFLs are denoted by $C1$, $C2$, $C3$, $C4$ from left to right, respectively. Similarly, the IFFLs are denoted by $I1$, $I2$, $I3$, $I4$ from left to right, respectively. Here, X, Y, Z are TFs, proteins or genes, the edges are the regulatory interactions between nodes, $\rightarrow$ represents activation, $\dashv$ means repression, X is input, Z is output, X regulates the production of Y, Z and X, Y co-regulate the production of Z through AND, OR or other logic gates. (**b**) Detailed regulation processes in the C1 circuit, where D_x, D_y, D_z denote genes, X, Y, Z are proteins, X_n, Y_n are polymers. Similarly, in the other circuits, one only needs to change the corresponding regulatory interactions. Reprinted from ref. [47], with permission from Elsevier

and touch), where the signal discrimination is closely relative to the background signal. All these results certainly enhance our understanding for the fundamental functions of the FFL motifs in biological networks. It sheds some light on the inner relation between the network topological structures and their corresponding dynamical functions. It also indicates us that the dynamical properties of network motifs may significantly contribute to biological network organization [58].

However, some researchers [65] insist that it is often difficult to gain significant insights into biological function by simply considering the connection architecture of single gene network, or its decomposition into simple structural motifs. That is, network motif structures cannot completely determine their biological functions because these motifs occur less frequently in the very complex biochemical networks. Moreover, most network motifs are often embedded in large biological systems, where there exist various inputs in motifs. For example, under various input conditions [65], the bi-fan motif can exhibit a wide range of dynamical responses to different inputs [1, 2, 30, 32].

Different from the conclusion in Reference [65], most researchers think that there exist some inherent relations between network structures and its dynamical functions. The investigation of motif structures is the first step to reveal the inherent law how small network motifs can form the overall complex networks. For example, Yang and Kuznetsov explored the oscillatory mechanisms in a merged artificial GRN [66]. Also, they discussed the emerging effects of the original relaxation oscillator network [67] and the original repressilator network [36].

To resolve the problem that whether structures can determine its biological functions, robustness [68, 69] or sensitivity analysis [70–72] is an effective tool.

It is well known that the sensitivity analysis approach has been widely applied into various disciplines as a good measure of robustness. It includes input / output sensitivity, parameter sensitivity [71], species sensitivities [72], sensitivity of sensitivities [73], and so on. However, most of the above sensitivity analysis methods are local concepts. That is, these sensitivities depend on the given initial values and nominal parameter values of a specific dynamical system. It means that it is inappropriate to determine whether a certain function of a biochemical network is robust or not as a whole. Therefore, it is very necessary yet important to develop a global parameter sensitivities (GPS) analysis approach.

The following sections aim to develop a novel GRPS and a GRIOS analysis approaches, and then apply it to analyze the FFLs in GRNs. To begin with, some known GPS methods will be briefly reviewed. Saltelli et al. introduced a global sensitivity analysis approach in [74, 75]. A random sampling high dimensional model representation (RS-HDMR) algorithm was further investigated in [76–78]. Feng et al. [79] studied the optimal design of artificial genetic circuits by using RS-HDMR. Also, they applied the proposed approach to estimate the optimal molecular species for perturbing and monitoring a simulated biochemical reaction network [80]. Moreover, the parameter sensitivity analysis can be used in various biochemical networks, such as elucidating the behavior of models, predicting the effect of interventions in medicine, biotechnology and metabolic control analysis, model calibration [70], model reduction in signal transduction pathways, performing reliable system identification [73], and so on. Motivated by the abovementioned various problems, the subsequent sections focus on the following key issues: (1) Develop a simple and intuitive approach to perform the GRPS analysis for some biological systems; (2) Explore the GRPS properties of different FFL configurations in GRNs with different logic gates; (3) Investigate the relationship between GRPS and the other properties, such as biological abundance, noise characteristics, fold change detection, and so on. (4) Propose the new index called GRIOS, and investigate the GRIOS properties of the FFLs in GRNs.

3.4.2 Methods for Parameter Sensitivities Analysis

3.4.2.1 Local Relative Parameter Sensitivities

To begin with, one briefly reviews a local parameter sensitivities method [70], which is different from the so-called direct differential method in [71, 81].

Consider the following dynamical system:

$$\frac{dx}{dt} = f(x, \lambda), \tag{3.23}$$

where $x = (x_1, x_2, \cdots, x_n)^T \in R^n$ is the vector of species concentrations, $\lambda = (\lambda_1, \cdots, \lambda_m)^T \in R^m$ is the parameters of interest. Let the steady state of x be

$x^* = (x_1^*, x_2^*, \cdots, x_n^*)^T$, where x^* is a function of λ. Then, one has

$$f(x^*(\lambda), \lambda) = 0.$$

Differentiating the above equation with respect to λ yields

$$\frac{\partial f}{\partial x^*}\frac{\partial x^*}{\partial \lambda} + \frac{\partial f}{\partial \lambda} = 0.$$

Suppose that $\frac{\partial f}{\partial x^*}$ is invertible, solving the above equation and one gets

$$\frac{\partial x^*}{\partial \lambda} = -\left[\frac{\partial f}{\partial x^*}\right]^{-1}\frac{\partial f}{\partial \lambda}. \tag{3.24}$$

Since the unit of the above absolute sensitivities is related to the unit of λ_i, it is not suitable to compare the sensitivities for different parameters. For example, let λ_i be a degradation rate in Eq. (3.23), λ_i has unit min^{-1}, then the absolute sensitivities in Eq. (3.24) has unit $mol.L^{-1}.min^{-1}$. The relative sensitivities are defined as the ratios of relative changes of output with respect to each parameter. They are described by

$$s_{ij} = \frac{\partial x_i^*/x_i^*}{\partial \lambda_j/\lambda_j} = \frac{\lambda_j}{x_i^*}\frac{\partial x_i^*}{\partial \lambda_j}, \tag{3.25}$$

where $i = 1, 2, \cdots, n$ and $j = 1, 2, \cdots, m$. If the above formula equals to C for a specific nominal value of λ_j, then the 1% increment of λ_j will lead to a $C\%$ increment in x_i^* near the nominal value of λ_j. The bigger absolute value of C, the more sensitive of x_i^* with respect to λ_j.

However, the relative sensitivity introduced in (3.25) is also a local concept. This is because the specific nominal value of λ_j is still needed for the calculation of the relative sensitivity. It should be pointed out that the system's dynamical behavior over a wide range of operating conditions has to be considered under some circumstances, such as the model construction and interpretation within a wide range of parameter values, predicting the effect of interventions in medicine and biotechnology, and so on. Therefore, it is very necessary to develop global sensitivities methods to resolve the abovementioned problems.

3.4.2.2 A Traditional GPS Method: RS-HDMR

Note that GPSs measures are generally based on various statistical tools. It is well known that the GPSs are widely used in the metabolic control analysis and signal transduction pathways for reducing the model complexity and performing the reliable system identification [73]. In the following, we consider the sensitivities of steady state of Eq. (3.23) with respect to each parameter. Let $x^* = y(\lambda)$, where $y(\lambda) = (y_1(\lambda), y_2(\lambda), \cdots, y_n(\lambda))^T$. According to the RS-HDMR method [76–78], to derive the sensitivities of $y_k(\lambda)$ with respect to each parameter, $y_k(\lambda)$ is decomposed as follows:

$$y_k(\lambda) = y_{k(0)} + \sum_{i=1}^{m} y_{k(i)}(\lambda_i) + \sum_{1 \le i < j \le m}^{m} y_{k(i,j)}(\lambda_i, \lambda_j) + \cdots + y_{k(1,2,\cdots,m)}(\lambda_1, \lambda_2, \cdots, \lambda_m). \tag{3.26}$$

Here, the above decomposition is a finite hierarchical correlated function expansion [76–78]. And all parts in Eq. (3.26) are independent with each other.

Next, one can select a range for each parameter λ_j in Eq. (3.23) and then randomly choose thousands of $\lambda_j^{(i)} (i = 1, 2, \cdots, N)$ in its range. Thus the sensitivities of y_k with respect to λ_j are described by

$$s_{k(j)} = \left[\frac{1}{N} \sum_{i=1}^{N} (y_{k(j)}(\lambda_j^{(i)}))^2 - (\frac{1}{N} \sum_{i=1}^{N} y_{k(j)}(\lambda_j^{(i)}))^2 \right]^{1/2} / w_{kj}, \tag{3.27}$$

where w_{kj} is a weight factor, which normalizes the absolute standard deviation of $y_{k(j)}(\lambda_j^{(i)})$. Hereafter,

$$w_{kj} = \frac{1}{N} \sum_{i=1}^{N} y_{k(j)}(\lambda_j^{(i)}).$$

Similarly, the combinational sensitivities $s_{k(i,j)}$ of y_k with respect to λ_i, λ_j can be calculated by using $y_{k(i,j)}$ in Eq. (3.26). Thus, the total sensitivity of y_k with respect to all parameters of interest is given by

$$S_k = \sum_{i=1}^{m} s_{k(i)} + \sum_{1 \le i < j \le m}^{m} s_{k(i,j)} + \cdots + s_{k(1,2,\cdots,m)}, \tag{3.28}$$

where the first term on the right hand is the effect of single parameter λ_i with $i = 1, 2, \cdots, m$; the second term is the cooperative influence of λ_i and λ_j with $i, j = 1, 2, \cdots, m$. Although the relevant Matlab's toolbox-SBTOOLBOX [82] is available, it is also difficult to precisely calculate the sensitivity of y_k with respect to each parameter. Therefore, it is also very necessary to simplify the above GPS approach.

3.4.2.3 The New Global Relative Parameter Sensitivities Approach

The above RS-HDMR method can be seen as the variance-based method in some senses [76–78], however, our current method can be seen as the mean-based relative sensitivities method. The GRPS approach can be summarized as follows [47]:

Step 1: For a given mathematical model with m parameters, one can determine the possible range of each parameter from the existing experimental data or references.

Step 2: Randomly select N sets of parameter values for each parameter in its range according to a probability distribution, such as the uniform distribution or normal distribution [83]. Then, one can pair these samples into N sets of m dimensional parameter vectors without repetitions. From [83], the Latin's hypercube sampling method [84] can well perform this process.

Step 3: According to Eq. (3.25), we perform the local relative parameter sensitivity $s_k^{(i)} = \left(s_{k1}^{(i)}, \cdots, s_{km}^{(i)}\right)^T$ $(i = 1, 2, \cdots, N)$ of $y_k(\lambda)$ with respect to each set of parameters.

Step 4: Define the GRPS of $y_k(\lambda)$ with respect to each parameter as the average of N sets of local relative sensitivities:

$$GRPS_k = \frac{1}{N}\sum_{i=1}^{N} s_k^{(i)}. \tag{3.29}$$

Hereafter, $s_k^{(i)}$ and $GRPS_k$ have the same dimension m. Note that the above average value does not rely on the specific nominal values of parameters. In fact, it reflects the overall sensitivities of the model to its parameters. And the total sensitivities of $y_k(\lambda)$ with respect to all parameters are defined by

$$TS_k = \sum_{i=1}^{m} |GRPS_k(i)|. \tag{3.30}$$

Moreover, the ranges or bounds of local relative sensitivities as well as the variance of these N sets of local relative sensitivities can be used as the accessorial indexes to measure the GRPS properties.

Compared with the traditional RS-HDMR method, The GRPS approach is very simple. However, the GRPS method is a little conservative since the definition of the GRPS requires that the signs of s_{ij} are unchanged for all parameter sets of interest. In many real-world biological systems, the above condition is easily satisfied since all s_{ij}s are either non-positive or non-negative for all parameter values. Sometimes, the big numbers may "eat" the small ones. In this case, the variance of local relative sensitivities is very large. Therefore, the GRPS approach is reliable under the condition without too large variance.

3.4.3 *Global Relative Parameter Sensitivities of the FFLs*

3.4.3.1 Mathematical Models for the FFLs in GRNs

Figure 3.6 shows the eight basic types of FFLs, where X is an input node and Z is an output node. Hereafter, X regulates the production of Y, the combination of TFs X, Y regulates the production of Z, and the regulatory relations are either an activation or a repression. The regulation of Z by X, Y can be through the AND, OR, XOR, SUM logics, and so on. The following analysis will focus on AND and OR logics.

Take the C1 configuration with AND logic as an example, for the expression of Y, X polymerizes to act as an active TF, then the complex X_n combines with the promoter set of gene D_y. When $D_y X_n$ is formed, the transcription and translation processes will begin. Thus, the biochemical reactions are described by

$$
\begin{aligned}
nX &\rightleftharpoons X_n, \\
D_y + X_n &\rightleftharpoons D_y X_n, \\
D_y X_n &\rightarrow D_y X_n + M_y, \\
M_y &\rightarrow M_y + Y, \\
M_y &\rightarrow \phi, \\
Y &\rightarrow \phi,
\end{aligned}
$$

where the last two reactions represent the degradation of mRNA M_y and protein Y. For the production of protein Z, the transcription will begin after $D_z X_n Y_n$ is formed. For simplification, suppose that either Y_n binds to $D_z X_n$ to produce $D_z X_n Y_n$ or X_n binds to $D_z Y_n$ to produce $D_z X_n Y_n$. That is, the binding to the promoter region of D_z is without order. And the reactions are given by

$$
\begin{aligned}
nY &\rightleftharpoons Y_n, \\
D_z + X_n &\rightleftharpoons D_z X_n, \\
D_z + Y_n &\rightleftharpoons D_z Y_n, \\
D_z X_n + Y_n &\rightleftharpoons D_z X_n Y_n, \\
D_z Y_n + X_n &\rightleftharpoons D_z X_n Y_n, \\
D_z X_n Y_n &\rightarrow D_z X_n Y_n + M_z,
\end{aligned}
$$

$$M_z \rightarrow M_z + Z,$$

$$M_z \rightarrow \phi,$$

$$Z \rightarrow \phi.$$

For an OR gate, either D_zX_n, D_zY_n, or $D_zX_nY_n$ may transcript to produce the output protein Z. Note that the binding manner determines the above production process of Z. In this case, the above biochemical reactions only need to be slightly modified from the binding manner. Similarly, the biochemical reactions can be obtained for the other configurations.

After assuming fast reactions, such as polymerizations and combination reactions, the transcription processes, to reach the chemical equilibriums quickly. Using some dimensionless transformations, the mathematical model of the FFLs is finally described by

$$\begin{cases} \frac{dy}{dt} = \alpha_1 f(x, k_1) - \beta_1 y, \\ \frac{dz}{dt} = \alpha_2 G(x, k_2; y, k_3) - \beta_2 z, \end{cases} \tag{3.31}$$

where x, y, z are the concentrations of X, Y, Z, respectively. Here, $k_i (i = 1, 2, 3)$ are dissociation constants, n is the Hill coefficient with the same value in the above biochemical reactions, α_i is the dimensionless maximum transcription rate of protein i with $i = Y, Z$, and β_i is the degradation rate of protein i with $i = Y, Z$. According to [37], one can derive the regulation function for an activator from the above reactions. It is described by

$$f(x, k_1) = \frac{x^n}{k_1^n + x^n}.$$

Similarly, the regulation function for a repressor is given by

$$f(x, k_1) = \frac{k_1^n}{k_1^n + x^n}.$$

Hereafter, $G(x, k_2; y, k_3)$ is a gate function which can be AND, OR, SUM, or the hybrid of AND and OR logic [1]. The gate function of an AND gate is $G(x, k_2; y, k_3) = f(x, k_2) f(y, k_3)$. For an OR gate, suppose that two TFs can be exclusively binding to the promoter region. Thus, one has $G(x, k_2; y, k_3) = f_c(x; k_2, y, k_3) + f_c(y; k_3, x, k_2)$, where the activator is given by

$$f_c(x; k_2, y, k_3) = \frac{(x/k_2)^n}{1 + (x/k_2)^n + (y/k_3)^n},$$

and the repressor is described by

$$f_c(x; k_2, y, k_3) = \frac{1}{1 + (x/k_2)^n + (y/k_3)^n}.$$

For the other binding manners, such as independent binding, cooperative binding [44], one only need to slightly modify the denominator parts of $G(x, k_2; y, k_3)$.

3.4.3.2 The GRPS of the FFLs

This subsection focuses on the GRPS of the FFLs based on the above GRPS procedure. From Eq. (3.31), for the eight basic types of the FFLs, one can easily write out their dynamical equations. According to Eq. (3.25) and the four steps of GRPS analysis, one can obtain the relative parameter sensitivities. For example, for C1 FFL with AND cis-regulatory input at Z, the dynamic equation of concentration change is described by

$$\begin{cases} \frac{dy}{dt} = \frac{\alpha_1 x^n}{k_1^n + x^n} - \beta_1 y, \\ \frac{dz}{dt} = \frac{\alpha_2 x^n y^n}{(k_2^n + x^n)(k_3^n + y^n)} - \beta_2 z. \end{cases} \tag{3.32}$$

Let the right hand side of Eq. (3.32) to be zeros and solving the related equations, the steady states of the proteins y, z can be obtained as

$$\begin{cases} y^* = \frac{\alpha_1 x^n}{\beta_1 (k_1^n + x^n)}, \\ z^* = \frac{\alpha_2 x^n y^{*n}}{\beta_2 (k_2^n + x^n)(k_3^n + y^{*n})}. \end{cases} \tag{3.33}$$

Here, there are seven parameters: $\alpha_1, \beta_1, \alpha_2, \beta_2, k_1, k_2, k_3$. In the following, one only investigates the relative sensitivities of steady-state output z^* with respect to each parameter. And the relative sensitivities equations of steady state z^* with respect to these parameters are described by

$$\begin{cases} \frac{\alpha_1}{z^*}\frac{\partial z^*}{\partial \alpha_1} = \frac{\alpha_1}{z^*}\frac{\partial z^*}{\partial y^*}\frac{\partial y^*}{\partial \alpha_1} = \frac{nk_3^n}{k_3^n + y^{*n}}, \\ \frac{\beta_1}{z^*}\frac{\partial z^*}{\partial \beta_1} = \frac{\beta_1}{z^*}\frac{\partial z^*}{\partial y^*}\frac{\partial y^*}{\partial \beta_1} = -\frac{nk_3^n}{k_3^n + y^{*n}}, \\ \frac{\alpha_2}{z^*}\frac{\partial z^*}{\partial \alpha_2} = \frac{\alpha_2}{z^*}\frac{x^n y^{*n}}{\beta_2 (k_2^n + x^n)(k_3^n + y^{*n})} = 1, \\ \frac{\beta_2}{z^*}\frac{\partial z^*}{\partial \beta_2} = -\frac{\beta_2}{z^*}\frac{\alpha_2 x^n y^{*n}}{\beta_2^2 (k_2^n + x^n)(k_3^n + y^{*n})} = -1, \\ \frac{k_1}{z^*}\frac{\partial z^*}{\partial k_1} = \frac{k_1}{z^*}\frac{\partial z^*}{\partial y^*}\frac{\partial y^*}{\partial k_1} = -\frac{n^2 k_1^n k_3^n}{(k_1^n + x^n)(k_3^n + y^{*n})}, \\ \frac{k_2}{z^*}\frac{\partial z^*}{\partial k_2} = -\frac{nk_2^n}{k_2^n + x^n}, \\ \frac{k_3}{z^*}\frac{\partial z^*}{\partial k_3} = -\frac{nk_3^n}{k_3^n + y^{*n}}. \end{cases} \tag{3.34}$$

Similarly, the relative parameter sensitivity equations of the other FFLs with different logic gates can be derived. According to Eq. (3.34), one can easily verify that all these local relative sensitivities are either positive or negative. It shows that our assumption on the signs of s_{ij} in the above subsection is reasonable. Hereafter, let $n = 2$ and the parameters $\alpha_1, \beta_1, \alpha_2, \beta_2, k_1, k_2, k_3$ are labeled as $1, 2, \cdots, 7$. From [2, 14, 37, 83], the sampling ranges of parameters are given by $\alpha_i \in [0.01, 10]$ and $\beta_i \in [0.01, 5]$ for $i = 1, 2$, and $k_j \in [0.001, 100]$ for $j = 1, 2, 3$. For each case, the GRPS is calculated based on the evaluation of 50,000 random samples.

3.4.3.3 The Global Relative Parameter Sensitivities of CFFLs

Firstly, we consider the GRPS of the four types of CFFLs with AND and OR gates. Let the concentration of the input signal x be 1 μmol.L^{-1}. Following the four steps of the proposed approach in Sect. 3.4.2.3, GRPS can be easily obtained by using Matlab. Figure 3.7 shows the GRPS for CFFLs with AND (a, c) and OR gates (b, d), where all parameters are sampled uniformly (a, b) or normally (c, d) in their ranges. Comparing the four panels in Fig. 3.7, it indicates us that the sampling distributions of parameters have no significant influence on the GRPS. Therefore, the following analysis will be mainly based on the uniform sampling.

According to Fig. 3.7, one can get some conclusions as follows: (1) The GRPS have a little difference for the different logic gates, however, the outputs of these four types of CFFLs are more sensitive under AND logic gate; (2) The C1 is the most sensitive to most parameters under different logic gates, especially for k_1. The C2 is very sensitive under AND gate, however, it is the least sensitive under OR gate. The C4 is the second most sensitive configuration under OR gate; (3) From Eq. (3.34), the GRPS are constants for the parameters α_2, β_2. And k_1 is a key parameter for the C1 under both AND gate and OR gate.

Table 3.5 shows all statistical indexes for the relative parameter sensitivities. Here, the sensitive bounds are derived analytically which only depend on the multimerization level n. According to Eq. (3.32), for the C1 FFL, when $\frac{y^*}{k_3} \gg 1$, $\frac{\partial z^*}{\partial \alpha_1}\frac{\alpha_1}{z^*} \approx 0$; when $\frac{y^*}{k_3} \ll 1$, $\frac{\partial z^*}{\partial \alpha_1}\frac{\alpha_1}{z^*} \approx n$. That is, the local relative sensitivities of z^* with respect to α_1 will range from 0 to n. And the GRPS of z^* with respect to α_1 is 1.9972. It means that the 1% increment of α_1 leads to the 1.9972% increment of steady state z^*. The variance 0.0031 implies that the local relative parameter sensitivities are mostly located around 1.9972 and the average value of GRPS is a reliable index. Similarly, the other data can be well explained. For the C3 FFL, the local relative sensitivities of z^* with respect to k_1 are $n^2/[(1+(k_3/y)^n)(1+(x/k_1)^n)]$ which tends to n^2 under the conditions of $k_3/y \ll 1$ and $x/k_1 \ll 1$.

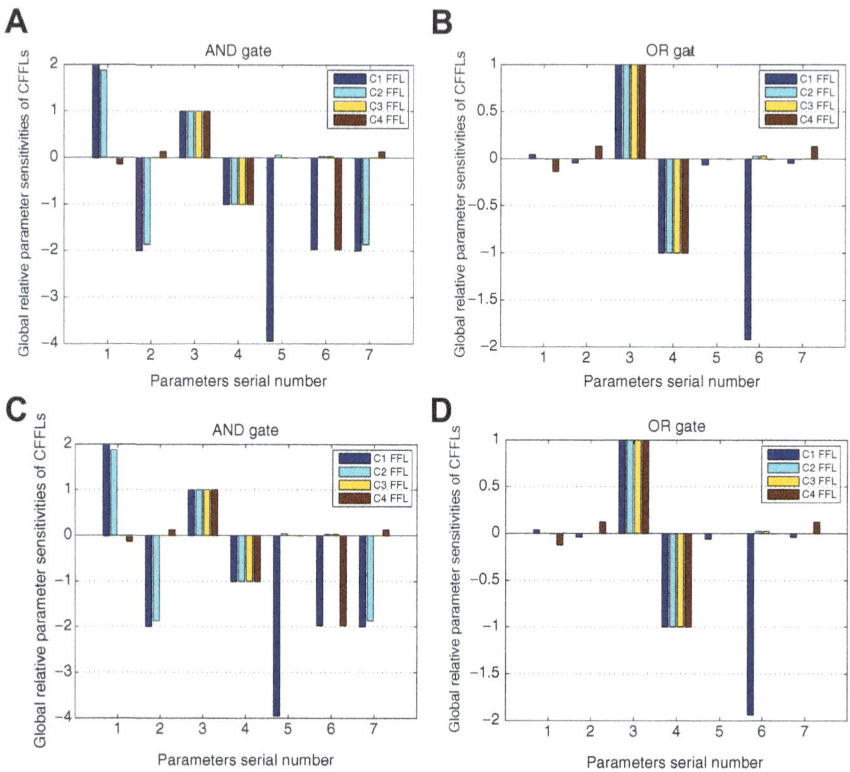

Fig. 3.7 The GRPS of the steady-state output z^* with respect to each parameter of CFFLs with AND and OR gates under different parameter sampling methods. (**a**, **b**) The GRPS of steady-state output z^* with respect to each parameter for CFFLs with AND and OR gates; Here, $x = 1$ and all parameters are uniformly sampled from their regions. (**c**, **d**) The GRPS of steady-state output z^* with respect to each parameter for CFFLs with AND and OR gates. Here $x = 1$ and all parameters are normally sampled from their regions. Reprinted from ref. [47], with permission from Elsevier

3.4.3.4 The Global Relative Parameter Sensitivities of ICFFLs

Similarly, the GRPS can be investigated for the four ICFFLs. Figure 3.8 shows the sensitivities under different logic gates. Our numerical results show that: (1) The GRPS have a little difference under the different logic gates for the ICFFLs; (2) The I1 and the I2 are the most insensitive to parameters under AND logic gate, however, the I1 and I3 are the most insensitive under OR gate. Also, the I4 is a sensitive loop under different logic gates; (3) The GRPS are constants for parameters α_2, β_2. k_1 is a key parameter for the I3 under AND gate, however, $\alpha_1, \beta_1, k_2, k_3$ are important parameters for the I4.

Table 3.5 Statistical characteristics of the GRPS for CFFLs with AND gate, $x = 1$, and parameters are uniformly sampled

Types	Parameters	Sensitive bounds	GRPS	Variances
$C1$	α_1	$[0, n]$	1.9972	0.0031
	β_1	$[-n, 0]$	−1.9972	0.0031
	k_1	$[-n^2, 0]$	−3.9332	0.1381
	k_2	$[-n, 0]$	−1.9687	0.0309
	k_3	$[-n, 0]$	−1.9972	0.0031
$C2$	α_1	$[0, n]$	1.8649	0.1442
	β_1	$[-n, 0]$	−1.8649	0.1442
	k_1	$[0, n^2]$	0.0615	0.1247
	k_2	$[0, n]$	0.0313	0.0309
	k_3	$[-n, 0]$	−1.8649	0.1442
$C3$	α_1	$[-n, 0]$	−0.0028	0.0031
	β_1	$[0, n]$	0.0028	0.0031
	k_1	$[0, n^2]$	0.0029	0.0048
	k_2	$[0, n]$	0.0313	0.0309
	k_3	$[0, n]$	0.0028	0.0031
$C4$	α_1	$[-n, 0]$	−0.1351	0.1442
	β_1	$[0, n]$	0.1351	0.1442
	k_1	$[-n^2, 0]$	−0.0024	0.0016
	k_2	$[-n, 0]$	−1.9687	0.0309
	k_3	$[0, n]$	0.1351	0.1442

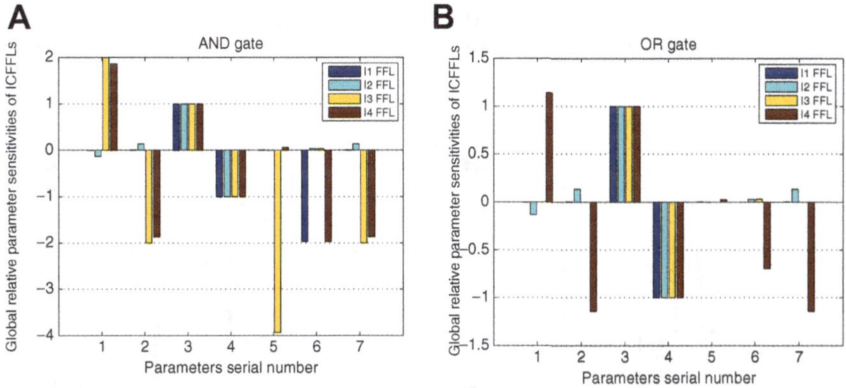

Fig. 3.8 The GRPS of steady-state output z^* with respect to each parameter for ICFFLs under AND and OR gates. Here, $x = 1$ and all parameter are uniformly sampled. Reprinted from ref. [47], with permission from Elsevier

3.4.3.5 The Effect of Input x on GRPS

Note that the input signal x of each configuration is assumed to be 1 in the above subsections. However, it is not the case in real-world applications. This subsection focuses on the evolution of GRPS with the changing of x. Figures 3.9 and 3.10 show the sensitivities of output z^* with respect to $\alpha_1, \beta_1, \alpha_2, \beta_2, k_1, k_2, k_3$ for $x \in [0.001, 100]$. Here, Fig. 3.9 shows the case for CFFLs with AND gate, and Fig. 3.10 shows the cases for the C1, C4 under the OR gate and the I1, I4 under the AND gate.

According to Figs. 3.9 and 3.10, the main conclusions are summarized as follows: (1) For any input signal concentration of x, C1 and C2 are the most sensitive to system parameters in CFFLs, however, I4 is the most sensitive in ICFFLs; (2) The GRPS with respect to $\alpha_1, \beta_1, k_1, k_2, k_3$ are monotone increasing or decreasing with the increasing or decreasing of x, respectively.

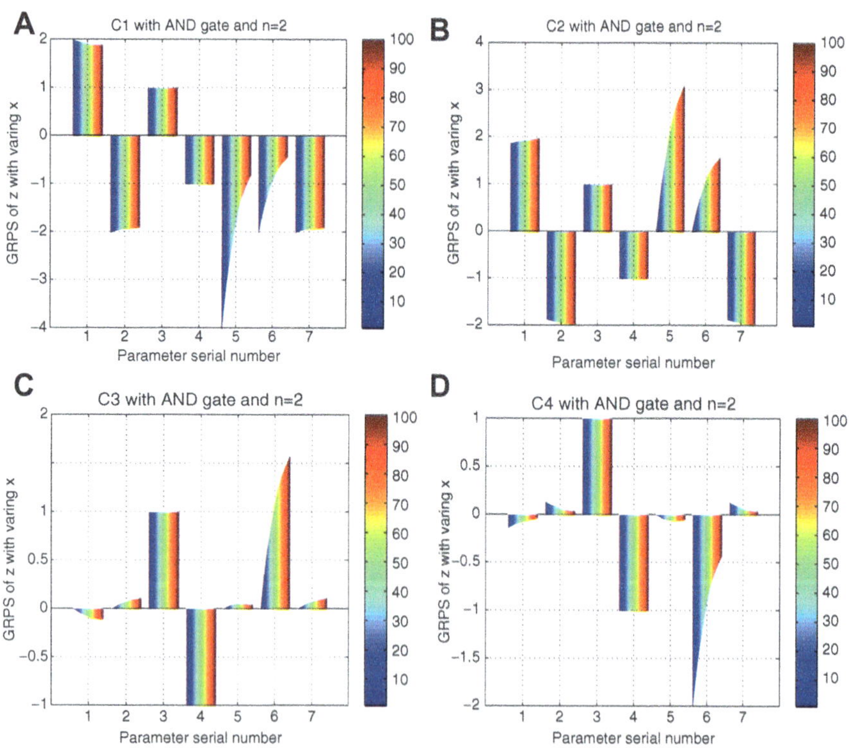

Fig. 3.9 The GRPS of the steady-state output z^* with respect to each parameter for CFFLs under AND gate. (**a–d**) Correspond to circuits C1–C4 respectively. Here, all parameters are uniformly sampled and the input x varies from 0.001 to 100. Colorbars correspond to different values of x, similarly hereinafter. Reprinted from ref. [47], with permission from Elsevier

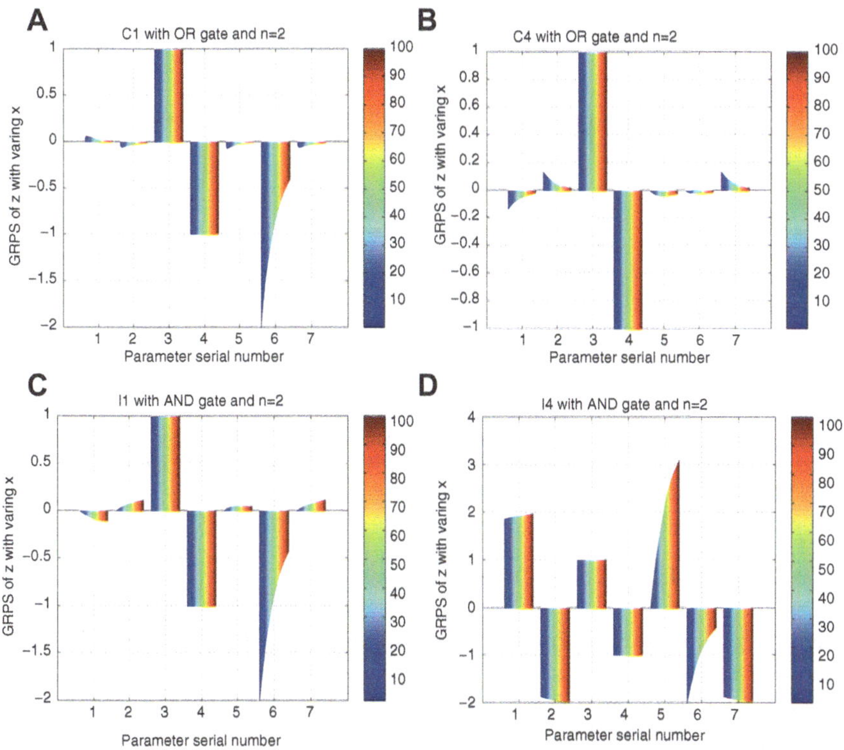

Fig. 3.10 The GRPS of the steady-state output z^* with respect to each parameter for C1 and C4 under OR gate (**a** and **b**) and the I1 and I4 under AND gate (**c** and **d**). All parameters are uniformly sampled and the input x varies from 0.001 to 100. Reprinted from ref. [47], with permission from Elsevier

3.4.3.6 The Effect of the Hill Coefficient *n* on the GRPS

It is well known that the Hill coefficient n describes how many monomers are assembled to form the polymer to bind to the gene's promoter sites. In the above subsections, we suppose that $n = 2$. However, the properties of the GRPS are similar for different n. For example, Fig. 3.11 shows the sensitivities of C1 and C4 in CFFLs under AND gate with $n = 4$. Compared with Fig. 3.9, n only changes the ranges of the GRPS. Moreover, for different FFL configurations, C1 is also the most sensitive circuit in CFFLs, however, C4 is not so sensitive. Furthermore, Fig. 3.11 can be well explained analytically as follow: (1) The sensitivities of z^* with respect to α_1 ranges from 0 to 4 for $n = 4$; (2) The sensitivities of z^* with respect to k_1 approximates to $-n^2 = -16$ under the conditions of $x/k_2 \ll 1$ and $y^*/k_3 \ll 1$.

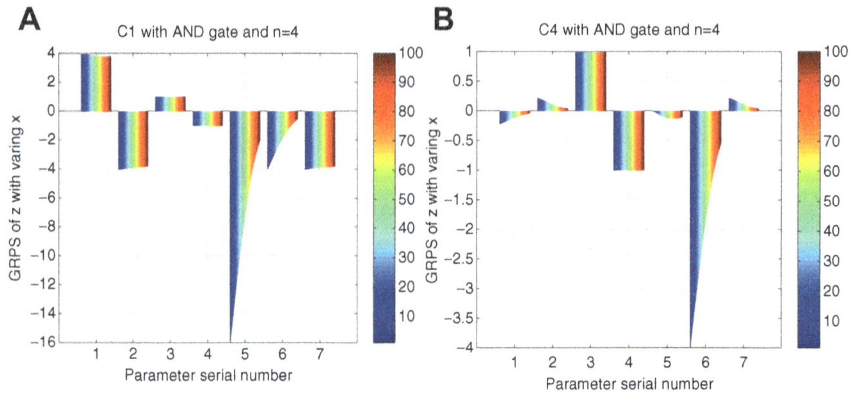

Fig. 3.11 The GRPS of the steady-state output z^* with respect to each parameter for C1 and C4 under AND gate and $n = 4$. All parameters are uniformly sampled and the input x varies from 0.001 to 100. Reprinted from ref. [47], with permission from Elsevier

3.4.3.7 RS-HDMR Versus GRPS on FFLs

This subsection aims to compare the traditional RS-HDMR method and the proposed approach in Sect. 3.4.2.3 based on the total global sensitivities for the FFLs. For simplicity, let $x = 1$. Table 3.6 shows all results for the eight basic types of configurations with AND and the OR gates.

According to Table 3.6, the calculating results of the proposed approach are mostly consistent with those of the RS-HDMR method except the slight difference for CFFLs with the AND gate. In detail, the order of total GRPS is $C1 > C2 > C4 > C3$ by using our proposed approach, however, the corresponding order of total GRPS is $C1 > C4 > C2 > C3$ by using the RS-HDMR method. For CFFLs with the OR gate, the above two methods have the same result of

Table 3.6 Total GRPS for FFLs under AND and OR gates by using the RS-HDMR method and the proposed approach. Here, $x = 1$ and all parameters are uniformly sampled, TS denotes the total GRPS derived from the proposed method, and RS-HDMR denotes the total GPS derived from the RS-HDMR method

Types	TS: AND	RS-HDMR: AND	TS: OR	RS-HDMR: OR
$C1$	13.8936	124.4566	4.1166	22.2208
$C2$	7.6875	9.4234	2.0335	4.1288
$C3$	2.0425	4.1342	2.0424	4.1342
$C4$	4.3764	23.1824	2.4056	4.2208
$I1$	3.9799	23.2738	2.0111	4.0972
$I2$	2.4391	4.2614	2.4340	4.2606
$I3$	11.9562	74.8487	2.0314	4.1301
$I4$	9.6248	62.4686	6.1455	9.0776

$C1 > C4 > C3 > C2$. For IFFLs, the above two approaches also have the same results of $I3 > I4 > I1 > I2$ under the AND gate and $I4 > I2 > I3 > I1$ under the OR gate, respectively. Therefore, our proposed approach is very effective for detecting the differences of sensitivities among different configurations of FFLs.

Although the RS-HDMR method is necessary for considering the cooperative effect of two or more parameters, the proposed GRPS approach is more effective for comparing the total sensitivities between different models or circuits. Furthermore, our method can also identify the first-order global sensitivities. However, it includes the cooperative effects of the other model parameters.

3.4.4 GRPS and Biological Functions of the FFLs

3.4.4.1 GRPS and Biological Abundance of FFLs

According to [2, 40], for the eight basic types of FFLs, there exists relative abundance in the TRNs of *E. coli* and in *S. cerevisiae*, as shown in Fig. 3.12. From [40], C1 and I1 are the most abundant in these two networks, which take up more than one-third of the total FFL configurations. Using the GRPS approach, it reveals that the most abundant C1 and C2 configurations are the most sensitive to its parameters for CFFL with AND gate, however, the conclusions are contrary for ICFFLs. Moreover, it implies that the more sensitive of the output to its parameters, the less robustness of the circuit's output to parameter variation. Also, I1 is insensitive in ICFFLs, which means that the output of this configuration is more robust to its parameter variation. Similarly, the most abundant circuit is also the most sensitive one for CFFLs. According to [2], the FFLs are most likely to be network motifs in sensory transcriptional network. To cope with the environmental fluctuations, it requires some of the FFLs to be sensitive to parameter variations.

3.4.4.2 Relations Between GRPS and Noise Characteristics

According to [41], the noise characteristics of the FFLs with AND and OR logics had been further investigated, where the Langevin formalism and the MC method based on the Gillespie algorithm were used. They calculated the variances around the mean protein levels in the steady states of the FFLs as noise. For the case of CFFLs with AND gate, the most abundant C1 is the least noisy circuit, another less noisy one is C2, while C3 and C4 are the most noisy. For ICFFLs with AND gate, I4 is the least noisy, the second least noisy circuit is I3 (see Figs. 2–4 of [41]), while the most abundant I1 and I2 in ICFFLs are the most noisy.

To compare between noise characteristics and the GRPS properties, one models the FFLs in a corresponding stochastic format and implements simulations by using the Gillespie stochastic algorithm [10]. Reaction probabilities or propensities of the synthesis of Y, Z are modeled by using the Hill-type rate expressions in the Gillespie

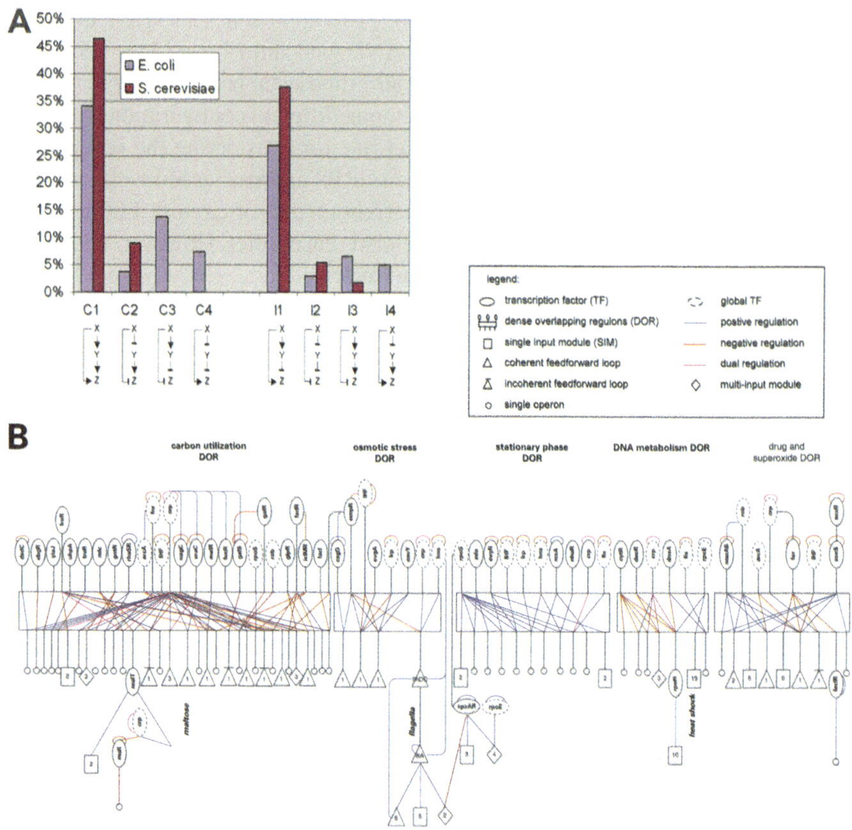

Fig. 3.12 Abundance of the FFLs in the TRNs of *E. coli* and *S. cerevisiae* (**a**) and part of the GRN for *E. coli* and motifs in the network (**b**). Panel A is reprinted from ref. [40], with permission from Elsevier; Panel (**b**) is reprinted by permission from Springer, ref. [32]

algorithm. Note that this method has been used to compare between deterministic models and stochastic models for some specific genetic systems [18, 19, 85]. Table 3.7 shows the stochastic reactions and its propensities for C1 with AND logic gate. Here, N_i are the molecular numbers for $i = x, y, z$, Ω is the system volume, and the other parameters have the same meaning as in Eq. (3.32). The detailed reaction processes for the Gillespie algorithm are omitted here and replaced by only four reactions with propensity functions scaled from the right hand side of Eq. (3.32). Moreover, the deterministic and stochastic models for the circadian rhythms circuit have been further explored by using this approach [19]. It should be pointed out that the above approach has the similar results with those of the regular stochastic models. Furthermore, this method has also been used in investigating the dynamical relations between deterministic ODE model and stochastic model for an interlinked positive and negative feedback loops of transcriptional regulation of CREB proteins [85].

Table 3.7 Stochastic model for C1 with AND gate

Reactions	Propensity functions	Increment of the molecular numbers
$\emptyset \xrightarrow{a_1} Y$	$a_1 = \frac{\alpha_1 \Omega N_x^n}{(k_1\Omega)^n + N_x^n}$	$(0, 1, 0)^T$
$Y \xrightarrow{a_2} \emptyset$	$a_2 = \beta_1 N_y$	$(0, -1, 0)^T$
$\emptyset \xrightarrow{a_3} Z$	$a_3 = \frac{\alpha_2 \Omega N_x^n N_y^n}{[(k_2\Omega)^n + N_x^n][(k_3\Omega)^n + N_y^n]}$	$(0, 0, 1)^T$
$Z \xrightarrow{a_4} \emptyset$	$a_4 = \beta_2 N_z$	$(0, 0, -1)^T$

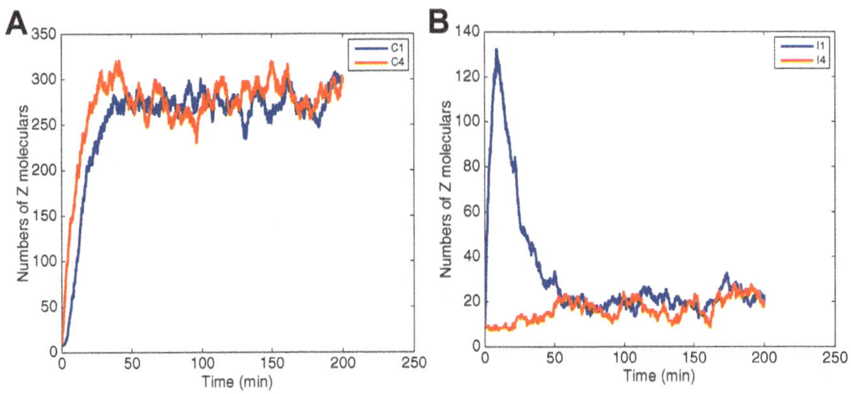

Fig. 3.13 Stochastic simulation results for C1 and C4 in the FFLs (**a**), I1 and I4 (**b**) under AND gates by using the Gillespie stochastic simulation algorithm with propensity functions similar to those that are listed in Table 3.7 for C1. Reprinted from ref. [47], with permission from Elsevier

Figure 3.13 shows the stochastic simulation results for the C1, C4 and the I1, I4 under AND gate, where the initial values N_x, N_y, N_z of the proteins X, Y, Z are 1000, 20, 10. Here, all parameters are taken as follows: $\Omega = 50, \alpha_1 = 1, \alpha_2 = 1.2, \beta_1 = 0.05, \beta_2 = 0.1, k_1 = 5, k_2 = 20, k_3 = 5$. Note that the results derived from this approach are similar to the results in [41]. The total number of Z in the C4 fluctuates more fiercely than in the C1. That is, the C1 is less noisy and the I1 is more noisy than the I4.

Compared this result with our findings on the total GRPS, the noise characteristics of the FFLs are mostly consistent with the total GRPS. That is, the less noisy FFL is more globally sensitive to parameters. For the CFFL, our results reveal that the C1 and C2 with AND gates are the most sensitive to parameters. However, for the case of the ICFFLs, the I3 and I4 are the most sensitive circuits. Under the OR gate, similar results can be obtained. The GRPS property combined with the noise characteristics can well explain the reason why some FFL configurations are the most frequently appeared ones. They are either the least noisy (e.g., C1) or the most robust to parameter perturbations (e.g., I1).

3.4.4.3 GRPS and Fold-Change Detection

Many sensory systems (e.g., vision and hearing) show a response that is proportional to the fold-change in the stimulus relative to the background, a feature related to the Weber's law. For a circuit with fold-change function, a response that depends on the fold-change in the input signal, and not on its absolute level [44, 63, 64]. For detailed explanations of the fold-change detection, one can see Fig. 3.14. The I1 circuit has been found to act as the fold-change detector, as shown in Fig. 3.14b. The circuit with fold-change detection function can buffer stochastic, genetic, and environmental variations [86, 87].

In the year 2009, Goentoro experimentally investigated the Wnt signaling pathway of the *Xenopus laevis*. They found that the gene β-catenin can act as a fold-change detector, and this circuit is very robust to parameter fluctuations. In fact, through the theoretical investigations in reference [44, 63], a sufficient condition for the onsetting of the fold-change detection function is the robustness of the system to parameter variations. Therefore, the function of I1 requires this configuration to be less sensitive to parameter variation [44, 63, 64].

3.4.5 Global Relative Input–Output Analysis of the FFLs

3.4.5.1 A GRIOS Index

Ref. [47] introduces a method to perform GRPS of a dynamical system, which is based on large-scale sampling and the average of local relative parameter sensitivity. Similar to GRPS, the input–output sensitivities [48] can be defined globally and performed as follows.

The first step is to compute the local relative input–output sensitivity, taking system (3.31) as an example, the local relative input–output sensitivity [70] can be defined as

$$s = \frac{\partial z^*/z^*}{\partial x/x} = \frac{x}{z^*}\frac{\partial z^*}{\partial x}, \tag{3.35}$$

where z^* represents the steady states of z, which can be derived from $\alpha_1 f(x, k_1) - \beta_1 y^* = 0$ and $\alpha_2 G(x, k_2; y^*, k_3) - \beta_2 z^* = 0$. $\frac{\partial z^*}{\partial x}$ can also be derived by taking derivative with respect to x in the above two equations, where one should only take y^* and z^* as implicit functions of x.

The second step is to determine a biology reasonable range for the input x, and also soundable parameter values, which can be derived from online databases or existing references [88]. After that, we sample N sets of x values, if there are many inputs, then one can use the Latin hypercube sampling method [84] to sample the inputs values.

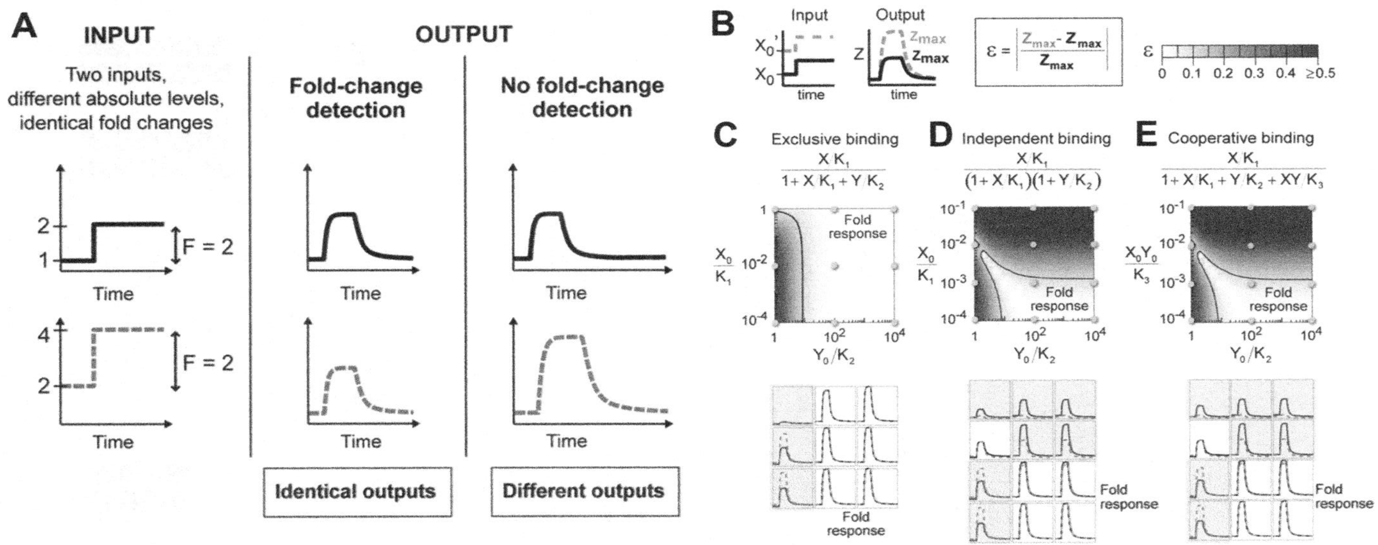

Fig. 3.14 The I1 can act as a fold-change detector. (**a**) Fold-change detector and non-fold-change detector. (**b–e**) I1 can act as a fold-change detector under a wide range of parameters. (**c–e**) show three detailed designs of the input functions for the promoters of *Z*, in which binding of *X* and *Y* is exclusive (**c**), independent (**d**), or cooperative (**e**). Reprinted from ref. [44], with permission from Elsevier

The next step is to compute local relative input–output sensitivity at each sampling point. One denotes the local relative input–output sensitivity index at the $i'th$ sampling point as

$$s_i = \frac{x^{(i)}}{z^*}\frac{\partial z^*}{\partial x^{(i)}}, (i = 1, 2, \cdots, N).$$

Finally, similar to the idea in Ref. [47], the GRIOS [48] can be defined as

$$GRIOS = \frac{1}{N}\sum_{i=1}^{N}|s_i|. \tag{3.36}$$

Here, the average is based on the absolute values at each sampling point, which is because for different x, s_i may be positive or negative; And we note that this average is meaningful, since one only concerns the relative quantity change, without considering whether the change is increment or decrement. The variance of these N sets of local relative input–output sensitivities can be used to measure the reliability of the GRIOS (3.36).

In the following analysis, by referring to Ref. [37], we set parameters $\alpha_i = 1, \beta_i = 1, k_1 = k_2 = 0.1, k_3 = 0.5, n = 2$, unless otherwise noted. The range of the input is set as $x \in [0.0001, 100]$, for each circuit, $N = 50,000$ sets of different x values will be sampled from this range.

3.4.5.2 GRIOS of the FFLs

By following the method in Sect. 3.4.5.1, the GRIOS of CFFLs and IFFLs undere AND and OR logic gates are plotted in Fig. 3.15; where, the length of these bars represent the GRIOS for each circuit.

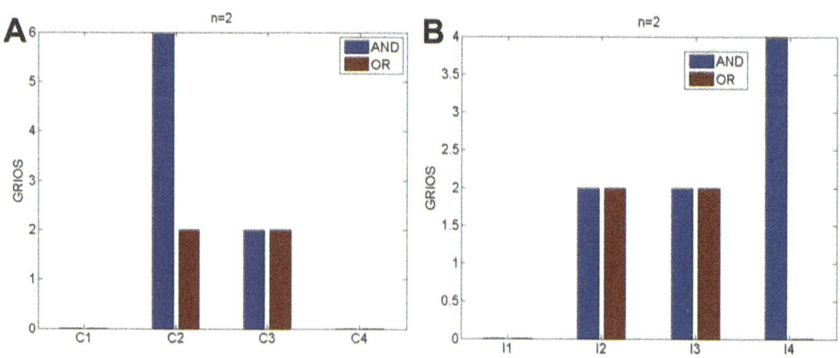

Fig. 3.15 GRIOS of (**a**) CFFLs and (**b**) IFFLs under AND logic gate and OR logic gate respectively. Reprinted from ref. [48]

From Fig. 3.15, one can easily find that no matter under AND or OR logic gates, C1 and I1 are always the least GRIOS configurations in CFFLs and IFFLs, respectively. And another least GRIOS circuit for CFFLs is C4, while for IFFLs, under the OR gate, I4 is also the least GRIOS one, but invalid under AND gate. Table 3.8 shows the variances of 50,000 sets of local relative input–output sensitivities for these eight types of FFLs under AND and OR gates, which corresponds to Fig. 3.15. From Table 3.8, one can see that these variances are all quite small, which illustrates that the averages are quite reliable to measure the GRIOS for these circuits.

Figures 3.16 and 3.17 show the evolution of the GRIOS with respect to system parameters $\alpha_1, \beta_1, k_1, k_2, k_3$. When we consider the evolution of the GRIOS with respect to one parameter, the other parameters are fixed as that mentioned in the above section. In Fig. 3.16a, the evolution of the GRIOS is considered for $\alpha_1 \in [0.01, 5]$, these four panels show the cases for CFFLs and IFFLs under AND and OR logic gates, respectively. From these figures, one can easily find that for CFFLs, the C1 and C4 are the least GRIOS, further, with the increasing of α_1, the C1 becomes less GRIOS than the C4; while for IFFLs, the I1 is the least GRIOS circuit under AND gate, but under OR gate, the I1 and the I4 are both very insensitive, and the I1 becomes more sensitive than the I4 with the increasing of α_1, whereas the absolute increment of GRIOS for I1 is still very small for $\alpha_1 \in [0.01, 5]$. Figure 3.16b shows the evolution of the GRIOS for $\beta_1 \in [0.01, 2]$, Fig. 3.16c, d and Fig. 3.17 show the case for $k_i \in [0.1, 10], i = 1, 2, 3$. These figures all demonstrate that, in the considered parameter ranges, for CFFLs, the C1 is quite insensitive compared with other coherent configurations, and for IFFLs, the I1 is quite insensitive, no matter under AND or OR logic gates.

We note that the evolutions of the GRIOS with respect to α_2, β_2 are not shown here, which is because that the GRIOS is irrelevant with α_2, β_2. One can understand this point from the following mathematical derivation. From Eqs. (3.31) and (3.32), one has the local relative input–output sensitivity as

$$s = \frac{x \partial z^*}{z^* \partial x} = \frac{x}{G}\left(\frac{\partial G}{\partial x} + \frac{\partial G}{\partial y^*}\frac{\partial y^*}{\partial x}\right).$$

Here, G denotes the function $G(x, k_2; y^*, k_3)$. Obviously, s is independent of α_2 and β_2.

Figure 3.18 shows the GRIOS for these eight types of FFLs under AND and OR logic gates for $n = 4$, with other parameters the same as that in Figs. 3.16

Table 3.8 Variances of the local relative input–output sensitivities that correspond to Fig. 3.15

Types	AND	OR	Type	AND	OR
$C1$	0.0244	0.0010	$I1$	0.0021	0.0005
$C2$	0.0394	0.0035	$I2$	0.0062	0.0050
$C3$	0.0021	0.0036	$I3$	0.0040	0.0076
$C4$	0.0070	0.0009	$I4$	0.0172	0.0002

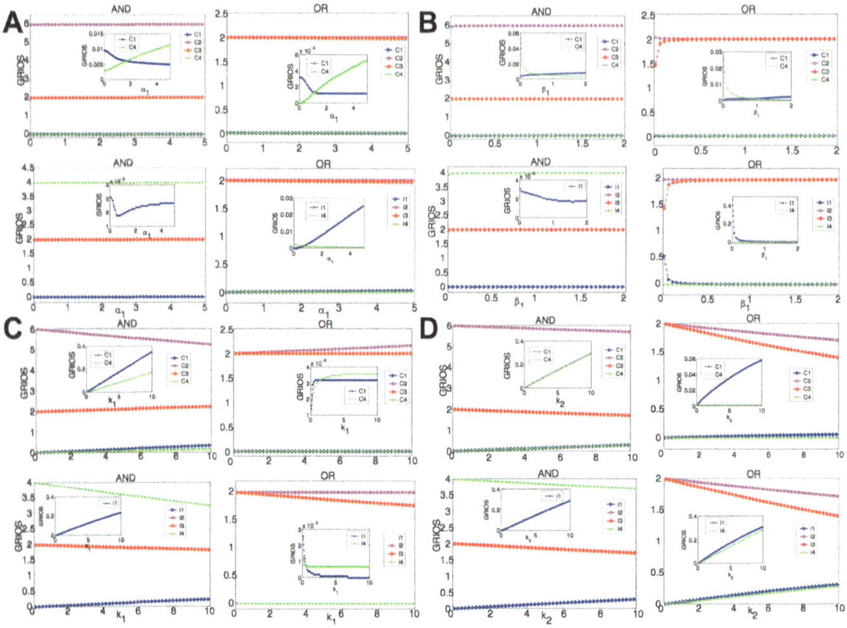

Fig. 3.16 Evolution of the GRIOS for CFFLs and IFFLs with respect to parameters α_1 (**a**) β_1 (**b**) k_1 (**c**) and k_2 (**d**). Reprinted from ref. [48]

and 3.17. Where, similar conclusions can be derived as that in Figs. 3.16 and 3.17. This indicates that the cooperative coefficient n has no destructive effect on our previous observations.

3.4.5.3 GRIOS of the FFLs Versus Its Structural and Functional Characteristics

As it has been shown in Ref. [40] that the C1 and I1 take up more than one-third of all the CFFLs and IFFLs configurations in real-world networks from *E.coli* and *S. cerevisiae*, respectively. Therefore, leading to the question that why the FFLs can become network motif and why the C1 and I1 are so frequently appeared in CFFLs and IFFLs. Existing works [37–63] have found that these FFLs can serve as important response detectors for various input signals, and some of them play an important role in sporulation formation in *Bacillus subtilis* [62] and so on, therefore, rationalizing why the FFLs can be network motif in so many real-world biological networks; Further from Ghosh and coworkers [41], the C1 circuit is the least noisy in the CFFLs, which can well explain why the C1 is so abundant in the CFFL configurations, but the I1 circuit cannot be explained from the perspective of intrinsic noise. Fortunately, our recent works found that the output of the I1 is the least globally relative sensitive to perturbations of its parameters, which indicates us

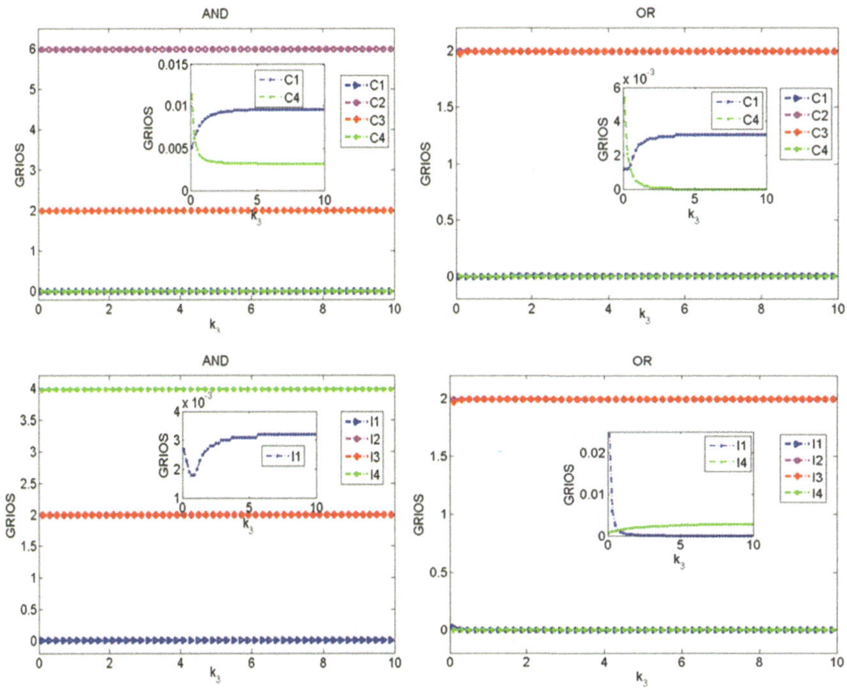

Fig. 3.17 Evolution of the GRIOS for CFFLs and IFFLs with respect to parameter k_3. Reprinted from ref. [48]

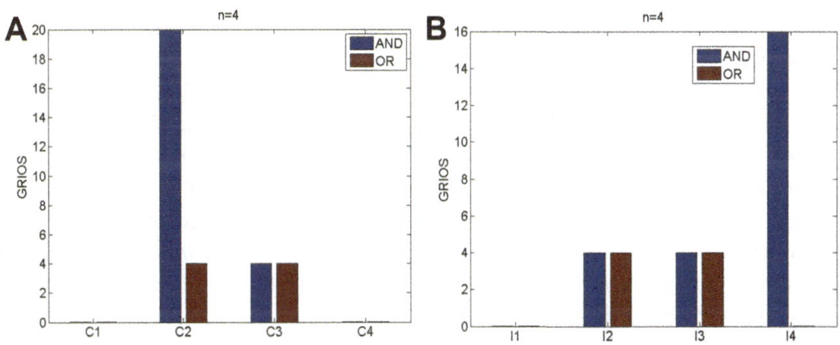

Fig. 3.18 GRIOS of (**a**) CFFLs and (**b**) IFFLs under AND and OR logic gates for $n = 4$. Reprinted from ref. [48]

that the I1 can robustly perform its functions in fluctuating environments, therefore, rationalizing why the I1 is so frequently appeared.

The above observations reveal that the output of the C1 and I1 are quite insensitive to various levels of inputs, which indicate that the C1 and I1 are input insensitive, that is to say, they can operate under different inputs robustly, therefore, maybe this can further explain why the C1 and I1 are so abundant, and most importantly, these FFLs are network motif in sensory related transcriptional networks [1, 2]. Such as, the I1 can act as a fold-change detector [44, 63], a very interesting phenomenon that relates to our taste, feeling, visual, smell, and so on, we note that this function immanently demands the I1 to be less sensitive to parameter variations and various input levels.

3.4.6 Summary

This section explores the relationships among the structures, dynamics, and functions of the genetic FFL circuits. We introduce two global sensitivity measures to analyze the feed-forward genetic circuits. The first measure is called GRPS and the second one is called GRIOS. Investigations reveal that the outputs of some frequently appeared FFL configurations are insensitive to its parameter variations and inputs, rationalizing why these configurations can be selected during evolution.

3.5 The Coupled Positive and Negative Feedback Genetic Circuits

Positive and negative feedbacks are two basic feedback mechanisms. It is reported that positive feedbacks can lead to multi-stability, act as signal amplifier, enhance the sensitivity of a system response to signals. Positive feedbacks also play an important role in cell memory [89–93], control the onset of some autosomal dominant diseases, such as autosomal dominant polycystic kidney disease and mature onset diabetes of youth [93], whereas negative feedbacks can lead to oscillations [35, 94]. However, in real-world biological systems, positive feedbacks and negative feedbacks always couple together to form coupled positive and negative feedback genetic circuits (CPNFGCs). Why these CPNFGCs are so frequently appeared in real-world systems? This is because the coupled circuits have their structural and functional advantages. In this section, we will overview the dynamics and functions of a simple CPNFGCs in the GRNs.

3.5.1 Related Works and Motivations

Network motifs with special dynamical behaviors have attracted the attentions of many researchers in the past decades, these motifs can either exhibit bistable switch, oscillation, excitability, or other special phenomenons [44, 94–97], which can be used to explain many biological phenomenon and functions, such as circadian rhythms [96, 97], biology memory [89], cell communication [98], cell differentiation [99, 101], and so on.

As we all know, negative feedbacks have the potential to generate sustained oscillation [94], while positive feedbacks can result in multi-stability [89, 94–102]. Recently, some coupled positive and negative feedback genetic motifs (see Fig. 3.19a) have been extensively investigated [103–111], and they found that these negative feedback oscillators become more robust and tunable when coupled by a positive one [103, 105]; and one can achieve a widely tunable frequency and near-constant amplitude for a negative feedback oscillator's coupled by a positive one [103].

There are different models to describe these CPNFGCs, such as the Hill kinetics [104, 107–109], RREs [105]; CMEs or CLEs, and so on. Whereas, the CMEs are well deemed as an accurate modeling approach, but at the expense of considerable large systems of equations and exhaustive computation time. Fortunately, Gillespie and followers proposed some stochastic simulation algorithms to accurately simulate biochemical reactions, which are equivalent to solve the chemical master equations, such as the direct Gillespie algorithm [10], the τ leap method [11, 112], the hybrid method [12]. On the other hand, time delays are ubiquitous in biological processes, correspondingly, stochastic simulation algorithms with delays [13, 15, 16] are also widely investigated. Although these algorithms can accurately describe systems behaviors, deterministic ODEs or delay differential equations (DDEs) are still widely used, raising the question that what is the relationships between stochastic models and its deterministic counterparts.

Gonze and coworkers firstly investigated the relationships between deterministic model and its stochastic counterparts of a circadian rhythm oscillator in *Drosophila* [19, 113–115], and they found that generally similar conclusions can be derived from these different models, but they only considered the cases when the system can display oscillation. Following, Hao et al. [109] investigated a minimal model of a CPNFGC, which has two components and are composed by transcriptional relations between two CREB proteins. In [109], stochastic dynamics near bifurcation points of the deterministic systems are also considered, providing some insights on the possible dynamics of the CREB regulatory motifs. While for delayed cases, Ref. [109] has not been further discussed and compared, and the developed stochastic model was also not considered in [109]. Moreover, excitable dynamics were also not discussed in their work.

Stimulated by the abovementioned problems, we will investigate and compare among different models for a three-component CPNFGC. Issues that will be explored include: (1) Bifurcation analysis for the deterministic ODE model and the DDE model, we will clarify the role of time delays on system dynamics, and the effect of bifurcation points on stochastic models; (2) How to simulate the undeveloped delayed stochastic models? (3) For both un-delayed and delayed cases, we explore the differences between the deterministic models and its stochastic counterparts. Especially, we will clarify the role of intrinsic noise inherently in stochastic simulations on the stochastic dynamics.

CPNFGCs exist in a wide range of biological systems [103, 106, 107], such as the MAPK/PKC system, the CREB system [109], the *yeast* galactose utilization network [34], just to name a few. The CPNFGC investigated in the following subsections is shown in Fig. 3.19b, dynamical behaviors for the deterministic system of this circuit have been previously partly discussed in Ref. [107]. In Fig. 3.19b, the TF X activates the expression of genes G_y, G_z, whereas Y activates and z represses the expression of the gene G_x; regulation relationships between genes G_x, G_y constitute the positive feedback loop, while G_x, G_z make the negative feedback loop. We denote D_x, D_y, D_z as the free promoter sites of the genes G_x, G_y, G_z respectively.

In the following, the deterministic ODEs will be solved by the Runge–Kutta method in Matlab, bifurcation analysis of the ODEs will be performed by Matcont [116], Oscill8 [117], bifurcation analysis of the DDEs is performed by DDE-BIFTOOL [118, 119], stability analysis as well as some bifurcation diagrams will be performed in XPPAUT [120]. Stochastic simulations of stochastic models are performed in Matlab. For the un-delayed stochastic models, the direct Gillespie algorithm [10] will be used, while for the delayed undeveloped case, a new method based on the Gillespie algorithm will be proposed, the related stochastic simulation results will be discussed in detail.

3.5.2 *Mathematical Models*

3.5.2.1 Deterministic Models: Without Time Delay

Basic biochemical reactions for the CPNFGC that is shown in Fig. 3.19b are listed in Table 3.9. $[D_x^T]$, $[D_y^T]$, $[D_z^T]$ are constants, which denote the total concentrations of these three genes G_x, G_y, G_z; X_2, Y_2, Z_2 are dimers, $D_xY_2, D_xZ_2, D_yX_2, D_zX_2$ are promoter-protein complexes. Hereinafter, we suppose X_2 and Y_2 exclusively bind to the promoters of gene G_x.

Assume that the fast reactions quickly reach their steady states, one has $[X_2] = K_1[X]^2$, $[D_yX_2] = K_1K_2[D_y][X]^2$, $[D_zX_2] = K_1K_3[D_z][X]^2$, $[Y_2] = K_4[Y]^2$, $[Z_2] = K_5[Z]^2$, $[D_xY_2] = K_4K_6[D_x][Y]^2$, $[D_xZ_2] = K_5K_7[D_x][Z]^2$; Here, $[M]$ denotes the concentration of the species M; $K_i = k_i/k_{-i}, (i = 1, 2, \cdots, 7)$ are dissociation constants; k_i, k_{-i} are forward and backward reaction rates for each revisable reactions in the left column of Table 3.9.

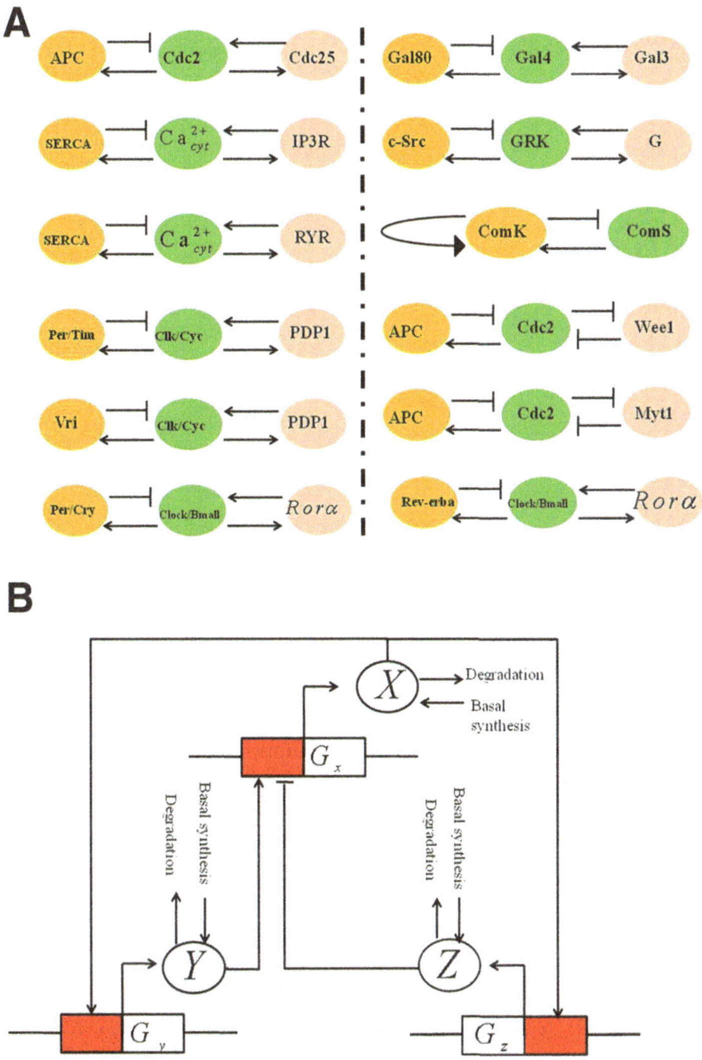

Fig. 3.19 The coupled positive and negative feedback genetic circuit. (**a**) Some real-world CPNFGCs [34]. (**b**)The detailed regulation processes in a CPNFGC. Genes G_x, G_y, G_z dominate the production of proteins X, Y, Z. The TF X activates the expression of genes G_y, G_z, whereas Y activates and Z represses the expression of gene G_x; Regulation relationships between genes G_x, G_y constitute the positive feedback loop, while G_x, G_z make the negative feedback loop. $\rightarrow$: activation; $\dashv$: repression. Each kind of proteins suffers from a degradation process and a basal synthesis process. ©[2011] IEEE. Reprinted, with permission, from ref. [21].

Further, for simplicity, we combine the transcription processes and translation processes as a unit process, we note that this simplification cannot change the conclusions and models too much. Mathematical model for the evolution of protein

Table 3.9 Biochemical reactions and reaction rates for the GRN that is shown in Fig. 3.19b

Fast reactions	Reaction rates
$2X \rightleftharpoons X_2$	K_1
$X_2 + D_y \rightleftharpoons D_y X_2$	K_2
$X_2 + D_z \rightleftharpoons D_z X_2$	$K_3 = K_2$
$2Y \rightleftharpoons Y_2$	$K_4 = K_1$
$2Z \rightleftharpoons Z_2$	$K_5 = K_1$
$Y_2 + D_x \rightleftharpoons D_x Y_2$	$K_6 = K_2$
$D_x + Z_2 \rightleftharpoons D_x Z_2$	$K_7 = K_2$
Slow reactions	Reaction rates
$D_x Y_2 \xrightarrow{t_1} D_x Y_2 + X$	$t_1 = 1\,\text{min}^{-1}$
$D_y X_2 \xrightarrow{t_2} D_y X_2 + Y$	$t_2 = 2\,\text{min}^{-1}$
$D_z X_2 \xrightarrow{t_3} D_z X_2 + Z$	$t_3 = 0.5\,\text{min}^{-1}$
$X \xrightarrow{d_X} \phi$	$d_X = 0.3\,\text{min}^{-1}$
$Y \xrightarrow{d_Y} \phi$	$d_Y = 0.2\,\text{min}^{-1}$
$Z \xrightarrow{d_Z} \phi$	$d_Z = 0.025\,\text{min}^{-1}$
$\phi \xrightarrow{r_i} i, (i = X, Y, Z)$	$r_i = (0.03, 0.2, 0.05)$ M/min
Conservation laws	
$[D_y] + [D_y X_2] = [D_y^T]$;	$[D_z] + [D_z X_2] = [D_z^T]$;
$[D_x] + [D_x Y_2] + [D_x Z_2] = [D_x^T]$	

concentrations under the exclusive binding assumption can be derived as

$$\begin{cases} [\dot{X}] = \frac{t_1[D_x^T]K_4K_6[Y]^2}{1+K_4K_6[Y]^2+K_5K_7[Z]^2} - d_X[X] + r_X, \\ [\dot{Y}] = \frac{t_2[D_y^T]K_1K_2[X]^2}{1+K_1K_2[X]^2} - d_Y[Y] + r_Y, \\ [\dot{Z}] = \frac{t_3[D_z^T]K_1K_3[X]^2}{1+K_1K_3[X]^2} - d_Z[Z] + r_Z. \end{cases} \tag{3.37}$$

After some algebraic substitutions, that are $x = \sqrt{K_1K_2}[X]$, $y = \sqrt{K_4K_6}[Y]$, $z = \sqrt{K_5K_7}[Z]$, $\alpha_1 = t_1[D_x^T]\sqrt{K_1K_2}$, $\alpha_2 = t_2[D_y^T]\sqrt{K_4K_6}$, $\alpha_3 = t_3[D_z^T]\sqrt{K_5K_7}$, $r_1 = r_X\sqrt{K_1K_2}$, $r_2 = r_Y\sqrt{K_4K_6}$, $r_3 = r_Z\sqrt{K_5K_7}$, $d_1 = d_X$, $d_2 = d_Y$, $d_3 = d_Z$, dimensionless model to describe the concentration evolutions of proteins X, Y, Z can be written as the following ODE system:

$$\begin{cases} \dot{x} = \frac{\alpha_1 y^2}{1+y^2+z^2} - d_1 x + r_1, \\ \dot{y} = \frac{\alpha_2 x^2}{1+x^2} - d_2 y + r_2, \\ \dot{z} = \frac{\alpha_3 x^2}{1+x^2} - d_3 z + r_3. \end{cases} \tag{3.38}$$

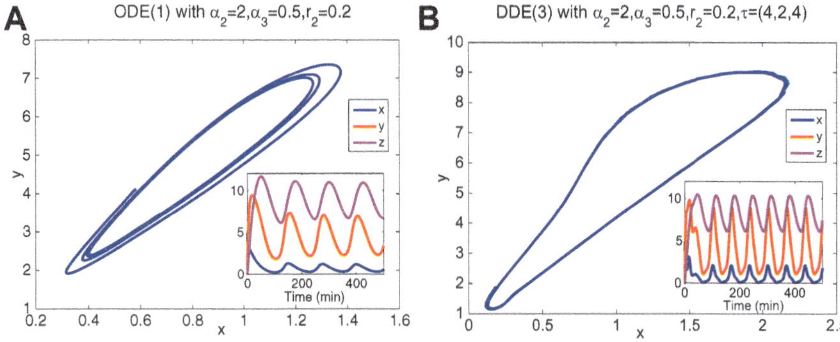

Fig. 3.20 Time evolutions as well as phase portrait in the $x - y$ plane for protein concentrations modeled as the ODE system (3.38) (**a**) and the DDE system (3.40) (**b**). Here $\alpha_2 = 2, \alpha_3 = 0.5, r_2 = 0.2$, time delays for the DDE (3.40) are supposed to be $\tau = (4, 2, 4)^T$, the history values of $(x, y, z)^T$ for $t < 0$ are chosen as $[5, 0, 0]^T$

We note here that, only the exclusive binding manner was considered, for other binding manners, mathematical models are similar to Eq. (3.38), only the first equation should be slightly modified as

$$\dot{x} = \frac{\alpha_1 y^2}{1 + y^2 + z^2 + \theta y^2 z^2} - d_1 x + r_1.$$

The values of θ determine different binding manners; $\theta = 0$ represents exclusive binding of Y_2, Z_2 to the promoter site D_x of gene G_x, $\theta = 1$ represents independent binding, whereas for other $\theta > 0 (\theta \neq 1)$, it can be seen as cooperative binding. For simplicity, we mainly consider the case of exclusive binding, other cases can be easily studied according to the same line. Figure 3.20a shows the oscillation behavior of the system (3.38) as well as phase portrait in the $x - y$ plane under appropriate parameters.

Steady states of system (3.38) can be derived by letting the right hand side of Eq. (3.38) to be zero; Local stability of the steady states $(x^*, y^*, z^*)^T$ can be analyzed through the linearization of system (3.38) around $(x^*, y^*, z^*)^T$ as

$$\dot{\tilde{X}} = A\tilde{X}, \tag{3.39}$$

where $\tilde{X} = (\tilde{x}, \tilde{y}, \tilde{z})^T = (x - x^*, y - y^*, z - z^*)^T$, and

$$A = \begin{bmatrix} -d_1 & \frac{2\alpha_1 y^*(1+z^{*2})}{(1+y^{*2}+z^{*2})^2} & \frac{-2\alpha_1 y^{*2} z^*}{(1+y^{*2}+z^{*2})^2} \\ \frac{2\alpha_2 x^*}{(1+x^{*2})^2} & -d_2 & 0 \\ \frac{2\alpha_3 x^*}{(1+x^{*2})^2} & 0 & -d_3 \end{bmatrix}.$$

The eigenvalues of matrix A determine the local stability of the steady state $(x^*, y^*, z^*)^T$. If there are eigenvalues with positive real parts, then the state $(x^*, y^*, z^*)^T$ is unstable, and it is stable if all the eigenvalues have negative real parts [71].

3.5.2.2 Deterministic Models with Time Delays

Time delays are ubiquitous in natural systems, gene regulation processes can also be seen as delayed systems. As we known that, gene transcription and translation can be seen as delayed processes, which are because that the transport processes of species between cellular compartments are time-consuming. Each protein cannot begin to degrade once it is produced, but after a life time. Therefore, protein degradations can also be seen as delayed processes. Hereinafter, we consider delayed models with delayed transcription or delayed degradation. The DDE model with delayed transcriptions is described as

$$\begin{cases} \dot{x} = \alpha_1 \frac{y^2(t-\tau_2)}{1+y^2(t-\tau_2)+z^2(t-\tau_3)} - d_1 x + r_1, \\ \dot{y} = \alpha_2 \frac{x^2(t-\tau_1)}{1+x^2(t-\tau_1)} - d_2 y + r_2, \\ \dot{z} = \alpha_3 \frac{x^2(t-\tau_1)}{1+x^2(t-\tau_1)} - d_3 z + r_3. \end{cases} \tag{3.40}$$

The DDE model with delayed degradations is described as

$$\begin{cases} \dot{x} = \alpha_1 \frac{y^2}{1+y^2+z^2} - d_1 x(t-\tau_1) + r_1, \\ \dot{y} = \alpha_2 \frac{x^2}{1+x^2} - d_2 y(t-\tau_2) + r_2, \\ \dot{z} = \alpha_3 \frac{x^2}{1+x^2} - d_3 z(t-\tau_3) + r_3. \end{cases} \tag{3.41}$$

Here, $\tau_i (i = 1, 2, 3)$ denote time delays.

Figure 3.20b shows the time evolution of protein concentrations and phase portrait in the $x - y$ plane for the DDE system (3.41). Here, when time delays are considered, oscillation period becomes shorter, and amplitudes become wider, which demonstrates that time delays may enhance oscillation [13].

3.5.2.3 Stochastic Model Directly from the Deterministic ODE: The Undeveloped Case

Just like the method in Ref. [19, 47, 109], we first consider a stochastic model directly from the deterministic ODE system (3.38), where the right hand side of Eq. (3.38) is taken as birth-death processes. Table 3.10 lists the simplified reactions and the corresponding propensity functions, where there are 9 reaction channels, Ω denotes the system volume, $N_i, (i = x, y, z)$ represents the molecular numbers of the proteins X, Y, Z. Parameters are mainly revised from Ref. [65, 121].

3.5.2.4 Stochastic Model from Table 3.9: The Developed Case

The developed stochastic model from Table 3.9 is rewritten into Table 3.11. For simplicity, we omit the detailed processes of transcription, translation, protein folding, and so on, the transcription and translation processes for each kinds of proteins are denoted by a single reaction, see reactions 15–17 in Table 3.11. In Table 3.11, we also denote N_x, N_y, N_z as the species numbers of proteins X, Y, Z. The other symbols, such as D_x, $D_y X_2$, X_2, are all denoted as the molecular numbers. k_i, k_{-i} are chosen such that $K_i = k_i/k_{-i}$ satisfies: $K_2 = K_3$, $K_1 K_2 = K_4 K_6 = K_5 K_7$; The total concentration of each gene is scaled by the system volume and the basal production rate for each protein is also scaled by Ω, which are mainly for the purpose of guaranteeing comparable between the undeveloped model and the developed model. Since the total concentration of each gene is assumed to be 1 μmol/L, therefore, the total numbers of these genes are assumed to be Ω in the stochastic models.

We also note that some propensity functions in Table 3.11 have been divided by Ω, this is because that k_i is the deterministic reaction rate, for stochastic simulations, we must transform the deterministic reaction rate k_i into the stochastic reaction rate c_i; For the zero and the one order reactions, the stochastic reaction rate

Table 3.10 Stochastic model directly from the deterministic ODE system (3.38)

Reactions	Propensity functions	Increment of the molecular numbers
$\emptyset \xrightarrow{a_1} X$	$a_1 = \frac{\alpha_1 \Omega N_y^2}{\Omega^2 + N_y^2 + N_z^2}$	$(1, 0, 0)^T$
$X \xrightarrow{a_2} \emptyset$	$a_2 = d_1 N_x$	$(-1, 0, 0)^T$
$\emptyset \xrightarrow{a_3} X$	$a_3 = r_1 \Omega$	$(1, 0, 0)^T$
$\emptyset \xrightarrow{a_4} Y$	$a_4 = \frac{\alpha_2 \Omega N_x^2}{\Omega^2 + N_x^2}$	$(0, 1, 0)^T$
$Y \xrightarrow{a_5} \emptyset$	$a_5 = d_2 N_y$	$(0, -1, 0)^T$
$\emptyset \xrightarrow{a_6} Y$	$a_6 = r_2 \Omega$	$(0, 1, 0)^T$
$\emptyset \xrightarrow{a_7} Z$	$a_7 = \frac{\alpha_3 \Omega N_x^2}{\Omega^2 + N_x^2}$	$(0, 0, 1)^T$
$Z \xrightarrow{a_8} \emptyset$	$a_8 = d_3 N_z$	$(0, 0, -1)^T$
$\emptyset \xrightarrow{a_9} Z$	$a_9 = r_3 \Omega$	$(0, 0, 1)^T$

Table 3.11 Developed stochastic model and parameters. We assume the following conservation laws: $D_y + D_yX_2 = \Omega$; $D_z + D_zX_2 = \Omega$; $D_x + D_xY_2 + D_xZ_2 = \Omega$. k_i, k_{-i} are randomly chosen such that $K_i = k_i/k_{-i}$ satisfies: $K_2 = K_3$, $K_1K_2 = K_4K_6 = K_5K_7$. The total concentration of each gene is scaled by system volume and the basal production rate for each protein is also scaled by Ω

Labels	Reactions	Propensity functions	Parameter values
1	$2X \xrightarrow{k_1} X_2$	$b_1 = k_1 \times N_x \times (N_x - 1)/\Omega$	$k_1 = 10$
2	$X_2 \xrightarrow{k_{-1}} 2X$	$b_2 = k_{-1} \times X_2$	$k_{-1} = 400$
3	$X_2 + D_y \xrightarrow{k_2} D_yX_2$	$b_3 = k_2 \times X_2 \times D_y/\Omega$	$k_2 = 50$
4	$D_yX_2 \xrightarrow{k_{-2}} X_2 + D_y$	$b_4 = k_{-2} \times D_yX_2$	$k_{-2} = 1.25$
5	$X_2 + D_z \xrightarrow{k_3} D_zX_2$	$b_5 = k_3 \times X_2 \times D_z/\Omega$	$k_3 = 60$
6	$D_zX_2 \xrightarrow{k_{-3}} X_2 + D_z$	$b_6 = k_{-3} \times D_zX_2$	$k_{-3} = 1.5$
7	$2Y \xrightarrow{k_4} Y_2$	$b_7 = k_4 \times N_y \times (N_y - 1)/\Omega$	$k_4 = 8$
8	$Y_2 \xrightarrow{k_{-4}} 2Y$	$b_8 = k_{-4} \times Y_2$	$k_{-4} = 240$
9	$2Z \xrightarrow{k_5} Z_2$	$b_9 = k_5 \times N_z \times (N_z - 1)/\Omega$	$k_5 = 8$
10	$Z_2 \xrightarrow{k_{-5}} 2Z$	$b_{10} = k_{-5} \times Z_2$	$k_{-5} = 320$
11	$Y_2 + D_x \xrightarrow{k_6} D_xY_2$	$b_{11} = k_6 \times Y_2 \times D_x/\Omega$	$k_6 = 40$
12	$D_xY_2 \xrightarrow{k_{-6}} Y_2 + D_x$	$b_{12} = k_{-6} \times D_xY_2$	$k_{-6} = 4/3$
13	$Z_2 + D_x \xrightarrow{k_7} D_xZ_2$	$b_{13} = k_7 \times Z_2 \times D_x/\Omega$	$k_7 = 60$
14	$D_xZ_2 \xrightarrow{k_{-7}} Z_2 + D_x$	$b_{14} = k_{-7} \times D_xZ_2$	$k_{-7} = 1.5$
15	$D_xY_2 \xrightarrow{t_1} D_xY_2 + X$	$b_{15} = t_1 \times D_xY_2$	$t_1 = 1$
16	$D_yX_2 \xrightarrow{t_2} D_yX_2 + Y$	$b_{16} = t_2 \times D_yX_2$	$t_2 = 2$
17	$D_zX_2 \xrightarrow{t_3} D_zX_2 + Z$	$b_{17} = t_3 \times D_zX_2$	$t_3 = 0.5$
18	$X \xrightarrow{d_1} \emptyset$	$b_{18} = d_1 \times N_x$	$d_1 = 0.3$
19	$Y \xrightarrow{d_2} \emptyset$	$b_{19} = d_2 \times N_y$	$d_2 = 0.2$
20	$Z \xrightarrow{d_3} \emptyset$	$b_{20} = d_3 \times N_z$	$d_3 = 0.025$
21	$\emptyset \xrightarrow{r_1} X$	$b_{21} = r_1$	$r_1 = 0.03 \times \Omega$
22	$\emptyset \xrightarrow{r_2} Y$	$b_{22} = r_2$	$r_2 = 0.2 \times \Omega$
23	$\emptyset \xrightarrow{r_3} Z$	$b_{23} = r_3$	$r_3 = 0.05 \times \Omega$

$c_i = k_i$; for the second order reactions, $c_i = k_i/\Omega$. For detailed discussions about the deterministic modeling, stochastic modeling as well as the related topics, one can refer to Ref. [26] or that discussed in Section 3.1.

3.5.2.5 Stochastic Simulations

The CME is an accurate description of biochemical systems, but it is impossible to be solved for most practical problems due to exhaustive computation time. To overcome this problem, stochastic simulation algorithms are proposed, which are equivalent to solve the CMEs for these systems. Thus, the stochastic simulation algorithms provide practical methods for simulating the stochastic biochemical systems.

In the following sections, we note there are N species in the overall reaction systems with the initial molecular number $X(0) = (X_1(0), \ldots, X_N(0))^T$. In stochastic simulations, two key points are that when will the next reaction occur, and which reaction will it be? From the discussion of Ref. [10], the two points can be determined from two uniformly randomly generated numbers and the probability propensity functions $a_i(X(t))(i = 1, 2, \cdots, M)$ for each reaction, where $a_i(X(t))$ relates to the reactant molecular numbers of the $i'th$ reaction at time t.

For the un-delayed case, the direct Gillespie algorithm [10] as discussed in section 3.1 will be used. For the delayed stochastic simulation, it is noted that some delayed stochastic simulation methods have been introduced in Ref. [13, 15, 16], which are developed from the direct Gillespie algorithm. Different from these methods, for stochastic simulations of the undeveloped model with time delays, we propose a new method, which is also based on the direct Gillespie algorithm. Taking the DDE system (3.40) as an example, first of all, system (3.40) is rewritten into birth-death processes, which is similar to the ones in Table 3.10, but with

$$a_1 = \frac{\alpha_1 \Omega N_y^2(t-\tau_2)}{\Omega^2 + N_y^2(t-\tau_2) + N_z^2(t-\tau_3)}, \quad a_4 = \frac{\alpha_2 \Omega N_x^2(t-\tau_1)}{\Omega^2 + N_x^2(t-\tau_1)},$$

$$a_7 = \frac{\alpha_3 \Omega N_x^2(t-\tau_1)}{\Omega^2 + N_x^2(t-\tau_1)}.$$

Most parts of the stochastic simulation processes are similar to the direct Gillespie algorithm, except when one computes these propensity functions, a_1, a_4, a_7 are computed according to the history values of N_x, N_y, N_z. We note that, since the reaction time steps are randomly generated, therefore, these history values of N_x, N_y, N_z cannot be always exactly derived, one can only guarantee to call these history values at time points t_d, where $t_d = t(d1)$, and $d1$ satisfies $[v1, d1] = min(abs(t - t(i) + \tau))$ (A Matlab sentence that finds the closest history values $N_i(t-\tau)$). Obviously, this method is not an accurate one; however, it is numerically proved to be powerful to simulate stochastic models directly from the deterministic DDEs. The detailed algorithms can be found in Algorithm 12.

3.5.3 *Dynamical Analysis and Functions*

3.5.3.1 Bifurcation Analysis

Mathematical model without delays for the circuit as shown in Fig. 3.19b can be modeled as the ODE system (3.38). The DDE models are described in systems (3.40) and (3.41). Following, we perform dynamical analysis on these systems.

From Ref. [107], α_2, α_3 can be seen as the strength of the positive and negative feedbacks respectively. This model can exhibit mono-stability, bistability, excitability, and oscillation under different parameter settings. In the following discussions, we set $\alpha_1 = 1, d_1 = 0.3, d_2 = 0.2, d_3 = 0.025, r_1 = 0.03, r_3 = 0.05$. We will take α_2, α_3 as bifurcation parameters in the subsequent analysis.

From Fig. 3.21, when $r_2 = 0.0265$, one-parameter bifurcation diagram of α_2 with respect to x reveals that there are two saddle-node (SN) points and two Hopf bifurcation (HB) points, and the two SN points locate between two HB points, which reveals that when α_2 locates between the two SN points, the system has three steady states, but further by XPPAUT, one case is that the lowest steady state is stable (for $\alpha_3 = 0.3$), which is actually excitable; Another case is that the lowest steady states is partly stable (for $\alpha_3 = 0.4$), where either excitable or oscillation behaviors can be observed; the other case is that all these three states are unstable ($\alpha_3 = 0.5$), which can display oscillations. The region between the two HB points except the excitable region can display oscillation behaviors; while for $r_2 = 0.2$, there are only two HB points for $\alpha_3 = 0.4$ and $\alpha_3 = 0.5$, which means that only mono-stability and oscillation behaviors can be observed; From Fig. 3.22 of two-parameter bifurcation diagrams, when $r_2 = 0.0265$, bistability region, mono-stability region, oscillation region as well as excitability region can be easily found; whereas, for $r_2 = 0.2$, the parameter space are partitioned only into two regions by two kinds of dynamical behaviors: oscillation and mono-stable.

Figure 3.22c, d show two different kinds of oscillation behaviors for $r_2 = 0.0265$. When α_2, α_3 locate in region VI, for example, for $\alpha_2 = 1.97, \alpha_3 = 0.48$, the system has three USSs, that is, $(0.1940, 0.4898, 2.6964)^T$, $(0.2600, 0.7562, 3.2157)^T$, $(0.3880, 1.4213, 4.5122)^T$, they are saddle focus (the linearized matrix has two complex eigenvalues with positive real parts and one negative real eigenvalue), saddle node (one positive real eigenvalue and

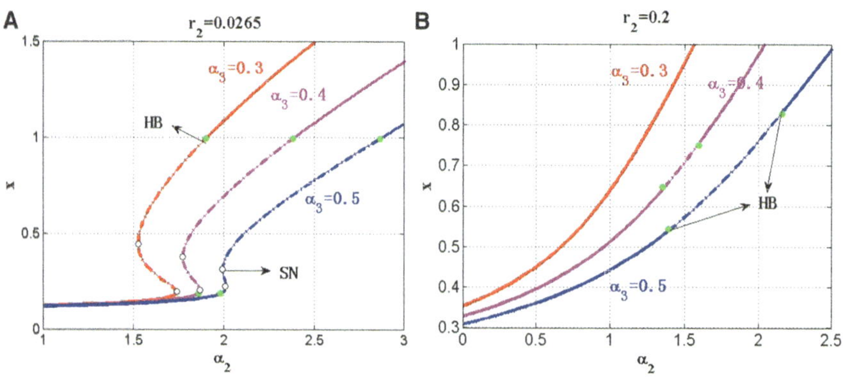

Fig. 3.21 One-parameter bifurcation diagrams for system (3.37). The two figures show the cases for $r_2 = 0.0265$ (**a**) and $r_2 = 0.2$ (**b**) respectively, where dashed-dotted segments denote the unstable steady states, and solid lines represent the stable steady states. SN: saddle-node point, HB: Hopf bifurcation point

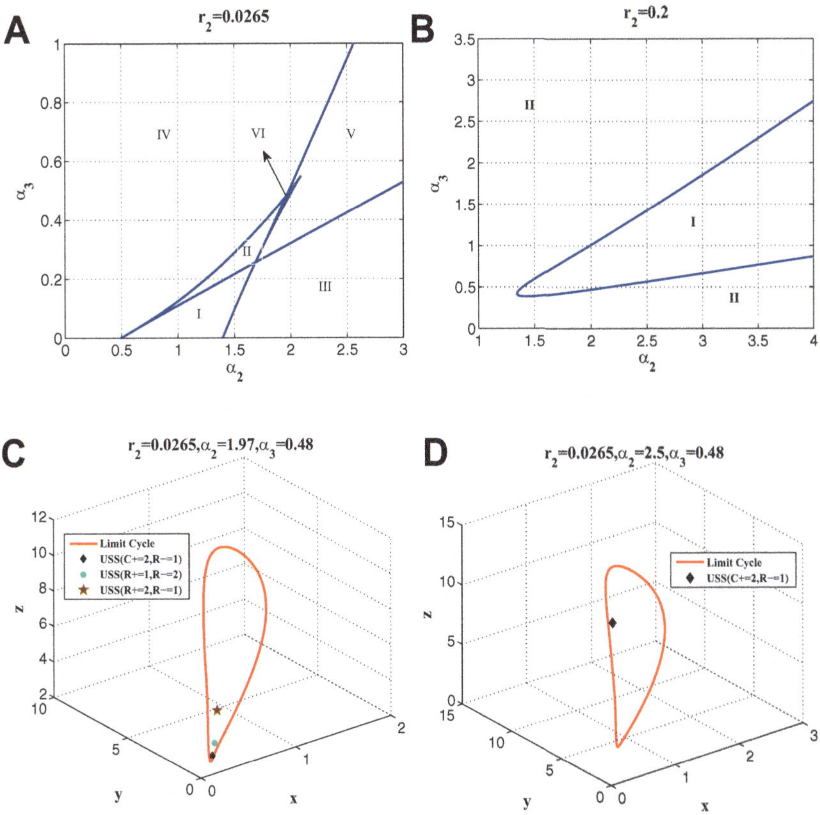

Fig. 3.22 Two-parameter bifurcation diagrams for system (3.38) and two different kinds of oscillation behaviors under $r_2 = 0.0265$. (**a**) $r_2 = 0.0265$. I: Bistable region; II: Excitable region; III: Mono-stable; IV: Mono-stable; VI: Oscillation with three unstable steady states (USSs) and V: Oscillation region with one USS. (**b**) $r_2 = 0.2$, I: Oscillation region, II: Mono-stable. Panels (**c**) and (**d**) show different oscillation behaviors for $r_2 = 0.0265$, the corresponding types of eigenvalues for the USSs are marked in the parentheses; $C+$: the number of complex eigenvalues with positive real parts, $R+$, $R-$ denote the number of positive, negative real eigenvalues respectively

two negative real eigenvalues for the linearized matrix), unstable node (two positive real eigenvalues and one negative real eigenvalue), respectively; While for parameters α_2, α_3 locate in region V, for example, for $\alpha_2 = 2.5, \alpha_3 = 0.48$, the oscillation behavior is accompanied by only one unstable saddle focus steady state $(0.8290, 5.2312, 9.8317)^T$.

Our results also show that the CPNFGC can be tunable among mono-stability, bistability, excitability, oscillation for small r_2, provided that the negative feedback strength α_3, the positive feedback strength α_2 are appropriately chosen. Interestingly, our result shows that there is an oscillation region with three steady states, which has not been reported in Ref. [107].

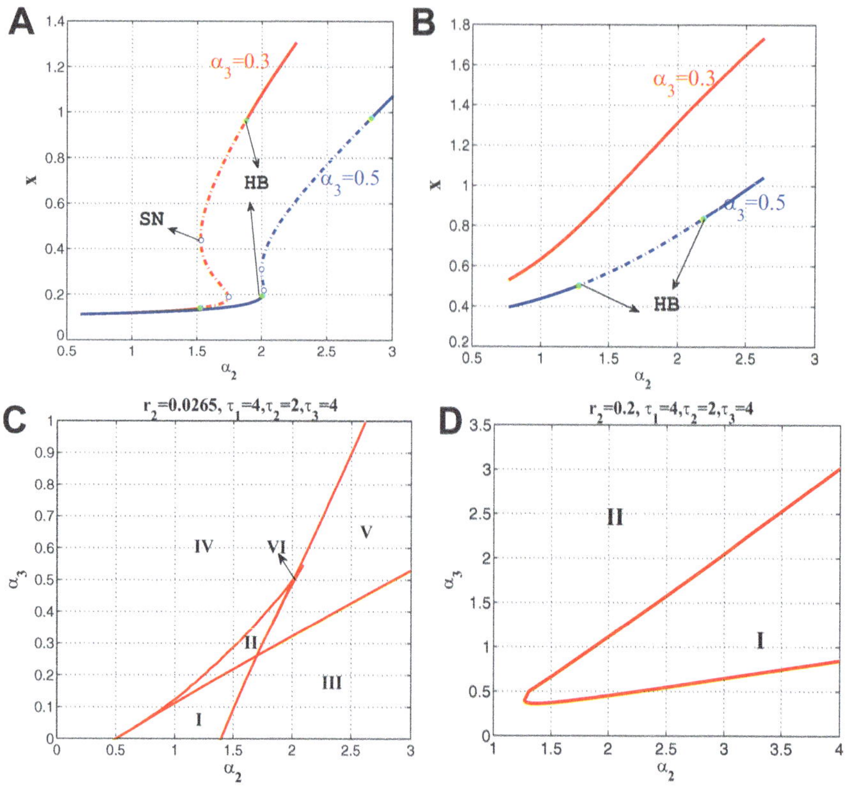

Fig. 3.23 One-parameter and two-parameter bifurcation diagrams for the DDE system (3.40). (**a**) The steady states x versus α_2 for $r_2 = 0.0265$, $\tau_1 = 4$, $\tau_2 = 2$, $\tau_3 = 4$; (**b**) The case for $r_2 = 0.2$. For panels (**a**) and (**b**), $\alpha_3 = 0.3, 0.5$ are shown. (**c**) The case for $r_2 = 0.0265$. I: Bistability region; II: Excitability region; III: Mono-stable; IV: Mono-stable; VI: Oscillation with three USSs and V: Oscillation regions with one unstable steady state. (**d**) The case for $r_2 = 0.2$, I: Oscillation region, II: Mono-stable. Where $\tau_1 = 4$, $\tau_2 = 2$, $\tau_3 = 4$ for both of the two panels

Using the DDE-BIFTOOL, bifurcation diagrams for the system (3.40) are shown in Fig. 3.23. In both figures, time delays are taken as $\tau_1 = 4$, $\tau_2 = 2$, $\tau_3 = 4$. Fig. 3.23a, b show one-parameter bifurcation of the steady states x versus α_2 with $\alpha_3 = 0.5, 0.3$, Fig. 3.23a shows the case for $r_2 = 0.0265$, while Fig. 3.23b shows the case for $r_2 = 0.2$. In Fig. 3.23a, for $\alpha_3 = 0.3$, there are two SN points, which indicates that there are three steady states when α_2 is taken between these two SN nodes, which is similar to the bifurcation diagrams for the ODE system (3.39). Figure 3.23c, d show the two-parameter bifurcation diagrams for the two cases in Fig. 3.23a, b, which are continued from the abovementioned two kinds of bifurcation points.

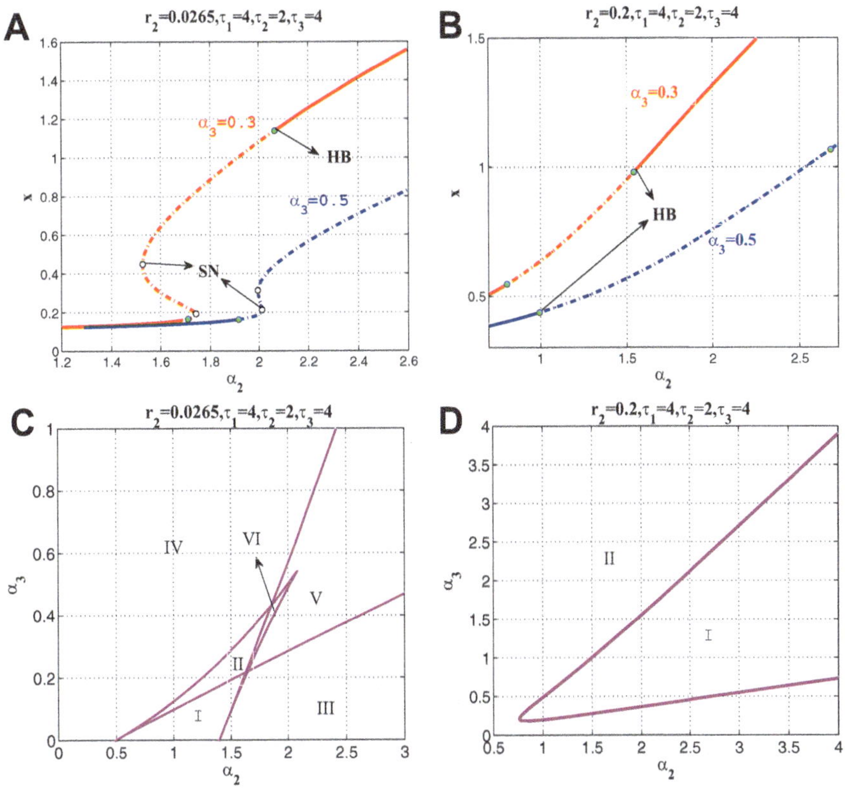

Fig. 3.24 One-parameter and two-parameter bifurcation diagrams for the DDE system (3.41). (**a**) The steady states x versus α_2 for $r_2 = 0.0265, \tau_1 = 4, \tau_2 = 2, \tau_3 = 4$; (**b**) The case for $r_2 = 0.2$. For panels (**a**) and (**b**), $\alpha_3 = 0.3, 0.5$ are shown. (**c**) The case for $r_2 = 0.0265$. I: Bistability region; II: Excitability region; III and IV: Mono-stability; VI: Oscillation with three USSs and V: Oscillation regions with one USS. (**d**) The case for $r_2 = 0.2$. I: Oscillation region, II: Mono-stable. Where $\tau_1 = 4, \tau_2 = 2, \tau_3 = 4$ for both panels

From Fig. 3.23, when $r_2 = 0.0265, \tau_1 = 4, \tau_2 = 2, \tau_3 = 4$, the parameter space (α_2, α_3) is divided into six regions, which are marked with $I - VI$, among which, I represents the bistability region, II represents the excitability region, V, VI are the oscillation regions, III and IV denote the mono-stability regions. For $r_2 = 0.2$, there are only two regions, with either mono-stable or oscillation behaviors.

Figure 3.24 shows the bifurcation diagrams for the DDE system (3.41), which are similar to the bifurcation diagrams for the DDE system (3.40), unnecessary details will be not given here.

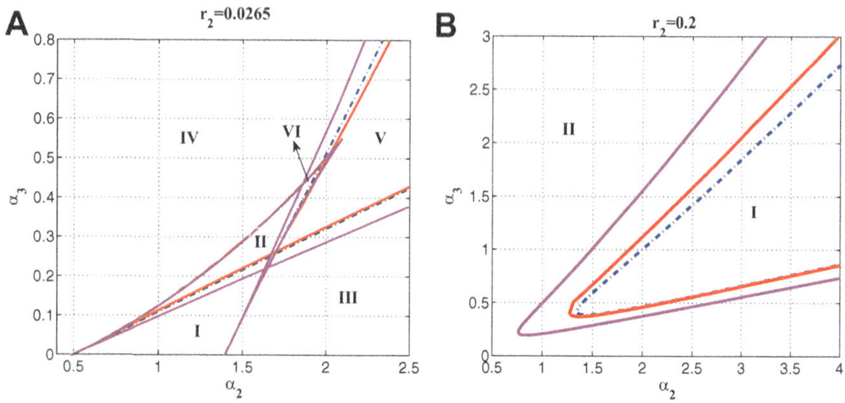

Fig. 3.25 Comparison between the two-parameter bifurcation diagrams of the ODE system (3.38), the DDE systems (3.40) and (3.41) under different r_2**s.** Dashed-dotted lines show bifurcations of the ODE system, while solid lines show the cases for the corresponding DDE systems, red lines show bifurcations of the DDE system (3.41), while bifurcations for the DDE system (3.40) are shown in magenta

Figure 3.25 shows the comparisons of two-parameter bifurcation diagrams between the ODE and the DDE systems, one can easily find that, under similar parameter conditions, for $r_2 = 0.0265$, the oscillation regions of the DDE system (3.41) are obviously larger than the ODE system (3.38) and the DDE system (3.40), this illustrates that time delays on the degradations can enhance or induce oscillation behaviors [13]; but oscillation region for the DDE system (3.40) is the smallest, which illustrates that time delays on transcriptions can shrink oscillation region, and further demonstrates that whether time delay can induce oscillations is context dependent; while for bistable regions, time delays on the degradations can shrink bistable regions, while delays on transcriptions can enlarge this region, which is an interesting phenomenon. However, the regions with three steady states are the same under the un-delayed and the delayed cases, which demonstrates that time delays cannot change the numbers of steady states, although the stability characteristic can be altered by time delays. For $r_2 = 0.2$, Fig. 3.25b shows that the two kinds of time delays can both induce oscillation, and the oscillation region for the DDE system (3.40) is broader than that for the DDE system (3.41).

Figure 3.26 shows the oscillation periods versus the varying of parameters α_2, α_3 under $r_2 = 0.0265$ and $r_2 = 0.2$. α_3 is fixed at 0.5 in panels a and c, while in panels b and d, α_2 is fixed at 2. From these four panels, one can see that delays in the degradation processes can effectively shorten the oscillation periods, while the transcription delays can lengthen the oscillation periods. One can also see that

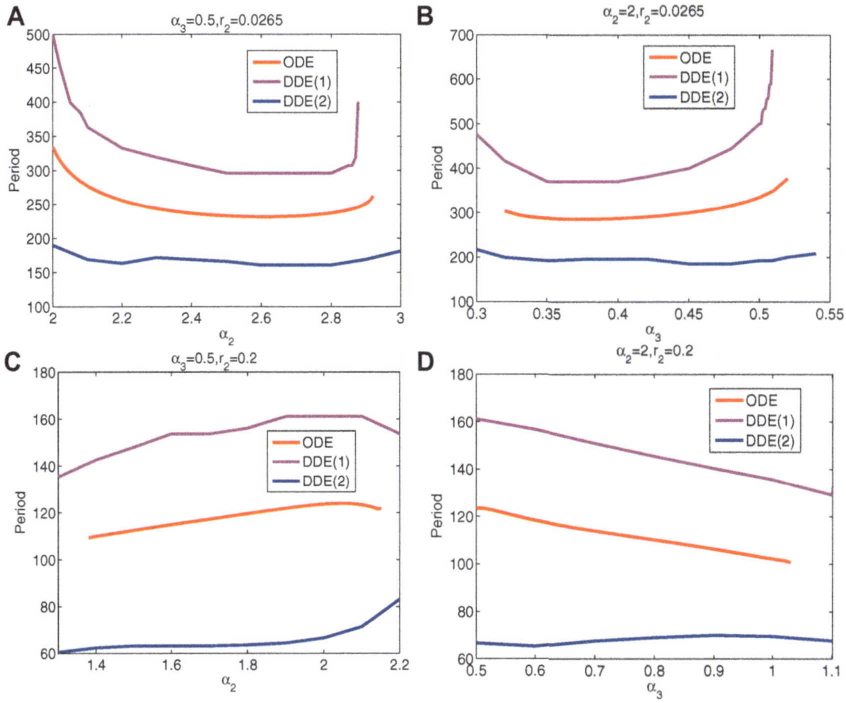

Fig. 3.26 Periods versus α_2, α_3 under $r_2 = 0.0265, r_2 = 0.2$, respectively. For the DDE systems, time delays are still fixed as $\tau_1 = 4, \tau_2 = 2, \tau_3 = 4$. Here, DDE(1) represents Eq. (3.40), and DDE(2) corresponds to Eq. (3.41)

for $\alpha_3 = 0.5, r_2 = 0.0265$, the system can oscillate in $\alpha_2 \in [2, 2.9]$, and the oscillation periods will first decrease and then increase with the increasing of α_2, for $\alpha_2 = 2, r_2 = 0.0265$, the system can oscillate in $\alpha_3 \in [0.32, 0.52]$, and the curve of the oscillation periods versus α_3 has similar behavior as panel A. For the case of $r_2 = 0.2$, the oscillation periods increase with α_2; but as to α_3, the periods almost linearly decrease with α_3. In fact, the phenomenon in panels C and D can be explained through the functional roles of the positive and negative feedbacks [122], the positive feedback can effectively stabilize a system, and the negative one can destabilize a system, for $r_2 = 0.2$, with the increasing of the positive feedback strength α_2, the oscillation periods become longer and longer, which indicates the system approximate to be stabilized; and with the increasing of the negative feedback strength α_3, the oscillation behaviors are enhanced, and therefore indicate the destabilizing role of the negative feedbacks. But for $r_2 = 0.0265$, no such conclusions can be observed, we guess that the mixture interactions of the positive and the negative feedbacks, the basal synthesis level as well as the time delays that cause this abnormal phenomenon.

3.5.3.2 Molecular Noise

System size Ω governs the size of fluctuations in the reaction systems [26], which has been discussed in many books and the references therein [26]. The Fokker–Planck approximation of the corresponding CMEs is just derived from the Kramers–Moyal expansion or the so-called Ω expansion. Figs. 3.27 and 3.28a–c show the stochastic simulations of the undeveloped model both without delays and with delays under different system volumes, where $\Omega = 1000, 100, 10$ are considered. From Figs. 3.27 and 3.28, we find that for small system volumes, noise in the reaction systems makes the evolution of protein numbers fluctuate abruptly, and the corresponding phase portraits are composed by messy dots. Similar conclusions have been derived by Gonze and colleagues [19, 115]. To quantitatively

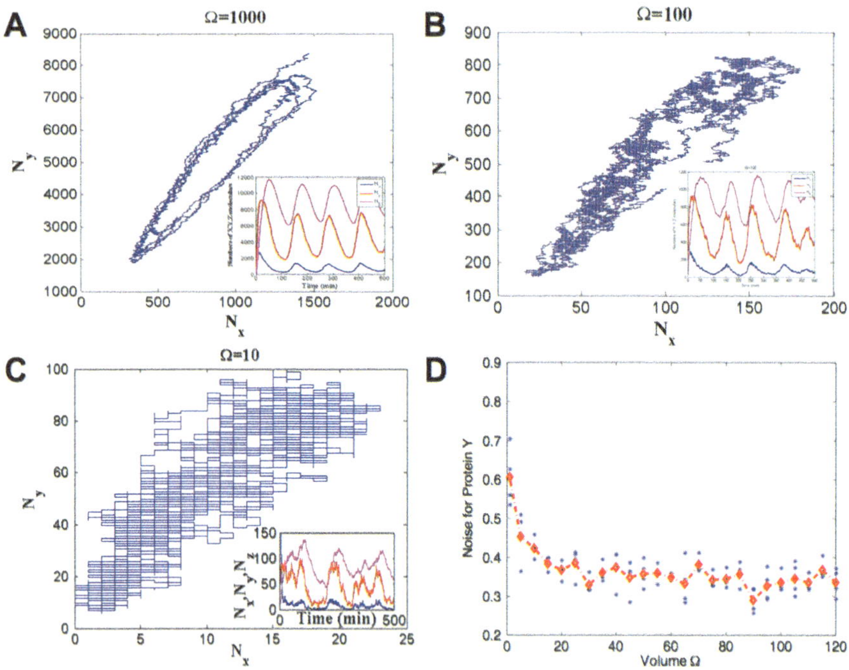

Fig. 3.27 Time evolutions as well as phase portraits in the $x-y$ plane for the protein numbers with the stochastic model that is shown in Table 3.10. Here, the system volumes are taken as $\Omega = 1000$ (**a**), 100 (**b**), 10 (**c**), respectively. Panel (**d**) shows the molecular noise versus the system volume

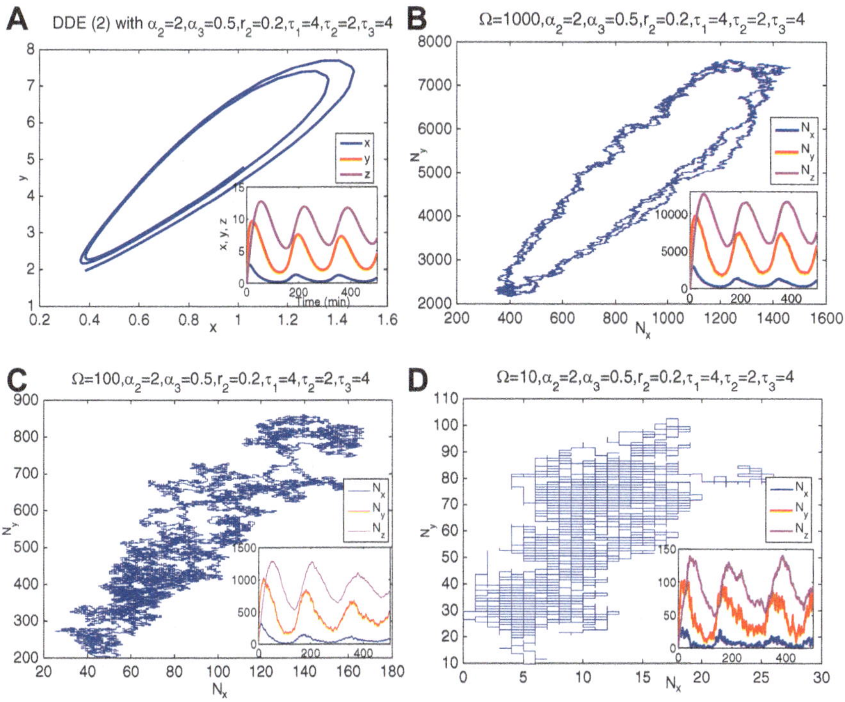

Fig. 3.28 Deterministic (**a**) as well as stochastic time evolutions, phase portraits in the **x-y** plane for the protein numbers with time delayed transcription $\boldsymbol{\tau} = (\mathbf{4, 2, 4})$. The system volumes are taken as $\Omega = 1000$ (**b**), 100 (**c**), 10 (**d**), respectively

describe this fact, just like in Ref. [123], we define noise of a stochastic variable N_x as

$$\eta = \frac{\sqrt{< (N_x - < N_x >)^2 >}}{< N_x >}, \tag{3.42}$$

here, $< \, . \, >$ denotes the time average. Figure 3.27d shows the noise η versus the system volume Ω for the stochastic models without delay, where the initial molecular numbers are taken as $(5\Omega, 0, 0)^T$, and parameters $\alpha_2 = 2, \alpha_3 = 0.5, r_2 = 0.2$. From this figure, one can easily derive the conclusion that, the molecular noise decreases with the increasing of the system volumes.

Figure 3.29 shows the stochastic simulation results for the developed stochastic models with the system volume $\Omega = 100$. Since there are 23 reaction channels and 10 species (due to the conservation laws, we only consider 10 of total 13 species in Table 3.11), the stochastic simulations for the developed stochastic models are very time consuming (hundreds or even thousands of hours are needed, which depends on Ω), therefore, we only show the case for $\Omega = 100$. For delayed

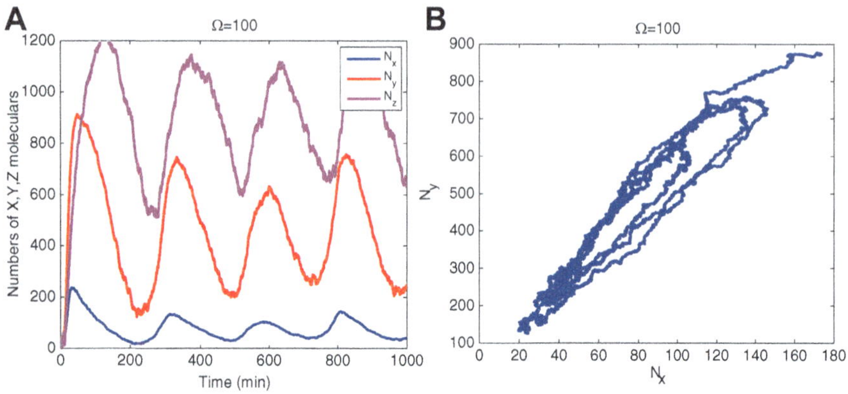

Fig. 3.29 Time evolutions (**a**) as well as phase portraits (**b**) in the **x-y** plane for the protein numbers from the developed stochastic model without delay. A set of parameters are randomly chosen and listed in Table 3.11, the system volumes are taken as $\Omega = 100$

models, short time stochastic simulation results show that similar conclusions can be derived as the un-delayed cases, we omit the detailed discussions and figures here. From Fig. 3.29, we see that the developed model can also display oscillation behavior similar to its undeveloped case and the simplified deterministic model, but it seems that the oscillation period for the developed model becomes longer, about twofold of the undeveloped case, which is mainly because that the undeveloped case and the deterministic models are based on the quasi equilibrium approximation of the developed case. Since the developed case and the undeveloped case can both display similar oscillation behaviors, and exhaustive computation time is needed for the developed case. In the following discussion, we mainly consider the relations between the deterministic models and its undeveloped stochastic counterparts, and special attentions are paid to the differences on the dynamical behaviors for the two classes of models.

3.5.3.3 Deterministic Versus Stochastic Dynamics for Parameters Near the Deterministic Bifurcation Points

We consider the effect of intrinsic noise inherently in the stochastic simulations on system dynamics near the bifurcation points, when we fixed $r_2 = 0.0265, \alpha_2 = 2.865872, \alpha_3 = 0.5$, there is a HB point for the system (3.38), Fig. 3.38 shows the deterministic as well as the stochastic simulation results for this case; When $\alpha_2 = 2.012, \alpha_3 = 0.5$, there is a SN point for the system (3.38), the simulation results are also shown in Fig. 3.30. For parameters locate near the HB point, the deterministic and the stochastic simulation results both show similar oscillation behaviors, and since there is only one steady state for the deterministic system (3.38), probability distribution of the protein number under stochastic simulation

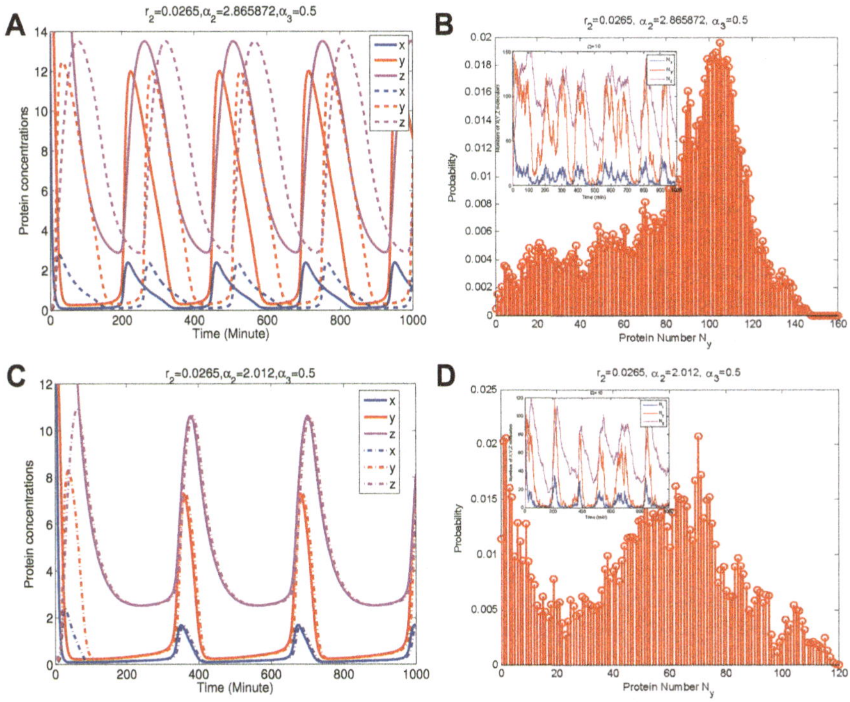

Fig. 3.30 Deterministic (**a**, **c**) versus stochastic simulations (**b**, **d**) for parameters near the bifurcation points. Where $\alpha_2 = 2.865872, \alpha_3 = 0.5$ for panels A and B, and $\alpha_2 = 2.012, \alpha_3 = 0.5$ for panels (**c**) and (**d**), for both cases, $r_2 = 0.0265$. The system volume for the stochastic models are taken as 10, the inset figures of the second and fourth panels show the time evolutions of the stochastic simulations, while the outside ones show distributions of the molecular numbers of protein Y

shows unimodal distribution. Furthermore, the deterministic oscillation curves near a HB point show dislocations of peaks under different initial values. While near the SN point, peak dislocations are not obvious. Moreover, the stochastic model shows shorter oscillation period than the deterministic model, and the probability distribution of the molecular numbers shows bimodal distribution, which is because that for $\alpha_2 = 2.012, \alpha_3 = 0.5$, there are three steady states. This can be explained as intrinsic noise-induced bimodality without deterministic bistability, very similar to the experimental observation in Ref. [124]. The related investigations can also be provided as another counterexample to the common thinking that bimodal population distribution can predict bistable gene expression in the deterministic model [125].

3.5.3.4 Deterministic Versus Stochastic Dynamics for Parameters Locating in the Deterministic Excitable Region

Excitability has been observed in a wide range of natural systems, such as lasers, chemical reactions, ion channels, neural systems, cardiovascular tissues and climate dynamics [126, 127], as well as gene regulatory circuits [110, 128, 129]. Common features to all of these excitable systems are in that there are three steady states, one is a "rest" state or vegetative growth state in gene regulatory circuits that dominate cell differentiation, which is stable; the other two states are unstable, and called "excited" state and "refractory" state (correspondingly, saddle and competence states in [110, 128]).

From Fig. 3.25, for $r_2 = 0.0265$, there is an excitable region labeled as region II. When $\alpha_2 = 1.7, \alpha_3 = 0.3$, system (3.38) is excitable, there are just three steady states for the system, one is stable, the other two are unstable. To further investigate this case, we explore system (3.38) analytically. Let the right hand side of system (3.38) to be zero and we consider its steady states:

$$\begin{cases} \frac{\alpha_1 y^2}{1+y^2+z^2} - d_1 x + r_1 = 0, \\ \frac{\alpha_2 x^2}{1+x^2} - d_2 y + r_2 = 0, \\ \frac{\alpha_3 x^2}{1+x^2} - d_3 z + r_3 = 0. \end{cases} \tag{3.43}$$

From Eq. (3.43), one has

$$y = \frac{1}{d_2}\left(\frac{\alpha_2 x^2}{1+x^2} + r_2\right); \; z = \frac{1}{d_3}\left(\frac{\alpha_3 x^2}{1+x^2} + r_3\right).$$

Substitute y, z into $f(x) = \frac{\alpha_1 y^2}{1+y^2+z^2} - d_1 x + r_1$, we have

$$f(x) = \frac{\alpha_1 d_3^2[(\alpha_2 + r_2)x^2 + r_2]^2}{d_2^2 d_3^2 (1+x^2)^2 + d_3^2[(\alpha_2 + r_2)x^2 + r_2]^2 + d_2^2[(\alpha_3 + r_3)x^2 + r_3]^2} - d_1 x + r_1.$$

The roots of equation $f(x) = 0$ are the steady state x values. To derive the roots, one has to solve the roots of the quintic function, it is difficult to be derived analytically, but one can obtain them numerically. From Fig. 3.39, the inset figure of the first panel shows that $f(x)$ crosses the straight line $g(x) = 0$ for three times. Therefore, there are three steady states for system (3.38). When $\alpha_2 = 1.7, \alpha_3 = 0.3$. One can easily judge that among these three steady states, one is stable, the other two are unstable, as noted by the black dot and the hollow circles, respectively.

By using the software package XPPAUT, one can easily analyze types of these singular points. For the stable steady states $(0.1680, 0.36583, 2.3294)^T$, the corresponding linearized matrix has two complex eigenvalues with negative real part and one negative real eigenvalue. For the unstable steady state $(0.23589, 0.58055, 2.6325)^T$, the corresponding linearized matrix has one positive

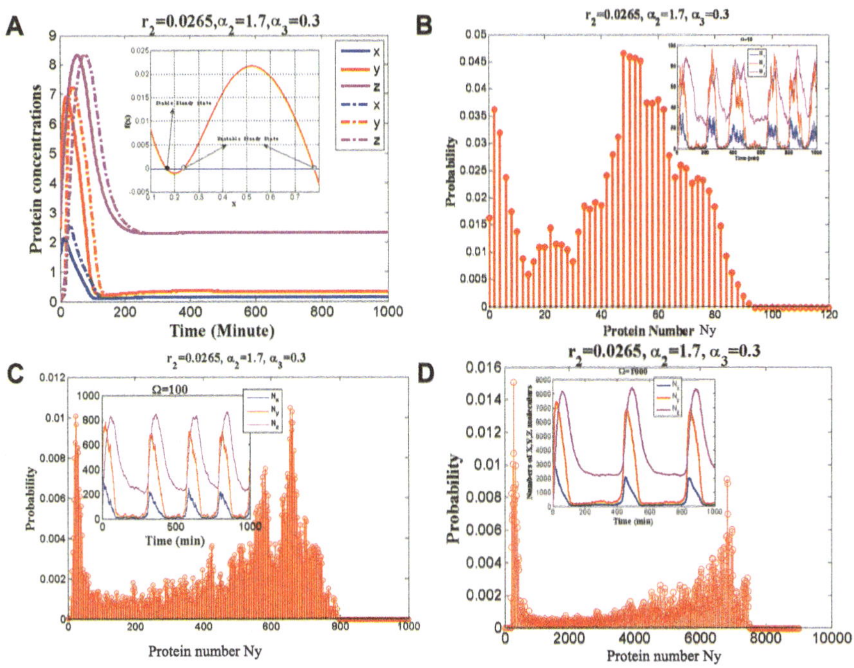

Fig. 3.31 Deterministic versus stochastic simulations for parameters locating in the deterministic excitable region. Here, $\alpha_2 = 1.7, \alpha_3 = 0.3$ and $r_2 = 0.0265$. (**a**) shows the deterministic time evolutions and three steady states for x, the stochastic time evolutions as well as the probability distributions of the molecular numbers for protein Y are shown in (**b–d**), where system volumes are taken as $\Omega = 10, 100, 1000$ respectively. ©[2011] IEEE. Reprinted, with permission, from ref. [21]

real eigenvalue and two negative ones. Therefore, it is a saddle node. For the second unstable steady state $(0.77928, 3.344, 6.5339)^T$, the corresponding linearized matrix has two complex eigenvalues with positive real part and one negative real eigenvalue, which is a saddle focus node.

From the stochastic simulation results of Fig. 3.31, we see that under different system volumes, stochastic oscillations can always be observed from the stochastic simulations. The probability distributions of the molecular numbers of protein Y all show bimodal distributions, which can be also provided as an example to illustrate that noise can induce bimodal population distribution without deterministic bistability [124]; with the increasing of system volumes, stochastic oscillation periods become longer; bimodal distributions of the protein numbers demonstrate that the oscillation behavior for this excitable region has switchable feature [130], which is very similar to the so-called periodic switch phenomenon in [131]. From Ref. [110], this may be explained as intrinsic noise-induced stabilization of the unstable saddle state, deterministically, the saddle node steady state is unstable, and deterministic time evolution converges to its stable steady state, but as to the stochastic model,

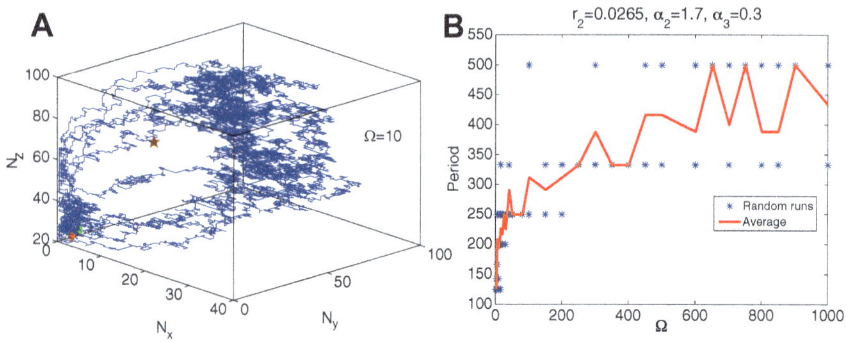

Fig. 3.32 The dwell time around the unstable saddle focus and the average switch periods versus system volumes. (**a**) The dwell time around the unstable saddle focus is considerable; (**b**) The average switch periods versus volumes for the stochastic model in Table 3.10 without delay. Here $\alpha_2 = 1.7, \alpha_3 = 0.3$, and $r_2 = 0.0265$

the intrinsic noise inherently in the stochastic model induce oscillations around the deterministic saddle focus node, and this oscillation increases the dwell time around this unstable steady state. Therefore, the unstable steady state is effectively stabilized. Figure 3.32a shows that the dwell time around the unstable saddle focus node is considerable for $\Omega = 10$, and Fig. 3.32b shows the periods of this switch-like oscillation versus the system volumes, one can see that the periods become longer and longer with the increasing of Ω.

3.5.3.5 Deterministic Versus Stochastic Dynamics for Parameters Locating in the Deterministic Bistable Region

From Fig. 3.25, for the case with $r_2 = 0.0265$, when parameters α_2, α_3 locate in region I, bistable behaviors can be onset. We take $\alpha_2 = 1.2, \alpha_3 = 0.1$, then system (3.38) is bistable, Fig. 3.33 shows some deterministic as well as stochastic simulation results. Where, the nullcline for x is drawn inset the first figure, and the figure outside is drawn under two different sets of initial state values. From the deterministic result, for two different sets of initial states, two different steady states are found, while under the observation of the undeveloped stochastic simulation, stochastic switch between the two deterministic steady states is presented, but the switch behavior can be only observed when the system volume is not so large. We note that, when the system volume is small, the molecular noise becomes important in evolution, this noise can drive the system switch between its two steady states [132], while for large system volumes, the molecular noise is comparatively low, and is not high enough to guarantee switch.

Figure 3.33c shows the probability distribution for protein Y when the system volume is set to be 10, where the time evolutions of these proteins can switch between the high and the low states, alternatively. Figure 3.33d shows the case

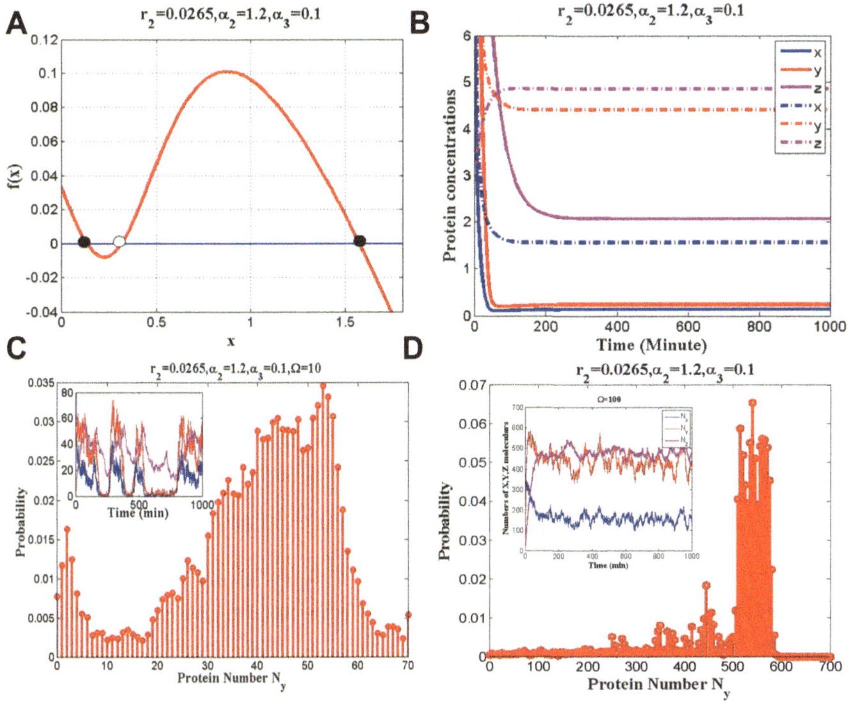

Fig. 3.33 Deterministic versus stochastic simulations for parameters locating in the bistable regions. Where $\alpha_2 = 1.2, \alpha_3 = 0.1$ and $r_2 = 0.0265$. Panels (**a**) and (**b**) show the nullclines for x and the deterministic time evolutions; Panels (**c**) and (**d**) show the probability distribution of the protein number N_y as well as the stochastic time evolutions for $\Omega = 10, 100$; where for $\Omega = 100$, bistable switch cannot be observed due to comparably low molecular noise. ©[2011] IEEE. Reprinted, with permission, from ref. [21]

for $\Omega = 100$, where the molecular numbers mainly fluctuate around one of its stable steady states. The stochastic switch behaviors can be observed as intrinsic noise-induced generation of new steady state, deterministically, system (3.38) can only converge to one of its steady states under a given initial condition, but due to intrinsic noise in stochastic models, a new steady state is generated and keeps the system transit between these two states [133].

As we have introduced, positive feedback has a role of stabilization, while negative feedback plays a role of destabilization, therefore, it is interesting to investigate the role of negative feedback in this bistable region. To measure the effect of negative feedback on the robustness of steady state, we introduce the concept of the first passage time [134–136]. The first passage time is defined as the time where the system first leaves a given domain. Figure 3.34 shows the first passage time for protein Y from the high state to the low steady state of the bistable switch versus negative feedback strength, where we note Y_{on} as the high state, and Y_{off} as the low state. Figure 3.34a shows the results for the un-delayed case, from

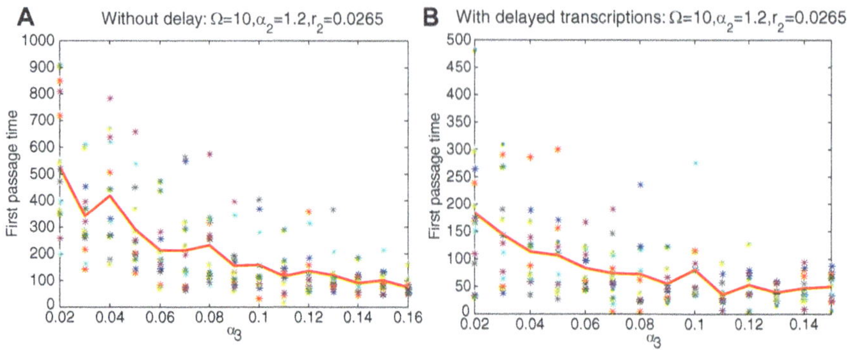

Fig. 3.34 The first passage time from Y_{On} to Y_{Off} of protein Y with respect to the negative feedback strength. The dots represent statistics data from the random stochastic simulations. A shows the case without delays, while panel B shows the case with degradation delays $\tau_1 = 4$, $\tau_2 = 2$, $\tau_3 = 4$. Here $\Omega = 10, \alpha_2 = 1.2, r_2 = 0.0265$, initial values for all simulations are taken as $(50, 0, 0)^T$

this panel, one can derive the conclusion that relatively strong negative feedback can effectively enforce the first passage of Y from its "on" to "off" state. This indicates that the stability of the high steady state becomes less and less robust to noise with the increasing of the negative feedback strength. It also indicates that the first passage time of a bistable switch can be controlled by an additional negative feedback. Figure 3.34b shows the cases for the stochastic models with degradation delays $\tau_1 = 4$, $\tau_2 = 2$, $\tau_3 = 4$, where similar conclusions can be derived as the un-delayed case. Moreover, compared between Fig. 3.34a, b, one can easily find that the first passage time for models with degradation delays is relative shorter than the un-delayed case, which is mainly due to the destabilization role of the time delays.

3.5.3.6 Deterministic Versus Stochastic Dynamics for Parameters Locating in the Deterministic Oscillation Region

To further investigate the dynamical behaviors for parameters locating in region VI, we take $\alpha_2 = 1.96, \alpha_3 = 0.47$, the nullcline for x is shown in Fig. 3.35a. Three steady states are all unstable, the state $(0.24514, 0.68802, 3.0657)^T$ has one positive and two negative eigenvalues, therefore, it is a saddle; the state $(0.1958, 0.49434, 2.6941)^T$ corresponds to two complex eigenvalues with positive real parts and one negative real eigenvalue; the state $(0.41585, 1.5774, 4.7718)^T$ has two positive real eigenvalues and one negative real one, correspondingly. From Fig. 3.35b, the deterministic time evolutions under different initial values show phase dislocations.

Figure 3.35c, d show the stochastic simulation results for $\Omega = 10, 100$. As to the stochastic simulations of the stochastic models, the stochastic oscillation behavior for small system volume has shorter oscillation period, with the increasing of the

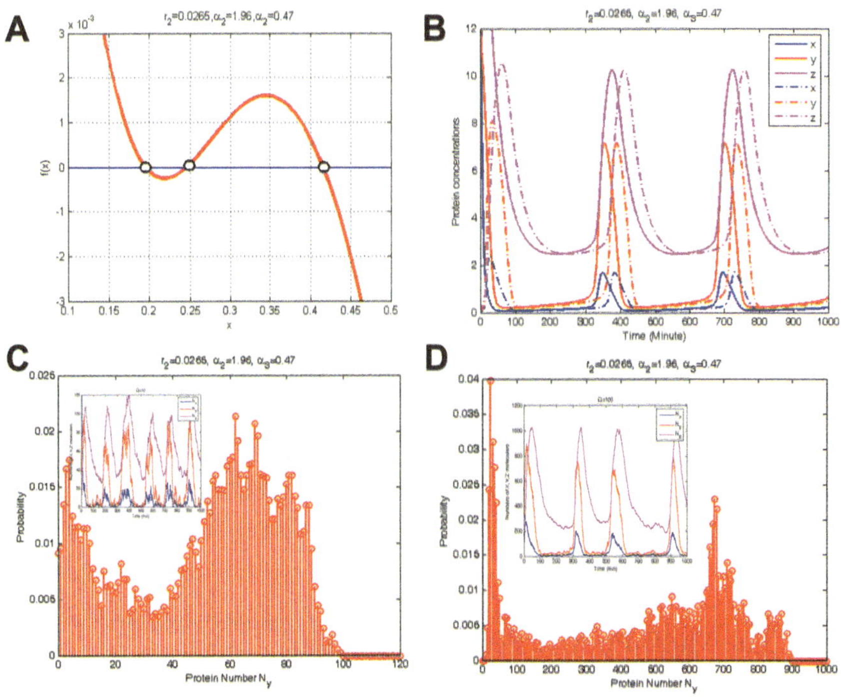

Fig. 3.35 Deterministic versus stochastic simulations for parameters locating in the oscillation region. Where $\alpha_2 = 1.96, \alpha_3 = 0.47$ and $r_2 = 0.0265$. (**a**) The nullcline for x; (**b**) The deterministic time evolutions, and the stochastic simulation results with system volume $\Omega = 10, 100$ are shown in panels (**c**) and (**d**), the probability distributions of the protein number N_y are shown outside and the time evolutions are shown as inset figures

system volume, the stochastic time evolutions approach the deterministic system better and better.

3.5.4 Summary

GRNs are research focus in the field of systems biology. These networks can be modeled both deterministically and stochastically. It is important yet difficult to establish biological reasonable mathematical models to simulate these circuits. In this section, we investigate a three-component CPNFGC, this circuit is firstly modeled deterministically by the Hill kinetics, and then the corresponding stochastic models have also been investigated based on the Gellispie's stochastic simulation method. Especially, a new stochastic simulation method is proposed to simulate the stochastic models directly revised from the deterministic models.

Through bifurcation analysis on the deterministic systems, interesting dynamical behaviors that this genetic circuit can display including: (1) When the system is modeled deterministically, it can display mono-stability, bistability, excitability, oscillation under different system parameters; (2) When parameters locate in deterministically mono-stable and oscillation regions, roughly similar results can be derived from the deterministic and the stochastic models; (3) While for parameters locating in the deterministically bistable and excitable regions, the corresponding stochastic simulations show that, the intrinsic noise inherently in the stochastic regime can induce bistable switch, periodic switch behavior, respectively. The onset of bistable switch can be seen as intrinsic noise-induced novel steady state and the transition between them, and periodic switch can be understood as intrinsic noise-induced stabilization of another unstable steady states and transitions; (4) When time delays are introduced in these two models, similar conclusions can be drawn as the un-delayed cases. The investigations provide a clear understanding of the relationships between different modeling regimes on this genetic circuit and others therein, potential real-world applications of the investigations are the engineering of synthetic circuits and so on.

3.6 The Multi-Positive Feedback Circuits

Multiple-positive feedback circuits are ubiquitous regulatory motifs in complex bio-molecular networks. A popular topic is why multiple-positive feedback mechanisms have been evolved and selected by organisms. To this end, a two-component dual-positive feedback genetic circuit is investigated, which consists of an auto-activation loop and a double negative feedback circuit. The auto-activation loop acts as an additional positive feedback loop (APFL), and our aim is to explore the functional characteristics of the APFL. Investigations reveal that the APFL can regulate the size of bistable region and the robust attractiveness of stable steady states (SSSs). It is also found that the APFL can regulate the GRIOS of the system. Furthermore, the APFL can tune the response speed, noise resistance, and stochastic switch behavior of the system, which makes it easy to realize functional tunability and robust decision-making. Therefore, rationalizing why multiple-positive feedback circuits so frequently appear in real-world biological systems. Potential applications of the associated investigations include the design of artificial genetic circuits, the modeling and model reduction for large-scale bio-molecular networks. This section is mainly based on our work [137] published in the year 2015.

3.6.1 Related Works and Motivations

Genetic circuits with special dynamical behaviors have been investigated extensively in the field of systems biology and synthetic biology, which can exhibit bistable switch [1, 2, 138–140], periodic switch [130], oscillation [103–109], excitability [128, 129], or even chaotic dynamics [94, 141–144]. It is well known that positive feedback with certain ultrasensitivity can result in multi-stability and hysteresis [89, 90, 145]. Positive feedback can act as a signal amplifier [146], a buffer for propagated noise improves the sensitivity of the system without a compromise in the ability to buffer propagated noise [147], and more interestingly, it can act as a biological memory module [89, 92], control the onset of autosomal dominant diseases, such as mature onset diabetes of youth and autosomal dominant polycystic kidney disease [93].

Although one positive feedback loop can already generate bistability, real-world functional biological circuits are always coupled by APFLs [106, 107, 111, 140, 148–153], such as the circuits that control mitotic trigger, *S. Cerevisiae* galactose regulation, p53 regulation, muscle cell fate specification, polarization of *yeast* cells [106]. Therefore, recently, some coupled positive feedback genetic circuits have been extensively investigated [106, 107, 111, 140, 148–153]. For example, in 2005, Brandman et al. investigated an interlinked fast and slow positive feedback loop, under mono-stable dynamics, researchers found that the response of the system became faster, at the same time, the system could effectively resist to noise in the upstream signaling system, therefore, the signal system became more robust [106]. Following the work [106], Zhang and coworkers [153] investigated the corresponding bistable switch behaviors in the similar positive feedback loop, and they found that the dual-time positive feedback loop could establish a balance between stimulus sensitivities and responsive noise-resistant. Smolen and coauthors [100] also discussed the similar circuits as that in [106] and [153], but the authors have paid more attention to internal noise resistance of the system. In 2009, through bifurcation analysis, Sriram et al. [92] investigated a mutual inhibition, a mutual activation and an interlocked mutual inhibition and activation circuit, and they found that the interlocked mutual inhibition and activation circuit has advantages in that, the mutual inhibition loop controls the switch behavior, while the mutual activation loop can enforce decision-making. Recently, Shi et al. [151, 152] have investigated the functional tunability and robustness of the multi-positive feedback circuits, where, in [151], the authors have considered the advantages of an auto-activation loop coupled by an additional toggle switch [20].

However, existing references [106, 153] are mainly concerned about the advantages of dual-time positive feedback loop over single fast or slow positive feedback loop in signal transduction pathways, the effect of the APFL strength on the robustness and tunability of bistable switch and signal processing were seldom considered, especially in genetic regulatory circuits. Thus, it is fascinating to explore performances of the APFL strength in multi-positive feedback genetic circuits.

Motivated by the abovementioned questions, and take a double negative circuit as a prototype, which is actually the well-known synthetic genetic circuit—the toggle switch [20], as shown in Fig. 3.36a. We add an auto-activation loop to the toggle switch system, and derive the representative two-component dual-positive feedback circuit, as shown in Fig. 3.36b. In Fig. 3.36c, S is an input stimulus, which can be seen as an inducer. The product of gene X acts as the output. We call the auto-

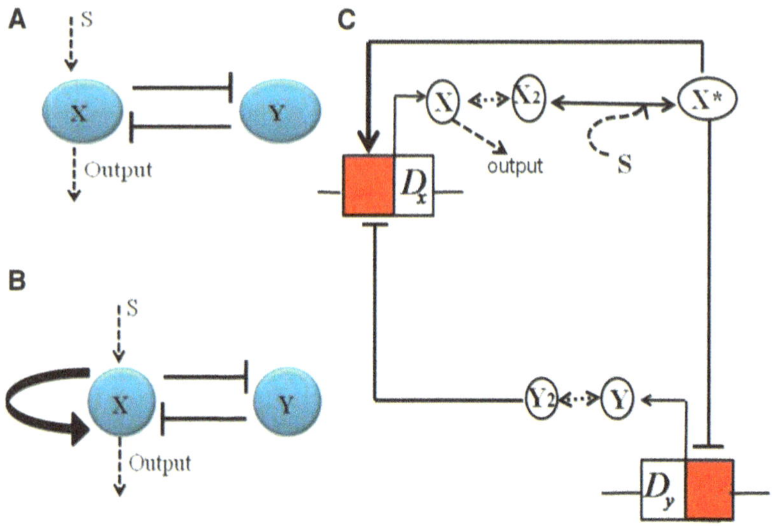

Fig. 3.36 Structures of the considered positive feedback circuits. (**a**) The negative–negative feedback genetic circuit: NN. (**b**) The negative–negative feedback genetic circuit with and APFL: NNP. (**c**) Detailed regulation processes in the NNP. Where S represents the input stimulus, protein X acts as the output. D_x, D_y are free promoter binding sites for genes X and Y. Reprinted by permission from Springer, ref. [137]

activation loop as the APFL, denote the double negative feedback circuit as NN, and denote the NN with an APFL as NNP. It is noted that the NNP is a representative multi-positive feedback circuit. For example, the NNP can be seen as a functional counterpart of the three-component interlocked mutual inhibition and activation circuits [92].

3.6.2 Mathematical Models

For the circuit NNP, D_x, D_y are free promoter binding sites for genes or proteins X and Y. X_2 and Y_2 are protein dimers of X and Y. We suppose the inducer S can catalyze the dimer X_2 into its active form X^*, and then X^* acts as a TF (Fig. 3.36c). We further assume that the regulation of gene expression is mediated by the protein dimers X^* and Y_2. Since X^* and Y_2 can both regulate the expression of gene X, one assumes that the TFs X^* and Y_2 competitively bind [150] to the gene X, i.e., the binding of one type of dimer excludes the binding of the other type. Detailed biochemical reactions for the NNP are summarized in Table 3.12, where D_x^T, D_y^T denote the total promoter binding sites for genes X and Y, which are assumed to be constants. $K_i = k_i^+/k_i^-(i = 1, 2, \cdots, 6)$ denote the dissociation constants, k_i^+, k_i^- are the forward and the backward reaction rates for the $i'th$ revisable reaction.

Table 3.12 Biochemical reactions, conservation laws and related parameters in the NNP circuit

Fast reactions	Reaction rates	Dissociation constants
$X + X \rightleftharpoons X_2$	k_1^+, k_1^-	$K_1 = k_1^+/k_1^-$
$X_2 + S \rightleftharpoons X^*$	k_2^+, k_2^-	$K_2 = k_2^+/k_2^-$
$X^* + D_y \rightleftharpoons D_yX^*$	k_3^+, k_3^-	$K_3 = k_3^+/k_3^-$
$Y + Y \rightleftharpoons Y_2$	k_4^+, k_4^-	$K_4 = k_4^+/k_4^-$
$Y_2 + D_x \rightleftharpoons D_xY_2$	k_5^+, k_5^-	$K_5 = k_5^+/k_5^-$
$D_x + X^* \rightleftharpoons D_xX^*$	k_6^+, k_6^-	$K_6 = k_6^+/k_6^-$
Slow reactions		Reaction rates
$D_x \rightarrow D_x + X$		t_1
$D_y \rightarrow D_y + Y$		t_2
$D_xX^* \rightarrow D_xX^* + X$		t_3
$X \rightarrow \phi$		d_X
$Y \rightarrow \phi$		d_Y
Conservation laws		
$[D_x] + [D_xX^*] + [D_xY_2] = [D_x^T]$		
$[D_y] + [D_yX^*] = [D_y^T]$		

$t_i (i = 1, 2, 3)$ are the transcription rates. d_X, d_Y denote the degradation rates. $[M]$ represents the concentration of species M.

Generally speaking, on the one hand, the synthesis reactions, decomposition reactions, and dimerization reactions are much faster than the translation and degradation reactions. On the other hand, the mRNA life time is usually in the time range of minutes, whereas the protein life time is in the range of hours, thus, the dynamics of mRNA molecules are much faster than those of proteins. Therefore, in the following, we assume the synthesis reactions, the decomposition reactions, and the dimerization reactions quickly reach chemical equilibrium, and we ignore the equations for mRNAs. By the law of mass action, and combined with the conservation equations, dynamical model to describe the concentration evolution of proteins X and Y in the NNP can be easily established. Actually, from Table 3.12 and by the law of mass action, differential equations to describe the concentration evolutions of species in the fast equations can be obtained as follows:

$$\begin{cases} \frac{d[X_2]}{dt} = k_1^+[X]^2 - k_1^-[X_2] - k_2^+[X_2][S] + k_2^-[X^*], \\ \frac{d[X^*]}{dt} = k_2^+[X_2][S] - k_2^-[X^*] - k_3^+[X^*][D_y] + k_3^-[D_yX^*] \\ -k_6^+[X^*][D_x] + k_6^-[D_xX^*], \\ \frac{d[D_yX^*]}{dt} = k_3^+[X^*][D_y] - k_3^-[D_yX^*], \\ \frac{d[Y_2]}{dt} = k_4^+[Y]^2 - k_4^-[Y_2] - k_5^+[Y_2][D_x] + k_5^-[D_xY_2], \\ \frac{d[D_xY_2]}{dt} = k_5^+[Y_2][D_x] - k_5^-[D_xY_2], \\ \frac{d[D_xX^*]}{dt} = k_6^+[X^*][D_x] - k_6^-[D_xX^*]. \end{cases} \tag{3.44}$$

Assume fast reactions can quickly reach their chemical equilibrium, one has $[X_2] = K_1[X]^2, [X^*] = K_2[X_2][S], [D_yX^*] = K_3[X^*][D_y], [Y_2] = K_4[Y]^2, [D_xY_2] = K_5[Y_2][D_x], [D_xX^*] = K_6[X^*][D_x]$. Further considering the conservation laws: $[D_x]+[D_xX^*]+[D_xY_2] = [D_x^T]$ and $[D_y]+[D_yX^*] = [D_y^T]$, one derives

$$[D_x] = \frac{[D_x^T]}{1 + K_4K_5[Y]^2 + K_1K_2K_6[S][X]^2}, \tag{3.45}$$

$$[D_xX^*] = \frac{K_1K_2K_6[S][X]^2[D_x^T]}{1 + K_4K_5[Y]^2 + K_1K_2K_6[S][X]^2}, \tag{3.46}$$

$$[D_y] = \frac{[D_y^T]}{1 + K_1K_2K_3[S][X]^2}. \tag{3.47}$$

From the slow reactions, one has

$$\begin{cases} \frac{d[X]}{dt} = t_1[D_x] + t_2[D_xX^*] - d_X[X], \\ \frac{d[Y]}{dt} = t_3[D_y] - d_Y[Y]. \end{cases} \tag{3.48}$$

Replace $[D_x]$, $[D_xX^*]$, $[D_y]$ in Eq. (3.48) with Eqs. (3.45)–(3.47), thus one derives Eq. (3.49).

$$\begin{cases} \frac{d[X]}{dt} = \frac{t_1[D_x^T]+t_3[D_x^T]K_1K_2K_6[S][X]^2}{1+K_4K_5[Y]^2+K_1K_2K_6[S][X]^2} - d_X[X], \\ \frac{d[Y]}{dt} = \frac{t_2[D_y^T]}{1+K_1K_2K_3[S][X]^2} - d_Y[Y]. \end{cases} \tag{3.49}$$

Let

$$\sqrt{K_1K_2K_3}[X] = x, \sqrt{K_4K_5}[Y] = y,$$

$$[S] = s, \sqrt{K_1K_2K_3t} \to t,$$

$$t_1[D_x^T] = \alpha_1, t_3[D_x^T] = \alpha_3, t_2[D_y^T] = \alpha_2,$$

$$K_6/K_3 = \theta, \sqrt{(K_4K_5)/(K_1K_2K_3)} = \varepsilon,$$

$$d_X/\sqrt{K_1K_2K_3} = d_1, d_Y/\sqrt{K_4K_5} = d_2, \alpha_3\theta = \eta,$$

then, one can derive the simplified mathematical model for the NNP, which is described as

$$\begin{cases} \frac{dx}{dt} &= \frac{\alpha_1+\eta s x^2}{1+\theta s x^2+y^2} - d_1 x, \\ \frac{dy}{dt} &= \varepsilon\left(\frac{\alpha_2}{1+sx^2} - d_2 y\right). \end{cases} \tag{3.50}$$

Set $K_6 = 0$, $t_2 = 0$ in Table 3.12 and Eq. (3.49), or equivalently, set $\theta = 0$, $\eta = 0$ in Eq. (3.50), one derives the dimensionless model for the NN:

$$\begin{cases} \frac{dx}{dt} &= \frac{\alpha_1}{1+y^2} - d_1 x, \\ \frac{dy}{dt} &= \varepsilon\left(\frac{\alpha_2}{1+sx^2} - d_2 y\right). \end{cases} \tag{3.51}$$

For simplicity, Eqs. (3.50) and (3.51) can be reorganized as the following form:

$$\begin{cases} \frac{dx}{dt} &= f(x, y, s) - d_1 x, \\ \frac{dy}{dt} &= \varepsilon(g(x, s) - d_2 y). \end{cases} \tag{3.52}$$

Where $f(x, y, s) = \alpha_1/(1 + y^2)$ for the NN, and $f(x, y, s) = (\alpha_1 + \eta s x^2)/(1 + \theta s x^2 + y^2)$ for the NNP, $g(x, s) = \alpha_2/(1 + sx^2)$ for both circuits.

In the dimensionless model (3.50), x, y represent the concentrations of the proteins X and Y, s represents the concentration of the input signal S. $\alpha_i (i = 1, 2)$ represent the dimensionless transcription rates. $d_i (i = 1, 2)$ denotes the degradation rates for proteins X, Y. It is noted that in Eq. (3.50), the term $f(x, y, s)$ is the tradeoff production rate of gene X, where the sub-item $\alpha_1/(1 + \theta s x^2 + y^2)$ corresponds to the repression of gene Y, it decreases with the increasing of y. The sub-item $\eta s x^2/(1+\theta s x^2+y^2)$ corresponds to the contribution of the auto-activation loop, it increases with the increasing of x. When $y >> x$ and the concentration of x is very low, $f(x, y, s)$ tends to be 0, that is, the expression of X tends to be totally inhibited by Y. When $x >> y$ and the concentration of y is very low, then $f(x, y, s)$ approximates to α_3, that is, the expression of X will be mainly controlled by the auto-activation loop.

In the mathematical model (3.50), parameters $\varepsilon, \theta, \eta$ are with interesting biological meanings. ε is proportional to the ratio between the dissociation constants for Y and X, it can reflect biological activities, or production and transportation efficiency of proteins X and Y. Thus, it can be seen as a timescale. For example, larger ε value indicates faster dissociation dynamics for Y than for X. θ controls the ratio of X^* binding to D_x and D_y, therefore, can be seen as whether there is an APFL. If $\theta = 0$ (namely $K_6 = 0$), then the NNP degenerates into the NN. η equals to $\alpha_3\theta$, since α_3 controls the production efficiency of the APFL, therefore, η can be seen as the APFL strength. Hereinafter, we take $d_1 = d_2 = 0.5, \eta = 1, \varepsilon = 1, s = 1$ as nominal parameter values, unless otherwise notated, and we will clarify the functional characteristics of the APFL strength.

3.6.3 Dynamical Analysis and Functions

3.6.3.1 The APFL Strength Can Tune the Size of the Bistable Region

Traditionally, the transcription rates or degradation rates are frequently taken as bifurcation parameters [106, 107, 111, 140, 148–153]. In what follows, we take α_1, α_2 and d_1, d_2 as two sets of bifurcation parameters, and consider the effect of the APFL strength η on bifurcation diagrams. To compare with the circuit NN, we also show the case with $\eta = 0, \theta = 0$. To investigate the effect of η, we choose several values for η in the two-parameter bifurcation diagrams, which include $\eta = 0, 0.1, 0.5, 1, 1.5, 2$.

Figure 3.37 shows two-parameter bifurcation diagrams for the related circuits. In Fig. 3.37a, we have taken α_1, α_2 as bifurcation parameters. The unbounded regions located in the acute angle areas can display bistable behavior. With the increasing of η, the size of bistable region for the NNP becomes larger and larger, which indicates that the APFL strength η can effectively tune the bistable region. The line corresponding to $\eta = 0, \theta = 0$ shows bifurcation diagram for the NN. The NN can display bistable behavior in quite a large region. However, the additional positive feedback loop in the NNP increases the tunability of the size of bistable region. Figure 3.37b shows the two-parameter diagram in the $d_1 - d_2$ plane. Here, we have fixed $\alpha_1 = 2, \alpha_2 = 5$. The bounded leaf-shaped regions denote bistable areas. From Fig. 3.37b, similar phenomenon can be observed as that in Fig. 3.37a, that is, the APFL strength η can effectively enlarge the bistable region. More intuitively, it also indicates us that the size of the bistable region for the NNP can be freely tuned by the APFL strength.

Positive feedbacks are empirically regarded as a prerequisite for bistability in genetic circuits [89, 145]. Biological systems can use positive feedback mechanisms to realize differentiation, memory and hysteresis and other functions. From the two-

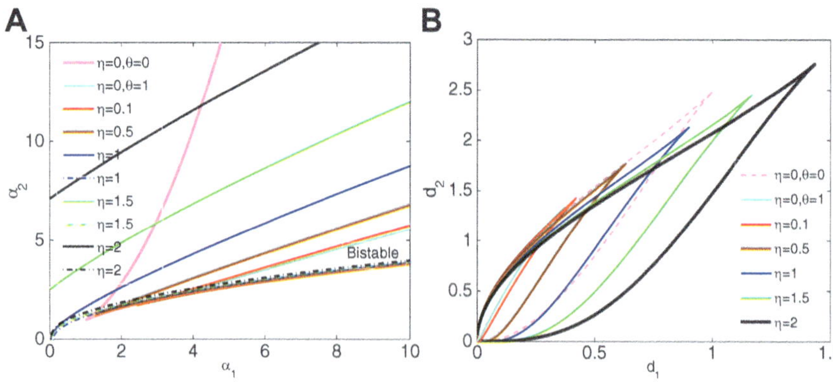

Fig. 3.37 Two-parameter bifurcation diagrams. (**a**) α_1, α_2 are bifurcation parameters. (**b**) d_1, d_2 are bifurcation parameters. Reprinted by permission from Springer, ref. [137]

parameter bifurcation diagrams for the dual-positive feedback genetic circuit, we realize that stronger APFL strength always indicates larger bistable areas. Therefore, the APFL can act as a tunable regulation component for the size of bistable region. The bistable region tunability of the APFL is crucial for decision-making of organisms during differentiation or cell communication.

3.6.3.2 The APFL Can Tune the Attractiveness of the Stable Steady States

Since the two genes in the NN repress each other's expression, it must be one at its high expression state, the other at its low state [1]. A natural arising question is what's the effect of the APFL strength on the attractiveness of stable steady states (SSSs) [154, 155]. To quantitatively investigate the problem, one sets $\alpha_1 = 5.2, \alpha_2 = 3$, and considers the APFL strength $\eta = 0, 0.1, 0.5, 1, 1.5, 2$. Similarly, when $\eta = 0, \theta = 0$, the NNP degenerates into the NN. With the above parameters, the circuits can all display bistable behaviors. One randomly takes 5000 cells with random initial values, then proportion of cells, which are ultimately attracted to the two steady states, are summarized in Table 3.13. One notes that similar conclusions can be derived under other parameters. The proportions can reveal the relative importance of the basin of attraction, as well as the robustness of the SSSs [154, 155].

To include the possible effect of initial values, we take three groups of the initial values $x(0), y(0)$ randomly from the intervals $[0, 2]$, $[0, 5]$ and $[0, 11]$. With the three different sets of initial values, there are two SSSs for the NN and NNP, which are labeled as SSS1 and SSS2 in Table 3.13. From Table 3.13, we see that values of the two SSSs for x, y are one at a high state, while the other at a low state.

Table 3.13 Attractiveness of the two SSSs for the NN and NNP. Here, $\alpha_1 = 5.2, \alpha_2 = 3$

Circuits	η	θ	SSSs	$[0, 2]$	$[0, 5]$	$[0, 11]$
NN	0	0	$SSS1 : (10.368, 0.055)$	96.22%	86.34%	84.28%
			$SSS2 : (0.353, 5.336)$	3.78%	13.66%	15.72%
NNP	0	1	$SSS1 : (1.643, 1.621)$	89.24%	80.00%	77.64%
			$SSS2 : (0.350, 5.344)$	10.76%	20.00%	22.36%
	0.1	...	$SSS1 : (1.766, 1.456)$	91.64%	82.46%	80.14%
			$SSS2 : (0.352, 5.339)$	8.36%	17.54%	19.86%
	0.5	...	$SSS1 : (2.208, 1.021)$	95.84%	87.48%	84.90%
			$SSS2 : (0.358, 5.319)$	4.16%	12.52%	15.10%
	1	...	$SSS1 : (2.803, 0.677)$	98.20%	89.20%	88.22%
			$SSS2 : (0.367, 5.291)$	1.80%	10.80%	11.78%
	1.5	...	$SSS1 : (3.503, 0.452)$	99.76%	91.90%	90.06%
			$SSS2 : (0.376, 5.256)$	0.24%	8.10%	9.94%
	2	...	$SSS1 : (4.306, 0.307)$	100.00%	93.64%	90.76%
			$SSS2 : (0.388, 5.215)$	0.00%	6.36%	9.24%

Furthermore, the APFL strength in the NNP can tune the attractiveness of the two SSSs. As one can see from Table 3.13, with the increasing of η, the amount of cells with high output state is significantly increased. Under the three sets of initial conditions, the high output state for the NN takes up $96.22\%, 86.34\%, 84.28\%$. For the NNP with $\eta \in [0, 2]$, the proportions range from 89.24% to 100% for $x(0), y(0) \in [0, 2]$, from 80.00% to 93.64% for $x(0), y(0) \in [0, 5]$, and from 77.64% to 90.76% for $x(0), y(0) \in [0, 11]$, respectively. The statistical results in Table 3.13 are a little different under different ranges of initial values. When $x(0), y(0)$ are randomly taken from $[0, 11]$, the proportions of the high output states are smaller than that $x(0), y(0) \in [0, 2]$ and $x(0), y(0) \in [0, 5]$, but similar statistical rules can be drawn.

The APFL can tune the attractiveness of the SSSs, which indicates that the APFL can increase flexibility and controllability of organisms. If one takes a steady state as a decision, then the data in Table 3.13 indicates that strong APFL strength can enhance decision-making, and promote robustness of the decision to the initial states. This is especially important for biological processes. For example, in cellular differentiation, this mechanism can be used to robustly promote the amount of certain functional cells.

3.6.3.3 The APFL Can Change the Global Relative I/O Sensitivities

In this section, one investigates the effect of the APFL on the GRIOS. By setting the right hand side of Eq. (3.52) to be zero, that is

$$\begin{cases} f(x^*, y^*, s) - d_1 x^* = 0, \\ g(x^*, s) - d_2 y^* = 0. \end{cases} \tag{3.53}$$

One can derive the steady states for system (3.52). By differentiating Eq. (3.53) with respect to s, one derives the sensitivity of the steady states x^*, y^* with respect to the input s:

$$\begin{cases} \frac{\partial x^*}{\partial s} = -\left(\frac{\partial f}{\partial s} + \frac{1}{d_2}\frac{\partial f}{\partial y^*}\frac{\partial g}{\partial s}\right)\left(\frac{\partial f}{\partial x^*} - d_1 + \frac{1}{d_2}\frac{\partial f}{\partial y^*}\frac{\partial g}{\partial x^*}\right)^{-1}; \\ \frac{\partial y^*}{\partial s} = \frac{1}{d_2}\left(\frac{\partial g}{\partial x^*}\frac{\partial x^*}{\partial s} + \frac{\partial g}{\partial s}\right). \end{cases} \tag{3.54}$$

Similar to the works in [47], the local relative I/O sensitivity can be computed as

$$\frac{\partial \ln x^*}{\partial \ln s} = \frac{s}{x^*}\frac{\partial x^*}{\partial s}.$$

By averaging over all the absolute values of the local relative I/O sensitivities under different s, one derives the so-called GRIOS [48]. For $s \in [0.01, 2]$, we show the GRIOS for the circuit NNP, which is depicted in Fig. 3.38, where, we have considered the cases with $\eta = 0, 0.1, 0.5, 1, 1.5, 2, 2.5, 3, 3.5$ and $\eta = 4$.

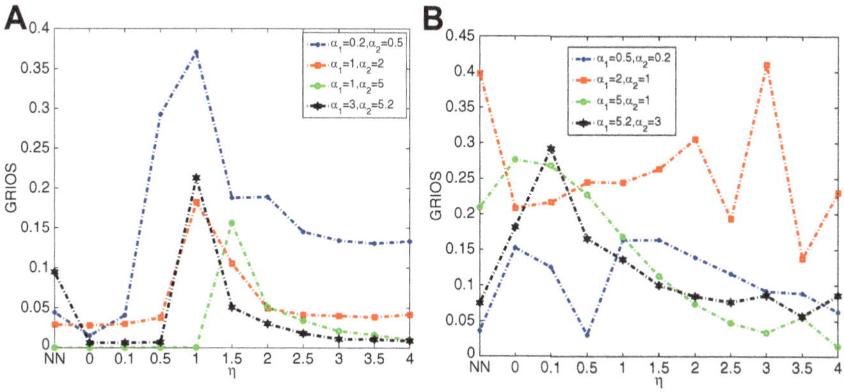

Fig. 3.38 Evolutions of the GRIOS under different APFL strength. Here, cases under eight sets of different α_1, α_2 values are shown. Reprinted by permission from Springer, ref. [137]

To investigate whether α_1, α_2 can affect our conclusions, we randomly choose eight sets of parameter values for α_1, α_2. The first four sets are $\alpha_1 = 0.2, \alpha_2 = 0.5$; $\alpha_1 = 1, \alpha_2 = 2$; $\alpha_1 = 1, \alpha_2 = 5$; $\alpha_1 = 3, \alpha_2 = 5.2$. By exchanging the values between α_1 and α_2, we derive the second four sets of parameters. The system can display bistable behaviors under some parameters, while it can display mono-stable behaviors under some other parameters. For bistable system, we average over the GRIOS for the two stable states as the GRIOS of the system (For bistable system, the trends of the GRIOS for the two SSSs have no much differences, data are not shown). It is noted that in Fig. 3.38, in order to compare with the cases without the APFL, one also considers the circuit NN. Steady states x^*, y^* are computed from Eqs. (3.51) and (3.52) with arbitrary random initial conditions.

From Fig. 3.38, roughly similar conclusions can be drawn under different sets of α_1, α_2 values. The NN has certain GRIOS. For the NNP, with the increasing of the APFL strength in the interval [0, 4], the GRIOS curves first increase and then decrease in Fig. 3.38a and for the case $\alpha_1 = 5.2, \alpha_2 = 3$ in Fig. 3.38b, while for the other cases in Fig. 3.38b, there exists optimal η, where the GRIOS can achieve its maximum value. Therefore, Fig. 3.38 indicates that there exists an optimal APFL strength η, under which, the NNP can achieve the highest GRIOS. More interestingly, the GRIOS of the NNP can be tuned through the APFL strength, to realize less or more global sensitive than the original circuit NN. This tunability increases the adaptability of the NNP to environmental stimulus. Therefore, from the perspective of the GRIOS, it is easy to understand why positive feedback circuits are always coupled by the APFLs.

3.6.3.4 Functional Characteristics of the APFL on Noisy Signal Processing

Living organisms evolve under noisy environments, therefore, their abilities to cope with noisy signals are critical for evolution and natural selection. It is very interesting to clarify advantages of the APFL strength on noisy signal processing.

To investigate the effect of the APFL strength on the response to noisy signal, one generates a noisy input s, as shown in Fig. 3.39a. We set $\alpha_1 = 5.2, \alpha_2 = 3$, and randomly take initial values for the system. In the following, when we consider the effect of the APFL strength, the timescale is taken as $\varepsilon = 1$, and cases for $\eta = 0, 0.1, 1, 2$ will be considered, while we consider the effect of timescale, we fix $\eta = 1$, and take $\varepsilon = 0.1, 1, 10, 100$ for investigation. Two response curves of the circuit under $\eta = 1, \varepsilon = 100$ and $\eta = 2, \varepsilon = 1$ are shown in Fig. 3.39b. Figure 3.39c, d show the normalized response curves for the NNP under different APFL strength and timescale, respectively, where, for the sake of easy comparison and observation, the outputs curves have been normalized through being divided by their steady state values.

From Fig. 3.39a, b, we can see that the magnitude of the input signal is about 1.5, while the magnitudes of the outputs for the two cases in Fig. 3.39b are all no more than 0.3, which indicates that the NNP tends to dampen the magnitude of the input

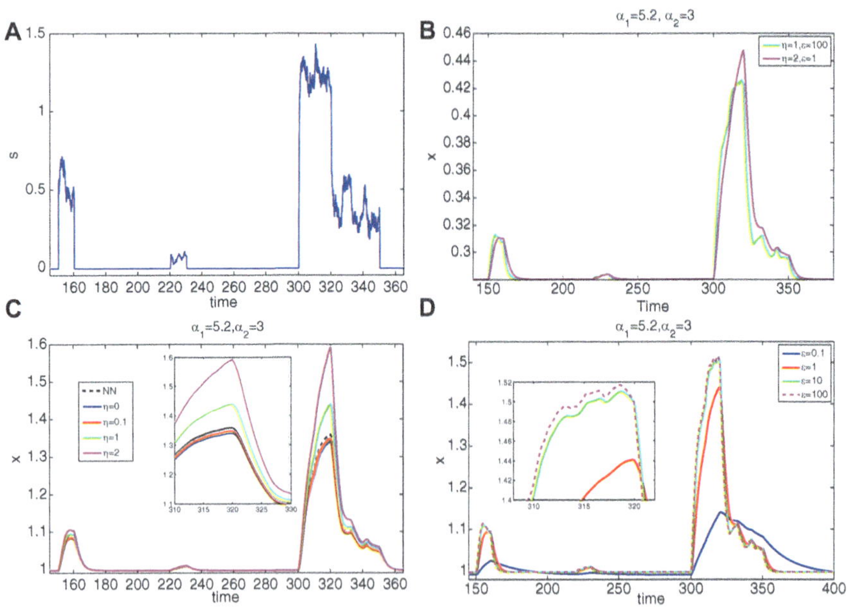

Fig. 3.39 Noisy signal and system response. (**a**) A noisy input signal. (**b**) Two response curves. (**c**) Normalized response curves of the circuit NNP to noisy signal under different APFL strength. (**d**) Normalized response curves of the circuit NNP to noisy signal under different timescale. Here, $\alpha_1 = 5.2, \alpha_2 = 3$. Reprinted by permission from Springer, ref. [137]

signal, large changes in s can lead to relatively smaller changes in the output. The two curves in Fig. 3.39b also indicate that large timescale can promote the response speed of the system. From Fig. 3.39c, d, we find that the NNP can detect weak noisy signal, the detection ability seems better with strong APFL strength and large timescale. Furthermore, the response of the NNP becomes rapider with stronger APFL strength and larger timescale. The curves in Fig. 3.39d also indicate that under slow timescale ε, the output of the NNP can better filter out fluctuations in the input, that is, the system becomes relatively more robust to the input fluctuations. In fact, robust noise-resistant of the slow dynamics can be interpreted as a low-pass filter [153], the "low-pass filter" can reject transient fluctuations in the input signal. With the decreasing of ε, the y loop becomes slower and slower, and high frequency of the input noise can be filtered out better and better. Therefore, the APFL strength and time scale can regulate the noise-resistant of the multi-positive feedback system.

In a word, the APFL and timescale can jointly regulate the positive feedback system to make appropriate response to the noisy inputs. Through the two mechanisms, the NNP system can realize the tunability of response speed and noise-resistant. In real-world biological systems, an organism can recur to this property to make appropriate decisions under noisy inputs. For example, from Pigliucci et al. [156], in times of stress, it is advantageous for an organism to increase variability. From this research, the variability can be increased through the increasing of η and timescale ε. Whereas, when an organism has well adapted to the environment, it is better to decrease variability. For the circuit NNP, one can lower the output variability by decreasing the APFL strength and slowing down the timescale. Therefore, the investigations indicate a possible way of input noise control [157].

3.6.3.5 Effect of the APFL on Stochastic Bistable Switch

As indicated by bifurcation diagrams in Fig. 3.37, the two circuits NN and NNP can all display bistable dynamical behaviors. To investigate the effect of the APFL on the stochastic bistable switch, one investigates the stochastic models directly from the deterministic ODE model (3.50), where the terms on the right hand side of Eq. (3.50) are transformed into birth-death processes. It is noted that this method has been applied in some investigations, it has been proved that roughly similar conclusions can be derived as the detailed stochastic models [18, 19, 47, 109, 157]. The stochastic models are shown in Table 3.14, where, Ω represents the system volume,

Table 3.14 Stochastic model directly from the deterministic model

Reactions	Propensity functions	Increment of the molecular numbers
$\emptyset \rightarrow X$	$p_1 = \Omega f(X/\Omega, Y/\Omega, S/\Omega)$	$(1, 0)^T$
$X \rightarrow \emptyset$	$p_2 = d_1 X$	$(-1, 0)^T$
$\emptyset \rightarrow Y$	$p_3 = \Omega \varepsilon g(X/\Omega, S/\Omega)$	$(0, 1)^T$
$Y \rightarrow \emptyset$	$p_4 = d_2 \varepsilon Y$	$(0, -1)^T$

X, Y, S denote the molecular numbers, $f(x, y, s)$ and $g(x, s)$ in Eq. (3.50) are transformed into the propensity functions by substituting x, y, s with X/Ω, Y/Ω and S/Ω respectively. In the following, the stochastic model in Table 3.14 will be investigated by stochastic simulations, which will be performed by the Gillespie's direct stochastic simulation algorithm [10].

We set $\alpha_1 = 5.2, \alpha_2 = 3, \Omega = 10, S = 10$, and randomly take the initial molecular numbers $X(0)$, $Y(0)$, we perform stochastic simulations under different random states of the random number generator. Since first switch time and switch frequency [135, 158] in finite time intervals can well quantify the intrinsic noise-induced bistable switch behavior, we collect the two statistical indexes under different random states. Figure 3.40a, b show the first switch time and the switch frequency versus the APFL strength respectively. Figure 3.40c, d consider the effect of timescale on the bistable switch, where the APFL strength $\eta = 0.5$ is fixed in each simulation. In Fig. 3.40, we have considered the first switch time from the low to the high output states as a measure. One shape of points denotes an independent simulation result under a random state, the dash-dotted lines represent the average curves. It is noted that when there are no switching events in the finite time interval

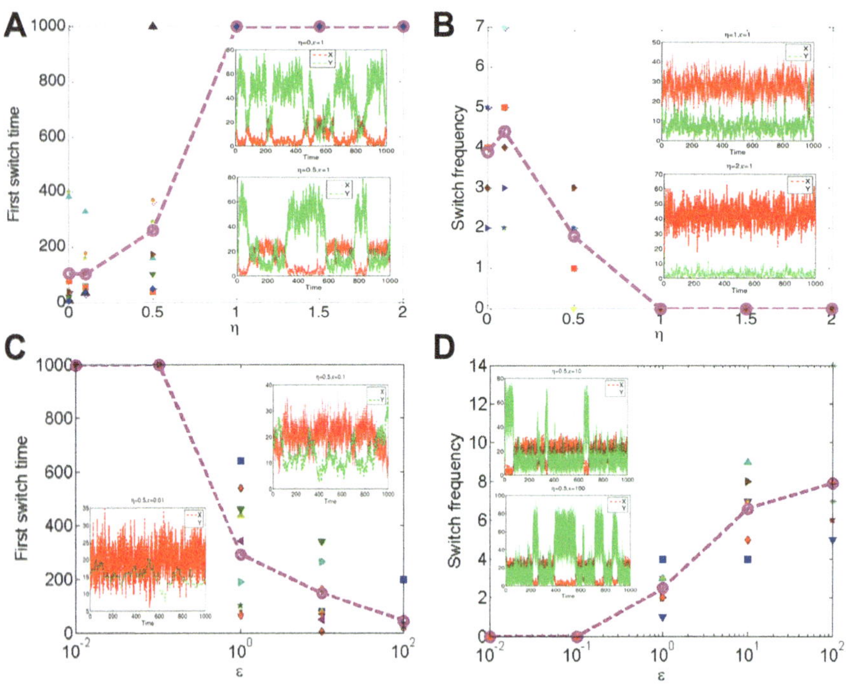

Fig. 3.40 Stochastic dynamics of the molecular numbers in the circuit NNP. (**a**) First switch time versus the APFL strength. (**b**) Switch frequency versus the APFL strength. (**c**) First switch time versus the timescale. (**d**) Switch frequency versus the timescale. Reprinted by permission from Springer, ref. [137]

[0, 1000], the first switch time is setted as 1000 and the switch frequency is taken as 0. The inset figures show random simulation runs of stochastic evolution of the molecular numbers with the different APFL strength.

From Fig. 3.40a, b, one can draw the conclusion that, averagely speaking, with the increasing of the APFL strengths, the first switch time is delayed, and fewer and fewer switches in the interval [0, 1000] can be observed. With enough strong APFL strengths, the NNP cannot switch between two SSSs, which indicate that strong APFL strength can strengthen cell decisions and persist cell memory. From Fig. 3.40c, d, reverse conclusions can be derived as compared with Fig. 3.40a, b. With the increasing of timescale, the first switch time can be effectively shortened and the switch frequency can be enhanced. Under slow timescale, the circuit NNP cannot display bistable switch, which is mainly because that, the dynamics of protein Y is slow with small ε. As we have demonstrated in the former section, the toggle switch loop can act as a low-pass filter under small ε, the high frequency of noise in the NNP can be partly filtered out by this loop, the low frequency noise cannot overcome double-well potentials to induce bistable switch, therefore, no switch behavior can be observed. The findings can be intuitively derived from the inset figures. We noted that the results also correspond to our observations in Sect. 3.6.3.2, where we have found that the APFL strength can enhance cell decision-making from the perspective of the attractiveness of the SSSs, the enhancement of cell decision-making just corresponds to the reductions of the switch frequency and the prolongation of the first switch time.

The above findings are biologically meaningful, especially in cell differentiation. A differentiated cell can keep a long-term memory of differentiated event, which requires the cell to slow down the time for the second round of state transition. In Ref. [92], Sriram et al. found that in interlocked mutual inhibition and activation positive feedback loop, the mutual inhibition loop controls the switch behavior, and the mutual activation loop can enforce decision. Here, the APFL plays a similar role as the mutual activation loop in the interlocked feedback circuit. Strong APFL strengths can decrease the switch frequency and increase the first switch time, which indicates cell decisions have been strengthened. Since the toggle switch loop controls the frequency of noise, therefore, it controls the noise-induced switch behavior. In a word, one can conclude that the APFL strength and timescale can jointly regulate the organism to make appropriate decision, and an organism with this mechanism can easily realize functional tunability and adaptability.

3.6.4 Summary

Genetic circuits are important cellular signal processing components [1, 140, 148]. Feedback is a key mechanism in cellular signal processing, therefore, genetic feedback circuits have been popular topics in the field of systems biology and synthetic biology. The two-component multi-positive feedback circuits are representative positive feedback circuits. In this section, we model a representative two-component dual-positive feedback circuit as an ODE system, and the ODE system can well reflect the APFL strength and timescale differences between the two positive feedback loops.

Functional roles of the APFL in the two-component dual-positive feedback circuit have been investigated. One finds that the APFL strength can tune the size of bistable region, stronger APFL strength always indicates larger size of bistable region. The APFL strength can effectively change the attractiveness of the two SSSs. The stronger the APFL strength is, the larger the attractiveness point of the high output state will be, therefore, it indicates that the APFL can strengthen cell decision-making. The APFL can also tune the GRIOS, with the increasing of the APFL strength, there exists an optimal APFL strength to realize the highest GRIOS, and the highest GRIOS for the NNP is higher than the GRIOS for the NN, which indicates that the GRIOS of the NNP can be freely tunable through the APFL strength. From the response curves of the circuit under noisy input, one finds that the APFL strength and timescale can jointly regulate the positive feedback system to respond more rapidly or tune noise resistance, which guarantees the adaptation of the circuit to environmental fluctuation. Finally, stochastic molecular evolution simulations also show the characteristics of the APFL in enhancing decision-making and realizing functional tunability.

Further issues that are deserved to be investigated include, whether the observations still stand for water when time delays [159, 160] are incorporated in the NNP system. Another question is what's the role of the APFL when the double negative feedback loop is substituted by the double positive feedback one. We will discuss these questions in our future works. However, the current investigations provide a clear understanding of the existence of multiple-positive feedback loops from the perspective of the APFL strength and timescale, and may have potential implications in the design of artificial genetic circuits, modeling and model reduction for large-scale genetic networks and so on.

3.7 Exploring Simple Bio-molecular Networks with Specific Functions

3.7.1 Motivations

Growing evidences have suggested the existence of the design principles that unify the organization of diverse circuits across all organisms. For example, it has been shown that there are recurrent network motifs linked to particular functions, such as temporal expression programs, reliable cell decisions and robust and tunable biological oscillations [1, 2, 83]. These findings suggest an intriguing hypothesis: despite the apparent complexity of cellular networks, there might only be a limited number of network topologies that are capable of robustly executing any particular biological function. Some topologies may be more favorable because of fewer parameter constraints. Other topologies may be incompatible with a particular function. The thorough understanding of a circuit function-topology map would be invaluable for synthetic biology, providing a manual for how to robustly engineer biological circuits that carry out a target function. In this section, we introduce the work by Ma et al. [83] and Zhang et al. [142], which have devoted to massive exploration of simple circuits with predefined biological functions.

3.7.2 Exploring Enzymatic Regulatory Networks with Adaption

As early as the year 2009, Ma et al. [83] have computationally explored the full range of simple enzyme circuit architectures that are capable of executing one critical and ubiquitous biological behavior—adaptation. They explored if there are finite solutions for achieving adaptation. Adaptation refers to the system's ability to respond to a change in input stimulus then return to its pre-stimulated output level, even when the change in the input persists. Adaptation is commonly used in sensory and other signaling networks to expand the input range that a circuit is able to sense, to more accurately detect changes in the input, and to maintain homeostasis in the presence of perturbations.[1]

A mathematical description of adaptation is diagrammed in Fig. 3.41a. Two characteristic quantities are defined to measure the adaption: the circuit's sensitivity to input change and the precision of adaptation.

$$Sensitivity = \left|\frac{(O_{peak} - O_1)/O_1}{(I_2 - I_1)/I_1}\right|. \tag{3.55}$$

$$Precision = \left|\frac{(O_2 - O_1)/O_1}{(I_2 - I_1)/I_1}\right|^{-1}. \tag{3.56}$$

Here, O_{peak}, O_1, O_2, I_1, I_2 are shown in Fig. 3.41a. If the system's response returns exactly to the prestimulus level (infinite precision), it is called the perfect adaptation. Examples of perfect or near perfect adaptation range from the chemotaxis of bacteria and neutrophils, osmo-response in *yeast*, to the sensor cells in higher organisms and calcium homeostasis in mammals.

To explore all possible network topologies that are capable of robust adaptation, by restricting to enzymatic nodes, Ma et al. [83] enumerated all possible three-node network topologies and study their adaptation properties over a range of kinetic parameters (Fig. 3.41b, c). Among the three nodes, one node that receives input, a second node that transmits output, and a third node that can play diverse regulatory

[1] Reprinted from CELL, 138, Ma, W., Trusina, A., El-Samad, H., Lim, W.A., Tang, C., Defining network topologies that can achieve biochemical adaptation, 760–773, Copyright (2009), with permission from Elsevier.

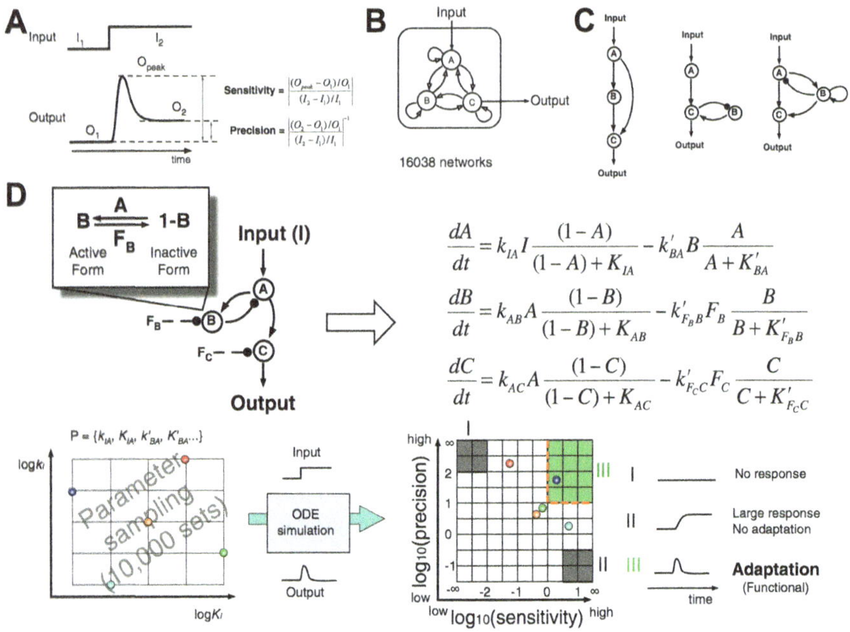

Fig. 3.41 Searching topology space for adaptation circuits. (**a**) Input–output curve defining adaptation. (**b**) Possible directed links among three nodes. (**c**) Illustrative examples of three-node circuit topologies. (**d**) Illustration of the analysis procedure for a given topology. Reprinted from ref. [83], with permission from Elsevier

roles. It is noted that although most biological circuits are likely to have more than three nodes, many of these cases can probably be reduced to these simpler frameworks, given that multiple molecules often function in concert as a single virtual node. There is a total of 16,038 possible three-node topologies that contain at least one direct or indirect causal link from the input node to the output node. For each topology, they sampled a wide range of parameter space (10,000 sets of network parameters) and characterized the resulting behavior in terms of the circuit's sensitivity to input change (Eq. (3.55)) and its ability to adapt (Eq. (3.56)). A total of $16{,}038 \times 10{,}000 \approx 1.6 \times 10^8$ different circuits. This search resulted in an exhaustive circuit function map, which have been used to extract the core topological motifs essential for adaptation. Overall, they revealed that despite the importance of adaptation in diverse biological systems, there are only a finite set of solutions for robustly achieving adaptation. These findings may provide a powerful framework in which to enhance our understanding of complex biological networks.

3.7.2.1 Searching for Circuits Capable of Adaptation

By using the two indicators as defined in Eqs. (3.55) and (3.56) and by constraining the search to three-node networks, Ma et al. [83] performed a coarse-grained network search for circuits capable of adaptation. They limited their research to enzymatic regulatory networks and modeled network linkages using the MM rate equations. Each node in the model network has a fixed total concentration that can be interconverted between two forms (active and inactive) by other active enzymes in the network or by basally available enzymes. For example, a positive link from node A to node B indicates that the active state of enzyme A is able to convert enzyme B from its inactive to active state (see Fig. 3.41d). If there is no negative link to node B from the other nodes in the network, they assumed that a basal (nonregulated) enzyme would inactivate B. They used ODEs to model these interactions, characterized by the MM constants (K_M's) and catalytic rate constants (k_{cat}'s) of the enzymes. Implicit in their analysis are assumptions that the enzyme nodes operate under the MM kinetics and that they are noncooperative (Hill coefficient = 1).

The circuit's sensitivity and adaptation precision can be mapped on the two-dimensional sensitivity versus precision plot (Fig. 3.41d). They defined a particular circuit architecture/parameter configuration to be "functional" for adaptation if its behavior falls within the upper-right rectangle in this plot (the green region in Fig. 3.41d)—these are circuits that show a strong initial response ($sensitivity > 1$) combined with strong adaptation ($precision > 10$). In most of their simulations they gave a non-zero initial input ($I_1 = 0.5$) and then changed it by 20% ($I_2 = 0.6$). The functional region corresponds to an initial output change of more than 20% and a final output level that is not more than 2% different from the initial output. Nonfunctional circuits fall into other quadrants of this plot, including circuits that show very little response (upper-left quadrant) and circuits that show a strong response but low adaptation (lower-right quadrant). For any particular circuit architecture, they focused on how many parameter sets can perform adaptation—a circuit is considered to be more robust if a larger number of parameter sets yield the behavior defined above.

To identify the network requirements for adaptation, two different but complementary approaches were taken. In the first approach, the authors searched for the simplest networks that are capable of achieving adaptation, limiting themselves to networks containing three or fewer links. They find that all circuits of this type that can achieve adaptation fall into two architectural classes: negative feedback loop with a buffering node (NFBLB) and incoherent FFL with a proportioner node (IFFLP). In the second approach, they searched all possible 16,038 three-node networks (with up to nine links) for architectures that can achieve adaptation over a wide range of parameters. These two approaches converge in their conclusions: the more complex robust architectures that emerge are highly enriched for the minimal NFBLB and IFFLP motifs. In fact, all adaptation circuits contain at least one of these two motifs. The convergent results indicate that these two architectural motifs present two classes of solutions that are necessary for adaptation.

3.7.2.2 Identifying Minimal Adaptation Networks

They started by examining the simplest networks capable of achieving adaptation (defined as $sensitivity > 1$ and $precision > 10$) for any of their parameter sets. For networks composed of only two nodes (an input-receiving node A and output-transmitting node C, with no third regulatory node), there are 4 possible links and 81 possible networks, none of which is capable of achieving adaptation for the considered parameter space.

Next, they examined minimal 3-node topologies with only three or fewer links between nodes (maximally complex 3-node topologies contain nine links). None of the 2-link, 3-node networks were capable of adaptation—the minimal number of links for this to be functional is three. The simplest topologies capable of adaptation, under at least some parameter sets, are eleven 3-node, 3-link networks. These network architectures are listed in Fig. 3.42 along with examples of the distribution of sensitivity/precision behaviors for the 10,000 parameter sets that were searched. An architecture is considered capable of adaptation if this distribution extends into the upper-right quadrant (high sensitivity, high precision). The common features of the networks capable of adaptation are either a single negative feedback loop or a single incoherent FFL. Here, a negative feedback loop is defined as a topology whose links, starting from any node in the loop, lead back to the original node with the cumulative sign of regulatory links within the loop being negative. An incoherent FFL is defined as a topology in which two different links starting from the input-receiving node both end at the output-transmitting node, with the cumulative sign of the two pathways having different signs (one positive and one negative). The first row of Fig. 3.42 shows several examples of 3-link, 3-node networks capable of adaptation; the second row shows related counter parts that cannot achieve adaptation. Overall incoherent FFLs appear to perform adaptation more robustly than negative feedback loops—they are capable of higher sensitivity and higher precision as indicated by the larger distribution of sampled parameters that lie in the upper-right corner of the sensitivity/precision plot.

While it is not surprising that positive feedback loops cannot achieve adaptation (Fig. 3.42a), it is interesting to note that negative feedback loop topologies differ widely in their ability to perform adaptation (Fig. 3.42a, lower panel). Notably, there is only one class of simple negative feedback loops that can robustly achieve adaptation. In this class of circuits, the output node must not directly feedback to the input node. Rather, the feedback must go through an intermediate node (b), which serves as a buffer.

Among FFLs (Fig. 3.42b), coherent FFL is clearly very poor at adaptation (Fig. 3.42b, lower panel). The three incoherent FFLs in Fig. 3.42b also differ drastically in their performance. Of these, only the circuit topology in which the output node C is subject to direct inputs of opposing signs (one positive and one negative) appears to be highly preferred. As will be seen later, the reason this architecture is preferred is because the only way for an incoherent FFL to achieve robust adaptation is for node B to serve as a proportioner for node A—i.e., node B is

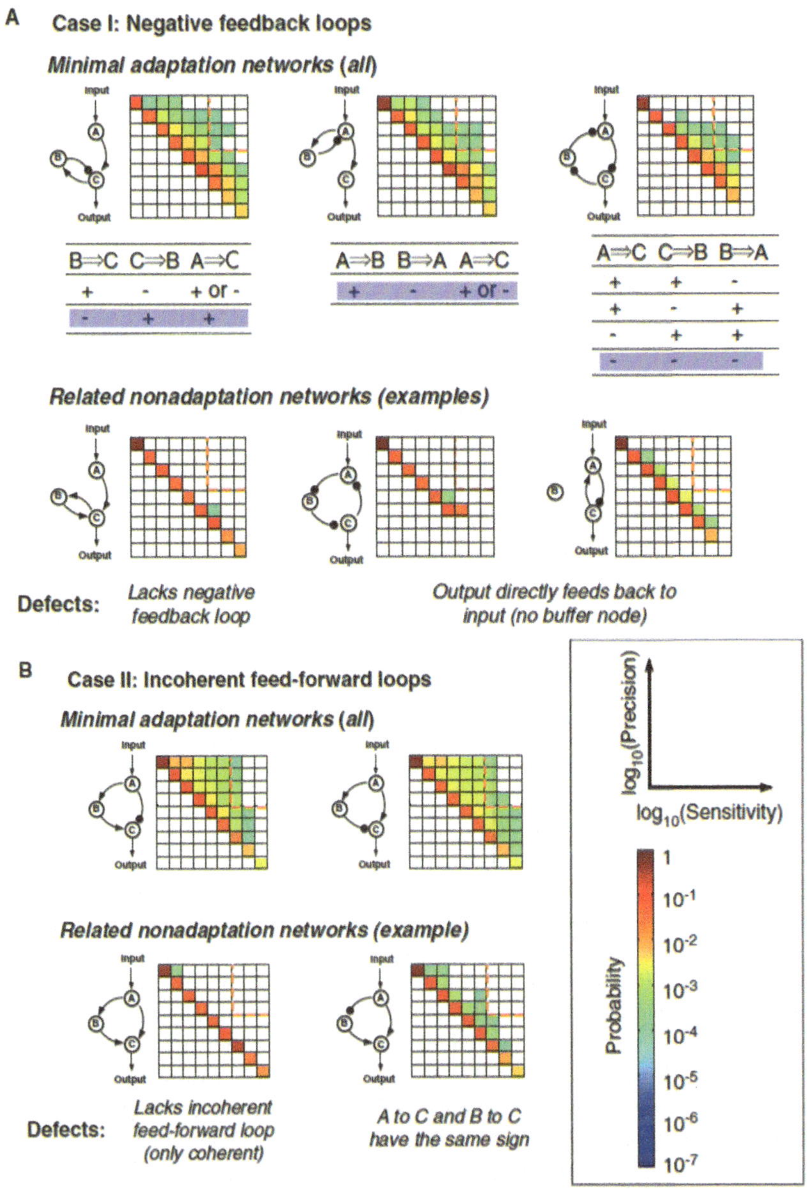

Fig. 3.42 Minimal networks (≤3 links) capable of adaptation. (**a**) Adaptive networks composed of negative feedback loops. Three examples of adaptation networks are shown in the upper panel. three examples of nonadaptive networks are shown in the lower panel. (**b**) Adaptive networks composed of incoherent FFLs. The only two minimal adaptation networks in this case are shown in the upper panel. Examples of nonadaptive networks are shown in the lower panel. Reprinted from ref. [83], with permission from Elsevier

activated in proportion to the activation of node A and to exert opposing regulation on node C.

3.7.2.3 Key Parameters in Minimal Adaptation Networks

Two major classes of minimal adaptive networks emerge from the above analysis: one type of negative feedback circuits and one type of incoherent FFLs. Why are these two classes of minimal architectures capable of adaptation? In this subsection, their underlying mechanisms, as well as the parameter conditions that must be met for adaptation will be investigated.

3.7.2.4 Negative Feedback Loop with a Buffer Node

The NFBLB class of topologies has multiple realizations in 3-node networks (Fig. 3.42a), all featuring a dedicated regulation node B that functions as a buffer. Considering a negative feedback loop between regulation node B and output-transmitting node C, the mechanism by which this NFBLB topology adapts and achieves a high sensitivity can be unraveled by the analysis of the following kinetic equations:

$$\begin{cases} \frac{dA}{dt} = Ik_{IA}\frac{1-A}{1-A+K_{IA}} - F_A k'_{F_A A}\frac{A}{A+K'_{F_A A}}, \\ \frac{dB}{dt} = Ck_{CB}\frac{1-B}{1-B+K_{CB}} - F_B k'_{F_B B}\frac{B}{B+K'_{F_B B}}, \\ \frac{dC}{dt} = Ak_{AC}\frac{1-C}{1-C+K_{AC}} - Bk'_{BC}\frac{C}{C+K'_{BC}}, \end{cases} \tag{3.57}$$

where F_A and F_B represent the concentrations of basal enzymes that carry out the reverse reactions on nodes A and B, respectively (they oppose the active network links that activate A and B). In this circuit, node A simply functions as a passive relay of the input to node C; the circuit would work in the same way if the input were directly acting on node C (just replacing A with I in the third equation of Eq. (3.57)). Analyzing the parameter sets that enabled this topology to adapt indicates that the two constants K_{CB} and $K'_{F_B B}$ (The MM constants for activation of B by C and inhibition of B by the basal enzyme) tend to be small, suggesting that the two enzymes acting on node B must approach saturation to achieve adaptation. Indeed, it can be shown that in the case of saturation, this topology can achieve perfect adaptation. Under saturation conditions, i.e., $(1 - B) >> K_{CB}$ and $B >> K'_{F_B B}$, the rate equation for B can be approximated by the following:

$$\frac{dB}{dt} = Ck_{CB} - F_B k'_{F_B B}. \tag{3.58}$$

The steady-state solution is

$$C^* = F_B k'_{F_B B}/k_{CB}, \tag{3.59}$$

which is independent of the input level I. The output C of the circuit can still transiently respond to changes in the input (see the first and the third equations in Eq. (3.57)) but eventually settles to the same steady state determined by Eq. (3.59). Note that Eq. (3.58) can be rewritten as

$$\begin{aligned} \frac{dB}{dt} &= k_{CB}(C - C^*), \\ B &= B^*(I_0) + k_{CB}\int_0^t (C - C^*)d\tau. \end{aligned} \tag{3.60}$$

Thus, the buffer node B integrates the difference between the output activity C and its input-independent steady-state value. Therefore, this NFBLB motif, node C ⊣ node B → node C, implements integral control—a common mechanism for perfect adaptation in engineering. All minimal NFBLB topologies use the same integral control mechanism for perfect adaptation.

The parameter conditions required for more accurate adaptation and higher sensitivity can also be visualized in the phase planes of nodes B and C (Fig. 3.43a). The nullclines for nodes B and C ($dB/dt = 0$ and $dC/dt = 0$, respectively) are shown for two different input values. For this topology, only the C nullcline (red curve) depends explicitly on the input through A (Eq. (3.57)). The B nullcline (black curve) does not depend on A. The steady state of the system is given by the intersection of the B and C nullclines. Thus, the change in steady state for any input change is only determined by the movement of the input-dependent C nullcline (e.g., dashed red curve in Fig. 3.43a). The adaptation precision is therefore directly related to the flatness of the B nullcline near the intersection of the two nullclines. The smaller the dependence of C on B, the smaller the adaptation error. One way to achieve a small dependence of C on B, or equivalently a sharp dependence of B on C, in an enzymatic cycle is through the zeroth order ultrasensitivity, which requires the two enzymes regulating the node B to work at saturation. This is precisely the condition leading to Eq. (3.58). All NFBLB minimal topologies have similar nullcline structures and their adaptation is related to the zeroth order ultrasensitivity in a similar fashion.

The ability of the network to mount an appropriate transient response to the input change before achieving steady-state adaptation depends on the vector fields $(dB/dt, dC/dt)$ in the phase plane (green arrows, Fig. 3.43). A large response, corresponding to sensitive detection of input changes, is achieved by a large excursion of the trajectory along the C-axis. This in turn requires a large initial $|dC/dt|$ and a small initial $|dB/dt|$ near the prestimulus steady state. For this class of topologies, this can be achieved if the response time of node C to the input change is faster than the adaptation time. The response time of node C is set by the first term in the dC/dt equation—faster response would require a larger k_{AC}. The timescale for adaptation is set by the equation for node B and the second term of the equation

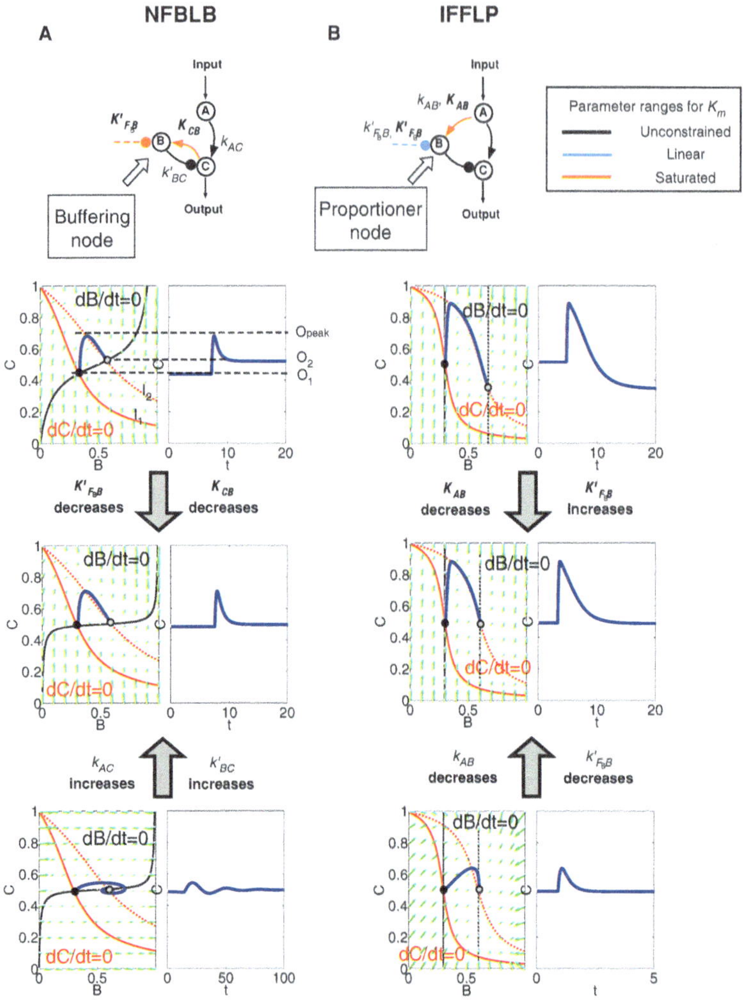

Fig. 3.43 Phase diagram and nullcline analysis of representative networks from the two classes of minimal adaptive topologies. (**a**) Phase planes of the variables (**b**) and (**c**) for a NFBLB topology. (**b**) Phase planes for an IFFLP topology. For details of the parameter settings, one can refer to the original paper. Reprinted from ref. [83], with permission from Elsevier

for node C—slower adaptation time would require a smaller k'_{BC}/K'_{BC} and/or a slower timescale for node B. This illustrates an important uncoupling of adaptation precision and sensitivity: once the MM constants are tuned to achieve operation in the saturated regimes, the timescales of the system can be independently tuned to modulate the sensitivity of the system to input changes.

3.7.2.5 Incoherent FFL with a Proportioner Node

The other minimal topological class sufficient for adaptation is the IFFLP (Fig. 3.42b). In an incoherent FFL, the output node C is subject to two regulations both originating from the input but with opposing cumulative signs in the two pathways. There are two possible classes of incoherent FFL architectures, but only one is able to robustly perform adaptation (Fig. 3.42b, upper panel): the functional architectures all have a "proportioner" (node B) that regulates the output (node C) with the opposite sign as the input to C.

The IFFLP topology achieves adaptation by using a different mechanism from that of the NFBLB class. Rather than monitoring the output and feeding back to adjust its level, the FFL "anticipates" the output from a direct reading of the input. Node B monitors the input and exerts an opposing force on node C to cancel the output's dependence on the input. For the IFFLP topology shown in Fig. 3.43b, the kinetic equations are as follows:

$$\begin{cases} \frac{dA}{dt} = Ik_{IA}\frac{1-A}{1-A+K_{IA}} - F_A k'_{F_A A}\frac{A}{A+K'_{F_A A}}, \\ \frac{dB}{dt} = Ak_{AB}\frac{1-B}{1-B+K_{AB}} - F_B k'_{F_B B}\frac{B}{B+K'_{F_B B}}, \\ \frac{dC}{dt} = Ak_{AC}\frac{1-C}{1-C+K_{AC}} - Bk'_{BC}\frac{C}{C+K'_{BC}}. \end{cases} \tag{3.61}$$

The adaptation mechanism is mathematically captured in the equation for node C: if the steady-state concentration of the negative regulator B is proportional to that of the positive regulator A, the equation determining the steady-state value of C, $dC/dt = 0$, would be independent of A and hence of the input I. In this case, the equation for node B generates the condition under which the steady-state value B^* would be proportional to A^*: the first term in dB/dt equation should depend on A only and the second term on B only. The condition can be satisfied if the first term is in the saturated region $((1 - B) >> K_{AB})$ and the second in the linear region $(B << K'_{F_B B})$, leading to

$$B^* = A^* k_{AB} K'_{F_B B}/(F_B k'_{F_B B}). \tag{3.62}$$

This relationship, established by the equation for node B, shows that the steady-state concentration of active B is proportional to the steady-state concentration of active A. Thus, node B will negatively regulate node C in proportion to the degree of pathway input. This effect of node B acting as a proportioner node of A can be graphically gleaned from the plot of the B and C nullclines (Fig. 3.43b). In this case, maintaining a constant C^* requires the B nullcline to move the same distance as the C nullcline in response to an input change. Here again, the sensitivity of the circuit (the magnitude of the transient response) depends on the ratio of the speeds of the two signal transduction branches: $A \rightarrow C$ and $A \rightarrow B \dashv C$, which can be independently tuned from the adaptation precision.

3.7.2.6 Exploration of All Possible 3-Node Networks: An NFBLB or IFFLP Architecture is Necessary for Adaptation

The above analyses focused on minimal (less than or equal to 3 links) 3-node networks and identified simple architectures that are sufficient for adaptation. But are these architectures also necessary for adaptation? In other words, are the identified minimal architectures the foundation of all possible adaptive circuits, or are there more complex higher-order solutions that do not contain these minimal topologies? To investigate this question, the authors expanded their study to encompass all possible 3-node networks, each with combinations of up to nine intra-network links. Again, for each network architecture, they sampled 10,000 possible parameter sets. Figure 3.44a shows a comprehensive map of the functional space, expressed as the distribution of all topologies and all sampled parameter sets on the sensitivity/precision plot. Only the regions above the diagonal are occupied, since by definition, sensitivity cannot be lower than adaptation error. The vast majority of the circuits lie on the diagonal where $sensitivity = 1/error$. This very common functional behavior is simply a monotonic change of the output in response to the input change, a hallmark of a direct, nonadaptive signal transduction response. The distribution plot quickly drops off away from the diagonal as the number of circuits with increasing sensitivity and/or adaptation precision drops. Overall, only 0.01% of all 1.6×10^8 possible architecture/parameter sets fall within the upper-right corner of the plot in Fig. 3.44a—i.e., those circuits that can achieve both high sensitivity and high adaptation precision. Defining overrepresentation topologies as the topology that are mapped to this region more than 10 times when sampled with 10,000 parameter sets. There are 395 out of 16,038 such topologies.

Analysis of these 395 robust topologies shows that they are overrepresented with feedback and FFLs. Strikingly, all 395 topologies contain at least one NFBLB or IFFLP motif (or both) (Fig. 3.44b). These results indicate that at least one of these motifs is necessary for adaptation.

3.7.2.7 Motif Combinations Can Improve Adaptation

Comparing the sensitivity/precision distribution plot of all networks (Fig. 3.44a) with that of the minimal networks (Fig. 3.42), it is clear that some of the more complex topologies occupy a larger functional space than the minimal topologies. Ma et al. [83] further investigated what additional features can improve the functional performance in these networks. They separated the 395 adaptation networks into the two categories, NFBLB and IFFLP. They then clustered the networks within each category using a pairwise distance between networks. The results, shown in Figs. 3.44c, d, clearly indicate the presence of common structural features (subclusters) in each category, some of which are shown on the righthand side in the figure. One striking feature shared by some of the more complex adaptation networks in the NFBLB category is a positive self-loop on the node B in the case where the other regulation on B is negative. This type of topology, with a saturated

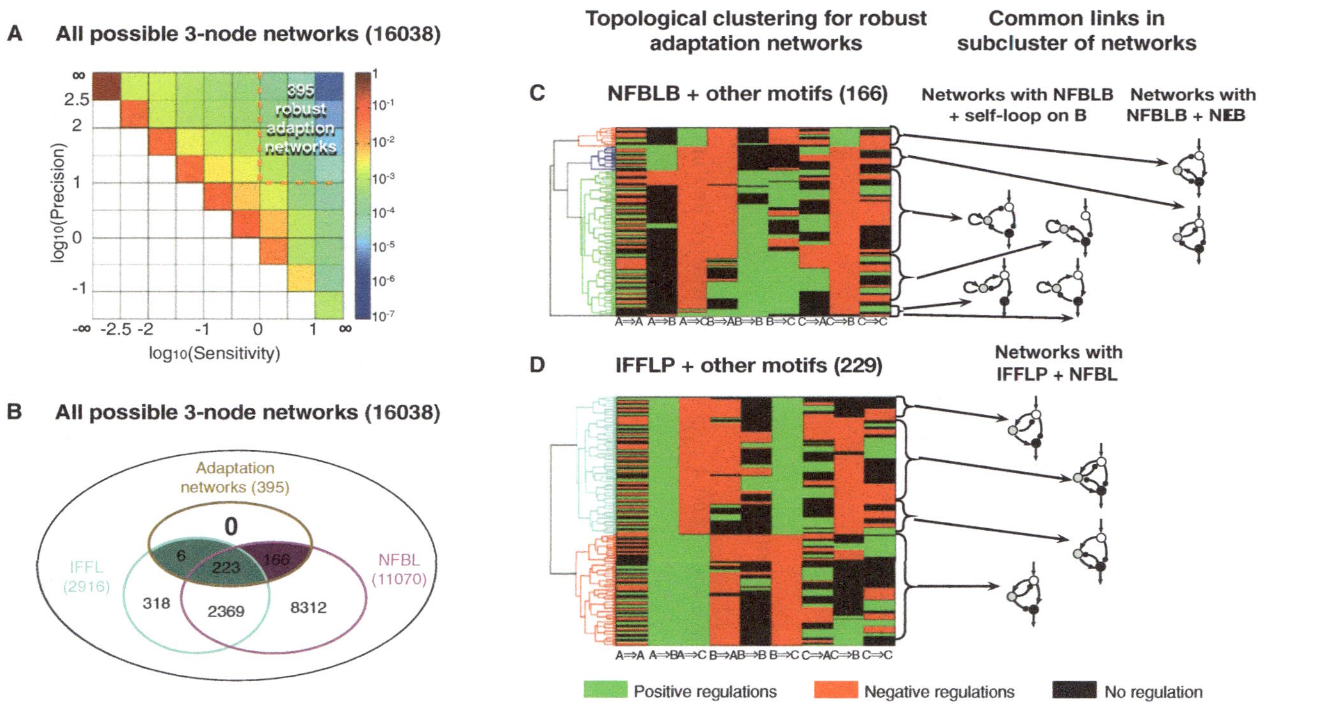

Fig. 3.44 Searching the full circuit-space for all robust adaptation networks. (**a**) The probability plot for all 16,038 networks with all the parameters sampled. 395 networks are overrepresented in the functional region shown by the orange rectangle. (**b**) Venn diagram of networks with three characters: adaptive, containing negative FBL, and containing incoherent FFL. (**c**) Clustering of the adaptation networks that belong to the NFBLB class. The network motifs associated with each of the subclusters are shown on the right. (**d**) Clustering of adaptation networks that belong to the IFFLP class. For details of the parameter settings, one can refer to the original paper. Reprinted from ref. [83], with permission from Elsevier

positive self-loop on B and linear negative regulation from other nodes, implements a special type of integral control to achieve perfect adaptation—here the $Log(B)$, rather than B itself, is the integrator. Another common feature of the more complex networks, which is present in both categories, is the presence of additional negative feedback loops that go through node B. They reported that this feature also enhances the performance—the networks with more such negative feedback loops have larger Q values (defined as the number of sampled parameter sets that yield the target adaptation behavior) than the minimal networks.

3.7.3 Exploring GRNs with Chaotic Behavior

In the year 2012, Zhang et al. [142] systematically investigated chaotic motifs (subnetworks with chaos) in GRNs. In this section, we will briefly introduce this work.

3.7.3.1 GRNs and Mathematical Models

Low dimensional GRNs with 3 and 4 nodes were considered and modeled as the following ODEs:

$$\frac{dp_i}{dt} = f_i(p) - p_i. \tag{3.63}$$

Here, p_i is the expression level of gene i, $0 < p_i < 1$. $p = (p_1, p_2, \cdots, p_N)^T$, $f_i(p) = A_i(p)$ for active regulation only, $f_i(p) = R_i(p)$ for repressive regulation only and $f_i(p) = A_i(p)R_i(p)$ for joint regulation.

$$A_i(p) = \frac{act_i^h}{act_i^h + K^h}, \ R_i(p) = \frac{K^h}{rep_i^h + K^h}, \tag{3.64}$$

and

$$act_i = \sum_{j=1}^{N} \alpha_{ji} p_j, rep_i = \sum_{j=1}^{N} \beta_{ji} p_j, (i, j = 1, 2, \cdots, N). \tag{3.65}$$

The elements in the adjacency matrix $\alpha = (\alpha_{ij})_{N\times N}$, $\beta = (\beta_{ij})_{N\times N}$ determine the network structure of the system, which are defined in such a way that $0 < \alpha_{ij} < 1$ if gene i activates gene j, $0 < \beta_{ij} < 1$ if gene i inhibits gene j and $\alpha_{ij} = \beta_{ij} = 0$ for no regulation of gene i by gene j. For convenience, all the corresponding elements in the matrix α or β in Fig. 3.45 are set to be 1. $act_i(rep_i)$ represents the sum of active (repressive) TFs to node i. The regulatory interactions of genes are

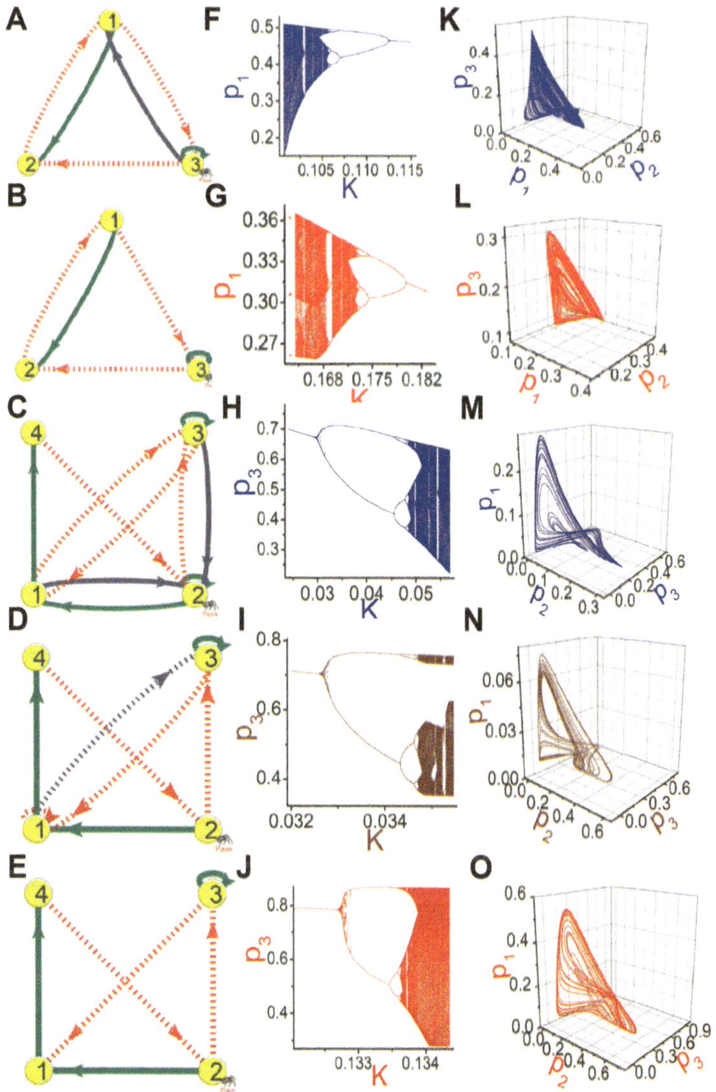

Fig. 3.45 Three-node and 4-node chaotic motifs. (**a–e**). Two 3-node and three 4-node chaotic GRNs; (**f–j**). Bifurcation diagrams for the circuits that are shown in panels (**a–b**); (**k–o**). Chaotic attractors. For details of the parameter settings, one can refer to the original paper. Reprinted from ref. [142]

represented by the Hill functions with the cooperativity exponent h ($1.0 < h < 10.0$) and the activation coefficient K ($0.005 < K < 0.30$), characterizing many real genetic systems. The model is of no-delay monotonous regulation. The leakage transcription rate is neglected, and the degradation rates of different proteins are supposed to be identical.

Each GRN has 1×10^6 samples from random network structures, parameter distributions, and initial variable conditions. The asymptotic states are finally recorded. The states are classified into three different types: steady states, periodic oscillations, and chaotic motions.

3.7.3.2 Conditions and Indicators for Chaos

Structure Requirements for Chaos

Some necessary conditions for chaotic motifs can be reasonably expected. The first crucial requirement of chaos is the competition between two or more oscillatory modes. It is more or less known that feedback loops may represent different oscillatory modes in the GRNs. Therefore, the existence of at least two feedback loops in the GRNs, generating different oscillatory modes in competition, must be fulfilled by chaotic GRNs. Moreover, self-sustained oscillation is also a basic condition for chaos. Since the absolutely necessary condition for self-sustained oscillation of the GRNs is existence of, at least, a negative feedback loop (NFL), and this is the second condition for chaos. Therefore, all the chaotic motifs may possess the following two characteristic features: (1) All these motifs have two feedback loops (the fewest feedback loops for chaotic motifs); (2) At least one of the two loops is NFL. Zhang et al. [142] exhaustively searched all 3-node and 4-node GRNs with minimal (irreducible) interactions satisfying the above two conditions, and eventually find 19 distinct 3-node and 86 distinct 4-node candidates of chaotic motifs.

Quantitative Analysis of Chaotic Motifs

Zhang et al. [142] applied the concept of dominant phase-advanced driving (DPAD) and extended the method of DPAD to DPAD time fraction (DTF). DPAD is a dynamical method that can find the strongest cross driving of the target node at any time instant when a system is in an oscillatory state, and DTF (ranged from 0 to 1) describes the driving contributions of all cross interactions to the target node during some given long period. The max value of DTF is one, which corresponds to there is only a single cross interaction to the target node; if the value is nearly zero, it means the driving effect of the given interaction to the target node is very weak. For the detail of these concepts, please refer to reference [142]. All the chaotic motifs discovered in Fig. 3.45 can be well explained by the driving relationships where the cross interactions are quantitatively described by DTFs. For details of the applications of the DTF in detecting chaotic behaviors, one can refer to the original paper by Zhang et al. [142].

3.7.3.3 Main Results

In order to search for chaos, the authors considered three ways, (1) network structures are randomly generated but to make sure that each node in the network has at least one input and at least one output (all the nodes are linked together). All the parameters $\alpha(\beta)$, K and h are randomly given within their ranges. (2) To enlarge the chaotic samples, oscillatory network structures are used to search for chaos. $\alpha(\beta)$ of the corresponding networks are remained, K and h are randomly given. 3) Chaotic samples are searched among the 105 two-loop structures with $\alpha(\beta)$, K and h given randomly. The authors found that chaos can only appear through competitions between different oscillatory modes with rivaling intensities. Conditions required for chaotic GRNs are found to be very strict, which make chaotic GRNs extremely rare. Chaotic motifs are explored as the simplest few-node structures capable of producing chaos, and serve as the intrinsic source of chaos of random few-node GRNs. Several optimal motifs causing chaos with atypically high probability are figured out. Moreover, a number of special oscillators can never produce chaos. These structures bring some advantages on rhythmic functions and may help us understand the robustness of diverse biological rhythms. The methods of DPAD and DTF were proposed to quantitatively identify chaotic motifs and to explain the origin of chaotic behaviors in GRNs.

3.7.4 Summary

In this section, we have introduced two works on the large-scale searching of simple bio-molecular circuits with predefined functions. The first work computationally searched all possible three-node enzyme network topologies to identify enzymatic regulatory networks that could perform adaptation. It was found that only two major core topologies emerge as robust solutions: a negative feedback loop with a buffering node and an incoherent FFL with a proportioner node. Minimal circuits containing these topologies are, within proper regions of parameter space, sufficient to achieve adaptation. More complex circuits that robustly perform adaptation ll contain at least one of these topologies at their core. The second work discussed the searching of three-node and four-node chaotic motifs in GRNs. It was founded that: (1) chaos can only appear through competitions between different oscillatory modes with rivaling intensities. Conditions required for chaotic GRNs are found to be very strict, which make chaotic GRNs extremely rare. (2) Chaotic motifs are explored as the simplest few-node structures capable of producing chaos, and serve as the intrinsic source of chaos of random few-node GRNs. Several optimal motifs causing chaos with atypically high probability are figured out. (3) Moreover, researchers discovered that a number of special oscillators can never produce chaos. These structures bring some advantages on rhythmic functions and may help us understand the robustness of diverse biological rhythms. (4) The methods of DPAD and DPAD time fraction are proposed to quantitatively identify chaotic motifs and to explain the origin of chaotic behaviors in GRNs.

The related works clarified some design roles of bio-molecular circuits, these design rules provide a framework for functionally classifying complex natural networks and a manual for engineering networks. Except the two mentioned works, some other works have also performed large-scale searching of simple bio-molecular networks with specific functions, for example, Ye et al. [161] have searched network motif family that can regulate multistep cell fate transitions with restricted reversibility.

3.8 Discussions and Conclusions

In this chapter, we discussed the topics of the modeling and analysis of simple network motifs in bio-molecular networks, where we pay our attention to explore the relationships among structure, dynamics, and functions of network motifs. Only the GRNs and enzymatic networks were considered in this chapter, it is noted that network motifs can be described by different mathematical models according to its backgrounds, it is interesting to explore the discussed functions of network motifs in other types of networks [61]. As a conclusion, we summarized several structures of bio-molecular networks and their functions in Table 3.15.

Finally, we noted that except the network motifs discussed in the above sections, there are many other motifs that have attracted many researchers, such as the auto-activation loop [135], the bi-fan motif [1, 2]. For details, one can refer to the related references therein, we give no detailed discussions on these motifs.

Table 3.15 Associations between common biochemical elements and dynamic behaviors

Biochemical elements	Networks	Favored behaviors
Positive feedback	GRN; signal transduction	Multi-stability; oscillations possible only if supply of biosynthetic precursors is rate limiting; State-dependent responses [100]; act as signal amplifier; enhance the sensitivity of a system response to signals
Negative feedback	GRN; signal transduction	Oscillations [100]
Biochemical time delays	GRN; signal transduction	complex responses to stimuli; oscillations; "Echoes"; chaos [100, 142]
Competition between gene activation and repression	GRN	Optimal stimulus frequencies for extrema in transcription rates; chaos [100, 142]
Protein oligomerization	GRN	Multi-stability; oscillations; optimal stimulus frequencies [100]
Stochastic fluctuations in molecule numbers	GRN	Destabilization of some steady states; randomness in responses to stimuli; random outcomes of processes dependent on genetic regulation [100]
CFFL	GRN; signal transduction	Steady-state logic is sensitive to inputs; sign-sensitive delay; cooperativity enhancement for some input [37]
IFFL	GRN; signal transduction	Steady-state logic is sensitive to inputs; pulse generator; sign-sensitive accelerator [37]
CFFL 1	GRN; signal transduction	Abundant in real systems; the least noisy among the 3-node FFLs [41]; globally sensitive to parameter variations [47]; not sensitive to input [48]
IFFL 1	GRN; signal transduction	Abundant in real systems; fold-change detection [44]; globally resistant to parameter variations [47]; not sensitive to input [48]
NFL with a buffering node	Enzymatic networks	Adaption [83]
IFFL with a proportioner node	Enzymatic networks	Adaption [83]
Coupled positive and negative feedback circuits	GRN	Multi-stability; oscillations; excitability; chaos etc.

References

1. Alon, U.: Network motifs: theory and experimental approaches. Nat. Rev. Genet. **8**, 450–461 (2007)
2. Alon, U.: An Introduction to systems biology: design principles of biological circuits. Chapman, Hall, CRC (2007)
3. Karlebach, G, Shamir, R.: Modelling and analysis of gene regulatory networks. Nat. Rev.: Mol. Cell Bio. **9**, 770–780 (2008)
4. Bornholdt, S.: Less is more in modeling large genetic networks. Science **310**, 449–450 (2005)

5. McAdams, H.H., Arkin, A.: Stochastic mechanisms in gene expression. Proc. Natl. Acad. Sci. USA. **94**, 814–819 (1997)
6. Brandman, O., Ferrell, J.E. Jr., Li, R., Meyer, T.: Positive feedback loops drive reliable cell decisions. Science **310**, 496–498 (2005)
7. Kauffman, S.A.: Metabolic stability and epigenesis in randomly connected nets. J. Theor. Biol. **22**, 437–467 (1969)
8. Li, F., Long, T., Lu, Y., Ouyang, Q., Tang, C.: The *yeast* cell-cycle network is robustly designed. Proc. Natl. Acad. Sci. USA. **101**, 4781–4786 (2004)
9. Maáyan, A., Sherry, L., Jenkins, S., et al.: Formation of regulatory patterns during signal propagation in a mammalian cellular network. Science **309**, 1078–1083 (2005)
10. Gillespie, D.T.: Exact stochastic simulation of coupled chemical reactions. J. Phys. Chem. **81**, 2340–2361 (1977)
11. Gillespie, D.T.: Approximate accelerated stochastic simulation of chemically reacting systems. J. Chem. Phys. **115**, 1716–1733 (2001)
12. Salis, H., Kaznessis, Y.: Accurate hybrid stochastic simulation of a system of coupled chemical or biochemical reactions. J. Chem. Phys. **122**(5), 054103 (2005)
13. Bratsun, D, Volfson, D., Tsimring, L.S., Hasty J.: Delay-induced stochastic oscillations in gene regulation. Proc. Natl. Acad. Sci. USA. **102**, 14593–14598 (2005)
14. Zhou, T.: Stochastic dynamics in biological systems. Science Publishing House, Beijing, China (2009). (In Chinese)
15. Barrio, M., Burrage, K., Leier, A., Tian, T.: Oscillatory regulation of Hes1: discrete stochastic delay modelling and simulation. PLoS Comput. Biol. **2**, e117 (2006)
16. Cai, X.: Exact stochastic simulation of coupled chemical reactions with delays. J. Chem. Phys. **126**, 124108 (2007)
17. Chen, L., Wang, R., Zhou, T., Aihara, K.: Noise-induced cooperative behavior in a multi-cell system. Bioinformat. **21**, 2722–2729, (2005)
18. Wang, P., Lü, J., Wan, L., Chen, Y.: A stochastic simulation algorithm for biochemical reactions with delays, IEEE Int. Conf. Syst. Biol. (ISB2013) **Aug. 23–25**, 109–114, Huangshan, China (2013)
19. Gonze, D., Halloy, J., Goldbeter, A.: Deterministic versus stochastic models for circadian rhythms. J. Biol. Phys. **28**, 637–653 (2002)
20. Gardner, T., Cantor, C., Cantor, J.: Construction of a genetic toggle switch in *Escherichia coli*. Nature **403**, 339–342 (2000)
21. Wang, P., Lü, J., Zhang, Y., Ogorzalek, M.J.: Intrinsic noise induced state transition in coupled positive and negative feedback genetic circuit. IEEE Int. Conf. Syst. Biol. **Sep.2–4**, 356–361 (2011)
22. Kloeden, P.E., Platen, E., Schurz, H.: Numerical solution of SDE through computer experiments. Springer-Verlag Berlin Heidelberg (1994)
23. Kloeden, P.E., Platen, E.: Numerical solution of stochastic differential equations. Springer-Verlag Berlin Heidelberg (1992)
24. Mil'shtein, G.N.: Approximate integration of stochastic differential equations. Theor. Probab. Its Appl. (In Russian). **19**, 583–588 (1974)
25. Mackevicius, V., Introduction to stochastic analysis (ISTE). Wiley (2011)
26. Wolkenhauer, O.: Systems biology: dynamic pathway modelling. www.sbi.uni-rostock.de, book to be pressed (2009)
27. Ainsworth, S.: Michaelis-Menten kinetics–steady-state enzyme kinetics. Macmillan Education UK (1977).
28. Hill, A.V.: The possible effects of the aggregation of the molecules of hemoglobin on its dissociation curve. J. Physiol. **40**, iv–vii (1910)
29. Butcher, J.C.: Numerical methods for ordinary differential equations. New York: John Wiley & Sons (2008)
30. Milo, R., Shen-Orr, S., Itzkovitz, S., Kashtan, N., Chklovskii, D., Alon, U.: Network motifs: simple building blocks of complex networks. Science **298**, 824–82 (2002)

31. Reigl, M., Alon, U., Chklovskii, D.B.: Search for computational modules in the *C. elegans* brain. BMC Biol. **2**, 25 (2004)
32. Shen-Orr, S.S., Milo, R., Mangan, S., Alon, U.: Network motifs in the transcriptional regulation network of *Escherichia coli*. Nat. Genet. **31**, 64–68 (2002)
33. Lee, I.T., Rinaldi, N.J., Robert, F., Odom, et al.: Transcriptional regulatory networks in *saccharomyces cerevisiae*. Science **298**, 799–804 (2002)
34. Kim, J., Yoon, Y., Cho, K.: Coupled feedback loops from dynamic motifs of cellular networks. Biophys. J. **94**, 359–365 (2008)
35. Sneppen, K., Krishna, S., Semsey, S.: Simplified models of biological networks. Annu. Rev. Biophys. **39**, 43–59 (2010)
36. Elowitz, M., Leibler, S.: A synthetic oscillatory network of transcriptional regulators. Nature **403**, 335–338 (2000)
37. Mangan, S., Alon, U.: Structure and function of the feed-forward loop network motif. Proc. Natl. Acad. Sci. USA. **100**, 11980–11985 (2003)
38. Mangan, S., Zaslaver, A., Alon, U.: The coherent feed-forward loop serves as a sign-sensitive delay element in transcription networks. J. Mol. Biol. **334**, 197–204 (2003)
39. Kalir, S., Mangan, S., Alon, U.: A coherent feed-forward loop with a SUM input function prolongs flagella expression in *Escherichia coli*. Mol. Syst. Biol. **1**, E1-E6 (2005)
40. Mangan, S., Itzkovitz, S., Zaslaver, A., Alon, U.: The incoherent feed-forward loop accelerates the response time of the gal system of *Escherichia coli*. J. Mol. Biol. **356**, 1073–1081 (2006)
41. Ghosh, B., Karmakar, R., Bose, I.: Noise characteristics of feed forward loops. Phys. Biol. **2**, 36–45 (2005)
42. Kaplan, S., Bren, A., Dekel, E., Alon, U.: The incoherent feed-forward loop can generate non-monotonic input functions for genes. Mol. Syst. Biol. **4**, 203 (2008)
43. Kittisopikul, M., Süel, G.M.: Biological role of noise encoded in a genetic network motif. Proc. Natl. Acad. Sci. USA. **107**, 13300–13305 (2010)
44. Goentoro, L., Shoval, O., Kirschner, M.W., Alon, U.: The incoherent feed-forward loop can provide fold-change detection in gene regulation. Mol. Cell **36**, 894–899 (2009)
45. Dekel, E., Mangan, S., Alon, U.: Environmental selection of the feed-forward loop circuit in gene-regulation networks. Phys. Biol. **2**, 81–88 (2005)
46. Kim, D., Kwon, Y.K., Cho, K.H.: The biphasic behavior of incoherent feed-forward loops in biomolecular regulatory networks. BioEssays **30**, 1204–1211 (2008)
47. Wang, P., Lü, J, Ogorzalek, M.J.: Global relative parameter sensitivities of the feed-forward loops in genetic networks. Neurocomput. **78(1)**, 155–165 (2012)
48. Wang, P., Lü, J., Zhang, Y., Ogorzalek, M.J.: Global relative input-output sensitivities of the feed-forward loops in genetic networks, Proc. 31th Chin. Control Conf. (CCC2012) **July 20–22**, 7376–7381, Hefei, China, (2012)
49. Guo, D., Li, C.: Stochastic and coherence resonance in feed-forward-loop neuronal network motifs. Phys. Rev. E **79** (2009)
50. Kashtan, N., Itzkovitz, S., Milo, R., Alon, U.: Efficient sampling algorithm for estimating subgraph concentrations and detecting network motifs. Bioinformat. **20**,1746–1758 (2004)
51. Wuchty, S., Oltvai, Z.N., Barabási, A.L.: Evolutionary conservation of motif constituents in the yeast protein interaction network. Nat Genet. **35**,176–179 (2003)
52. Yan, X., Han, J.: gSpan: Graph-based substructure pattern mining. IEEE Int. Conf. Data Mining **Dec. 9–12**, 721–724 (2002)
53. Koyutürk, M., Grama, A., Szpankowski, W.: An efficient algorithm for detecting frequent subgraphs in biological networks. Bioinformat. **20**, i200–i207 (2004)
54. Wernicke, S., Rasche, F.: FANMOD: a tool for fast network motif detection. Bioinformat. **22**,1152–1153 (2006)
55. Schreiber, F., Schwöbbermeyer, H.: Mavisto: A tool for the exploration of network motifs. Bioinformat. **21**,3572–3574 (2005)
56. Zhou, J., Lu, J., Lü, J.: Adaptive synchronization of an uncertain complex dynamical network. IEEE Trans. Automat. Contr. **51(4)**, 652–656 (2006)

57. Lü, J., Yu, X., Chen, G., Cheng, D.: Characterizing the synchronizability of small-world dynamical networks. IEEE Trans. Circ. Syst. I **51(4)**, 787–796 (2004)
58. Prill, R.J., Iglesias, P.A., Levchenko, A.: Dynamic properties of network motifs contribute to biological network organization. PLoS Biol. **3(11)**, e343 (2005)
59. macía, J., Solé, R., Widder, R.: Specialized or flexible feed-forward loop motifs: a question of topology. BMC Syst. Biol. **3**, 84 (2009)
60. Wall, M. E., Dunlop, M.J., Hlavacek, W.S.: Multiple functions of a feed-forward-loop gene circuit. J. Mol. Biol. **349**, 501–514 (2005)
61. Sontag, E.D.: Remarks on feedforward circuits, adaptation, and pulse memory. IET Syst. Biol. **4**, 39–51 (2010)
62. Eichenberger, P., Fujita, M., Jensen, S.T., et al.: The program of gene transcription for a single differentiating cell type during sporulation in *Bacillus subtilis*. PLoS Biol. **2**, e328 (2004)
63. Shoval, O., Goentoro, L., Hart, Y., Mayo, A., Sontag, E. D., Alon, U.: Fold-change detection and scalar symmetry of sensory input fields. Proc. Natl. Acad. Sci. USA. **107**, 15995–16000 (2010)
64. Goentoro, L., Kirschner, M.W.: Evidence that fold-change, and not absolute level, of β-catenin dictates Wnt signaling. Mol. Cell **36**, 872–884 (2009)
65. Ingram, P.J., Stumpf, M.P., Stark, J.: Network motifs: structure does not determine function. BMC Genom. **7**,108 (2006)
66. Yang, D., Kuznetsov, A.: Characterization and merger of oscillatory mechanisms in an artificial genetic regulatory network. Chaos **19** (2009)
67. Kuznetsov, A., Kaern, M., Kopell, N.: Synchrony in a population of hysteresis-based genetic oscillators. SIAM J. Appl. Math. **65**, 392–425 (2004)
68. Stelling, J., Sauer, U., Szallasi, Z., Doyle, F.J., Doyle, J.: Robustness of cellular functions. Cell **118**, 675–685 (2004)
69. Wang, P., Lü, J., Yu, X.: Colored noise induced bistable switch in the genetic toggle switch systems. IEEE/ACM Trans. Comput. Biol. Bioinformat. **12**(3), 579–589 (2015)
70. Ingalls B.: Sensitivity analysis: from model parameters to system behaviour. Essays Biochem. **45**, 177–193 (2008)
71. Khalil, H.K.: Nonlinear systems (third edition). Publishing House of Electronics Industry, Beijing, China (2007)
72. Shinar, G., Alon, U., Feinberg, M.: Sensitivity and robustness in chemical reaction networks. SIAM J. Appl. Math. **69**, 977–998 (2009)
73. Alberghina, L., Weaterhoff, H.V.: Systems biology: definitions and perspectives (topics in current genetics). Springer-Verlag, Berlin, Heidelberg (2005)
74. Saltelli, A., Tarantola, S., Campolongo, F., Ratto, M.: Sensitivity analysis in practice. Wiley, West Sussex, England (2004)
75. Saltelli, A., Annoni, P., Azzini, I., Campolongo, F., Ratto, M., Tarantola, S.: Variance based sensitivity analysis of model output. Design and estimator for the total sensitivity index. Comput. Phys. Commun. **181**, 259–270 (2010)
76. Soboĺ, I.M.: Global sensitivity indices for nonlinear mathematical models and their Monte Carlo estimates. Math. Comput. Simul. **55**, 271–280 (2001)
77. Li, G., Rosenthal, C., Rabitz, H.: High dimensional model representations. J. Phys. Chem. A **105**, 7765–7777 (2001)
78. Li, G., Wang, S. W., Rabitz, H., Wang, S., Jaffe, P.: Global uncertainty assessments by high dimensional model representations (HDMR). Chem. Engin. Sci. **57**, 4445–4460 (2002)
79. Feng, X., Hooshangi, S., Chen, D., Li, G., Weiss, R., Rabitz, H.: Optimizing genetic circuits by global sensitivity analysis. Biophys. J. **87**, 2195–2202 (2004)
80. Feng, X., Rabitz, H.: Optimal identification of biochemical reaction networks. Biophys. J. **86**, 1270–1281 (2004)
81. Varma, A., Morbidelli, M., Wu, H.: Parametric sensitivity in chemical systems. Cambridge Univ. Press, Cambridge, UK (2005)
82. Schmidt, H., Jirstrand, M.: Systems biology toolbox for Matlab: a computational platform for research in systems biology. Bioinformat. **22**, 514–515 (2006)

83. Ma, W., Trusina, A., El-Samad, H., Lim, W.A., Tang, C.: Defining network topologies that can achieve biochemical adaptation. Cell **138**, 760–773 (2009)
84. Iman, R.L., Davenport, J.M., Zeigler, D.K.: Latin hypercube sampling (program user's guide). Albuquerque, NM: Sandia Labs (1980)
85. Song, H., Smolen, P., Av-Ron, E., Baxter, D.A., ByrneJ. H.: Dynamics of a minimal model of interlocked positive and negative feedback loops of transcriptional regulation by cAMP-response element binding proteins. Biophys. J. **92**, 3407–3424 (2007)
86. Lü, J., Chen, G.: A time-varying complex dynamical network model and its controlled synchronization criteria. IEEE Trans. Automat. Contr. **50**, 841–846 (2005)
87. Liu, H., Lu, J., Lü, J., Hill, D.J.: Structure identification of uncertain general complex dynamical networks with time delay. Automatica **45**, 1799–1807 (2009)
88. Moran, U., Phillips, R., Milo, R.: SnapShot: key numbers in biology. Cell **141**, 1261–1262 (2010)
89. Xiong, W., Ferrell, J.E. Jr.: A positive-feedback-based bistable "memory module" that governs a cell fate decision. Nature **426**, 460–465 (2003)
90. Snoussi, E.H.: Necessary conditions for multistationary and stable periodicity. J. Biol. Syst. **6**, 3–9 (1998)
91. Ozbudak, E.M., Thattai, M., Lim, H.N., Shraiman, B. I., Van Oudenaarden, A.: Multistability in the lactose utilization network of *Escherichia coli*. Nature **427**, 737–740 (2004)
92. Sriram, K., Soliman, S., Fages, F.: Dynamics of the interlocked positive feedback loops explaining the robust epigenetic switching in *Candida albicans*. J. Theor. Biol. **258**, 71–88 (2009)
93. Bosl, W.J., Li, R.: The role of noise and positive feedback in the onset of autosomal dominant diseases. BMC Syst. Biol. **4**, 93 (2010)
94. Novak, B., Tyson, J.J.: Design principles of biochemical oscillators. Nat. Rev. Mol. Cell Biol. **9**, 981–991 (2008)
95. Tyson, J.J., Albert, R., Goldbeter, A., Ruoff, P., Sible, J.: Biological switches and clocks. J. R. Soc. Inter. **5**, S1-S8 (2008)
96. Moore-Ede, M.C., Sulzman, F.M, Fuller, C.A. : The clocks that time us: physiology of the circadian timing system. Harvard Univ. Press, Cambridge, MA (1982)
97. Danino, T., Mondragón-Palomino, O., Tsimring, L., Hasty, J.: A synchronized quorum of genetic clocks. Nature **463**, 326–330 (2010)
98. Zhou, T., Chen, L., Aihara, K.: Molecular communication through stochastic synchronization induced by extracellular fluctuations. Phys. Rev. Lett. **95**, 178103 (2005)
99. Laurent, M., Kellershohn, N.: Multistability: a major means of differentiation and evolution in biological systems. Trends Biochem. Sci. **24**, 418–422 (1999)
100. Smolen, P., Baxter, D.A., Byrne, J.H.: Mathematical modeling of gene networks. Neuron **26**, 567–580 (2000)
101. Smolen, P., Baxter, D., Byrne, J.: Interlinked dual-time feedback loops can enhance robustness to stochasticity and persistence of memory. Phys. Rev. E **79** (2009)
102. Dubnau, D., Losick, R.: Bistability in bacteria. Mol. Microbiol. 61, 564–572(2006)
103. Tsai, T.Y., Choi, Y.S., Ma, W., Pomerening, J.R., Tang, C., Ferrell, Jr. J.E.: Robust, tunable biological oscillations from interlinked positive and negative feedback loops. Science **321**, 126–129 (2008)
104. Purcell, O., Savery, N.J., Grierson, C.S., di Bernardo, M.: A comparative analysis of synthetic genetic oscillators. J. Roy. Soc. Inter. **7**, 1503–1524 (2010)
105. Stricker, J., Cookson, S., Bennett, M. R., Mather, W. H., Tsimring, L. S., Hasty J.: A fast, robust and tunable synthetic gene oscillator. Nature **456**, 516–519 (2008)
106. Brandman, O., Ferrell, Jr. JE., Li, R., Meyer, T.: Interlinked fast and slow positive feedback loops drive reliable cell decisions. Science **310**,496–498 (2005)
107. Tian, X., Zhang, X., Liu, F., Wang, W.: Interlinking positive and negative feedback loops creates a tunable motif in gene regulatory networks. Phys. Rev. E **80**, 011926 (2009)
108. Hasty, J., Dolnik, M., Rottschäfer, V., Collins, J.J.: Synthetic gene network for entraining and amplifying cellular oscillations. Phys. Rev. Lett. **88**, 148101 (2002)

109. Song, H., Smolen, P., Av-Ron, E., Baxter, D.A., Byrne, J.H.: Dynamics of a minimal model of interlocked positive and negative feedback loops of transcriptional regulation by cAMP-response element binding proteins. Biophys. J. **92**, 3407–3424 (2007)
110. Turcotte, M., Garcia-Ojalvo, J., Süel, G.M.: A genetic timer through noise-induced stabilization of an unstable state. Proc. Natl. Acad. Sci. USA. **105** 15732–15737 (2008)
111. Marucci, L., Barton, D.A.W., Cantone, I., Ricci, M.A., Cosma, M.P., Santini, S., di Bernardo, D., di Bernardo, M: How to turn a genetic circuit into a synthetic tunable oscillator, or a bistable switch. PLoS One **4**, e8083 (2009)
112. Gillespie, D.: The chemical Langevin equation. J. Chem. Phys. **113**, 297–306 (2000)
113. Goldbeter, A.: A model for circadian oscillation in the *Drosophila* period protein (PER). Proc. Roy. Soc. Lond. B. **261**, 319–324 (1995)
114. Goldbeter, A.: Biochemical oscillations and cellular rhythms. the molecular bases of periodic and chaotic behaviour. Cambridge Univ. Press, Cambridge, UK (1996)
115. Gonze, D., Goldbeter, A.: Circadian rhythms and molecular noise. Chaos **16**, 026110 (2006)
116. Dhooge, A., Govaerts, W., Kuznetsov, Y.A.: MATCONT: A Matlab package for numerical bifurcation analysis of ODEs. ACM Trans. Math. Softw. **29**, 141–164 (2003)
117. http://oscill8.sourceforge.net/.
118. Engelborghs, K., Luzyanina, T., Roose, D.: Numerical bifurcation analysis of delay differential equations using DDE-BIFTOOL. ACM Trans. Math. Softw. **28**, 1–21 (2002)
119. Engelborghs, K., Luzyanina, T., Samaey, G.: DDE-BIFTOOL v. 2.00: a Matlab package for numerical bifurcation analysis of delay differential equations. Report TW 330, Department of Computer Science, K.U. Leuven, Leuven, Belgium, (2001). Available from http://www.cs.kuleuven.be/~twr/research/software/delay/ddebiftool.shtml
120. Ermentrout, B.: Simulating, analyzing, and animating dynamical systems: a guide to XPPAUT for researchers and students. SIAM: Softw. Environm. Tools (2002)
121. Sneppen, K., Zocchi, G.: Physics in molecular biology. Cambridge Univ. Press, The Edinburgh Building, Cambridge, UK (2005)
122. Cinquin, O., Demongeot, J.: Roles of positive and negative feedback in biological systems. C.R. Biologies. **325**, 1–11 (2002)
123. Thattai, M., van Oudenaarden, A.: Intrinsic noise in gene regulatory networks. Proc. Natl. Acad. Sci. USA. **98**, 8614–8619 (2001)
124. To, T.L.,Maheshri, M.: Noise can induce biomodality in positive transcriptional feedback loops without bistability. Science **327**, 1142–1145 (2010)
125. Keller, A.D.: Model genetic circuits encoding autoregulatory transcriptional factor. J. Theor. Biol. **172**,169–185 (1995)
126. Gammaitoni, L., Hänggi, P., Jung, P., Marchesoni, F.: Stochastic resonance. Rev. Modern Phys. **70**, 223–287 (1998)
127. Lindner, B., Garcia-Ojalvo, J., Neiman, A., Schimansky-Geier, L.: Effects of noise in excitable systems. Phys. Rep. **392**, 321–424 (2004)
128. Süel, G., Garcia-Ojalvo, J., Liberman, L., Elowitz, M.: An excitable gene regulatory circuit induces transient cellular differentiation. Nature **440**, 545–550 (2006)
129. Rue, P., Garcia-Ojalvo, J.: Gene circuit designs for noisy excitable dynamics. Math. Biosci. **231**, 90–97 (2011)
130. Strelkowa, N., Barahona, M.: Switchable genetic oscillator operating in quasi-stable mode. J. Roy. Soc. Inter. **7**, 1071–1082 (2010)
131. Gonze, D.: Coupling oscillations and switches in genetic networks. Biosyst. **99**, 60–69 (2010)
132. Wang, P., Lü, J., Ogorzalek, M.J.: Synchronized switching induced by colored noise in the genetic toggle switch systems coupled by quorum sensing mechanism. Proc. 30th Chin. Contr. Conf.(CCC2011), Yantai, China, July **22–24**, 6605–6609 (2011).
133. Horsthemke, W., Lefever, R.: Noise-induced transitions: theory and applications in physics, chemistry and biology. Springer, New York (1984)
134. Risken, H.: The Fokker-Plank equation: methods of solution and application. Springer, Berlin (1992)

135. Zheng, X., Yang, X., Tao, Y.: Bistability, probability transition rate and first-passage time in an autoactivating positive-feedback loop. PLoS One **6**, e17104 (2011)
136. Masoliver, J., West, B.J., Lindenberg, K.: Bistability driven by Gaussian colored noise: First-passage times. Phys. Rev. A **35**, 3086–3094 (1987)
137. Wang, P., Zhang, Y., Lü, J., Yu, X.: Functional characteristics of additional positive feedback in genetic circuits. Nonlinear Dyn. **79**(1), 397–408 (2015)
138. Tyson, J., Novák, B.: Functional motif in biochemical reaction networks. Annu. Rev. Phys. Chem. **61**, 219–240 (2010)
139. Wang, J., Zhang, J., Yuan, Z., Zhou, T.: Noise-induced switches in network systems of the genetic toggle switch. BMC Syst. Biol. **1**, 50 (2007)
140. Shah, N., Sarkar, C.: Robust network topologies for generating switch-like cellular responses. PLoS Comput. Biol. **7**, e1002085 (2011)
141. Chen, A.: Modeling a synthetic biological chaotic system: relaxation oscillators coupled by quorum sensing. Nonlinear Dyn. **63**,711–718 (2011)
142. Zhang, Z., Ye, W., Qian, Y., Zheng, Z., Huang, X., Hu, G.: Chaotic motifs in gene regulatory networks. PLoS One **7(7)**, e39355 (2012)
143. Wang, P., Zhang, Y., Tan, S., Wan, L.: Explicit ultimate bound sets of a new hyperchaotic system and its application in estimating the Hausdorff dimension. Nonlinear Dyn. **74**, 133–142 (2013)
144. Wang, P., Li, D., Wu, X., Lü, J., Yu, X.: Ultimate bound estimation of a class of high dimensional quadratic autonomous dynamical systems. Int. J. Bifur. Chaos **21**, 2679–2694 (2011)
145. Ozbudak, E., Thattai, M., Lim, H., Shraiman, B., Van Oudenaarden, A.: Multistability in the lactose utilization network of *Escherichia coli*. Nature **427**, 737–740 (2004)
146. Shibata, T., Fujimoto, K.: Noisy signal amplification in ultrasensitive signal transduction. Proc. Natl. Acad. Sci. USA. **102**, 331–336 (2005)
147. Hornung, G., Barkai, N.: Noise propagation and signaling sensitivity in biological networks: a role for positive feedback. PLoS Comput. Biol. **4(1)**, e8 (2008)
148. Ferrell, J.E. Jr: Self-perpetuating states in signal transduction: positive feedback, double-negative feedback and bistability. Curr. Opinion Chem. Biol. **6**, 140–148 (2002)
149. Mitrophanov, A., Groisman, E.: Positive feedback in cellular control systems. BioEssays **30**, 542–555 (2008)
150. Banerjee, S., Bose, I.: Functional characteristics of a double positive feedback loop coupled with autorepression. Phys. Biol. **5(4)**, 046008 (2008)
151. Shi, C., Zhou, T., Yuan, Z.: Functional tunability of biological circuits from additional toggle switches. IET Syst. Biol. **7(5)**, 126–134 (2013)
152. Shi, C., Li, H., Zhou, T.: Architecture-dependent robustness in a class of multiple positive feedback loops. IET Syst. Biol. **7(1)**, 1–10 (2013)
153. Zhang, X., Cheng, Z., Liu, F., Wang, W.: Linking fast and slow positive feedback loops creates an optimal bistable switch in cell signaling. Phys. Rev. E **76(3)**, 031924 (2007)
154. Li, F., Long, T., Liu,Y., Ouyang, Q., Tang, C.: The *yeast* cell-cycle is robustly designed. Proc. Natl. Acad. Sci. USA. **101**, 4781–4786 (2004)
155. Wang, G., Du, C., Chen, H., Simha, R., Rong, Y., Xiao, Y., Zeng, C.: Process-based network decomposition reveals backbone motif structure. Proc. Natl. Acad. Sci. USA. **107**, 10478–10483 (2010)
156. Pigliucci, M., Murren, C. J. : Genetic assimilation and a possible evolutionary paradox: can macroevolution sometimes be so fast as to pass us by? Evolution Int. J. Org. Evolution **57**, 1455–1464 (2003)
157. Raser, J. M., O'Shea, E. K.: Noise in gene expression: origins, consequences, and control. Science **309**, 2010–2013 (2005)
158. Masoliver, J., West, B.J., Lindenberg, K.: Bistability driven by Gaussian colored noise: first passage times. Phys. Rev. A **35**, 3086–3094 (1987)

159. Wang, P., Lü, J.: Control of genetic regulatory networks: opportunities and challenges. Acta Automat. Sin. **39**, 1969–1979 (2013) (In Chinese)
160. Liu, H., Yan, F., Liu, Z.: Oscillatory dynamics in a gene regulatory network mediated by small RNA with time delay. Nonlinear Dyn. **76(1)**, 147–159 (2013)
161. Ye, Y., Kang, X., Bailey, J., Li, C., Hong, T.: An enriched network motif family regulates multistep cell fate transitions with restricted reversibility. PLoS Comput. Biol. **15(3)**, e1006855 (2019)

Chapter 4
Modeling and Analysis of Coupled Bio-molecular Circuits

Abstract This chapter discusses the dynamical and statistical analysis of some coupled circuits. Simple network motifs are building blocks of complex biological networks, and they can be coupled together to consist of complex networks. The dynamical and statistical analysis of coupled circuits is a key step to understand the complex life phenomenon. In this chapter, some works on the merged genetic circuits and genetic circuits coupled by the quorum sensing mechanism will be introduced.

4.1 Backgrounds

Regulatory molecular networks have numerous pharmacological and medical applications. Simple biological circuits function as basic building blocks of complex biological systems. The investigation on these simple circuits is the first step to understanding the life phenomena at the system level. In Chap. 3, we have introduced some works on the mathematical and dynamical analysis of simple genetic circuits. Simple circuits can be coupled together in real-world systems to play functional roles. In this chapter, on one hand, since genetic oscillators play a fundamental role in biological systems, which are found in many biological processes, such as apoptotic, metabolic, circadian rhythms, cell cycle, and morphogenic pathways, we will first introduce a work on a composite genetic oscillator [1]. On the other hand, we will introduce one of our works on the coupled bistable genetic toggle switch systems [2, 3]. The investigations on the composite or coupled genetic circuits are another important step to understanding the real-world biological systems.

J. Lü, P. Wang, *Modeling and Analysis of Bio-molecular Networks*,
https://doi.org/10.1007/978-981-15-9144-0_4

4.2 Dynamical Analysis of a Composite Genetic Oscillator

4.2.1 Related Works and Motivations

In the literature, several simple synthetic genetic oscillators have been designed [4–14]. Among which, the hyteresis-based relaxation oscillator [14] and the repressilator [6] are two well-known ones. The two simple oscillator networks are shown in Fig. 4.1a, b, respectively. Up to now, the oscillatory mechanisms and the role of oscillations in these regulatory networks have not been fully understood. Actually, oscillations in genetic circuits may be due to different mechanisms. For example, the hyteresis-based relaxation oscillator is susceptible to the Andronov–Hopf bifurcation (AHB), the saddle node on invariant circle (SNIC) bifurcation, or the homoclinic bifurcation (HCB), whereas the repressilator is susceptible to the AHB [1].

In the year 2009, Yang et al. [1] explored two oscillatory mechanisms: the hysteresis-based relaxation oscillator and the repressilator. They combined these mechanisms into one regulatory network, the composite oscillator is shown in Fig. 4.1c. In the mathematical models of the composite oscillator, only two parame-

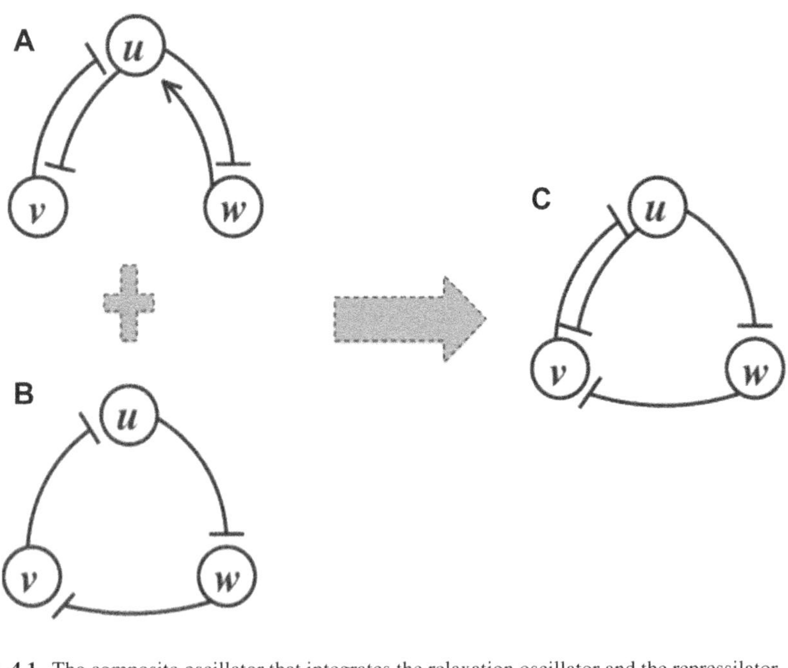

Fig. 4.1 The composite oscillator that integrates the relaxation oscillator and the repressilator. (**a**) The original relaxation oscillator network. Hammerheads show repression; arrows show activation. (**b**) The original repressilator network. (**c**) The composite oscillator. Reprinted from ref. [1], with permission from AIP Publishing

ters, the strength of an additional regulatory connection and the timescale separation for one of the variables, control the transition from one mechanism to the other. The investigations support a qualitative difference between the two oscillatory mechanisms, but in the parameter space, they found a single oscillatory region, suggesting that the two mechanisms support each other. The authors examined interactions in a basic population: that is, a pair of the composite oscillators. The authors found that the relaxation oscillation mechanism is much more resistant to oscillatory death as the cells are diffusively coupled in a population. Additionally, stationary pattern formation has been found to accompany the relaxation oscillation but not the repressilator mechanism. These properties may guide the identification of oscillatory mechanisms in complex natural regulatory networks. In the following, we briefly overview this work.

4.2.2 Mathematical Models

4.2.2.1 The Hysteresis-Based Oscillator

For the hysteresis-based oscillator proposed in Ref. [14], the circuit can be mathematically modeled as follows:

$$\begin{cases} \frac{du}{dt} &= \frac{\alpha_1}{1+v^n} + \frac{\alpha_3 w^\beta}{1+w^\beta} - u, \\ \frac{dv}{dt} &= \frac{\alpha_2}{1+u^\gamma} - v, \\ \frac{dw}{dt} &= \epsilon\left(\frac{\alpha_4}{1+u^\gamma} - w\right). \end{cases} \tag{4.1}$$

Here, each variable describes a concentration: u and v stand for two proteins and w stands for a small molecule called autoinducer. All negative terms in the right-hand side represent degradation of the molecules. The nonlinear functions reflect synthesis of the molecules from the DNA controlled by regulatory elements called promoters. The system and the diagram present a highly simplified model of the oscillatory network. As such, synthesis of the proteins and the autoinducer is modeled as one-step reactions, whereas intermediate steps such as mRNA synthesis or binding of an effector to a promoter are assumed to be very fast and, therefore, they are not explicitly shown in the model.

The oscillatory mechanism in this system is based on driving the bistable subnetwork of mutually repressing proteins u and v (Fig. 4.1a) through a hysteresis range by the autoinducer. The autoinducer concentration is under a negative feedback from the hysteretic subnetwork, leading to sustained autonomous oscillations. The dynamics of the autoinducer is made slow by introducing a small parameter ($\epsilon << 1$) as a factor in the third equation. As a result, the oscillations are made of the relaxation type.

4.2.2.2 The Repressilator

Another synthetic oscillatory network is called the repressilator. The design replicates the idea of a ring oscillator and is implemented by cyclic repression of protein synthesis as shown in Fig. 4.1b. Equation (4.2)

$$\begin{cases} \frac{du}{dt} &= \frac{\alpha}{1+v^n} + \alpha_0 - u, \\ \frac{dv}{dt} &= \frac{\alpha}{1+w^n} + \alpha_0 - v, \\ \frac{dw}{dt} &= \frac{\alpha}{1+u^n} + \alpha_0 - w. \end{cases} \tag{4.2}$$

describes the behavior of the repressilator. u and v, as well as w, are protein concentrations. α_0 represents uncontrolled leaky gene expression from the promoters. The system has been shown to oscillate both in simulations and in experiments [6].

4.2.2.3 The Composite Oscillator

The repressilator and the relaxation oscillator networks share some similar motifs (Fig. 4.1a, b. The similarities make it possible to combine the two systems with a few changes. First, the uncontrolled part of the synthesis can be omitted $\alpha_0 = 0$, because it is assumed to be small and is not a part of the oscillatory mechanism. Second, the only activation connection in the relaxation oscillator $w \rightarrow u$ is replaced with the repression of v by w: $w \dashv v$, as shown in Fig. 4.1c. These connections play the same role because the repression of v ends up with the activation of u as these two components are mutually repressively coupled. Third, Yang et al. [1] introduced a parameter ϵ as a factor in the w equation into the repressilator (Eq. (4.3)) to be able to slow down the dynamics of this component and match the timescale separation used in the relaxation oscillator. The resulting network diagram and equations are shown in Fig. 4.1c and Eq. (4.3), respectively.

$$\begin{cases} \frac{du}{dt} &= \frac{\alpha_1}{1+v^n} - u, \\ \frac{dv}{dt} &= \frac{\alpha_1}{1+w^n} + \frac{\alpha_2}{1+u^n} - v, \\ \frac{dw}{dt} &= \epsilon\left(\frac{\alpha_1}{1+u^n} - w\right). \end{cases} \tag{4.3}$$

Dynamical analysis was performed via a differential equation simulator and numerical bifurcation analysis tool, XPPAUT [15].

4.2.3 *Dynamical Analysis of the Merged Genetic Oscillator*

The composite network as shown in Fig. 4.1c can display both relaxation oscillation and repressilator. For example, when one sets $\epsilon = 1, \alpha_2 = 0$, the composite network is a repressilator, while for $\epsilon = 0.01, \alpha_2 = 2$, the network displays relaxation oscillation. Thus, the network as shown in Fig. 4.1c contains both the relaxation oscillation mechanism and the repressilator mechanism as limiting cases. Two parameters control both oscillatory mechanisms: the strength of the u to v connection α_2 and the parameter of timescale separation ϵ. Therefore, the problems of characterization and comparison of the oscillatory mechanisms can be reduced to the exploration of a two-dimensional parameter space.

4.2.3.1 The Two Oscillatory Mechanisms Support Each Other

The authors investigated the transition between the two limiting cases as the parameters change. Several disjoint oscillatory regions may exist around the parameter values at which one of the pure oscillatory mechanisms is realized. This would suggest that the mechanisms cannot merge but rather compete with each other. Alternatively, a single parameter region where oscillations persist may include both limiting cases. This would suggest that the mechanisms can be combined or support each other to generate a common rhythm.

Figure 4.2c presents the bifurcation diagram for the composite oscillator on the (α_2, ϵ) plane with the parameter α_1 fixed at 5. The shaded oscillatory region spans values of parameters from the relaxation case at the α_2-axis to the repressilator case on the ϵ-axis. Most of the boundaries of the region correspond to the AHB of an equilibrium state [16]. The AHB1 branch separates the oscillatory region from the origin. The upper boundary from the same bifurcation type is denoted as AHB2. The rest of the oscillatory region's boundaries correspond to the interruption of the high-amplitude limit cycle by equilibrium states. The characteristic of the corresponding bifurcations is the growing period of the limit cycle as the parameter approaches the boundary. These bifurcations are of two types: the SNIC and the homoclinic bifurcation (HCB). Taken together, the bifurcation diagram suggests that the two oscillatory mechanisms support each other.

4.2.3.2 Oscillatory Mechanisms Are Distinct

There are clear distinctions between the two oscillatory mechanisms. The oscillatory region is separated from the origin, meaning that the repressilator mechanism cannot function for small ϵ. Similarly, the relaxation oscillator cannot function without the u to v connection $\alpha_2 = 0$. The authors reported that the loss of relaxation oscillations as α_2 decreases is caused by the loss of bistability in the (u, v) subnetwork. The repressilator mechanism requires the system to be truly 3D, that is, it requires all timescales to be similar. The distinctions between the two oscillatory mechanisms are that one requires bistability in a subnetwork and separation of timescales, whereas the other does not use bistability and requires all timescales to be similar. The authors further reported that the mechanisms of instability can be characterized by bifurcations that define boundaries of the oscillatory region: the repressilator is susceptible to the AHB only, whereas the SNIC and HCB are also possible for the relaxation oscillations and the mixed mechanism, respectively.

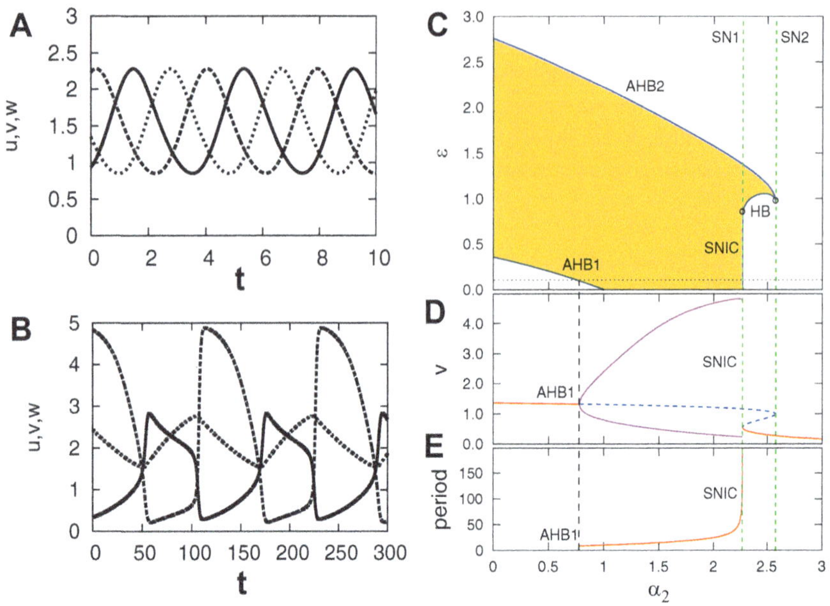

Fig. 4.2 Time series and oscillatory region in parameter space of the composite network (4.3). (**a**) Pure repressilator oscillations $\epsilon = 1, \alpha_2 = 0$. (**b**) Relaxation oscillations $\epsilon = 0.01, \alpha_2 = 2$. For both cases, $\alpha_1 = 5$ and $n = 3$. (**c**) Oscillations persist in the shaded region on the (α_2, ϵ) plane. (**d**) One-parameter diagram illustrating bifurcation transitions at the boundaries of the region at fixed $\epsilon = 0.1$. (**e**) The dependence of the period for the solution within the oscillatory region at the same fixed ϵ. Reprinted from ref. [1], with permission from AIP Publishing

4.2.4 Population Dynamics of Coupled Composite Oscillators

In the literature, there are different ways to couple the oscillators. All of them use an intercell signaling via the quorum sensing mechanism [17]. That is, they use the diffusion of a small molecule—the autoinducer. However, some works [11, 12] suggested addition of a subnetwork that provides coupling to each of the cells, whereas other works encapsulate the quorum sensing system into the regulatory network for the oscillator [14]. By following the method as proposed in [14], the mathematical models for the coupled composite oscillators can be established in Eq. (4.4).

$$\begin{cases} \frac{du_i}{dt} &= \frac{\alpha_1}{1+v_i^n} - u_i, \\ \frac{dv_i}{dt} &= \frac{\alpha_1}{1+w_i^n} + \frac{\alpha_2}{1+u_i^n} - v_i, \\ \frac{dw_i}{dt} &= \epsilon\left(\frac{\alpha_1}{1+u_i^n} - w_i\right) + d(w_e - w_i), \\ \frac{dw_e}{dt} &= \frac{d}{N}\frac{\rho}{1-\rho}\sum_{j=1}^{N}(w_j - w_e). \end{cases} \tag{4.4}$$

Here, w_e denotes the extracellular concentration of the autoinducer; N represents the number of coupled cells ($i = 1, 2, \cdots, N$); and the parameter d represents the permeability of the cell membrane to autoinducer molecules. The parameter ρ represents the population density, with $0 \leq \rho \leq 1$. $\rho = 0.5$ approximates the population in which half of the total volume is occupied by the cells.

Based on Eq. (4.4) and dynamical analysis, the authors reported that the repressilator oscillations will die out at moderate permeability; the synchronization of the repressilators can be re-established at strong coupling; and oscillations by the relaxation mechanism can persist for all values of the permeability. Moreover, the authors also found that the stationary pattern formation accompanies the relaxation oscillation mechanism; the synchronous oscillations and stationary patterns are resistant to noise.

For $N = 2$, Fig. 4.3 shows time series in different regimes for the noise intensity of about 4% of the maximal synthesis rate α_1 in the model. For all traces, the permeability was taken high to allow the comparison of synchronous repressilator oscillations and those by the hysteresis-based mechanism at the same coupling strength. Both types of oscillations remain highly synchronized as shown in Fig. 4.3b, c. On the other hand, regularity of repressilator oscillations (Fig. 4.3d) is greatly lost under the influence of noise, whereas hysteresis-based relaxation oscillations remain almost periodic (Fig. 4.3e). The frequencies of these oscillations are distinct (note the different timescales in Fig. 4.3d, e), and the frequency difference may contribute to the resistance to noise. Additionally, the amplitude of oscillations is significantly different. The lower amplitude effectively strengthens the influence of the noise on the repressilator oscillations and leads to the reduction in their regularity. The lower amplitude is tightly linked to the susceptibility of the repressilator mechanism to oscillatory death. By contrast, the robustness of the hysteresis-based oscillations to the influence of noise can be linked to their persistence under changes in parameters of the model. Additionally, the stationary pattern formed at the same permeability is also not significantly affected by the noise (Fig. 4.3f). Therefore, the reported synchronization and pattern formation properties are robust to noise of a small intensity.

4.2.5 Summary

The authors have combined the repressilator and the relaxation oscillator into one system so that only two parameters, the strength of an additional regulatory connection and the timescale separation for one of the variables from the rest of the system, control the transition from one mechanism to the other. Investigations support a qualitative difference between the oscillatory mechanisms: one requires the separation of timescales as well as bistability in a subnetwork achieved via the additional connection, whereas the other does not use bistability and requires all timescales to be similar. In the parameter space, though, the authors found a single oscillatory region that includes both relaxation oscillations and those by the

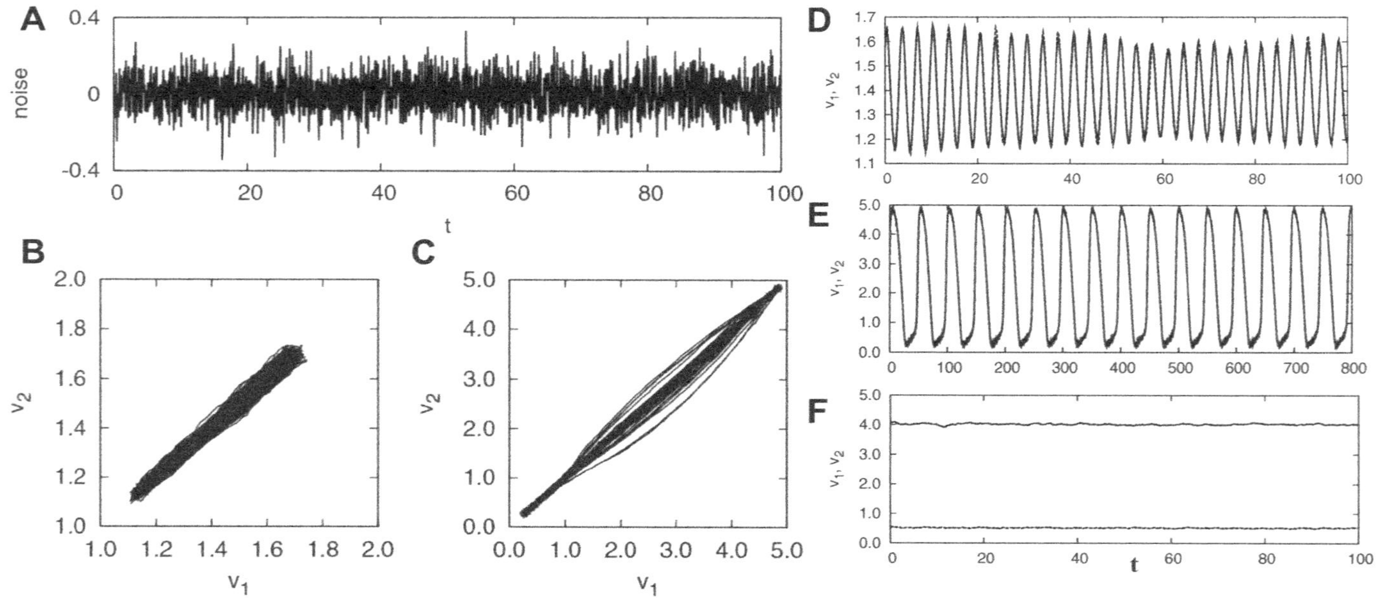

Fig. 4.3 Population dynamics coupled via the quorum sensing mechanism. (**a**) Noise introduced into the right-hand side of each equation. (**b**) and (**c**) The trajectory stays close to the diagonal in the v_1, v_2 projection, indicating synchronous oscillations under noise for the case of repressilator and hysteresis-based oscillations, respectively. (**d**) The repressilator oscillations display high irregularity. (**e**) The oscillations by hysteresis-based mechanism remain highly regular under noise of the same intensity. (**f**) The stationary pattern persists under the noise. In (**b**) and (**d**), $\epsilon = 3$ and $\alpha_2 = 0$. In (**c**) and (**e**), $\epsilon = 0.05$ and $\alpha_2 = 2$. In (**f**), $\epsilon = 3$ and $\alpha_2 = 2$. The permeability is the same in all cases $d = 20$. Reprinted from ref. [1], with permission from AIP Publishing

repressilator mechanism as limiting cases. This suggests that the two oscillatory mechanisms support each other in the composite network. The mechanisms may be characterized by bifurcations that define boundaries of the oscillatory region: the repressilator is susceptible to the AHB only, whereas the SNIC and HCB are also possible for the relaxation oscillations and the mixed mechanism, respectively.

The composite oscillator offers two ways of tuning its period, which may be relevant to distinct biological oscillators. First, the period can be increased by a transition to the relaxation case. Second, the period can also grow toward infinity without the timescale separation as the parameters approach a boundary of the oscillatory region. The growth is due to the type of bifurcation transitions at the boundary: SNIC or HCB. The proximity of this bifurcation is displayed as slowing at one point along the oscillatory solution, whereas in the relaxation case, long parts of the solution become slow. This suggests that the SNIC or HCB is more relevant to checkpoints along the cell division cycle [18], at which the progression may stop until the completion of the previous phase. Changes in the dynamics caused by interaction between cells will characterize the oscillatory mechanism. Therefore, synchronization and pattern formation properties may guide the identification of the oscillatory mechanisms in complex natural regulatory networks.

The merger of the oscillatory mechanism may be also interesting in light of other applications. In neuroscience, many types of neurons possess multiple oscillatory mechanisms [19]. The most well known is the spiking mechanism, but pharmacological blockade of spikes often reveals other types of oscillations [20, 21]. The neural oscillatory mechanisms share common variables and may function separately or together. The representation of observed oscillations as produced by a combination of the mechanisms may be instrumental for understanding the complex firing patterns and synchronization properties of neurons.

4.3 Modeling and Analysis of the Genetic Toggle Switch Circuit

Noise can induce various dynamical behaviors in nonlinear systems. White noise perturbed systems have been extensively investigated during the last decades. In gene networks, experimentally observed extrinsic noise is colored. As an attempt, this section investigates the genetic toggle switch circuit perturbed by colored extrinsic noise and with kinetic parameters. This section is mainly based on our work [2, 3]. Compared with white noise perturbed systems, we show that there also exists an optimal colored noise strength to induce the best stochastic switch behaviors in the single toggle switch, and the best synchronized switching in the networked systems, which demonstrate that noise induced optimal switch behaviors are widely in existence. Moreover, under a wide range of system parameter regions, we find there exist wider ranges of white and colored noise strengths to induce good switch and synchronization behaviors. Moreover, we found that white noise is beneficial for switch, and colored noise is beneficial for population synchronization. Our observations are very robust to extrinsic stimulus strength, cell density, and diffusion rate. Finally, based on the Waddington's epigenetic landscape and the Wiener–Khintchine theorem, physical mechanisms underlying the observations are interpreted. The investigations can provide guidelines for experimental design and have potential clinical implications in gene therapy and synthetic biology.

4.3.1 Related Works and Motivations

Bistable systems arise within biological systems ranging from the λ phage switch [22, 23] in bacteria to the cellular signal transduction pathways in mammalian cells [24–26]. Generally speaking, a positive feedback loop with certain ultra-sensitivity can result in multi-stability [27]. Under the excitation of external stimuli, bistable systems can switch between two distinct stable states but cannot rest on intermediate state [28]. Usually, bistable systems in the biological context involve in the generation of switch-like biochemical responses, the establishment of cell cycle oscillations and mutually exclusive cell cycle phases, the production of self-sustaining biochemical "memories" of transient stimuli, or the rapid lateral propagation of receptor tyrosine kinase activation [24–31]. Bistable systems are building blocks of large-scale networks, such as genetic networks and signaling cascades. Investigations on the dynamical properties and characteristics of bistable systems are therefore fascinating.

Noise plays a vital role in biological systems. Gene expression is a stochastic, or "noisy" process [32–43]. Noise arises from two ways. The inherent stochasticity of biochemical processes, such as transcription and translation generate intrinsic noise, and fluctuations in the amounts or states of other cellular components lead indirectly to variation in the expression of a particular gene, and thus represent extrinsic noise [32, 34]. Both intrinsic and extrinsic noises can be utilized by cells to perform certain biological functions. As an interesting biological phenomenon, noise induced switch or synchronized switching in a population of cells can be used to explain some biological behaviors, such as cell communication [44, 45], chemotaxis, cell differentiation [46, 47], and cell memory [27]. In 2006, Tian and Burrage [31] developed a novel stochastic model to study large-scale GRNs, that is, they introduced Poisson random variables into deterministic models described by ODEs. Two stochastic models for the toggle switch interfaced with either the SOS signaling pathway or a quorum sensing (QS) signaling pathway [40] were investigated, and bimodal population distributions were successfully observed in both cases, which is a typical feature of bistability. Zhou and coauthors [28, 37, 38] investigated the effects of noise on the single toggle switch and networked systems of the toggle switch interfaced by the QS and found that both intrinsic and extrinsic noises can induce switch behaviors, for the networked systems, called synchronized switching. Moreover, they found that there is an optimal noise strength to induce the best switch behavior. Furthermore, for networked systems, they found that the extracellular additive noise can lead to robust collective rhythm. However, most of the investigations supposed that the extrinsic noise is Gaussian white noise. Recently, Shahrezaei et al. [41], Rosenfeld et al. [43], and Lei [42] suggested that the extrinsic fluctuations have a nonzero average life time, namely, they are colored. White noise perturbed toggle switch systems have been extensively studied [28, 31, 35–39] in the last decades, while colored noise perturbed systems are seldom considered [3].

Additionally, due to random molecular collisions and thermal noise, reaction rates should be kinetic in real biochemical reaction processes. In 1995, Jansen [48] developed a stochastic simulation method based on the MC and the first reaction method [49] to simulate chemical reactions with kinetic reaction rates. Recently, Shahrezaei et al. [41] and Lei [42] investigated the stochasticity in gene expression by introducing fluctuations to deterministic ODEs through time-dependent reaction rates. Therefore, the investigation on the toggle switch systems with kinetic parameters will be more realistic.

Motivated by the mentioned problems, we investigate the genetic toggle switch system in *E.coli*, where both the single system and the networked systems will be considered under colored extrinsic noise and kinetic parameters. Figure 4.4a shows the single toggle switch system, where two genes $lacI$, cI produce proteins $LacI$ (Denoted by x) and λcI (Denoted by y). The two proteins act as TFs to repress the expression of the counterpart genes. Figure 4.4b shows the networked

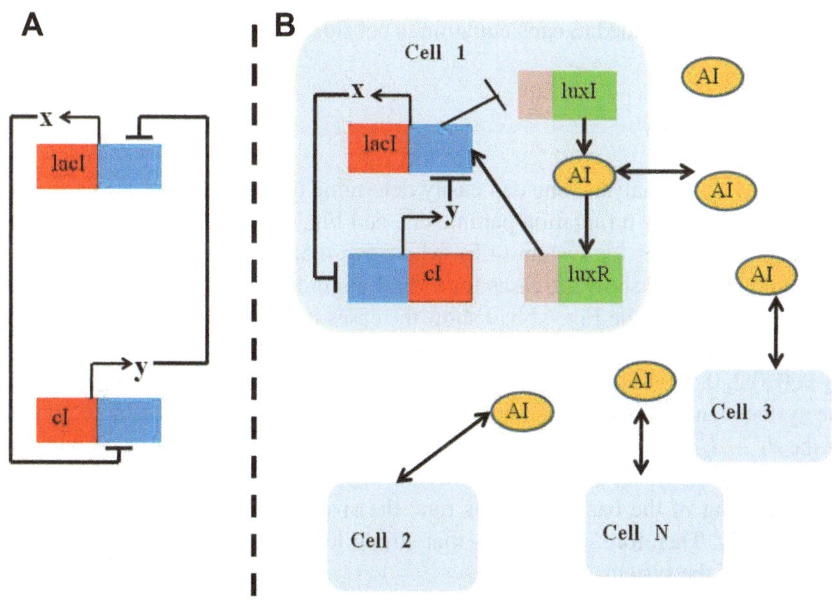

Fig. 4.4 Topological structures of the single toggle switch and the networked toggle switch systems. (**a**) The single toggle switch system. (**b**) The networked toggle switch systems coupled by the QS [28, 38, 50]. ©[2015] IEEE. Reprinted, with permission, from ref. [2]

toggle switch systems interfaced by the QS [11, 28, 31, 38, 50]. The QS signaling apparatus is incorporated into the toggle switch by placing the gene that encodes *LuxI* under the control of the protein *LacI*. *E. coli* can use this mechanism to respond to fluctuations in cell population density, producing and releasing chemical signal molecules, called autoinducers (AI) [40].

4.3.2 Modeling and Analysis of the Single Toggle Switch System

4.3.2.1 Deterministic Model

The modified deterministic ODE model for the toggle switch system from [5] is described as:

$$\begin{cases} \frac{dx}{dt} = \frac{\alpha_1}{1+y^{n_1}} - d_1 x + \gamma_1, \\ \frac{dy}{dt} = \frac{\alpha_2}{1+x^{n_2}} - d_2 y + \gamma_2, \end{cases} \tag{4.5}$$

where x and y represent the dimensionless concentrations of the repressors $LacI$ and λcI respectively. α_i, d_i, γ_i $(i = 1, 2)$ are dimensionless transcription rates, degradation rates, and basal synthesis rates, respectively. n_1, n_2 are the Hill coefficients. The main difference of model (4.5) from Ref. [5] lies in that the basal synthesis rates are added to each equation in consideration of the leaky expression.

4.3.2.2 Bistability

From bifurcation analysis, one can easily determine the bistablity of Eq. (4.5). We choose α_i and d_i as bifurcation parameters, and Fig. 4.5 shows the two-parameter bifurcation diagrams for system (4.5), where Fig. 4.5a, b show the cases for $n_i = 1.6$, and Fig. 4.5c, d show the cases for $n_i = 4$. Figure 4.5a, c show the bifurcations in $\alpha_1 - \alpha_2$ plane, while Fig. 4.5b, d show the cases in $d_1 - d_2$ plane. For $n_i = 1.6$, bifurcations under $\gamma_1 = \gamma_2 = 0, 0.1$ are plotted. For $n_i = 4$, bifurcations with $\gamma_1 = \gamma_2 = 0, 0.2, 0.5, 0.8$ are investigated. From Fig. 4.5, we can draw the conclusion that the system can all display bistability with the chosen parameters. Bistable regions in the $d_1 - d_2$ plane are bounded, and the sizes of bistable region shrink with the increasing of γ_i. Bistable regions in the $\alpha_1 - \alpha_2$ plane are all unbounded, but with the increasing of the basal synthesis rate, the sizes of bistable areas also reduce considerably. Therefore, it indicates that strong leaky expression can weaken the bistability of the system.

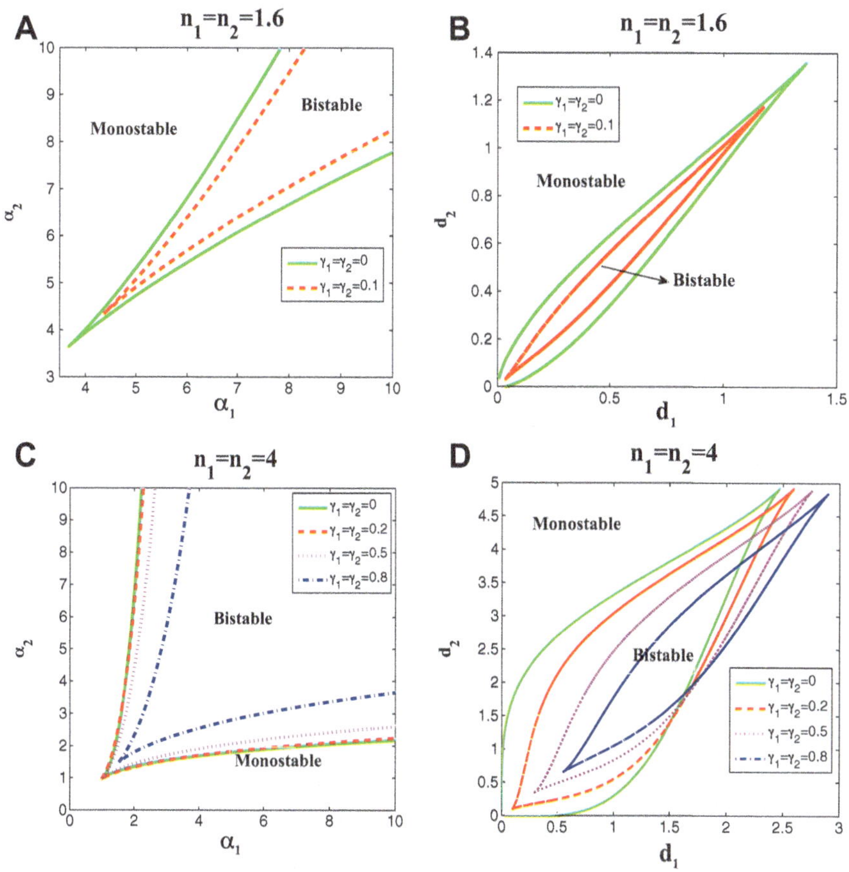

Fig. 4.5 Two-parameter bifurcation diagrams for system (4.5). (**a**) and (**c**) α_1 versus α_2 with fixed $d_1 = d_2 = 1$. (**b**) and (**d**). d_1 versus d_2 with fixed $\alpha_1 = \alpha_2 = 5$ for (**b**) and fixed $\alpha_1 = 2.5, \alpha_2 = 5$ for (**d**). In (**a**) and (**b**), $n_1 = n_2 = 1.6$, and bifurcations under $\gamma_1 = \gamma_2 = 0, 0.1$ are plotted. While (**c**) and (**d**) show the case for $n_1 = n_2 = 4$, and diagrams for $\gamma_1 = \gamma_2 = 0, 0.2, 0.5, 0.8$ are plotted. ©[2015] IEEE. Reprinted, with permission, from ref. [2]

4.3.2.3 Stochastic Model for the Single Toggle Switch System

For the single toggle switch system, dimensionless maximal transcription rates control the production rates of proteins, which are crucial in gene expression processes. The transcription rates have also been frequently used as bifurcation parameters in dynamical analysis of gene systems [5, 11, 28, 38, 50]. Therefore, we choose α_i as kinetic parameters and also incorporate the extrinsic fluctuations into system (4.5), and the stochastic mathematical model for the toggle switch system can be described as:

$$\begin{cases} \frac{dx}{dt} = \frac{\alpha_1(t)}{1+y^{n_1}} - d_1x + \gamma_1 + \zeta(t) + Asin(\Omega t), \\ \frac{dy}{dt} = \frac{\alpha_2(t)}{1+x^{n_2}} - d_2y + \gamma_2, \end{cases} \tag{4.6}$$

where $\alpha_i(t)(i = 1, 2)$ represents time-dependent maximal transcription rates. In the following, we suppose $\alpha_i(t) \sim N(\alpha_i, \sigma^2)$, where α_i is the average maximal transcription rate and σ^2 describes the fluctuation or the heterogeneity of the average rate [51]. $\zeta(t)$ represents the extrinsic noise.

For Gaussian white noise, it satisfies $\langle\zeta(t)\rangle = 0$; $\langle\zeta(t)\zeta(t')\rangle = D_{ext}\delta(t - t')$, where D_{ext} denotes the noise strength. For colored extrinsic noise, since the Ornstein–Uhlenbeck (OU) process is frequently used to consider colored noise in the context of genetic regulatory systems [42, 52], we simply generate an OU time series for consideration:

$$\frac{d\zeta(t)}{dt} = -\frac{\zeta(t)}{\tau} + \sqrt{2/\tau}\xi(t), \tag{4.7}$$

where $\zeta(t)$ is the colored noise, τ is the autocorrelation time. $\xi(t)$ is the Gaussian white noise that satisfies:

$$\langle\xi(t)\rangle = 0, \langle\xi(t)\xi(t')\rangle = D_{ext}\delta(t - t'),$$

which can be used to generate the colored one. Here, we call D_{ext} as the colored noise strength. It is easy to obtain from the Fourier transformation that the process defined by (4.7) satisfies:

$$\langle\zeta(t)\rangle = 0, \langle\zeta(t)\zeta(t')\rangle = D_{ext}exp(-|t - t'|/\tau).$$

$Asin(\Omega t)$ represents the periodical stimulus from outside, which has been widely used to investigate the stochastic resonance [53] and switch behaviors in bistable systems [28, 37, 38], as well as oscillator systems [54, 55]. In fact, it can be analogy to extrinsic day-light stimulus.

In the following, for the single toggle switch system, we choose $n_1 = n_2 = 1.6, \alpha_1 = \alpha_2 = 5, d_1 = d_2 = 1, \gamma_1 = \gamma_2 = 0$, and $A = 0.25, \Omega = 2\pi/400, \tau = 10$, which are mainly from [5, 28, 37]. Initial values for x, y will be taken as $x(0) = 3.4, y(0) = 0.64$. The SDE (4.6) will be solved by the Euler–Maruyama method [56].

4.3.3 Modeling the Networked Toggle Switch Systems

The sketch map of the networked genetic toggle switch systems is shown in Fig. 4.4b. The coupled network system combines two features: acting as a toggle switch and using an intracellular signaling mechanism to couple among cells [28]. The QS system enables cells to sense population density through a TF $LuxR$, which acts as a transcription activator of gene $lacI$ when AI binds to it [57, 58]. The AI is synthesized by the protein encoded by the gene $luxI$, which can diffuse across the cell membrane, resulting in two kinds of concentrations: the extracellular concentration of AIs in the environment and the AIs concentration in individual cells. The property of the QS mechanism has been successfully used in the design of some cellular communication systems, where collective rhythmic behaviors across a cell population can be achieved [12, 14, 44, 59, 60]. For theoretical analysis about the synchrony in the QS networks, one can refer to Ref. [14, 61].

Noted that in Ref. [28], both intrinsic and extrinsic noises have been considered in the system, and constant reaction rates are assumed, that is, the parameters are time-independent for all cells. Different from their model, we consider the synchronized switching behaviors under kinetic reaction parameters and extrinsic noise, where both white and colored extrinsic noises will be considered. Just as the single cell case, we only consider the time dependence of the transcription rates for each cell, and the other parameters are assumed to be time independent. The mathematical model for the QS network can be finally established as:

$$\begin{cases} \frac{dx_i}{dt} &= \frac{\alpha_{1i}(t)}{1+y_i^{n_1}} - d_1 x_i + \gamma_1 + \frac{\beta A_i}{1+A_i}, \\ \frac{dy_i}{dt} &= \frac{\alpha_{2i}(t)}{1+x_i^{n_2}} - d_2 y_i + \gamma_2, \\ \frac{dA_i}{dt} &= \varepsilon y_i - \mu A_i + k_i(A_e - A_i), \\ \frac{dA_e}{dt} &= \frac{Q}{N}\sum_{i=1}^{N}(A_i - A_e) - d_e A_e + \zeta(t) + Asin(\Omega t)\,. \end{cases} \tag{4.8}$$

which is similar to that in Ref. [28], with the difference on the kinetic parameters $\alpha_{ji}(t), j = 1, 2, i = 1, \ldots, N$. In system (4.8), x_i and y_i are the concentrations of $LacI$ and λCI of the $i'th$ cell, respectively. $\alpha_{ij}(t)$ is the maximal dimensionless transcription rate of protein $i(i = LacI, \lambda cI)$ in the $j'th$ cell at time t, and we suppose they are normally distributed, that is $\alpha_{ij}(t) \sim N(\alpha_i, \sigma^2)$, where α_i is the average transcription rate for $LacI(i = 1)$ and $\lambda cI(i = 2)$. σ^2 describes the fluctuation of time-dependent rate $\alpha_{ij}(t)$ and can be seen as the fluctuation of transcription processes with time. A_i is the concentration of intracellular signal molecules AI of the $i'th$ cell, whereas A_e is the concentration of AI in extracellular environment. β is the maximal contribution of saturating amounts of AI to the transcription of the $LacI$. k_i and Q measure the diffusion rates of AI across the cell membrane and cell density. ε and μ are the synthesis and degradation rates of the intracellular AI. d_e is the degradation rate of the extracellular AIs. $Asin(\Omega t) and \zeta(t)$ stand for the extracellular stimulus and environment perturbation, just as the one in the system (4.6).

In the following investigation, for system (4.8), we choose parameters $\alpha_1 = 2.5, \alpha_2 = 5, d_1 = d_2 = 1, \gamma_1 = \gamma_2 = 0.5, n_1 = n_2 = 4, d_e = 3, A = 0.08, \Omega = 2\pi/400, k = 10, \beta = 15, \varepsilon = 0.07, \mu = 1, Q = 0.5, \tau = 10$, unless otherwise noted, which are mainly from [28]. Furthermore, the initial values for each equation in system (4.8) are randomly chosen.

4.3.4 Statistical Measurements

To quantitatively evaluate the relative performance of the stochastic switch behaviors, the spectral amplification factor according to Ref. [28, 53] is verified to be a perfect measure, defined as:

$$\eta = 4A^{-2} \left|\langle e^{\vartheta \Omega t} M(t)\rangle\right|^2, \tag{4.9}$$

where

$$M(t) = \frac{1}{N}\sum_{i=1}^{N} x_i(t),$$

$\vartheta^2 = -1$, N represents cell numbers. As to the system (4.6), $N = 1, x_1 = x$ represents the concentration of protein $LacI$. While for system (4.8), x_i represents concentration of $LacI$ in the $i'th$ cell. $\langle . \rangle$ is the average over time. Noted that the bigger η, the better the switch behaviors.

For the networked toggle switch systems, more complex dynamical behaviors can be induced by noise. Noise may induce synchronization of the coupled cells and switch behaviors in each cell, and such behaviors are called synchronized switching [28]. For synchronized switching, the indicator introduced in (4.9) can characterize the switch behaviors. Synchronization phenomenon among cell population can be measured by the average synchronization error (ASE) [28, 38], defined as:

$$ASE = \left\langle \frac{1}{C_N^2} \sum_{i>j} \left[x_i(t) - x_j(t)\right]^2 \right\rangle, \tag{4.10}$$

where

$$C_N^2 = \frac{N(N-1)}{2}.$$

N represents the total number of cells. $\langle . \rangle$ also denotes the average over time. If all the coupled cells were synchronized, $ASE \approx 0$. Otherwise, ASE would be above zero.

4.3.5 Stochastic Switch in the Single Toggle Switch System

In the following, we investigate whether colored or white extrinsic noise can induce stochastic switch behaviors in system (4.6). We choose $\sigma^2 \in [0, 0.1]$, $D_{ext} \in [0, 0.5]$. Then the spectral amplification factor η computed in these intervals is shown in Fig. 4.6. Figure 4.6a shows the case with white extrinsic noise, while Fig. 4.6b shows the case with colored one. From both panels, one can easily find that at each fixed σ^2, there always exists an optimal D_{ext} that corresponds to the biggest η. That is, there exists an optimal extrinsic noise strength to induce the best switch behavior. Here, for $\sigma^2 = 0$, the observation under white extrinsic noise corresponds to the investigation in Ref. [37]. Moreover, by comparing Fig. 4.6a with Fig. 4.6b, one can see that, in the parameter space $[0, 0.5] \times [0, 0.1]$, when extrinsic noise is white, there exists a large area that makes $\eta > 70$, while it comes to colored noise, the area becomes much smaller. Therefore, we can draw the conclusion that white noise is beneficial for switch.

To illustrate the abovementioned observations more explicitly, we consider the case for $\sigma^2 = 0$ and $\sigma^2 = 0.05$, and Fig. 4.7 shows the curves η versus D_{ext}. It

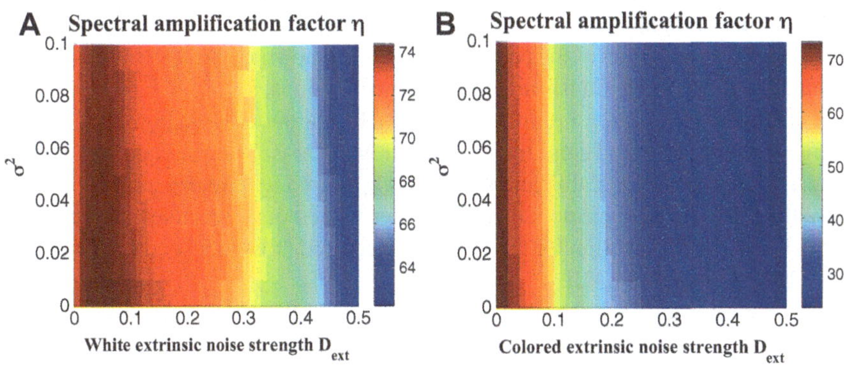

Fig. 4.6 The spectral amplification factor η versus σ^2 and extrinsic noise strength D_{ext}. (**a**) Extrinsic noise is white. (**b**) Colored extrinsic noise. Each figure is averaged over 100 simulation runs. ©[2015] IEEE. Reprinted, with permission, from ref. [2]

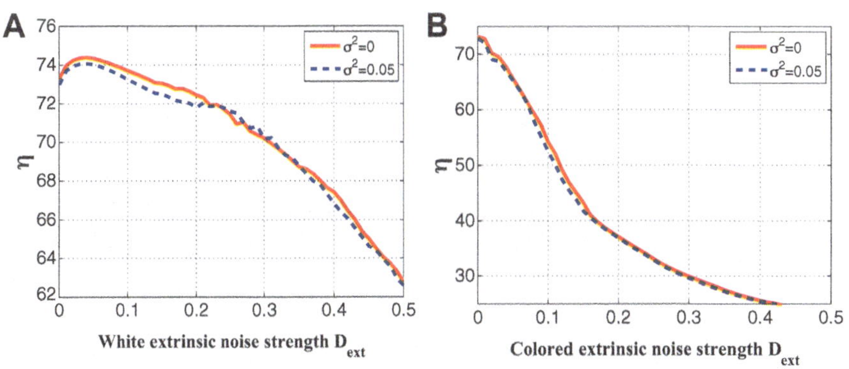

Fig. 4.7 The spectral amplification factor η versus D_{ext} for $\sigma^2 = 0$ and $\sigma^2 = 0.05$. (**a**) Extrinsic noise is white. (**b**) Colored extrinsic noise. Each curve is averaged over 100 simulation runs. ©[2015] IEEE. Reprinted, with permission, from ref. [2]

can be seen that, no matter under what kinds of extrinsic noise sources, there always exists an optimal D_{ext} to make η to reach its maximum value. The optimal D_{ext} under white extrinsic noise is around $D_{ext} = 0.04$, while it is 0.01 under colored noise. Moreover, η becomes smaller due to fluctuation in the reaction rates. Fig. 4.8 shows the concentration evolutions of the two proteins under different extrinsic noise strengths D_{ext} and variance σ^2. From Fig. 4.8, one can see that for D_{ext} near its optimal value, the switch behavior is more successive, whereas, for large D_{ext}, the switch behavior becomes irregular and blurred. Under the same initial conditions, the extrinsic white noise induced switch behaviors are better than that under the colored one.

Interestingly, from Fig. 4.8, it can be seen that the time evolutions of proteins x, y switch alternatively, one at a high state, and the other one must be at a low state, which is because the protein $LacI$ acts as an inhibitor, and it can repress the expression of gene cI. Therefore, if the concentration of $LacI$ is high, the expression of λcI must be low, and vice versa. The bistability of the system leads to the bimodal distribution of the protein numbers, as shown in Fig. 4.9. In Fig. 4.9, the molecule numbers are transformed from concentrations, since for *E.coli*, there are about 500 molecules each 1 μmol.L^{-1} [37, 55]. Furthermore, from Fig. 4.9, there is a longer tail under colored noise than that under white noise. In fact, under the colored noise and for 1000 cells, the number of cells appeared with high x protein number (>2400) is about 7 times more than that under white noise. We note that the longer right tail of the distribution for the case under colored noise is just corresponding to the observations in [41, 42].

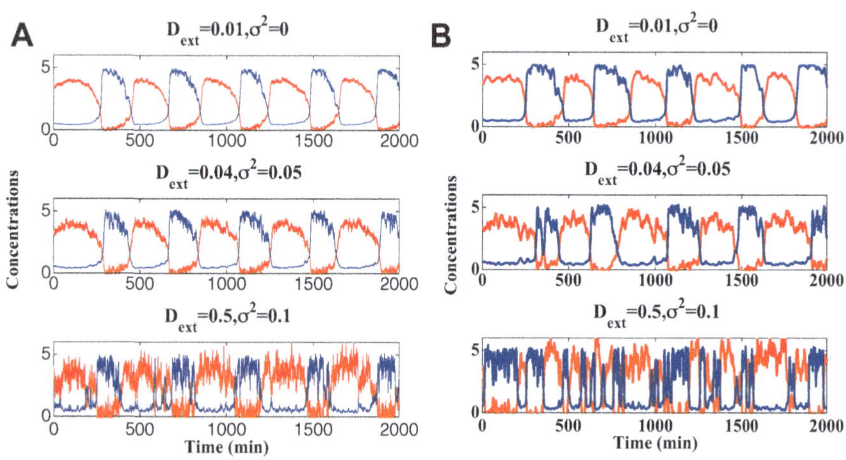

Fig. 4.8 Time evolutions of the protein concentrations under different D_{ext} and σ^2. (**a**) Extrinsic noise is white; (**b**) Colored extrinsic noise. Each figure is obtained from a random simulation run. ©[2015] IEEE. Reprinted, with permission, from ref. [2]

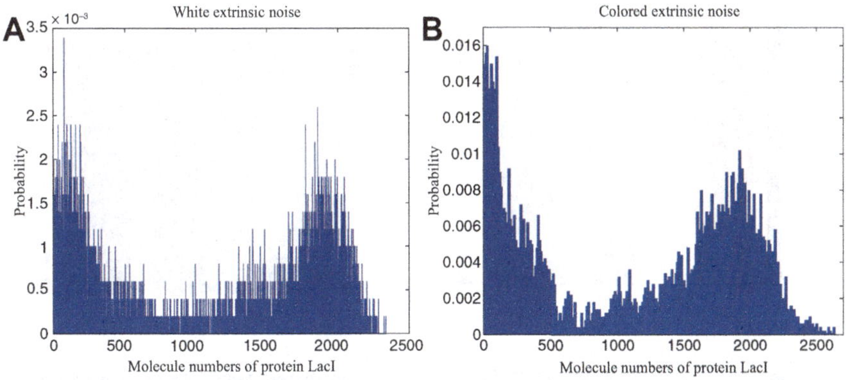

Fig. 4.9 Probability distributions of the molecule numbers for the protein $LacI$ from system (4.6). (**a**) Under white extrinsic noise. (**b**) Colored extrinsic noise. Where the systems can always display bimodal distributions under the two cases. Here $\alpha_i \sim N(5, 0.05)$, $D_{ext} = 0.04$. ©[2015] IEEE. Reprinted, with permission, from ref. [2]

4.3.6 *Synchronized Switching in Networked Toggle Switch Systems*

4.3.6.1 Feature Comparison Between White and Colored Noises Induced Synchronized Switching

Just like the discussions for the single cell case, we suppose transcription rate $\alpha_{1i}(t) \sim N(2.5, \sigma^2), \alpha_{2i}(t) \sim N(5, \sigma^2)$. We further assume $\sigma^2 \in [0, 0.1]$ and vary D_{ext} in the interval $[0, 0.1]$, and the spectral amplification factor η and the average synchronization error ASE versus σ^2 and D_{ext} are plotted in Fig. 4.10. Figure 4.10a, b show the cases under white extrinsic noise, and Fig. 4.10c, d show the cases under colored extrinsic noise.

From Fig. 4.10, under two different extrinsic noise sources, at each fixed σ^2, there exists an optimal D_{ext} to achieve the maximum η. Under white noise, the optimal D_{ext} is around 0.02, while it is 0.01 under colored noise. As one of the conclusions in Ref. [41], Shahrezaei and coauthors suggested that extrinsic fluctuations can affect the performance of genetic networks. Furthermore, colored noise can speed up typical network response time. In our investigation, network systems can quickly reach the best switch at smaller D_{ext} under colored noise, where the switch can also be seen as a response to extrinsic stimulus, therefore, in this sense, our investigations are coincident with the conclusion derived by Shahrezaei et al. Moreover, under colored noise, the overall values of η are relatively smaller than that under white noise. More interestingly, under colored noise, the area with $\eta > 200$ is much smaller than that under white noise. Therefore, one can derive a similar conclusion as that in the single cell case, white noise is beneficial for the switch behaviors.

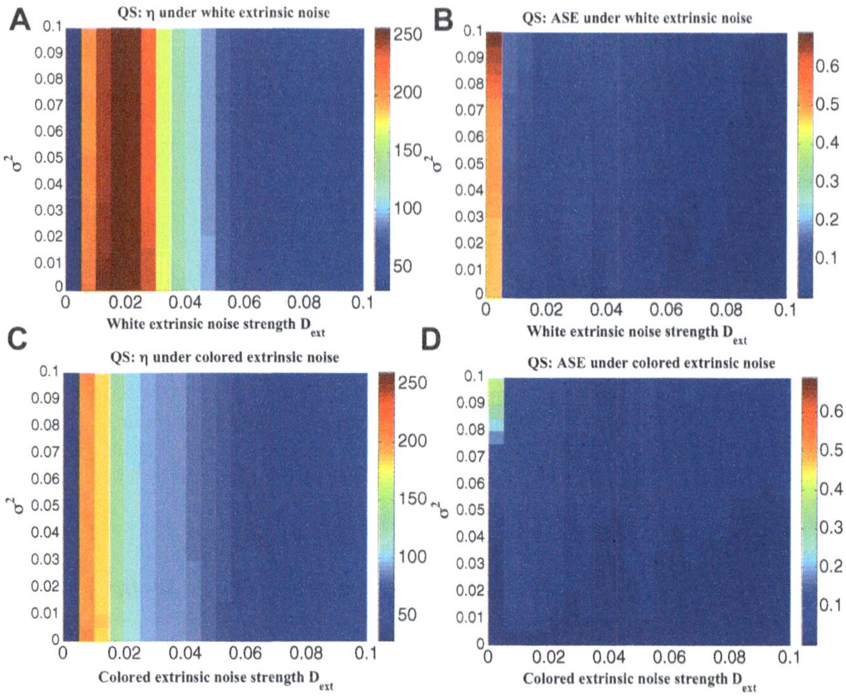

Fig. 4.10 The spectral amplification factor η, the average synchronization error ASE versus σ^2, and extrinsic noise strength D_{ext}. (**a**) and (**b**) White extrinsic noise. (**c**) and (**d**) Colored extrinsic noise. The coupled population of cells are $N = 100$. ©[2015] IEEE. Reprinted, with permission, from ref. [2]

The coupled toggle switch systems can realize synchronization under the driving of extrinsic noise. The overall values of the ASE under white noise are relatively bigger than that under colored noise. That is, for parameter D_{ext}, σ^2 in the area $[0, 0.1] \times [0, 0.1]$, the coupled system perturbed by colored noise may display better synchronization behaviors than the system perturbed by white extrinsic noise. Therefore, one concludes that colored noise is more prone to promote population synchronization.

4.3.6.2 Colored Noise Can Promote the Mean Protein Numbers

We choose $\sigma^2 = 0.05, D_{ext} = 0.01$ for both white and colored noise, then the probability distributions of the molecule numbers of $LacI$ for two randomly chosen cells are shown in Fig. 4.11, where the distributions show bimodal features. Compared with the case under white noise, the longer tail at the right hand side of the distribution map under colored noise is also obvious (Statistically, under

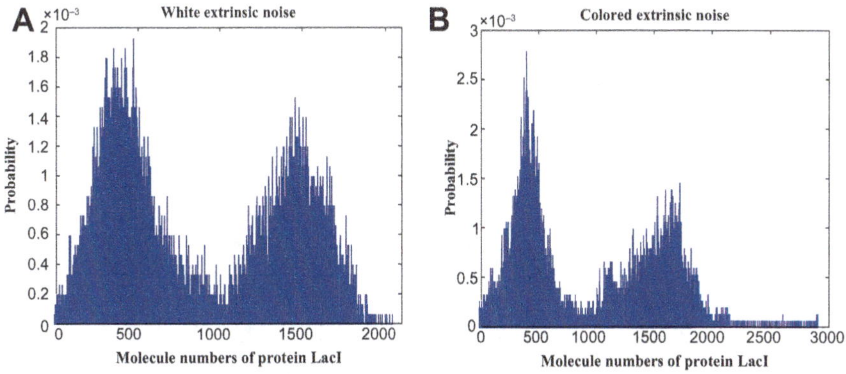

Fig. 4.11 For the coupled systems, bimodal probability distributions for the numbers of protein $LacI$ both under white (**a**) and colored noise (**b**) with $\sigma^2 = 0.05$, $D_{ext} = 0.01$. ©[2015] IEEE. Reprinted, with permission, from ref. [2]

colored noise and for 1000 cells, the number of cells appeared with high abundance of $LacI$ (>2000) is more than 6 times higher than that under white noise.).

4.3.6.3 Robustness of Synchronized Switching Against Parameter Perturbations

Several other parameters in system (4.8) may influence the phenomena observed in the above sections, such as the cell density Q, the diffusion rate k, as well as the extrinsic stimulus strength A. In the following, we investigate the robustness of the synchronized switching behaviors with respect to parameter perturbations.

Figure 4.12 shows η versus D_{ext} under $\sigma^2 = 0.05$. Each panel plots five curves, and each curve corresponds to different A, Q, or k. For Fig. 4.12a, b, we limit A in the interval $[0.02, 0.2]$ and find that under both noise sources, with the increasing of A, the curves η versus D_{ext} change from tight bell shape to fat bell shape to roughly horizontal line. Therefore, one can always observe that there exists an optimal D_{ext} for any $A \leq 0.1$, which corresponds to the biggest η, and indicates robust synchronized switch behaviors for $A \in [0.02, 0.1]$. However, for $A = 0.2$, the effect of D_{ext} is not so obvious as smaller A. Furthermore, for $A \leq 0.05$, the optimal η under colored noise is far larger than that under white noise, which indicates the optimal switch behaviors under colored noise are very sensitive to A. The phenomenon observed in the Sect. 4.3.6.1 thus may only hold for $A \in (0.05, 0.2)$. For Fig. 4.12c, d, the robustness of the synchronized switching with respect to Q is investigated, where we choose five Q values in the interval $[0.1, 1]$, and draw the curves η versus D_{ext}. From the two panels, it is found that the synchronized switching behaviors are very robust to Q, curves under different Q values are almost coincide with each other. The performance of the switch behaviors with respect to the intracellular diffusion rate k is investigated for $k \in [2, 12]$. Five

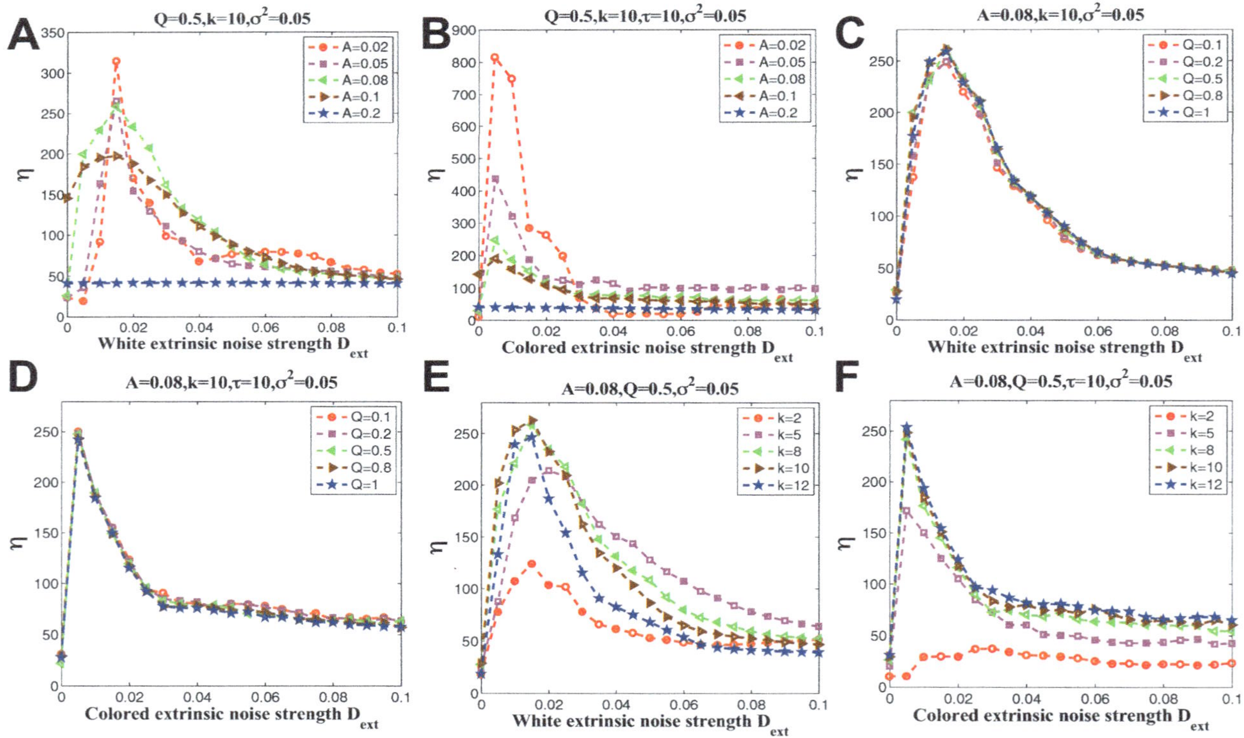

Fig. 4.12 Robustness of the synchronized switching behaviors with respect to parameters A, Q, k. (**a**, **c**), and (**e**) show the cases under white extrinsic noise, while (**b**, **d**), and (**f**) show the cases under colored noise. Under different noise sources, the curves of η versus D_{ext} are plotted. Each curve is averaged over 500 simulation runs. ©[2015] IEEE. Reprinted, with permission, from ref. [2]

cases are shown in Fig. 4.12e, f, from which one can see that the synchronized switching behaviors are also very robust to parameter k, with the increasing of k, roughly speaking, switch behaviors becomes better and better.

The corresponding curves of ASE versus D_{ext} are shown in Fig. 4.13. From Fig. 4.13, we see that for $D_{ext} \in (0.02, 0.1]$, the ASEs obtained from different parameters are almost all lower than 0.1, which indicates robust synchronization behaviors. For $D_{ext} \in [0, 0.02]$ and under white extrinsic noise, the ASEs obtained from many parameters are very high. Compared the corresponding D_{ext} regions with high ASE in Fig. 4.13, it further supports the conclusion that colored extrinsic noise is beneficial for population synchronization. It is noted that the robustness of the synchronized switching behaviors with respect to the other parameters can also be similarly investigated. Here, we only consider the effects of A, Q, k, since these parameters play crucial roles in the coupled systems.

4.3.6.4 Effect of Noise Autocorrelation Time

White noise is different from colored noise in that colored noise has nonzero autocorrelation time, and the corresponding autocorrelation function is not a δ function. Different length of autocorrelation time may contribute to performance diversities of the synchronized switching behaviors. To illustrate this point, we consider the effect of different autocorrelation times on the synchronized switching behaviors. Figure 4.14 shows the effect of the different autocorrelation times τ on the synchronized switching, where $\tau = 2, 5, 10, 30, 45$ are considered. Figure 4.14a shows the curves of η versus D_{ext}, while Fig. 4.14b shows the curves of ASE versus D_{ext}. From the two panels, one finds that the bigger τ, the relative smaller of the optimal D_{ext} to realize the best switch. For $D_{ext} \geq 0.02$ and large τ ($\tau = 30, 45$), η tends to be stable and with higher values. The ASE values are all very low and tend to decrease with the increasing of D_{ext}. As we have found that white noise is beneficial for the switch behaviors, colored noise is beneficial for the population synchronization, and we know white noise has zero autocorrelation time, combining these facts and our observations, we can conclude that noise autocorrelation time indeed contributes to the performance differences between white and colored extrinsic noises.

4.3.7 Physical Mechanisms of Bistable Switch

Combining the Waddington's epigenetic potential landscape theory [67, 68] and power spectral density analysis based on the Wiener–Khintchine theorem [69–71], hereinafter, we illustrate why white noise is beneficial for switch and colored noise is beneficial for population synchronization. To distinguish between white and colored noises, we denote $\zeta_w(t)$ as the Gaussian white noise and $\zeta_c(t)$ as the OU colored

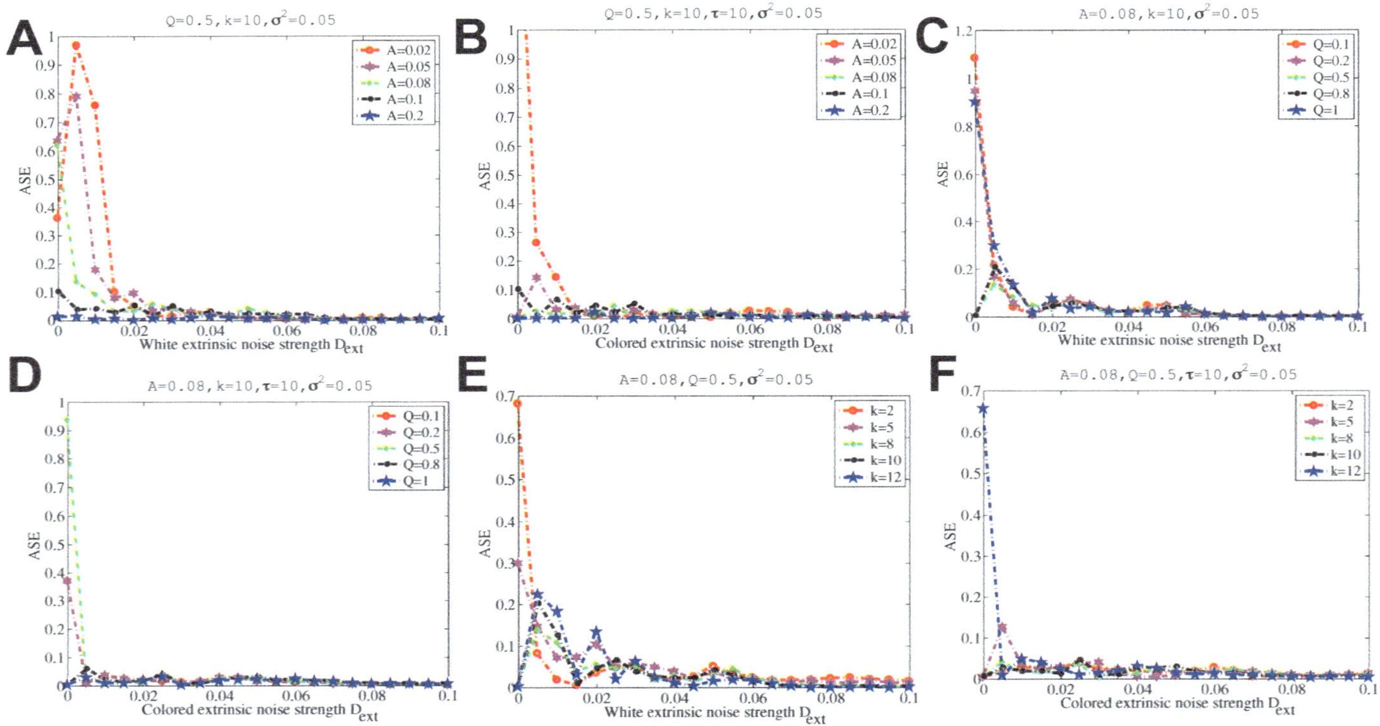

Fig. 4.13 Robustness of the synchronization behaviors with respect to parameters A, Q, k. (**a**, **c**), and (**e**) show the cases under white extrinsic noise, while (**b**, **d**), and (**f**) show the cases under colored noise. Under different noise sources, the curves of ASE versus D_{ext} are plotted. Each curve is averaged over 500 simulation runs. ©[2015] IEEE. Reprinted, with permission, from ref. [2]

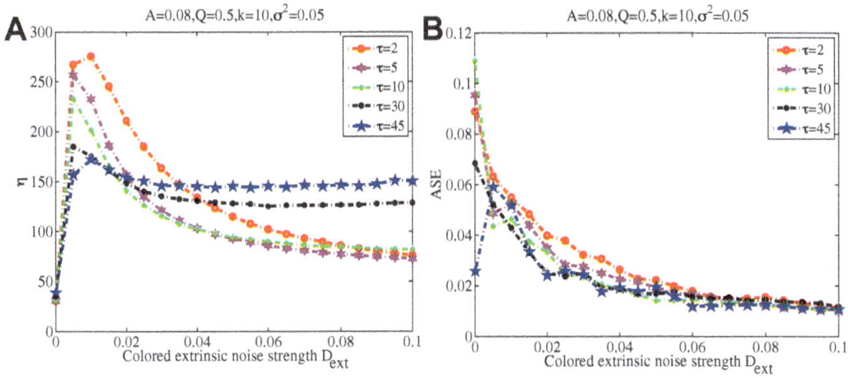

Fig. 4.14 The effect of noise autocorrelation time τ on the synchronized switching behaviors. Each curve is averaged over 1000 simulation runs. ©[2015] IEEE. Reprinted, with permission, from ref. [2]

noise. White noise $\zeta_w(t)$ has an autocorrelation function:

$$\langle \zeta_w(t)\zeta_w(t')\rangle = D_{ext}\delta(t-t'). \tag{4.11}$$

While colored noise $\zeta_c(t)$ has an autocorrelation function:

$$\langle \zeta_c(t)\zeta_c(t')\rangle = D_{ext}exp(-|t-t'|/\tau). \tag{4.12}$$

Based on the autocorrelation functions and the Wiener–Khintchine theorem, the spectral density $S_w(\omega)$, obtained by the Fourier transformation, can be calculated for white noise as:

$$S_w(\omega) = \int_{-\infty}^{+\infty} \langle \zeta_w(t)\zeta_w(0)\rangle exp(-i\omega t)dt = D_{ext}. \tag{4.13}$$

Obviously, white noise has constant spectral density D_{ext}. The spectral density $S_c(\omega)$ for the OU colored noise is:

$$S_c(\omega) = \int_{-\infty}^{+\infty} \langle \zeta_c(t)\zeta_c(0)\rangle exp(-i\omega t)dt = \frac{2D_{ext}\tau}{1+\omega^2\tau^2}, \tag{4.14}$$

which is not constant and depends on the autocorrelation time τ.

Based on the spectral density functions, the total power for white noise is:

$$P_w = \int_{-\infty}^{+\infty} S_w(\omega)d\omega = \infty, \tag{4.15}$$

which tends to be infinity. While for the colored noise, its total power is:

$$P_c = \int_{-\infty}^{+\infty} S_c(\omega)d\omega = \int_{-\infty}^{+\infty} \frac{2D_{ext}\tau}{1+\omega^2\tau^2}d\omega = 2\pi D_{ext}. \tag{4.16}$$

As $\tau \to 0$, the autocorrelation time of the colored noise tends to be zero, and it approximates to the Gaussian white noise, the spectral density $S_c(\omega)$ tends to be zero, and its total power is uncorrelated with τ and equals to $2\pi D_{ext}$.

Furthermore, intrinsic mechanisms of bistable switch can be explained from the physics point of view by the Waddington's epigenetic landscape theory. For system (4.5), the potential function $V(x, y)$ satisfies:

$$\begin{cases} \frac{\partial V}{\partial x} = -\frac{dx}{dt}, \\ \frac{\partial V}{\partial y} = -\frac{dy}{dt}, \end{cases} \tag{4.17}$$

which cannot be solved analytically, but fortunately, Bhattacharya and coworkers have developed a numerical method to derive the quasi-potential landscape. For detailed discussions, one can refer to their works [68]. By employing their methods, potential functions for the system (4.5) are drawn in Fig. 4.15, where Fig. 4.15a shows the case for $\alpha_1 = \alpha_2 = 5, d_1 = d_2 = 1, \gamma_1 = \gamma_2 = 0, n_1 = n_2 = 1.6$. Figure 4.15b shows the case for $\alpha_1 = 2.5, \alpha_2 = 5, d_1 = d_2 = 1, \gamma_1 = \gamma_2 = 0.5, n_1 = n_2 = 4$. For both cases, the landscapes have three local minima, which correspond to the three steady states of the system (4.5). Different stable states are separated by "ridges" that act as barriers.

After considering the noise perturbed single toggle switch system and multicellular system, system (4.5) becomes systems (4.6) and (4.8). As for systems (4.6) and (4.8), initially, the systems may stay around one of its stable steady states,

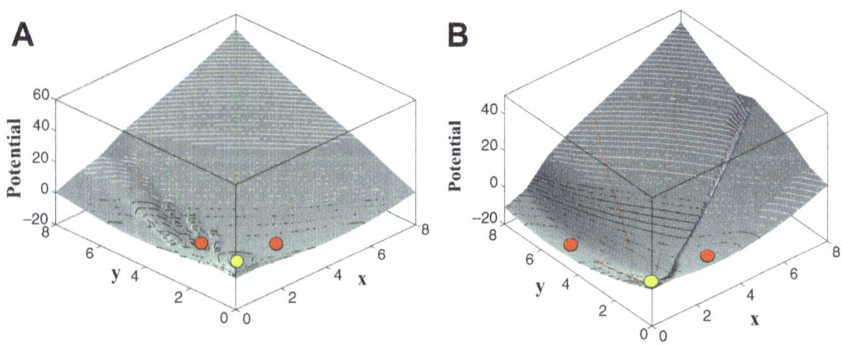

Fig. 4.15 Energy landscape of system (4.5). (**a**) $\alpha_1 = \alpha_2 = 5, d_1 = d_2 = 1, \gamma_1 = \gamma_2 = 0, n_1 = n_2 = 1.6$. (**b**) $\alpha_1 = 2.5, \alpha_2 = 5, d_1 = d_2 = 1, \gamma_1 = \gamma_2 = 0.5, n_1 = n_2 = 4$. Where red filled circles denote the stable steady states, while the yellow one represents unstable state in each figure (color online). ©[2015] IEEE. Reprinted, with permission, from ref. [2]

but due to extrinsic noise perturbation, under certain noise strength and with the accumulation of potential, the systems can overcome potential barriers and jump to another stable steady state. Due to ceaseless noise perturbation, the bistable systems can alternately jump between the two stable states, and therefore, stochastic switch behavior is onset.

Combining the power spectral density analysis and the Waddington's epigenetic landscapes, we can explain why white noise is beneficial for switch, and colored noise is beneficial for synchronization. From Eqs. (4.15) and (4.16), the total power of white noise is infinite, while colored noise has finite total power. Therefore, under similar conditions, white noise can more easily induce good switch behaviors than colored noise. Since the power of colored noise is limited, the coupled systems cannot fluctuate as fiercely as white noise; therefore, different cells in the coupled system can more easily achieve synchronization. Following a logical train of thought, for most cases, the colored noise perturbed networked system can display relatively better synchronization behaviors. Moreover, since too weak noise strength cannot overcome the potential barriers, while too strong noise perturbation can induce the system frequently transits between its two steady states, and switch becomes vague, there is an optimal extrinsic noise strength to induce the best switch behavior.

4.3.8 Some Further Issues

The kinetic rate parameters play an important role in our investigations, which make our conclusions more practical and general. On one hand, the assumption considers the real-world characteristics of gene regulatory systems. On the other hand, the effects of kinetic rate parameters on system dynamics can be seen as the robustness of the system against time-varying parameters. From the above research, too much fluctuation on parameters may be harmful to the switch behaviors (data not shown), and the population in the QS system is also difficult to reach synchronization.

The obtained conclusions are mainly based on numerical experiments and mechanism analysis. Due to the complexity of the networked systems, it is still difficult to theoretically prove the synchronized switching behaviors in the networked systems. As for the cases without noise perturbation, some mathematical results have been reported [14, 61]. It is expected that theoretical analysis could be carried out with the development of complex networks theory [72–77] and SDEs. Furthermore, we have considered the Gaussian white noise and colored noise based on the OU process, which are frequently used noise sources in the literature, especially in the context of biological systems. It is intriguing to explore the effect of the other types of noise sources, such as non-Gaussian white noise and colored noise generated by the other stochastic processes [52]. Furthermore, as another interesting future topic, it is interesting to further clarify some other differences between white and colored noise induced behaviors, such as the mean protein numbers. For simplicity, we define the mean number $< X >$ of protein x at high state as the average number at the "on" state (The cases for numbers greater than 1000 in Figs. 4.9 and 4.11). For the single and coupled toggle switch systems, under the same random number generators and parameters, for white and colored noise perturbed systems, the mean numbers of the protein x at high states are shown in Fig. 4.16, where each figure is averaged over 100 simulation runs. From Fig. 4.16, one observes that the colored noise perturbed systems tend to express high mean protein numbers. However, it is still an open topic to further verify the observations and its effect size in more details, such as considering the effect of possible numerical artifacts and the width of protein number distributions.

4.3.9 Summary

GRNs are a class of important real-world biological networks, which have attracted extensive attention in the field of systems biology and synthetic biology [62–66]. In this section, we have investigated colored noise induced switch behaviors in the genetic toggle switch systems, where a periodically stimulated single cell system and a networked system coupled by the QS mechanism have been studied. The investigated systems are incorporated with kinetic parameters, and both colored and white extrinsic noises have been investigated and compared. We find that there exists some optimal extrinsic noise strength to induce the best stochastic switch behaviors in the single toggle switch, the best synchronized switching behaviors in the networked system at each fixed reaction rate fluctuation level. For a large region of system parameters and under the same initial conditions, the switch behaviors under white extrinsic noise are prone to be better than that under the colored one. White noise is beneficial for the switch behaviors, while colored noise is more propitious to promote the coupled cells to reach population synchronization. The related findings are robust to many crucial parameters.

The above findings further extend and develop the conclusions in the existing investigations, which can be used to guide experimental design of artificial genetic circuits and artificial organisms. Moreover, the associated researches may have potential implications in gene therapy, since current gene therapy techniques are limited in that transfected genes are typically either in an "on" or "off" state. For the effective treatment of many diseases, the expression of a transfected gene needs to be regulated in some systematic fashion. Thus, the development of externally controllable noise-based switches for gene expression could have significant clinical implications [30].

4.4 Discussions and Conclusions

Except for the composite oscillator and the toggle switch system coupled by the QS mechanism, there are some other works on the coupling and merge of simple circuits. For example, switches (bistability) and oscillations (limit cycle)

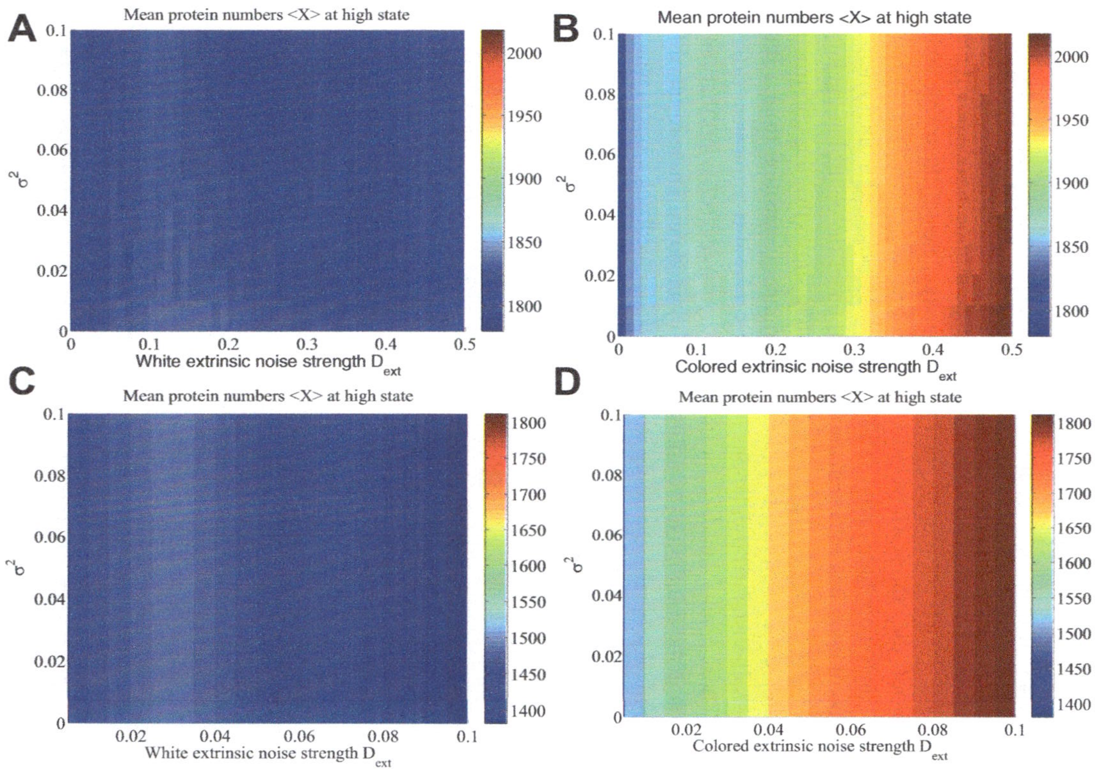

Fig. 4.16 Mean numbers $< X >$ of protein *LacI* at high states. (**a**) and (**b**) For white and colored extrinsic noise perturbed single system. (**c**) and (**d**). For the coupled systems. The coupled population of cells is $N = 100$. ©[2015] IEEE. Reprinted, with permission, from ref. [2]

are omnipresent in biological networks. Synthetic genetic networks producing bistability and oscillations have been designed and constructed experimentally. However, in real biological systems, regulatory circuits are usually interconnected and the dynamics of those complex networks are often richer than the dynamics of simple modules.

In the year 2010, Gonze coupled the genetic toggle switch and the repressilator, two prototypic systems exhibiting bistability and oscillations, respectively. They studied two types of coupling, as shown in Figs. 4.17 and 4.18, respectively. In the first type, the bistable switch is under the control of the oscillator (Fig. 4.17). Numerical simulation of this system allows people to determine the conditions under which a periodic switch between the two stable steady states of the toggle switch occurs. In addition, they showed how birhythmicity characterized by the coexistence of two stable small-amplitude limit cycles can easily be obtained in the system. In the second type of coupling, the oscillator is placed under the control of the toggle switch (Fig. 4.18). Numerical simulation of this system shows that this construction could, for example, be exploited to generate a permanent transition from a stable steady state to self-sustained oscillations (and vice versa) after a transient external perturbation. Those results thus describe qualitative dynamical behaviors that can be generated through the coupling of two simple network modules. These results differ from the dynamical properties resulting from interlocked feedback loops in which a given variable is involved at the same time in both positive and negative feedbacks. The proposed models may be of interest in synthetic biology, as they give hints on

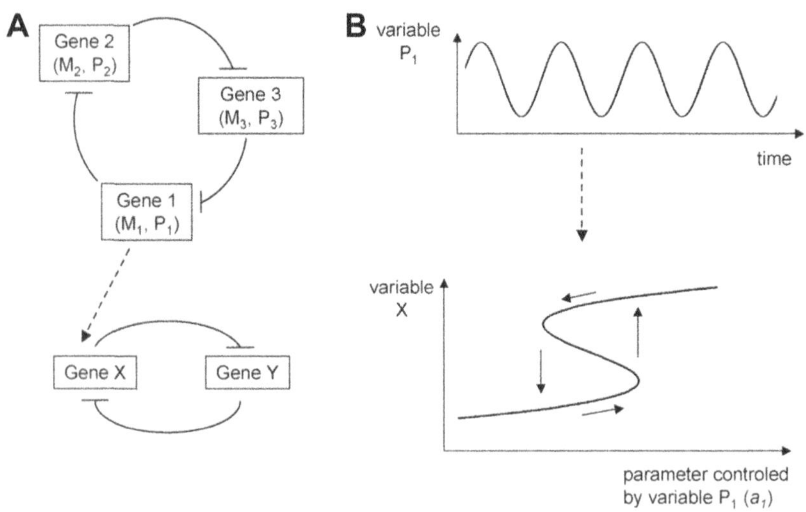

Fig. 4.17 Illustration of the first type of coupling: The repressilator controls the toggle switch. More specifically, one of the three proteins of the repressilator (here protein $P1$, associated with gene 1) enhances the expression of gene X of the toggle switch (panel **a**). As a result, it is expected that in some conditions the oscillations of the repressilator protein will induce a periodic switch from one steady state to the other (panel **b**). Reprinted from ref. [78], with permission from Elsevier

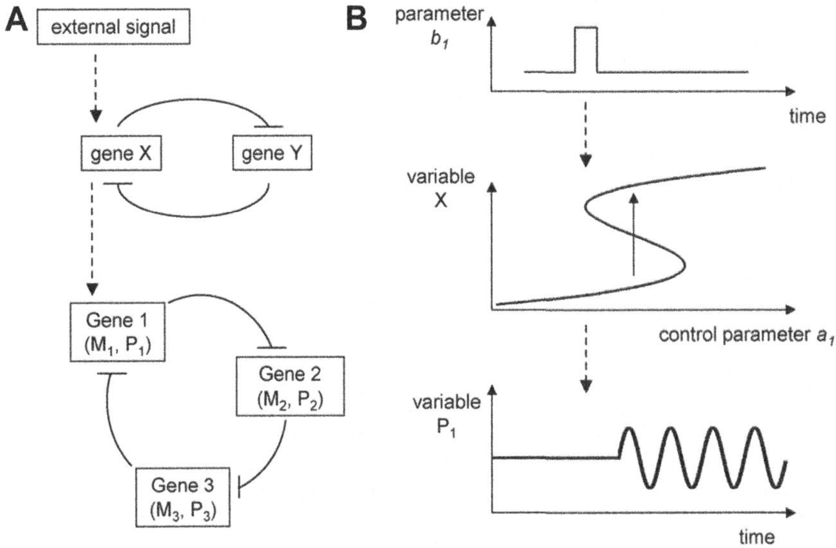

Fig. 4.18 Illustration of the second type of coupling: The toggle switch controls the repressilator. More specifically, gene X of the toggle switch enhances the expression of gene 1 of the repressilator (panel **a**). If the lower steady state of X corresponds to a non-oscillatory (i.e., steady) state of the repressilator, while the upper steady state is associated to oscillations, it should be able to induce (or stop) oscillations by a simple switch in the toggle switch. If a permanent switch could be induced by a transient perturbation, it would thus be able to induce permanent oscillations by a single transient external signal (panel **b**). Reprinted from ref. [78], with permission from Elsevier

how the coupling should be designed to get the required properties. For details of the analysis for the two types of coupling, one can refer to Gonze's work [78].

References

1. Yang, D., Li, Y., Kuznetsov, A.: Characterization and merger of oscillatory mechanisms in an artificial genetic regulatory network. Chaos **19**, 033115 (2009)
2. Wang, P., Lü, J., Yu, X.: Colored noise induced bistable switch in the genetic toggle switch systems. IEEE/ACM Trans. Comput. Biol. Bioinformat. **12**, 579–589 (2015)
3. Wang, P., Lü, J., Ogorzalek, M.J.: Synchronized switching induced by colored noise in the genetic toggle switch systems coupled by quorum sensing mechanism. Proc. of the 30th Chin. Control Confer. **July 22–24**, 6605–6609 (2011)
4. Atkinson, M.R., Savageau, M.A., Myers, J.T., Ninfa, A.J.: Development of genetic circuitry exhibiting toggle switch or oscillatory behavior in *Escherichia coli*. Cell **113**, 597–607 (2003)
5. Gardner, T.S., Cantor, C.R., Collins, J.J.: Construction of a genetic toggle switch in *Escherichia coli*. Nature **403**, 339–342 (2000)
6. Elowitz, M.B., Leibler, S.: A synthetic oscillatory network of transcriptional regulators. Nature **403**, 335–338 (2000)
7. Becskei, A., Serrano, L.: Engineering stability in gene networks by autoregulation. Nature **405**, 590–593 (2000)

8. Hasty, J., McMillen, D., Collins, J.J.: Engineered gene circuits. Nature **420**, 224–230 (2002)
9. Kholodenko, B.N.: Cell-signalling dynamics in time and space. Nat. Rev. Mol. Cell Bio. **7**, 165–176 (2006)
10. Tyson, J.J., Chen, K.C., Novak, B.: Sniffers, buzzers, toggles and blinkers: dynamics of regulatory and signaling pathways in the cell. Curr. Opin. Cell Biol. **15**, 221–231 (2003)
11. Garcia-Ojalvo, J., Elowitz, M.B., Strogatz, S.H.: Modeling a synthetic multicellular clock: repressilators coupled by quorum sensing. Proc. Natl. Acad. Sci. USA. **101**, 10955–10960 (2004)
12. McMillen, D., Kopell, N., Hasty, J., Collins, J.J.: Synchronizing genetic relaxation oscillators by intercell signaling. Proc. Natl. Acad. Sci. USA. **99**, 679–684 (2002)
13. Yamaguchi, S., Isejima, H., Matsuo, T., Okura, R., Yagita, K., Kobayashi, M., Okamura, H.: Synchronization of cellular clocks in the suprachiasmatic nucleus. Science **302**, 1408–1412 (2003)
14. Kuznetsov, A., Kærn, M., Kopell, N.: Synchrony in a population of hysteresis-based genetic oscillators. SIAM J. Appl. Math. **65**, 392–425 (2004)
15. Ermentrout, B.: Simulating, analyzing, and animating dynamical systems: a guide to XPPAUT for researchers and students. SIAM, Philadelphia (2002)
16. Andronov, A., Leontovich, E., Gordon, I., Maier, A.: Theory of bifurcations of dynamical systems on a plane, Israel Program for Sc. Translations, Jerusalem (1971)
17. You, L., Cox, R.S., Weiss, R., Arnold, F.H.: Programmed population control by cell–cell communication and regulated killing. Nature **428**, 868–871 (2004)
18. Nurse, P.: A long twentieth century of the cell cycle and beyond. Cell **100**, 71–78 (2000)
19. Hodgkin, A.L., Huxley, A.F.: A quantitative description of membrane current and its application to conduction and excitation in nerve. J. Physiol. **117**, 500 (1952)
20. Izhikevich, E.M.: Dynamical systems in neuroscience: the geometry of excitability and bursting. MIT press, Cambridge (2007)
21. Kuznetsov, A.S., Kopell, N.J., Wilson, C.J.: Transient high-frequency firing in a coupled-oscillator model of the mesencephalic dopaminergic neuron. J. Neurophysiol. **95**, 932–947 (2006)
22. Ptashne, M., Switch, A.G.: Phage lambda and higher organisms. Cell and Blackwell Scientific, Cambridge (1992)
23. Arkin, A., Ross, J., McAdams, H.H.: Stochastic kinetic analysis of developmental pathway bifurcation in phage λ-infected *Escherichia coli* cells. Genetics **149**, 1633–1648 (1998)
24. Ferrell, J.E., Machleder, E.M.: The biochemical basis of an all-or-none cell fate switch in *Xenopus oocytes*. Science **280**, 895–898 (1998)
25. Bhalla, U.S., Ram, P.T., Iyengar, R.: MAP kinase phosphatase as a locus of flexibility in a mitogen-activated protein kinase signaling network. Science **297**, 1018–1023. (2002)
26. Bagowski, C.P., Ferrell, J.E.: Bistability in the JNK cascade. Curr. Biol. **11**, 1176–1182 (2001)
27. Sriram, K., Soliman, S., Fages, F.: Dynamics of the interlocked positive feedback loops explaining the robust epigenetic switching in *Candida albicans*. J. Theor. Biol. **258**, 71–88 (2009)
28. Wang, J., Zhang, J., Yuan, Z., Zhou, T.: Noise-induced switches in network systems of the genetic toggle switch. BMC Syst. Biol. **1**, 50 (2007)
29. Kobayashi, H., Kærn, M., Araki, M., Chung, K., Gardner, T.S., Cantor, C.R., Collins, J.J.: Programmable cells: interfacing natural and engineered gene networks. Proc. Natl. Acad. Sci. USA. **101**, 8414–8419 (2004)
30. Hasty, J., Pradines, J., Dolnik, M., Collins, J.J.: Noise-based switches and amplifiers for gene expression. Proc. Natl. Acad. Sci. USA. **97**, 2075–2080 (2000)
31. Tian, T., Burrage, K.: Stochastic models for regulatory networks of the genetic toggle switch. Proc. Natl. Acad. Sci. USA. **103**, 8372–8377 (2006)
32. Swain, P.S., Elowitz, M.B., Siggia, E.D.: Intrinsic and extrinsic contributions to stochasticity in gene expression. Proc. Natl. Acad. Sci. USA. **99**, 12795–12800 (2002)
33. Pedraza, J.M., van Oudenaarden, A.: Noise propagation in gene networks. Science **307**, 1965–1969 (2005)

34. Kaern, M., Elston, T.C., Blake, W.J., Collins, J.J.: Stochasticity in gene expression: from theories to phenotypes. Nat. Rev. Genet. **6**, 451–464 (2005)
35. Lipshtat, A., Loinger, A., Balaban, N.Q., Biham, O.: Genetic toggle switch without cooperative binding. Phys. Rev. Lett. **96**, 188101 (2006)
36. Warren, P.B., ten Wolde, P.R.: Chemical models of genetic toggle switches. J. Phys. Chem. B **109**, 6812–6823 (2005)
37. Yuan, Z., Zhang, J., Zhou, T.: Noise-induced continuous switch. Sci. China Ser. B **37**, 446–452 (2007)
38. Zhou, T., Zhang, J., Wang, J., Yuan, Z.: Noise-induced synchronized switching of a multicellular system. Progress Biochem. Biophys. **35**, 929–939 (2008)
39. Lipshtat, A., Loinger, A., Balaban, N.Q., Biham, O.: Stochastic simulations of genetic toggle switch system. Phys. Rev. E **75**, 021904 (2007)
40. Miller, M.B., Bassler, B.L.: Quorum sensing in bacteria. Annu. Rev. Microbiol. **55**, 165–199 (2001)
41. Shahrezaei, V., Ollivier, J.F., Swain, P.S.: Colored extrinsic fluctuations and stochastic gene expression. Mol. Syst. Biol. **4**, 196 (2008)
42. Lei, J.: Stochasticity in single gene expression with both intrinsic noise and fluctuation in kinetic parameters. J. Theor. Biol. **256** 485–492 (2009)
43. Rosenfeld, N., Young, J.W., Alon, U., Swain, P.S., Elowitz, M.B.: Gene regulation at the single-cell level. Science **307**, 1962–1965 (2005)
44. Zhou, T., Chen, L., Aihara, K.: Molecular communication through stochastic synchronization induced by extracellular fluctuations. Phys. Rev. Lett. **95**, 178103 (2005)
45. Zhou, T., Zhang, J., Yuan, Z., Xu, A.: External stimuli mediate collective rhythms: artificial control strategies. PLoS One **2**, e231 (2007)
46. Laurent, M., Kellershohn, N.: Multistability: a major means of differentiation and evolution in biological systems. Trends Biochem. Sci. **24**, 418–422 (1999)
47. Smolen, P., Baxter, D.A., Byrne, J.H.: Mathematical modeling of gene networks. Neuron **26**, 567–580 (2000)
48. Jansen, A.P.J.: Monte Carlo simulations of chemical reactions on a surface with time-dependent reaction-rate constants. Comput. Phys. Commun. **86**, 1–12 (1995)
49. Gillespie, D.T.: Exact stochastic simulation of coupled chemical reactions. J. Phys. Chem. **81**, 2340–2361 (1977)
50. Zhou, T.: Stochastic dynamics of biological systems. Science publishing house, Beijing (2009) (In Chinese)
51. Zhang, H., Lu, L., Yan, X., Gao, P.: Effect of the population heterogeneity on growth behavior and its estimation. Sci. China Ser. C: Life Sci. **50**, 535–547 (2007)
52. Van Kampen, N. G.: Stochastic processes in physics and chemistry. North Holland, New York (2007)
53. Tessone, C.J., Mirasso, C.R., Toral, R., Gunton, J.D.: Diversity-induced resonance. Phys. Rev. Lett. **97**, 194101 (2006)
54. Kori, H., Mikhailov, A.S.: Entrainment of randomly coupled oscillator networks by a pacemaker. Phys. Rev. Lett. **93**, 254101 (2004)
55. Hasty, J., Isaacs, F., Dolnik, M., McMillen, D., Collins, J.J.: Designer gene networks: Towards fundamental cellular control. Chaos **11**, 207–220 (2001)
56. Higham, D.J.: An algorithmic introduction to numerical simulation of stochastic differential equations. SIAM Rev. **43**, 525–546 (2001)
57. Fuqua, C., Winans, S.C., Greenberg, E.P.: Census and consensus in bacterial ecosystems: the LuxR-LuxI family of quorum-sensing transcriptional regulators. Annu. Rev. Microbiol. **50**, 727–751 (1996)
58. Bassler, B.L.: How bacteria talk to each other: regulation of gene expression by quorum sensing. Curr. Opin. Microbiol. **2**, 582–587 (1999)
59. Danino, T., Mondragón-Palomino, O., Tsimring, L., Hasty, J.: A synchronized quorum of genetic clocks. Nature **463**, 326–330 (2010)

60. Bohn, A., García-Ojalvo, J.: Synchronization of coupled biological oscillators under spatially heterogeneous environmental forcing. J. Theor. Biol. **250**, 37–47 (2008)
61. Russo, G., Slotine, J.J.E.: Global convergence of quorum-sensing networks. Phys. Rev. E **82**, 041919 (2010)
62. Wang, P., Lü, J.: Control of genetic regulatory networks: opportunities and challenges. Acta Automat. Sin. **39**, 1969–1979 (2013) (In Chinese)
63. Mitra, S., Das, R., Hayashi, Y.: Genetic networks and soft computing. IEEE/ACM Trans. Comput. Biol. Bioinformat. **8**, 94–107 (2011)
64. Rottger, R., Ruckert, U., Taubert, J., Baumbach, J.: How little do we actually know? On the size of gene regulatory networks. IEEE/ACM Trans. Comput. Biol. Bioinformat. **9**, 1293–1300 (2012)
65. Todor, A., Dobra, A., Kahveci, T.: Characterizing the topology of probabilistic biological networks. IEEE/ACM Trans. Comput. Biol. Bioinformat. **10**, 970–983 (2013)
66. Wang, J., Huang, Y., Wu, F.X., Pan, Y.: Symmetry compression method for discovering network motifs. IEEE/ACM Trans. Comput. Biol. Bioinformat. **9**, 1776–1789 (2012)
67. Waddington, C.H.: The strategy of the genes. London: Allen. **86** (1957)
68. Bhattacharya, S., Zhang, Q., Andersen, M.E.: A deterministic map of Waddington's epigenetic landscape for cell fate specification. BMC Syst. Biol. **5**, 1 (2011)
69. Ma, J., Xiao, T., Hou, Z., Xin, H.: Coherence resonance induced by colored noise near Hopf bifurcation. Chaos **18**, 043116 (2008)
70. W-Couch II, L.: Digital and analog communications systems. Prentice Hall, New Jersey (2001)
71. Middleton, J.W., Chacron, M.J., Lindner, B., Longtin, A.: Firing statistics of a neuron model driven by long-range correlated noise. Phys. Rev. E **68**, 021920 (2003)
72. Wang, P., Lu, R., Chen, Y., Wu, X.: Hybrid modelling of the general middle-sized genetic regulatory networks. IEEE Int. Symp. Circ. Syst. (ISCAS'13) **May 19–23**, 2103–2106 (2013)
73. Wang, P., Yu, X., Lü, J.: Identification and evolution of structurally dominant nodes in protein-protein interaction networks. IEEE Trans. Biomed. Circ. Syst. **8**, 87–97 (2014)
74. Wen, G., Duan, Z., Chen, G., Yu, W.: Consensus tracking of multi-agent systems with Lipschitz-type node dynamics and switching topologies. IEEE Trans. Circ. Syst.-I **61**, 499–511 (2014)
75. Wang, P., Lü, J., Yu, X., Liu, Z.: Duplication and divergence effect on network motifs in undirected bio-molecular networks. IEEE Trans. Biomed. Circ. Syst. **9**, 312–320 (2015)
76. Wang, P., Lü, J., Yu, X.: Identification of important nodes in directed biological networks: a network motif approach. PLoS One **9**, e106132 (2014)
77. Wen, G., Hu, G., Yu, W., Chen, G.: Distributed consensus of higher order multiagent systems with switching topologies. IEEE Trans. Circ. Syst.-II **61**, 359–363 (2014)
78. Gonze, D.: Coupling oscillations and switches in genetic networks. Biosyst. **99**, 60–69 (2010)

Chapter 5
Modeling and Analysis of Large-Scale Networks

Abstract In the previous chapters, we have discussed the mathematical modeling and dynamical analysis of several simple circuits and coupled genetic circuits. Generally, the established models are ordinary differential equations or stochastic differential equations, which are always with the Michaelis–Menten or Hill forms. The ordinary or stochastic differential equation models are inappropriate for large-scale networks, due to complexity. In this chapter, we introduce some works on large-scale networks. The word "large-scale" is a relative concept. When we discuss discrete or continuous models for bio-molecular networks, we call networks with tens of nodes as large-scale networks. For hybrid discrete and continuous models and the percolating flow model, we discuss their applications in networks with hundreds or thousands of nodes.

5.1 Backgrounds

Real-world bio-molecular networks are large-scale ones, which contain hundreds to tens of thousands of nodes. It is often a difficult task to mathematically model these large-scale networks [1–13]. There are three main reasons. Firstly, for large networks, the traditional ODE or SDE models are often too complex, there are too many parameters to be estimated [12], and parameter estimation is a difficult task in biological systems. Secondly, too large ODEs or SDEs are difficult to be analyzed. The ODEs or SDEs always contain the Hill equations, and numerically solving of these equations is time-consuming. Thirdly, for large-scale networks, there are too many state variables (high dimensional), each variable is always affected by many other nodes, and it is difficult to distinguish which nodes are the main elements of the networks for certain behaviors and should be considered in great detail.

In the existing works, some large networks have been modeled and analyzed. In the following, we briefly overview some of the frequently explored networks. The first network is the *yeast* cell cycle network, which contains hundreds of nodes, as shown in Fig. 5.1. The complete *yeast* cell cycle network is very complex; however, the core network that controls the cell cycle only contains tens of nodes, and

J. Lü, P. Wang, *Modeling and Analysis of Bio-molecular Networks*,
https://doi.org/10.1007/978-981-15-9144-0_5

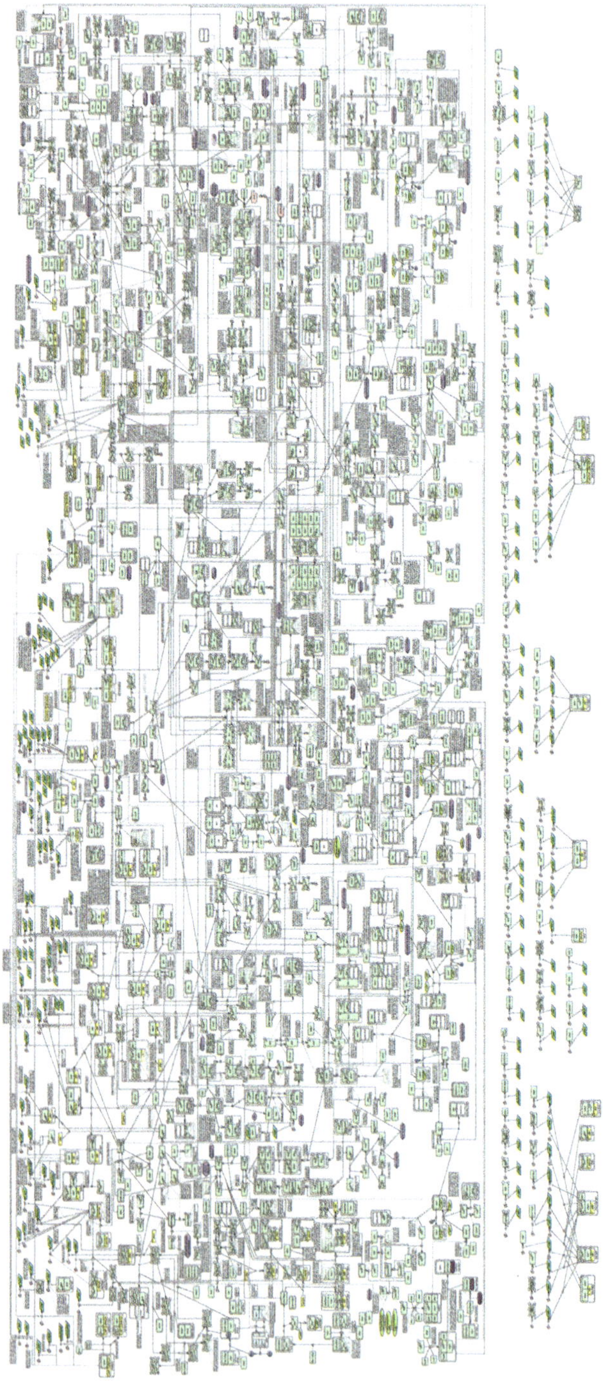

Fig. 5.1 A comprehensive molecular interaction map for the budding *yeast* cell cycle. A total of 880 species and 732 reactions are included. Copyright ©(2010) Wiley. Used with permission from Ref. [14]

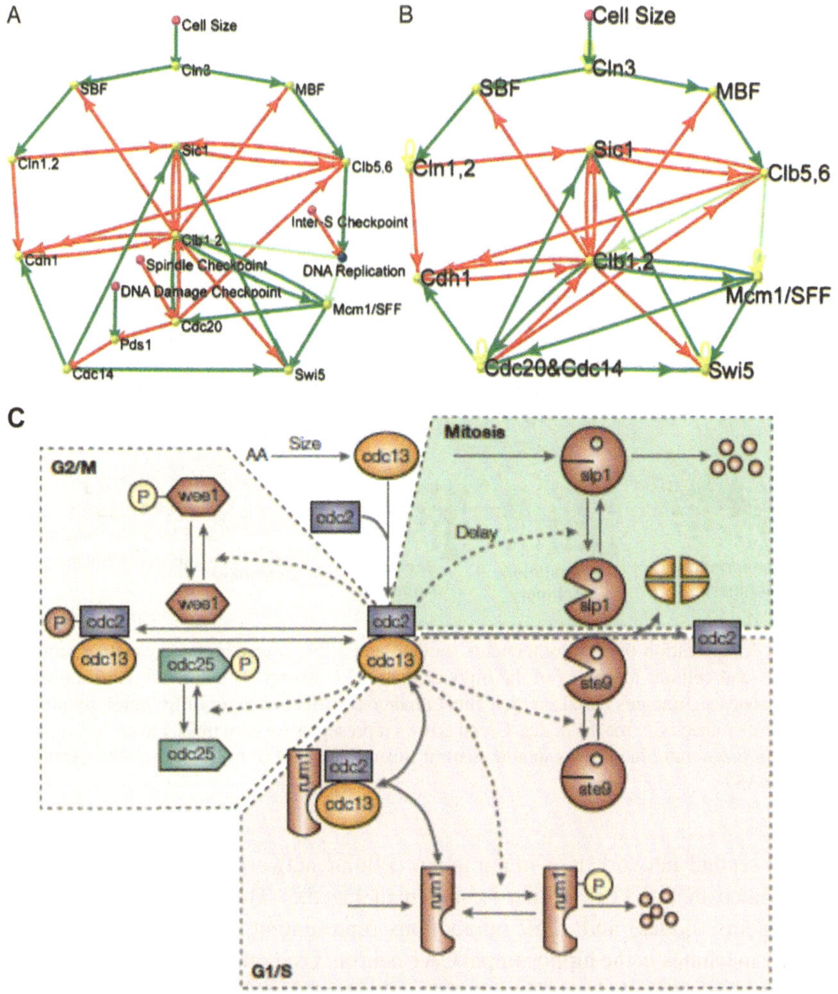

Fig. 5.2 The *yeast* cell cycle network. (**a**) The core cell cycle network for the budding *yeast*. (**b**) The simplified cell cycle network for the budding *yeast* with only one checkpoint "cell size". Reprinted from Ref. [11] (Copyright (2004) National Academy of Sciences, U.S.A.). (**c**) The cell cycle control system for the fission *yeast*. This system can be divided into three modules, which regulate the transitions from G1 into S phase, from G2 into M phase, and exit from mitosis. Reprinted by permission from Springer, Ref. [15]

therefore, it receives many attentions. The core cell cycle networks for the budding *yeast* and fission *yeast* are shown in Fig. 5.2. Figure 5.2a, b show the cell cycle network of the budding *yeast* and the simplified cell cycle network with only one checkpoint "cell size" [11]. Figure 5.2c represents the cell cycle control system in the fission *yeast*. This system can be divided into three modules, which regulate the transitions from G1 into S phase, from G2 into M phase, and exit from mitosis [15].

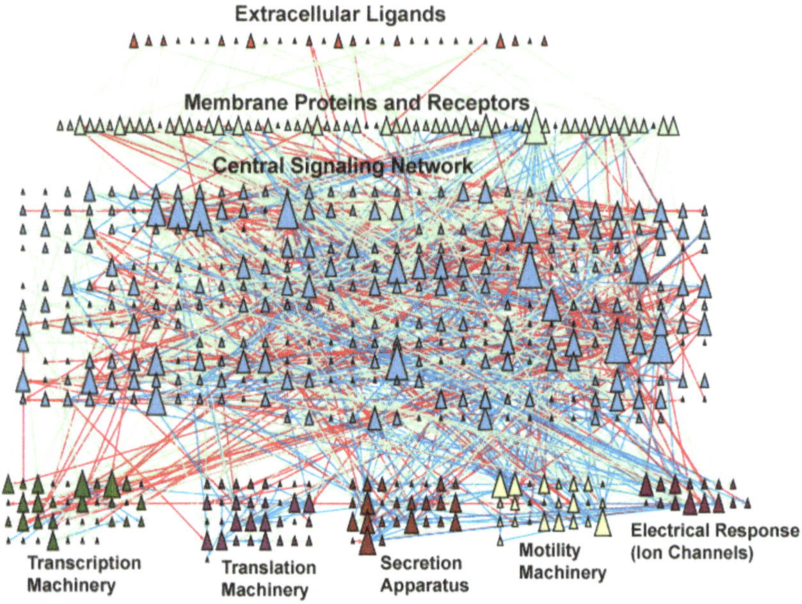

Fig. 5.3 Visualization of 545 components (nodes) and 1259 interactions representing signaling pathways and cellular machines in the hippocampal CA1 neuron. The network is visualized by placing nodes as triangles based on their functional compartments. Sizes of triangles demonstrate the level of connectivity for the nodes. Green arrows represent activation links, red arrows represent inhibition links, and blue arrows denote neutral links. Reprinted from Ref. [13], with permission from AAAS

The second network is a mammalian cellular network that was reported and investigated in Ref. [13], which is shown in Fig. 5.3. The network contains 545 components (nodes) and 1259 interactions representing signaling pathways and cellular machines in the hippocampal CA1 neuron. Using the graph theory methods, Máayan et al. [13] analyzed ligand-induced signal flow through the system. The specification of input and output nodes allowed people to identify functional modules. Networking resulted in the emergence of regulatory motifs, such as positive and negative feedback and FFLs, which process information. Key regulators of plasticity were highly connected nodes required for the formation of regulatory motifs, indicating the potential importance of such motifs in determining cellular choices between homeostasis and plasticity.

In the following sections, we will review and discuss the related works on the mentioned networks. Hereinafter, we firstly present the works based on the continuous ODE models. Then, we review works that are based on the discrete Boolean dynamical models. Finally, we overview and discuss some recent works on hybrid modeling of the bio-molecular networks.

5.2 Continuous Models for the *Yeast* Cell Cycle Network

5.2.1 Related Works and Motivations

For single component (node) biological systems, such as the single gene auto-activation and the single gene auto-repression circuits, the CMEs can be used to describe detailed molecular evolutions. For GRNs with several nodes, one can establish ODE or SDE models to quantitatively investigate deterministic dynamics in these systems, and the MM equations or the Hill equations are always used to model such highly nonlinear differential equation systems [16]. The single component circuits or circuits with several components are all simple regulatory circuits, and the general format of the continuous ODE models for these simple circuits can be described as follows:

$$\frac{dx_i}{dt} = \sum_{j=1}^{M} s_{ij} f_j(x_1, x_2, \ldots, x_N). \tag{5.1}$$

Here, x_i denotes protein concentration, s_{ij} represents the stoichiometric coefficient of species i in the j'th reaction [16], and f_j is the reaction rate of the j'th reaction, which is often highly nonlinear. In addition, for the middle-sized GRNs, Eq. (5.1) is always a very complex differential algebraic equation.

Hereinafter, we briefly discuss some related works based on the continuous ODEs for the fission *yeast* cell cycle network [17].

5.2.2 Dynamical Analysis

The fission *yeast* cell cycle network is a typical biological network, which has been extensively investigated during the last decades [17–27]. The cell cycle is a complex biological process, which involves lots of bio-molecule species and includes lots of biochemical reactions. Generally, the cell cycle can be separated into four stages, namely, G_1, S, G_2, and M phases. For more details of the cell cycle process, one can refer to Ref. [17, 27]. Hereinafter, we only pay attention to the investigations on the mathematical model of the fission *yeast* cell cycle. In the year 1997, Novak and Tyson [17] established an ODE to describe the control of DNA replication in the fission *yeast* cell cycle as follows:

$$
\begin{cases}
\frac{dx_1}{dt} = k_1 - (k_2 + k_{wee} + k_7 x_3)x_1 + k_{25} x_2 + (k_{7r} + k_4)x_6, \\
\frac{dx_2}{dt} = k_{wee} x_1 - (k_{25} + k_2 + k_7 x_3)x_2 + (k_{7r} + k_4)x_7, \\
\frac{dx_3}{dt} = k_3 - k_4 x_3 - \frac{k_p x_3 x_{13} \theta_2}{K_{mp} + x_3} - k_7 x_3 (x_1 + x_2) + (k_{7r} + k_2 + k_{2p})(x_6 + x_7) \\
\qquad -k_8 x_3 x_4 + (k_{8r} + k_{6p})x_5, \\
\frac{dx_4}{dt} = k_5 - (k_6 + k_8 x_3)x_4 + (k_{8r} + k_4)x_5, \\
\frac{dx_5}{dt} = k_8 x_3 x_4 - (k_{8r} + k_4 + k_{6p})x_5, \\
\frac{dx_6}{dt} = k_7 x_3 x_1 - (k_{7r} + k_4 + k_2 + k_{2p})x_6, \\
\frac{dx_7}{dt} = k_7 x_3 x_2 - (k_{7r} + k_4 + k_2 + k_{2p})x_7, \\
\frac{dx_8}{dt} = \frac{k_i (1 - x_8)\theta_1}{K_{mi} + 1 - x_8} - \frac{k_{ir} x_8}{K_{mir} + x_8}, \\
\frac{dx_9}{dt} = \frac{k_u x_8 (1 - x_9)}{K_{mu} + 1 - x_9} - \frac{k_{ur} x_9}{K_{mur} + x_9}, \\
\frac{dx_{10}}{dt} = \frac{k_{u2} (1 - x_{10})\theta_1}{K_{mu2} + 1 - x_{10}} - \frac{k_{ur2} x_{10}}{K_{mur2} + x_{10}}, \\
\frac{dx_{11}}{dt} = \frac{k_{wr} (1 - x_{11})}{K_{mwr} + 1 - x_{11}} - \frac{k_w x_{11} \theta_1}{K_{mw} + x_{11}}, \\
\frac{dx_{12}}{dt} = \frac{k_c (1 - x_{12})\theta_1}{K_{mc} + 1 - x_{12}} - \frac{k_{cr} x_{12}}{K_{mcr} + x_{12}}, \\
\frac{dx_{13}}{dt} = \mu x_{13},
\end{cases}
\tag{5.2}
$$

where $x_i (i = 1, \ldots, 13)$ are state variables, which correspond to the species concentrations, as shown in Table 5.1. Parameter values are shown in Table 5.2. It is noted that, in Table 5.1, $\theta_1 = x_1 + \beta x_2$ is the M-phase promoting factor and $\theta_2 = \theta_1 + \alpha x_4 + [Cig1]$ is the S-phase promoting factor. In the above model, when θ_2 crosses 0.1 from below, S phase is initiated. When $x_9 = [UbE]$ crosses 0.1 from above, the cell divides functionally, that is, x_{13} should be reset as $x_{13}/2$. 60 min after start, k_p is divided by 2, and at cell division, k_p is multiplied by 2. Parameters in Eq. (5.2) are considerable, which makes the analysis of this system to be a difficult task.

Table 5.1 Variables and symbols in Eq. (5.2)

Variables	Symbols	Variables	Symbols
x_1	$[Cdc13/Cdc2]$	x_2	$[Cdc13/P - Cdc2]$
x_3	$[Rum1]$	x_4	$[Cig2/Cdc2]$
x_5	$[Cig2/Cdc2/Rum1]$	x_6	$[Cdc13/Cdc2/Rum1]$
x_7	$[Cdc13/P - Cdc2/Rum1]$	x_8	$[IE]$
x_9	$[UbE]$	x_{10}	$[UbE2]$
x_{11}	$[Wee1]$	x_{12}	$[Cdc25]$
x_{13}	$[mass]$	θ_1	$x_1 + \beta x_2$
θ_2	$\theta_1 + \alpha x_4 + [Cig1]$	k_2	$V_{2p}(1 - x_9) + V_2 x_9$
k_6	$V_{6p}(1 - x_{10}) + V_6 x_{10}$	k_{wee}	$V_{wp}(1 - x_{11}) + V_w x_{11}$
k_{25}	$V_{25p}(1 - x_{12}) + V_{25} x_{12}$		

$[A]$ denotes the concentration of species A (Copyright (1997) National Academy of Sciences, U.S.A.)

Table 5.2 Parameter values in Eq. (5.2)

Parameters	Values	Parameters	Values
k_1	0.015 min^{-1}	k_{2p}	0.05 min^{-1}
k_3	0.09375 min^{-1}	k_4	0.1875 min^{-1}
k_5	0.00175 min^{-1}	k_{6p}	0 min^{-1}
k_7	100 min^{-1}	k_{7r}	0.1 min^{-1}
k_8	10 min^{-1}	k_{8r}	0.1 min^{-1}
k_c	1 min^{-1}	k_{cr}	0.25 min^{-1}
k_i	0.4 min^{-1}	k_{ir}	0.1 min^{-1}
k_p	3.25 min^{-1}	k_u	0.2 min^{-1}
k_{ur}	0.1 min^{-1}	k_{u2}	1 min^{-1}
k_{ur2}	0.3 min^{-1}	k_w	1 min^{-1}
k_{wr}	0.25 min^{-1}	V_2	0.25 min^{-1}
V_{2p}	0.0075 min^{-1}	V_6	7.5 min^{-1}
V_{6p}	0.0375 min^{-1}	V_{25}	0.5 min^{-1}
V_{25p}	0.025 min^{-1}	V_w	0.35 min^{-1}
V_{wp}	0.035 min^{-1}	μ	0.00495 min^{-1}
K_{mc}	0.1	K_{mcr}	0.1
K_{mi}	0.01	K_{mir}	0.01
K_{mp}	0.001	K_{mu}	0.01
K_{mur}	0.01	K_{mu2}	0.05
K_{mur2}	0.05	K_{mw}	0.1
K_{mwr}	0.1	α	0.25
β	0.05	$[Cig1]$	0

The fission *yeast* cells have two size requirements: one enforced at S-phase initiation and the other before entering M phase. In wild-type cells, the larger of the two size requirements (governing the G2/M transition) is operative; at birth, cells are already large enough to initiate DNA synthesis, so the G1/S size requirement is cryptic [28]. In $wee1^-$ cells (cells with mutant gene wee1), the G2/M size control is inoperative, permitting cells to divide at an abnormally small size, which brings the G1/S size control into play. That is, $wee1^-$ cells maintain balanced growth and division at about half the size of wild-type cells by operation of a size control mechanism over the Start transition. The model (Fig. 5.2b and Eq. (5.2)) is designed for $wee1^-$ mutants: it has no size control over the Cdc2 tyrosine phosphorylation reactions, and the rate constants for the $Wee1 + Mik1$-catalyzed reactions are chosen to represent a $wee1^-$ strain. A typical simulation of this model is illustrated in Fig. 5.4. G1 phase, during which Rum1 level is high and CDK activity is low, lasts for roughly 65 min, whereas $S + G2 + M$ phase (about 75 min in duration) is shorter than in wild-type cells (130 min) and incompressible. That is, if one decreases the specific growth rate, μ, the cycle time increases by lengthening G1 phase; the duration of $S + G2 + M$ is independent of μ. These are characteristic features of $wee1^-$ cells, whose division cycle is governed by a size requirement in G1 phase [28].

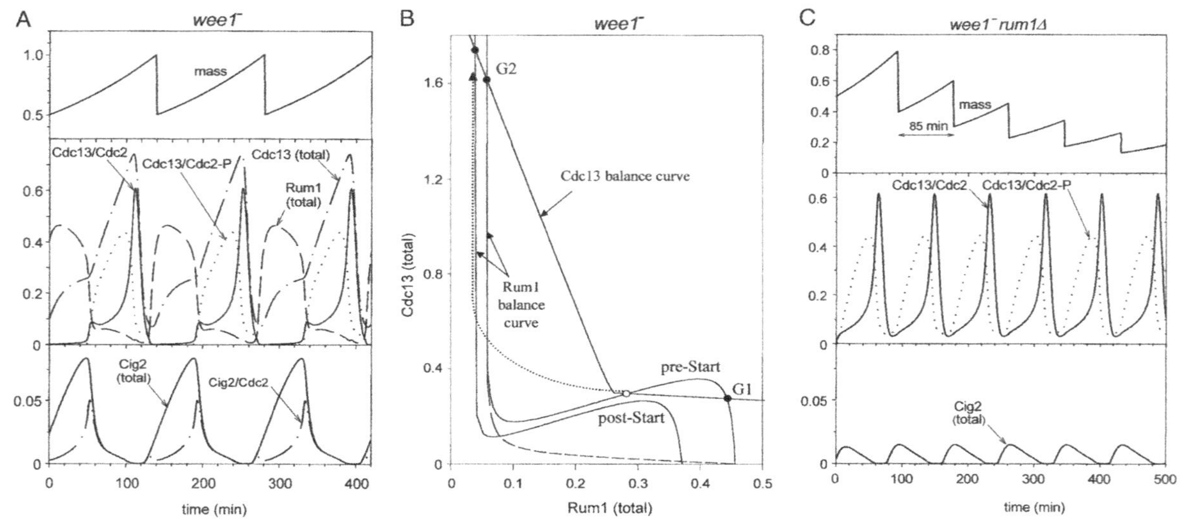

Fig. 5.4 Simulation results for the $wee1^-$ cell cycle. (**a**) The $wee1^-$ cell cycle. Simulated time courses for the major components. Parameters are given in Table 5.2. $Rum1(total) = R + G1R + G2R + PG2R$, $Cdc13(total) = G2K + G2R + PG2 + PG2R$, $Cig2(total) = G1K + G1R$. The cycle time is 140 min, identical to the mass doubling time. (**b**) Phase portrait of the Start transition. Only those steps involved in the synthesis, degradation, and interactions of Cdc13 and Rum1 are selected, thereby reducing the complete system to just two differential equations (about $G2T$ and RT, see Ref. [17] for details). The properties of this two-dimensional subsystem are conveniently portrayed as balance curves (solid lines) in the phase plane [29]. The dashed curve shows how the activity of Cdc13/Cdc2 is quenched as Rum1 accumulates. In pre-Start, the balance curves intersect in two stable steady states (the black dot labeled G_1 and G_2) and an intermediate unstable steady state (hollow circle). In post-Start, the balance curves intersect only in the G_2 steady state. As the cell grows, the Rum1 balance curves move down, causing the G_1 state to disappear (by a saddle-node bifurcation) and the system to proceed along the dotted trajectory to the G_2 state. (**c**) Unbalanced growth and division in $wee1^-rum1\Delta$ mutants. Parameter values are shown in Table 5.2, except $k_3 = 0$ (therefore, $R = G1R = G2R = PG2R = 0$). Both size controls, at G_1/S and G_2/M, are inoperative. There is no stable steady state (checkpoint) at which the cycle can pause to query cell size. Instead the control system executes autonomous (limit cycle) oscillations with a division time (85 min) shorter than the mass doubling time (140 min). Hence, cells get smaller each cycle. Reprinted from Ref. [17] (Copyright (1997) National Academy of Sciences, U.S.A.)

The simulation in Fig. 5.4a indicates that Cig2 is much less abundant than Cdc13. Nonetheless, cell size at Start, when wee12 cells commit to DNA synthesis, is determined equally by Cig2 and Cdc13, because $wee1^{-}cig2\Delta$ cells execute Start at about twice the size of $wee1^{-}cig2^{+}$ cells. Figure 5.4b illustrates that, in pre-Start, the two balance curves intersect at two stable steady states: a G_1 state with lots of Rum1 and little Cdc13-dependent kinase activity and a G_2 state with little Rum1 and lots of active Cdc13yCdc2 dimers. Figure 5.4c shows the double mutant strain wee1^{-} rum1Δ, the cycle of DNA replication and cell division proceeds more rapidly than the mass-doubling process, and cells get progressively smaller each cycle. Clearly, these cells lack any mechanism to coordinate cell growth and division.

Figure 5.5a simulates the endoreplication cycles in cdc13Δ. In this case, the interactions between Rum1 and Cig2 generate size-controlled oscillations of SPF, but cells never enter mitosis because MPF (Cdc13/Cdc2) is missing. The portrait in Fig. 5.5b shows how this cycle works. In G_1 phase, when cells are small, the Cig2 (total) and Rum1(total) balance curves intersect in a unique stable steady state with lots of Rum1 and total Cig2, but little Cig2-dependent kinase activity because most of the cyclin molecules are tied up in inactive trimers. However, as the cell grows, the rate of Rum1 phosphorylation by Cig2-dependent kinase increases and the Rum1 balance curve drops. Eventually, the G_1 steady state loses stability, Rum1 is degraded, and Cig2/Cdc2 dimers are unmasked, driving the cell into S phase (① in Fig. 5.5b). Then, Cig2 level drops ② because Rum1 was shielding Cig2 from proteolysis (notice that, although MPF50, UbE2 retains a basal activity of $V_{6'}$). The subsequent drop in Cig2-mediated phosphorylation of Rum1 allows Rum1 to make a comeback ③. As Rum1 accumulates, Cig2 is stabilized and reaccumulated ④. The authors assume that doubling the DNA content causes an increase in the phosphatase that opposes SPF in the phosphorylation of Rum1, which is equivalent to reducing the effective activity of SPF and is modeled by dividing k_p by 2. This brings the Rum1 balance curve back to the pre-Start position, and the system arrests at the G_1 checkpoint. Cell size must increase by another factor of two before Start is executed again. (During endoreplication, cells do not divide but their DNA-to-mass ratio oscillates over a two-fold range.)

Figure 5.5c illustrates the other case, $rum1^{OP}$, in which fission *yeast* cells undergo multiple rounds of DNA replication. As before, there is a G_1 steady state with lots of Rum1, which is destabilized by growth. In this case, however, because the cell contains excess Rum1, it must grow larger before it can eliminate the G_1 steady state. After Start, Rum1 is degraded, Cig2-dependent kinase activity rises, and the cell enters S phase. But even when Rum1 degradation is large (post-Start), the level of Rum1 remains high enough to squash Cdc13-dependent kinase activity; hence, the cell does not enter mitosis. There is enough SPF activity to trigger DNA synthesis but not enough MPF activity to trigger mitosis.

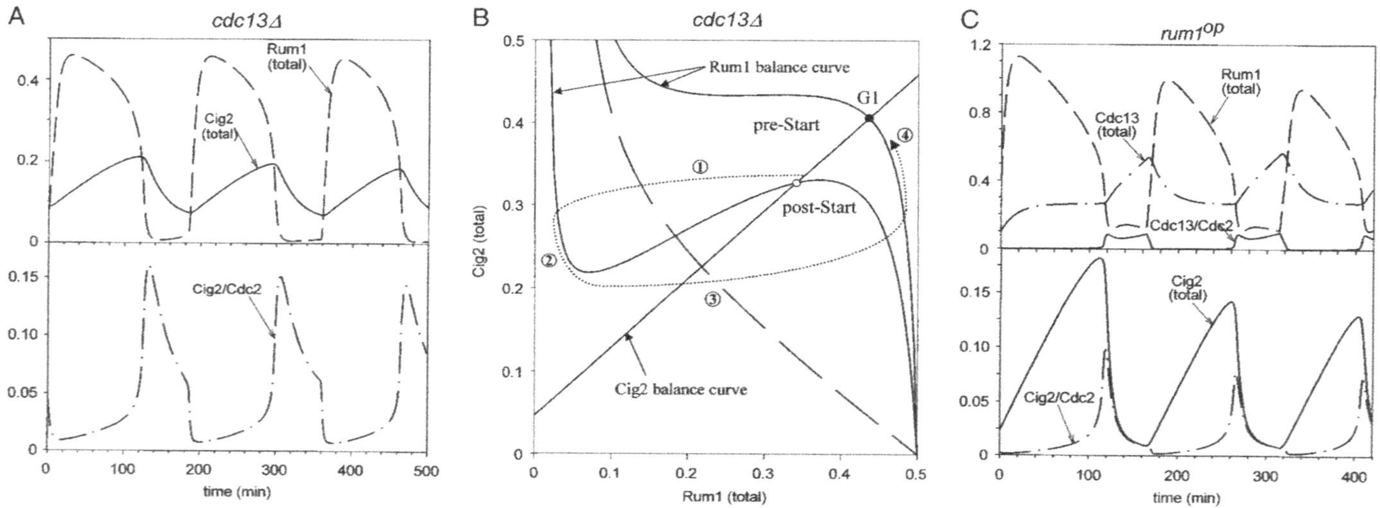

Fig. 5.5 (**a**) Endoreplication cycles in $cdc13\Delta$. Simulations with parameters as in Table 5.2, except $k_1 = 0$ (therefore, $G2K = PG2 = G2R = PG2R = 0$). Because cells never divide, k_p gets smaller by a factor of two each cycle of DNA replication, and cell mass must increase by a factor of two to compensate. (**b**) Phase portrait for the periodic execution of Start without intervening mitoses. Here, only those steps involved in the synthesis, degradation, and interactions of Cig2 and Rum1 (about $G1T$ and RT, see Ref. [17] for details) were considered, and their balance curves (solid lines) in the phase plane were plotted. The dashed curve shows how the activity of Cig2/Cdc2 is quenched as Rum1 accumulates. In pre-Start, the balance curves intersect in a unique stable steady state corresponding to a G_1 checkpoint (black dot). In post-Start, due to cell growth, the steady state (hollow circle) has become unstable by a Hopf bifurcation. Instead of going to a G_2 state, the trajectory (dotted curve) carries the cell back to G_1, as described in the text. (**c**) Endoreplication cycles in $rum1^{OP}$. Simulations with parameters as in Table 5.2, except $k3 = 0.375$ and $Cig1 = 0.07$. During endoreplication cycles, $Cig1$ can accumulate because it is not destroyed at anaphase, as normally it would be [30]. Reprinted from Ref. [17] (Copyright (1997) National Academy of Sciences, U.S.A.)

5.2.3 Summary

Continuous ODEs have been frequently used to model the biological systems. In this section, we have introduced a work reported by Novak and Tyson [17] in the year 1997. They established an ODE to describe the control of DNA replication in the fission *yeast* cell cycle. The authors proposed a molecular mechanism of "Start" control in *yeast*. Then, they compared the properties of the model in detail with the observed behavior of various mutant strains of fission *yeast*: $Wee1^-$ (size control at start), cdc13Δ and $rum1^{OP}$ (endoreplication), and $Wee1^-rum1\Delta$ (rapid division cycles of diminishing cell size). The authors discussed essential features of the mechanism that are responsible for characteristic properties of "Start" control in fission *yeast*.

5.3 Discrete Models for the *Yeast* Cell Cycle Network

Bio-molecular systems can also be described by discrete models, such as the Boolean network model. A Boolean network is a particular kind of sequential dynamical system, where time and states are discrete, i.e., both the set of variables and the set of states in the time series each have a bijection onto an integer series. Boolean networks are related to cellular automata. Usually, cellular automata are defined with a homogeneous topology, i.e., a single line of nodes, a square or hexagonal grid of nodes or an even higher-dimensional structure, but each variable (node in the grid) may take on more than two values (and hence not be Boolean). In this section, we mainly discuss several applications of the Boolean network model on the *yeast* cell cycle networks.

5.3.1 Related Works and Motivations

As early as the year 1969, Kauffman [10] firstly proposed the Boolean network model to describe the metabolic stability and epigenesis in GRNs. A Boolean network is a system of N binary-state nodes (representing genes) with K inputs to each node representing regulatory mechanisms. The two states (on/off) represent the status of a gene being active or inactive. The variable K is typically held constant, but it can also be varied across all genes. For simplicity, each gene is assigned, at random, K regulatory inputs from among the N genes. This gives a single random sample from the ensemble of possible networks of size N and either with connectivity K or with connectivities with some deviation around K. The state of a network at any point in time is given by the current states of all N genes. Thus, the size of the state space of any such network is $2N$. Simulation of Boolean networks is done in discrete time steps. The state of a node at time $t+1$ is computed by applying the Boolean function associated with the node to the state of its input nodes at time t. The sequence of states of the whole network starting from some initial state is called the trajectory of that state.

After the year of 1969, due to the relatively simplicity and parameter free of Boolean models, Boolean models had been extensively applied to investigate various biological systems. In the years 2004 and 2010, Li et al. [11] and Wang et al. [31] investigated the budding *yeast* cell cycle network, where Boolean network models are used. Following the work from Li et al. [11], in the year 2008, Davidich et al. [32] investigated the fission *yeast* cell cycle network. Hereinafter, we briefly review the works reported by Li et al. [11].

The cell cycle process, by which one cell grows and divides into two daughter cells, is a vital biological process, the regulation of which is highly conserved among the eukaryotes [33]. The process consists of four phases: G_1 (in which the cell grows and, under appropriate conditions, commits to division), S (in which the DNA is synthesized and chromosomes are replicated), G_2 (a "gap" between S and M), and M (in which chromosomes are separated and the cell divides into two). After the M phase, the cell enters the G_1 phase, hence completing a "cycle." The process has been studied in great detail in the budding yeast *Saccharomyces cerevisiae* [11, 32–34]. There are about 800 genes involved in the cell cycle process of the budding *yeast* [34]. However, the number of key regulators that are responsible for the control and regulation of this complex process is much smaller. Based on extensive literature studies, Li et al. [11] constructed a network of key regulators that are known so far, as shown in Fig. 5.2a.

There are four classes of members in this regulatory network: cyclins (Cln1, -2, and -3 and Clb1, -2, -5, and -6, which bind to the kinase Cdc28); the inhibitors, degraders, and competitors of the cyclin/Cdc28 complexes (Sic1, Cdh1, Cdc20, and Cdc14); TFs (SBF, MBF, Mcm1/SFF, and Swi5); and checkpoints (the cell size, the DNA replication and damage, and the spindle assembly). Green arrows in Fig. 5.2 represent positive regulations. For example, under rich nutrient conditions and when the cell grows large enough, the Cln3/Cdc28 will be "activated," which in turn activates (by phosphorylation) a pair of TF groups, SBF and MBF, which transcriptionally activate the genes of the cyclins Cln1 and -2 and Clb5 and -6, respectively. Red arrows in Fig. 5.2 represent "deactivation" (inhibition, repression, or degradation). For example, the protein Sic1 can bind to the Clb/Cdc28 complex to inhibit its function, Clb1 and -2 phosphorylates Swi5 to prevent its entry into the nucleus, whereas Cdh1 targets Clb1 and -2 for degradation. The cell cycle sequence starts when the cell commits to division by activating Cln3 (the START). The subsequent activity of Clb5 drives the cell into the S phase. The entry into and exit from the M phase are controlled by the activation and degradation of Clb2. After the M phase, the cell comes back to the stationary G_1 phase, waiting for the signal for another round of division. Thus, the cell cycle process starts with the "excitation" from the stationary G_1 state by the "cell size" signal and evolves back to the stationary G_1 state through a well-defined sequence of states.

5.3.2 Dynamical Analysis

In principle, the arrows in the network have very different time scales of action, and a dynamic model would involve various binding constants and rates [15, 19]. However, because in the cell cycle network, much of the biology seems to be reflected in the on–off characteristics of the network components, and people are mainly concerned here with the overall dynamic properties and the stability of the network. Li et al. established a simplified discrete dynamical model for the network, which treated the nodes and arrows as logic-like operations [11]. Thus, in the model, each node i has only two states, $S_i = 1$ and $S_i = 0$, representing the active and the inactive state of the protein, respectively. The protein states in the next time step are determined by the protein states in the present time step via the following rule:

$$S_i(t+1) = \begin{cases} 1, & \sum_j a_{ji} S_j(t) > \theta_i, \\ 0, & \sum_j a_{ji} S_j(t) < \theta_i, \\ S_i(t), & \sum_j a_{ji} S_j(t) = \theta_i. \end{cases} \tag{5.3}$$

Here, $S_i(t)$ denotes the state of protein i at time t, which can be 0 or 1, θ_i denotes the activation threshold for protein i. In Ref. [11], θ_i is set to be zero. a_{ji} represents network topology; if protein j can activate the expression of protein i, then $a_{ji} = 1$; if the regulation is repression, then $a_{ji} = -1$, and if there is no interaction between j and i, then $a_{ji} = 0$; a_{ii} can be -1, 1, and 0, depending on whether there is auto-degradation, auto-activation, or no self-regulation.

By using the dynamical model described above, Li et al. studied the time evolution of the protein states. First, they studied the attractors of the network dynamics by starting from each of the $2^{11} = 2048$ initial states in the 11-node network of Fig. 5.2b. It is reported that all of the initial states eventually flow into one of the seven stationary states (fixed points). Among the seven fixed points, there is one big fixed point attracting 1764, about 86% of protein states. Remarkably, this super stable state is the biological G1 stationary state. The advantage for a cell's stationary state to be a big attractor of the network is obvious: the stability of the cell state is guaranteed. Under normal conditions, the cell will be sitting at this fixed point, waiting for the signal for another round of division.

Next, the authors started the cell cycle process by "exciting" the G_1 stationary state with the cell size signal and observed that the system goes back to the G_1 stationary state. The temporal evolution of the protein states indeed follows the cell cycle sequence, going from the excited G1 state (the START) to S phase, G_2 phase, M phase, and finally to the stationary G_1 state. This is the biological trajectory or pathway of the cell cycle network.

To investigate the dynamical stability of this biological pathway, the authors studied the dynamic trajectories of all 1764 protein states that will flow to the G_1 fixed point. In Fig. 5.6, each of these protein states is represented by a dot, with the arrows between them indicating dynamic flows from one state to another. The biological pathway is colored in blue and so is the node representing the

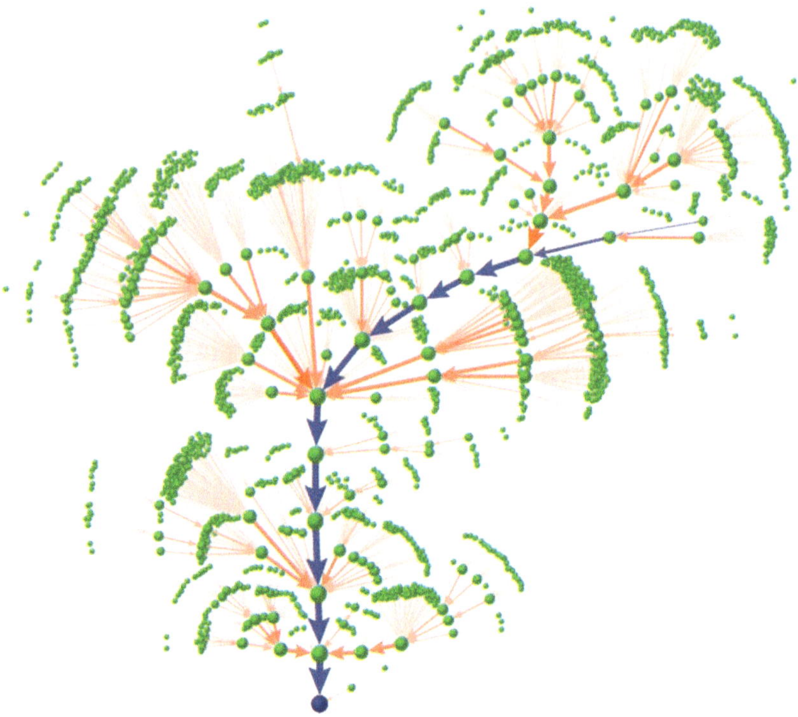

Fig. 5.6 Dynamical trajectories of the 1764 protein states (green nodes) flowing to the G_1 fixed point (blue node). Arrows between states indicate the direction of dynamic flow from one state to another. The cell cycle sequence is colored blue. The size of a node and the thickness of an arrow are proportional to the logarithm of the traffic flow passing through them. Reprinted from Ref. [11] (Copyright (2004) National Academy of Sciences, U.S.A.)

G_1 stationary state. One observes that the dynamic flow of the protein states is convergent to the biological pathway, making the pathway an attracting trajectory of the dynamics. With such a topological structure of the phase diagram of protein states, the cell cycle pathway is a very stable trajectory; it is very unlikely for a sequence of events, starting at the beginning (or at any other point) of the cell cycle process, to deviate from the cell cycle pathway. Interestingly, the topology of the converging trajectories as shown in Fig. 5.6 is reminiscent of the converging kinetic pathways in protein folding, where a protein sequence is facing the challenge of finding the unique native state among a huge number of conformations [35–37].

5.3.3 *Statistical Analysis*

5.3.3.1 Comparison with Random Networks

To investigate how likely a big fixed point and a converging pathway can arise by chance, Li et al. studied an ensemble of random networks that have the same numbers of nodes and links in each color as in the cell cycle network. They found that the random networks typically have more attractors (fixed points and limit cycles), with the average number being 14.28. The sizes of the basins of attraction in the random networks have a power-law distribution, as shown in Fig. 5.7a. The probability for a random network to have an attractor of a basin size B equal to or larger than that of the cell cycle network ($B \geq 1764$) is 10.34%.

To quantify the "convergence" of trajectories, the authors defined a quantity $w_n (n = 1, 2, \ldots, 2048)$ for each of the 2048 network states that measures the overlap of its trajectory with all other trajectories (Fig. 5.7c). Denote by $T_{j,k}$ the total traffic flow through the arrow $A_{j,k}$ that takes state j to k in one time step,

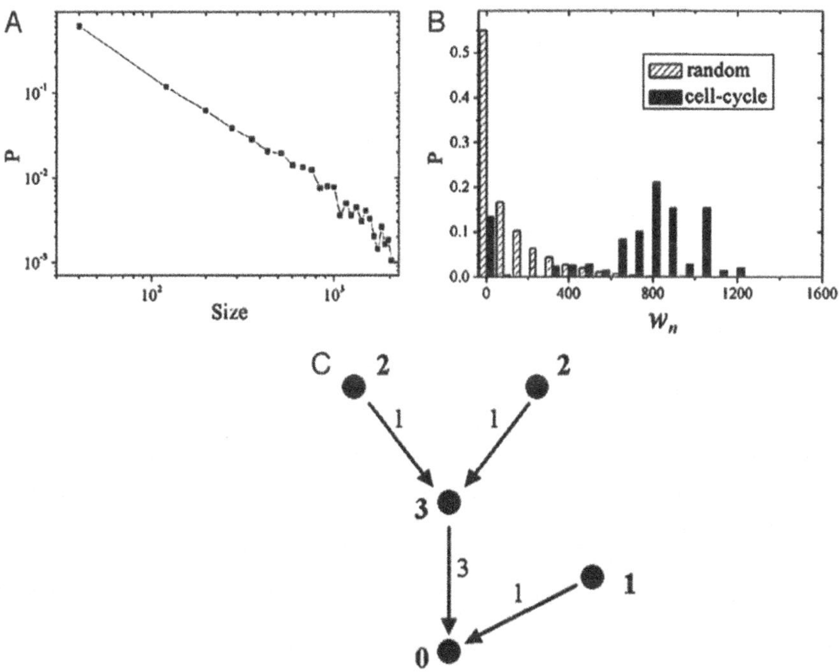

Fig. 5.7 Comparison with random networks. (**a**) Attractor size distribution of random networks. (**b**) w-Value distributions for the cell cycle network and for random networks. 10,000 random networks were used to generate the statistics. (**c**) Schematic illustration of the definition of w_n. The number next to an arrow indicates the total traffic through the arrow. The number next to a node is the w_n of the node. Reprinted from Ref. [11] (Copyright (2004) National Academy of Sciences, U.S.A.)

i.e., $T_{j,k}$ is the total number of trajectories starting from all network states that pass through $A_{j,k}$. If the trajectory from state n to its attractor has L_n steps, so that it consists L_n arrows $A_{k-1,k}$, $k = 1, 2, \ldots, L_n$.

$$w_n = \sum_{k=1}^{L_n} T_{k-1,k}/L_n.$$

The overall overlap of all trajectories in a network can be measured by $W =< w_n >$, where the average is over all network states. The normalized histogram of w_n for all network states is shown in Fig. 5.7b for both the cell cycle and the random networks. Without any significant overlap or convergence of trajectories and with a much shorter transient times to attractors, the random networks have their w distribution peaked at small w's, with an average $W = 124$. However, for the cell cycle network, the distribution is peaked at very large numbers ($W = 743$), indicating significant convergence of trajectories. The probability for a random network to have a $W \geq 743$ is only 0.25%.

5.3.3.2 Network Perturbations

The cell cycle network has two distinct dynamic properties compared with the random networks: it has a super fixed point and a converging pathway. What effects would perturbations of the network have on these properties? The authors perturbed a network by deleting an interaction arrow, adding a green or red arrow between nodes that are not linked by an arrow, or switching a green arrow to red and vice versa. The relative change in the size of the basin of attraction (B) for the biggest attractor, $\Delta B/B$, was then measured as a result of the perturbation. The distribution of $\Delta B/B$ is plotted in Fig. 5.8 for each kind of perturbation, respectively, along with those obtained from the ensemble of random networks. One observes that only a very small fraction of perturbations will eliminate the fixed point completely ($\Delta B/B = 1$). For most perturbations, the relative changes of the basin size are small. A similar behavior in the changes of the quantity W as results of the perturbations was also seen. Interestingly, this high "homeostatic stability" [38] is also evident in the ensemble of random networks of the same size (Fig. 5.8). In fact, it can be found that for random networks with the dynamic rule of Eq. (5.3), the homeostatic stability increases monotonically with the average number of arrows per node k, which is very different from the random Boolean network, where a "chaotic" phase with low homeostatic stability is seen for $k > k_c$ [38]. Existing studies suggested that either a SF Boolean network [39] or a genetic network with minimal frustration [40] would also lead to a more stable phase.

To examine the effects of these perturbations on the biological pathway itself, for each perturbed cell cycle network, the authors started at the START state and followed its time evolution. They found that under perturbation, a significant fraction of the trajectories reach the G_1 stationary state and the cell cycle sequence is by far the most probable trajectory (Fig. 5.9).

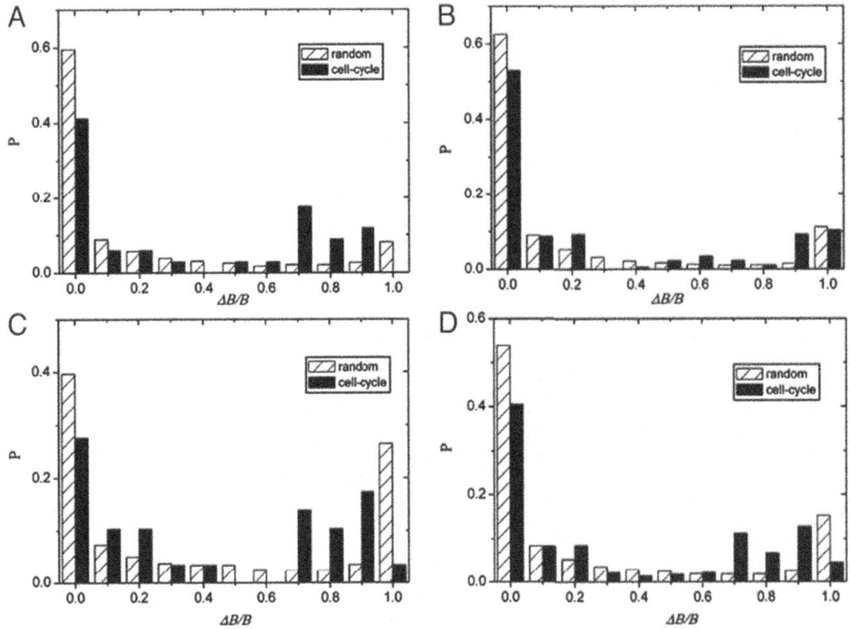

Fig. 5.8 The histogram of the relative changes of the size of the basin of attraction for the biggest fixed point with respect to network perturbations. The four panels are corresponding to 34 line deletions (**a**), 174 line additions (**b**), 29 red–green switchings (**c**), and the average of (**a**)–(**c**) (**d**). 1000 random networks were used to generate the statistics. Reprinted from Ref. [11] (Copyright (2004) National Academy of Sciences, U.S.A.)

5.3.4 Summary

Through a Boolean dynamic model for the *yeast* cell cycle network, it has been demonstrated that the *yeast* cell cycle network is robustly designed [11]. The biological states at the checkpoints are big attractors, and the biopathway is an attracting trajectory. These robust dynamical properties are also seen in the life cycle network of the budding *yeast*, suggesting that they may be common features of regulatory networks. The cell cycle network is rather stable against perturbations. Note that the network studied here is only a skeleton of a larger cell cycle network with many "redundant" components and interactions (e.g., any member of the G_1 cyclins can, to a large extent, perform the functions of other members). Thus, it is expected that the complete network is even more stable against perturbations. In some sense, biological systems have to be robust to function in complex (and very noisy) environments. More robust could also mean more evolvable, and thus more likely to survive; a robust "module" is easier to be modified, adapted, added-on, and combined with others for new functions and new environments [41]. Indeed, robustness may provide us with a handle to understand the profound driving force of evolution.

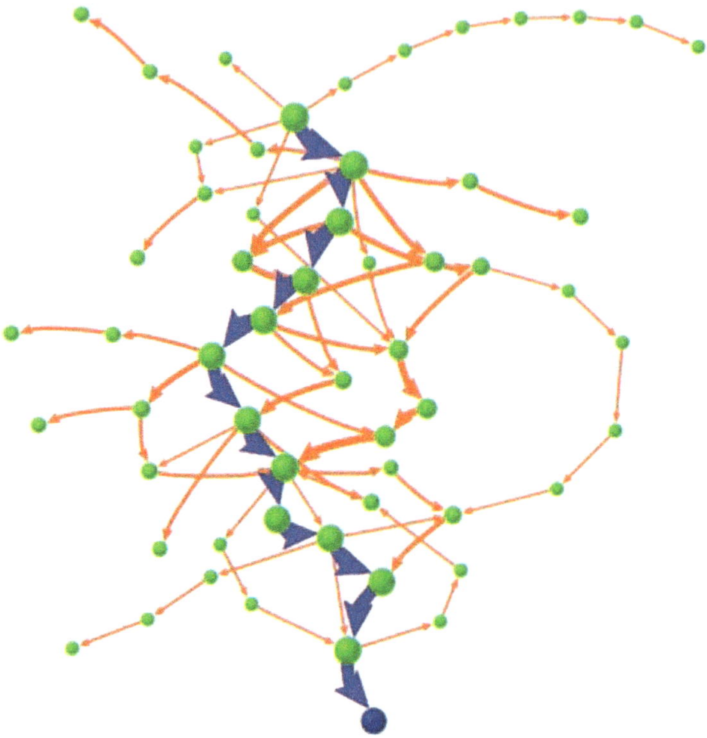

Fig. 5.9 Trajectories of the perturbed cell cycle network starting from the START. The trajectories from each kind of perturbations (34 from arrow deletions, 174 from arrow additions, and 29 from red–green switchings) are first superimposed on top of each other to form three groups. The three groups are then superimposed on top of each other with equal weights. The width of an arrow and the size of a node are proportional to the logarithm of the number of shared trajectories. The biological pathway is colored blue. The percentages of the perturbed networks that still evolve to the G_1 state from START are 41.2%, 57.4%, and 64.7% for arrow deletion, arrow addition, and color switching, respectively. Reprinted from Ref. [11] (Copyright (2004) National Academy of Sciences, U.S.A.)

By following the work by Li et al. [11], Wang et al. [31] have used a unique process-based approach to analyze the budding *yeast* and the fission *yeast* cell cycle networks (Fig. 5.10). They modeled the cell cycle network as the following Boolean network:

$$S_i(t+1) = \begin{cases} 1, & \sum_j a_{ji} S_j(t) > 0, \\ 0, & \sum_j a_{ji} S_j(t) < 0, \\ S_i(t), & \sum_j a_{ji} S_j(t) = 0. \end{cases} \tag{5.4}$$

Here, $A = (a_{ji})_{N \times N}$ is a $N \times N$ matrix encoding the network structure. The diagonal entries, a_{ii}, take the value -1 (self-degradation), 1 (self-activation), or

A

Time	Cln3	MBF	SBF	Cln1,2	Cdh1	Swi5	Cdc20/14	Clb5,6	Sic1	Clb1,2	Mcm1/SFF	Phase
t	s_1	s_2	s_3	s_4	s_5	s_6	s_7	s_8	s_9	s_{10}	s_{11}	
0	1	0	0	0	1	0	0	0	1	0	0	START
1	0	1	1	0	1	0	0	0	1	0	0	G1
2	0	1	1	1	1	0	0	0	1	0	0	G1
3	0	1	1	1	0	0	0	0	0	0	0	G1
4	0	1	1	1	0	0	0	1	0	0	0	S
5	0	1	1	1	0	0	0	1	0	1	1	G2
6	0	0	0	1	0	0	1	1	0	1	1	M
7	0	0	0	0	0	0	1	0	0	0	1	M
8	0	0	0	0	1	1	1	0	1	0	0	M
9	0	0	0	0	1	1	0	0	1	0	0	G1
10	0	0	0	0	1	0	0	0	1	0	0	G1
11	0	0	0	0	1	0	0	0	1	0	0	G1

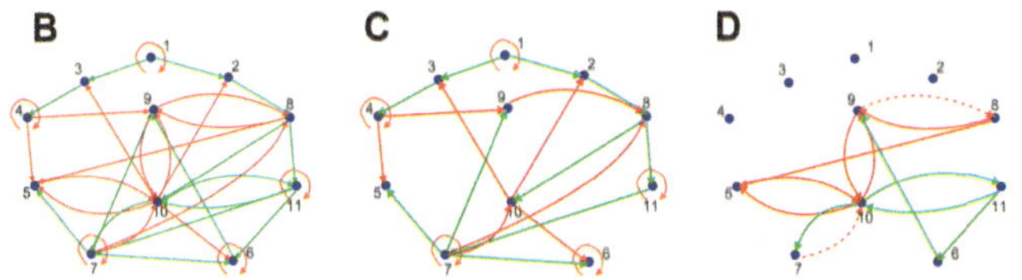

E

Time	SK	Cdc2/13	Ste9	Rum1	Slp1	Cdc2/13*	Wee1	Cdc25	PP	Phase
t	s_1	s_2	s_3	s_4	s_5	s_6	s_7	s_8	s_9	
0	1	0	1	1	0	0	1	0	0	START
1	0	0	0	0	0	0	1	0	0	G1/S
2	0	1	0	0	0	0	1	0	0	G2
3	0	1	0	0	0	0	0	1	0	G2
4	0	1	0	0	0	1	0	1	0	G2/M
5	0	1	0	0	1	1	0	1	0	G2/M
6	0	0	0	0	1	0	0	1	1	M
7	0	0	1	1	0	0	1	0	1	M
8	0	0	1	1	0	0	1	0	0	G1
9	0	0	1	1	0	0	1	0	0	G1

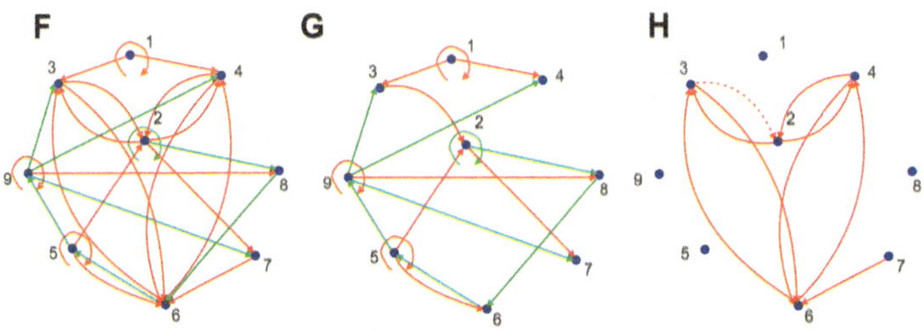

Fig. 5.10 Process-based network decomposition of the budding *yeast* and the fission *yeast* cell cycle networks. (**a**)–(**d**) are for the budding *yeast*. (**a**) The time course of the 11 nodes as a representation of the cell cycle process. (**b**) The full budding *yeast* cell cycle network. (**c**) The backbone subnetwork contained in the full network (**b**). (**d**) The supplemental edges are characterized by various feedback loops. The edges r_{98} and $r_{7,10}$ are shown as dashed lines because they are shared with the backbone. (**e**)–(**h**) are for the fission *yeast*. (**e**) The time course of the nine nodes as a representation of the cell cycle process. (**f**) The full fission *yeast* cell cycle network. (**g**) The backbone subnetwork contained in the full network (**f**). (**h**) The remaining edges are characterized by mutual inhibitive loops. The edge r_{32} is shown as dashed line because it is shared with the backbone. Reprinted from Ref. [31]

0 (no action). The nondiagonal entries, $a_{ji}(j \neq i)$, take the value $-\gamma$, 1, or 0, depending on whether node j inhibits, activates, or does not interact with node i. The parameter γ models the relative dominance of inhibition over stimulation. Because inhibition is dominant over stimulation for most bio-molecular interactions, one prefers $\gamma \geq 1$. Moreover, the network dynamics is usually not sensitive to the value of γ (the network topology is more important than the actual interaction strength).

Because the state variables $S(t)$ are known from the biological process (Fig. 5.10a, e), Eq. (5.4), for $t = 0, 1, \cdots, T - 1$, is used to infer the network connections to node i. The authors enumerated all minimal networks to identify which minimal network (backbone) is present in a given network. For example, there are 108,864 minimal networks that arise from analyzing the budding *yeast* cell cycle process (Fig. 5.10a), among which one and only one is contained in the budding *yeast* network. For the budding *yeast* and the fission *yeast* cell cycle networks, they found that each of these networks contains a giant backbone motif spanning all the network nodes that provides the main functional response. The backbone is in fact the smallest network capable of providing the desired functionality (Fig. 5.10c, g). Furthermore, the remaining edges in the network form smaller motifs whose role is to confer the stability properties rather than provide function (Fig. 5.10d, h). The process-based approach used in their analysis has additional benefits: it is scalable, analytic (resulting in a single analyzable expression that describes the behavior), and computationally efficient (all possible minimal networks for a biological process can be identified and enumerated). For the detailed investigations, one can refer to Ref. [31].

5.4 Percolating Flow Model for a Mammalian Cellular Network

For large-scale bio-molecular networks, pseudodynamics can be used to investigate the information propagation in chemical space. The following sections are mainly based on Ref. [13].

5.4.1 Related Works and Motivations

In the year 2005, Máayan et al. [13] used published research literature to identify the key components of signaling pathways and cellular machines and their binary interactions. Most components (about 80%) have been described in *hippocampal neurons* or related neuronal cells. Other components are from other cells but are included because they are key components in processes known to occur in *hippocampal neurons*, such as translation. The authors developed a system made of 545 components (nodes) and 1259 links (connections), as shown in Fig. 5.3. They used arbitrary but consistent rules to sort components into various groups. For instance, TFs are considered as part of the transcriptional machinery, although it may also be equally valid to consider them as the most downstream component of the central signaling network. Similarly, the AMPA receptor (AMPAR) channel is considered as part of the ion channels in the electrical response system, since its activity is essential to define the postsynaptic response, although it binds to and is activated by glutamate and hence can be also considered as a ligand-gated receptor channel in the plasma membrane.

The links were specified into three classes, namely, activating, inhibitory, or neutral. Neutral links do not specify directionality between two components and are mostly used to represent scaffolding and anchoring undirected or bidirectional interactions.

5.4.2 Dynamical Analysis

Components within mammalian cells interact with one another to form local networks that together form a single large network. This organization is essential for cellular components to coordinate their individual activities and achieve the cohesiveness needed for cellular functions. Information needs to flow between components in a continuous and organized manner. Determining how this flow of information occurs is a crucial step in understanding the functional organization of mammalian cells. This system of interacting cellular components based on phenotypic behavior allows people to analyze the flow of information between the components to identify the emergence of regulatory motifs that are capable of processing information as it flows through the network.

Understanding how the functional organization of cellular systems changes in response to information flow is an important goal in systems biology. For systems containing many components, obtaining an overview of the patterns of regulatory motifs and defining their interrelationships can provide a format for in-depth analysis of individual units using quantitative biochemical representations.

From data in the experimental literature, Máayan et al. [13] constructed a system of interacting cellular components involved in phenotypic behavior and used graph theory methods to analyze qualitative relationships between nodes (components) in a network. In signaling networks, activation is achieved as a response to a stimulus. Information propagates through the system by a series of coupled biochemical reactions to regulate components responsible for cellular phenotypic functions. Máayan et al. identified the regulatory features that emerge during such information flow in a simplified representation of a mammalian *hippocampal* CA1 neuron. Such neurons are capable of plasticity as defined by their ability to undergo long-term potentiation of synaptic responses.

The CA1 neuron is seen as a set of interacting components that make up a network of signaling pathways that connects to various cellular machines (Fig. 5.10a).

The authors identified various regulatory motifs in the network. They studied signal propagation resulting from ligand occupancy of receptors by building and analyzing a series of subnetworks that originate from nodes representing ligands. This is termed pseudodynamics because it represents propagation of reactions in chemical space rather than time series. Direct interaction between any two components, termed a link, consists of one or more underlying chemical reactions. Signal propagation from node to node is organized within the chemical space in units of steps. For any given node at step n, all immediate upstream nodes are at step $n - 1$ and all nodes positioned one link downstream are at step $n + 1$.

The authors counted the number of links per step as signals propagate from ligand–receptor interactions to their downstream effectors. The analysis of the emergent subnetworks for the different ligands showed a discernible pattern. For ligands that cause rapid, transient changes, early signal branching was extensive (that is, many links were formed in relatively few steps) (Fig. 5.11b). For ligands that cause permanent changes, such as FasL, which induces apoptosis, or ephrin, which alters neuronal morphology, there were fewer branches as the signal moved through the network. Between these extremes were many growth factors and ligands that bind G protein-coupled receptors (Fig. 5.11b). When the signal originating from any ligand had progressed through 15 steps, most of the network (nearly 1000 links) was engaged. This is a common property of large, highly connected directed graphs and in the considered case represents a very extensive propagation of the signal from the ligand analogous to prolonged receptor activation.

For any individual ligand, the whole network was never fully affected, with a few nodes (usually other ligands) with single directed outgoing interactions not engaged. The authors characterized the network that emerged as signals were traced through each step from three ligands that are key regulators of plasticity in hippocampal neurons: glutamate, norepinephrine (NE), and brain-derived neurotrophic factor (BDNF). Glutamate influenced more links and nodes in early stages than did NE or BDNF, which showed similar profiles. The subnetwork characteristics were similar. The authors analyzed the types of regulatory motifs formed as signals propagate from the ligands.

A motif is a group of interacting components capable of signal processing. Motifs such as positive feedback loops promote the persistence of signals and serve as information storage devices [42], whereas negative loops limit signal propagation through the network. In counting these motifs, the authors used all possible configurations of loops with three and four components (Fig. 5.11c). In their model, glutamate activates 25 feedback loops within five steps, whereas NE and BDNF require six steps to recruit 20 feedback loops (Fig. 5.11c). They determined the relative abundances of negative and positive feedback loops as signal propagated from glutamate, NE, or BDNF (Fig. 5.11d). Using a binomial test and the combinatorial possibilities arising from the ratios of positive to negative links, it appeared that there were more than the expected number of negative feedback loops compared with positive feedback loops for glutamate at early steps (<0.05). This indicates that the early steps may have built-in controls to limit signal propagation. As the signal progressed to later steps, positive feedback loops tended to be more

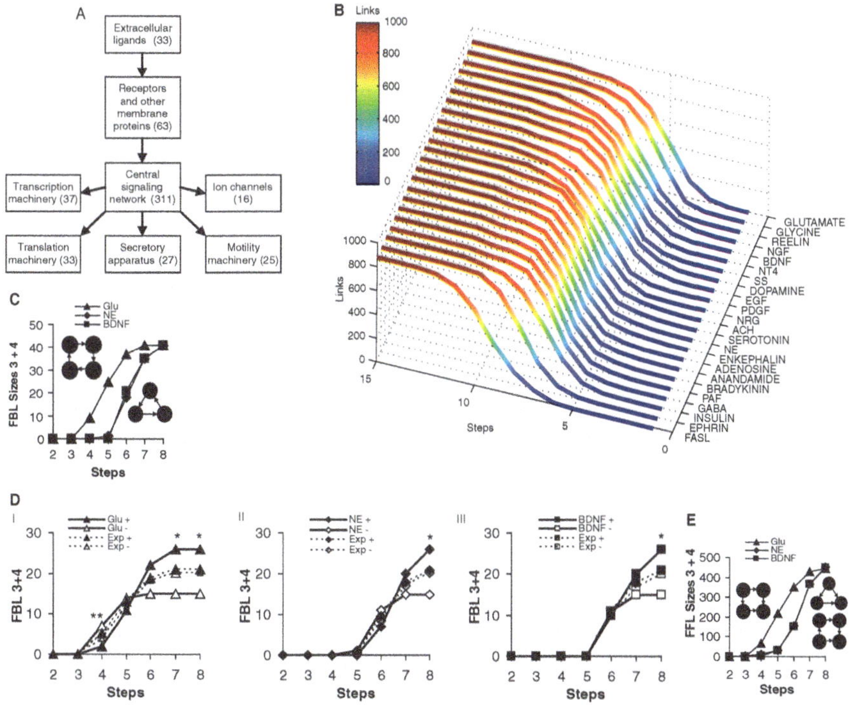

Fig. 5.11 Assembly of a function-based cellular network. (**a**) Block diagram of information flow from ligands to cellular machines. Nodes are associated with functional compartments. The counts of nodes are in parentheses. (**b**) Total links in subnetworks created in steps from ligands. The total number of links accumulated as a signal moves through the steps, downstream from various ligands, is compared. (**c**) The total number of 3- and 4-component feedback loops (positive and negative) in subnetworks with two to eight steps from Glu, NE, and BDNF. (**d**) The counts of positive and negative feedback loops in subnetworks from Glu (I), NE (II), and BDNF (III). The counts in the biological subnetworks (solid lines) are compared with the expected counts (dashed lines) developed using combinatorial probabilities for positive or negative feedback loops based on the number of total positive and negative links. ** : $P < 0.05$;* : $P < 0.08$; binomial test. (**e**) The counts of 3- and 4-component feed-forward motifs in subnetworks with two to eight steps from Glu, NE, and BDNF. Reprinted from Ref. [13], with permission from AAAS

abundant than expected for all three ligands ($P < 0.08$). These profiles suggest that weak or short-lived signals may not penetrate into the network because of the early barrier posed by the abundance of negative feedback loops. However, as the number of steps increases and positive feedback loops appear to be more abundant, signals should persist and be able to evoke a biological response. The authors also counted feed-forward motifs (Fig. 5.11e), scaffolds, and bi-fans. Scaffold motifs are sets of three nodes connected through three neutral links. Their presence may indicate a mechanism for local clustering and may represent the basis for spatial specification of information flow.

5.4.3 Statistical Analysis

The authors developed maps to represent the pseudodynamic regulatory topology that their analyses had identified. They plotted the density of motifs at each step for the three ligands, glutamate, NE, and BDNF by defining a "density of information processing" (DIP), which is the ratio of the additional new motifs (feedback loops, FFLs, and bi-fans) formed in each step to the increase in new links in each step multiplied by the grid coefficient. Mathematically, the DIP is defined as

$$DIP_i = \left(\frac{M_i - M_{i-1}}{L_i - L_{i-1}}\right) \times GC_i, \tag{5.5}$$

where $M_i = FBL3_i + FBL4_i + FFL3_i + FFL4_i + BIFAN_i$. M_i is the total number of feedback loops, FFLs, and bi-fan motifs, L_i is the total number of links, and i is the step. FBL3 and FBL4 are feedback loops of sizes 3 and 4, FFL3 and FFL4 are FFLs of sizes 3 and 4, and BIFAN is the bi-fan motifs of size 4. GC_i is the grid coefficient representing interconnectedness for the motifs, computed for the subnetwork at step i.

This measure can be used to identify the intensity and position in chemical space of information-processing activities. The DIP profile (Fig. 5.12a) at each step is plotted for the three different ligands through eight steps as signal propagates from receptors to cellular machines. Each DIP profile is distinctive, indicating that these represent differential stimulation of partially overlapping regions of the cellular network. Each ligand showed a "hot zone," where extensive information processing may occur. The authors also developed detailed maps to determine the position of the regulatory motifs in the chemical space of the network. For this, they specified the location of the nodes that participate in a motif between extracellular ligands and cellular machines on the basis of the shortest path lengths from the node within the motif to all extracellular ligands and to all components in the specified cellular machine. For each node, this yielded a "node location index" that measures the functional distance to each of the five cellular machines and to the ligands. They then identified the participation of these nodes in the various motifs. A parameter termed "motif location index" (MLI) was defined as the average of the location indices for the various nodes that comprise the motif in relation to the distance from the specified machine, defined as follows:

$$MLI = \frac{\sum_{i=1}^{n}\left(\frac{CPLM_i}{CPLM_i + CPLL_i}\right)}{n}, \tag{5.6}$$

where n is the size of the motif, $CPLM_i$ is the characteristic path length from a node within the motif to all other nodes in the cellular machine, and $CPLL_i$ is the characteristic path length from a node to all extracellular ligands. If a node is an extracellular ligand, $CPLL_i = 0$ for that node; if the node is in the plasma membrane, $CPLL_i = 1$. If a node belongs to a cellular machine, $CPLM_i = 0$

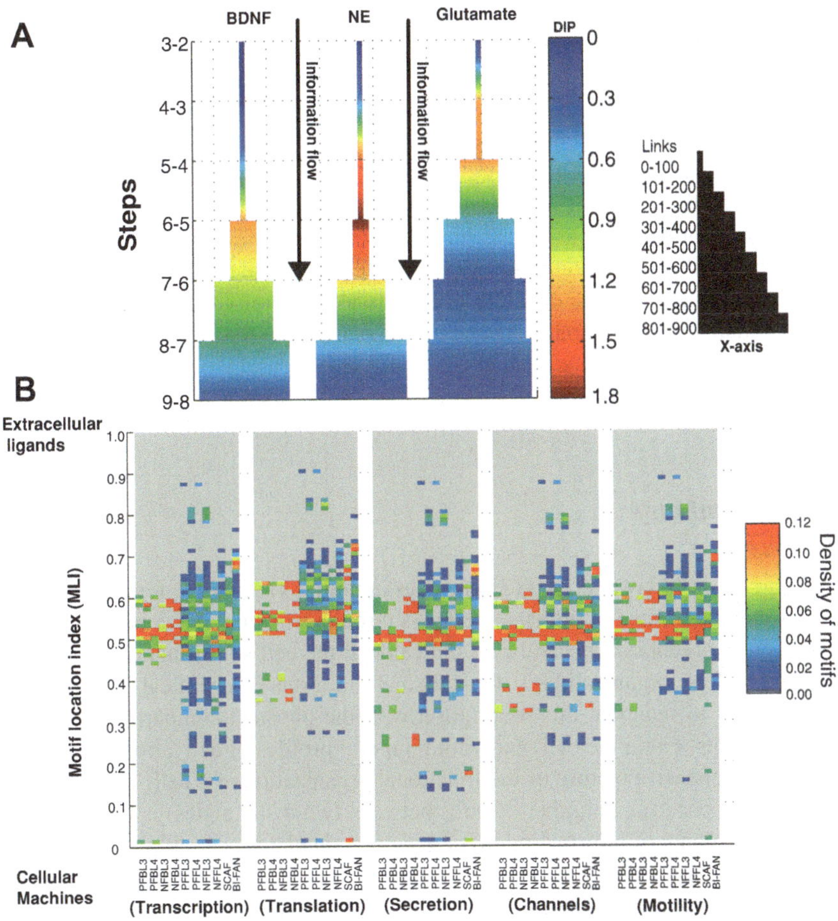

Fig. 5.12 Maps of the regulatory profiles within the cellular network. (**a**) DIP, representing the concentration of all motifs (shown in pseudocolor, with red indicating high density) and their local interconnectedness relative to the network size. DIP is plotted as a signal propagates vectorially through the network, as indicated by the downward arrow. The width of the bar represents the number of links engaged. (**b**) Relative distribution of regulatory motifs between ligands and cellular machines. PFBL and NFBL, 3- and 4-node positive and negative feedback loops, respectively; PFFL and NFFL, 3- and 4-node positive and negative FFLs; SCAF, 3-node scaffold motifs; BIFAN, 4-node bi-fan motifs. Reprinted from Ref. [13], with permission from AAAS

for that node. The counts of motifs were placed in 100 bins based on the computed MLI and normalized to the fraction of total motifs for each class of motifs. A MLI value of 0 represents location at the cellular machines where all the nodes that make up the motif are within the cellular machine; a value of 1.0 represents location at the ligand level.

Five maps corresponding to the different cellular machines were generated (Fig. 5.12b). These maps indicate the location of the various regulatory motifs between extracellular ligands and cellular machines. Both common and distinctive features are observed. When pathways from ligands to each of the cellular machines were considered, a higher density of regulatory motifs was found at the middle of the maps (note the red band in each map), indicating that a major portion of the information processing occurs at the "center" of the network.

Distinct patterns of motifs were observed upstream of the different cellular machines. In the direct upstream of the transcriptional machinery, feed-forward motifs were abundant. In contrast, for the translational machinery, the regulation was more distal. For the secretory machinery, feed-forward and feedback loops and scaffolds were found to be closer to the machine. For both the motility machinery and ion channels, regulation was largely concentrated in the center of the network (around $MLI + 0.5$). Some regulatory motifs are made of components within the cellular machines themselves.

5.4.4 Summary

The work from Máayan et al. [13] provides an initial view of the regulatory capabilities that are formed as information flows through a cellular network. The overall profile of emergent motifs can be an indicator of the cell's information-processing capability. Here, organization is defined solely in terms of chemical space. It will be necessary to integrate this description with the physical compartmentalization of components within the cell, and with temporal profiles of activation, to obtain a more comprehensive picture of the functional organization of a cell. Still, distance in chemical space (i.e., number of links between two distal nodes) is likely to be a major determinant of information processing that regulates phenotypic behavior.

The maps for individual ligands or cellular machines show distinct patterns of motifs. Combinations of ligands will likely produce many more patterns of connectivity. Thus, a cellular system may not be a single network but rather an ensemble of network configurations that are evoked by the stimuli-induced activation of various parts of the system. Identifying these network configurations and the functions they evoke is likely to provide more complete descriptions of how molecular interactions lead to cellular choices between homeostasis and plasticity.

5.5 A Hybrid Model for Mammalian Cell Cycle Regulation

5.5.1 Related Works and Motivations

The physiological behaviors of cells (growth and division, differentiation, movement, death, etc.) are controlled by complex networks of interacting genes and proteins, and a fundamental goal of computational cell biology is to develop dynamical models of these regulatory networks that are realistic, accurate, and predictive. Historically, these models have divided along two basic lines: deterministic or stochastic and continuous or discrete. The discrete Boolean models have severe limitations. First of all, metronomic time in Boolean models is unrelated to clock time in the laboratory, so the Boolean models cannot be compared to even the most basic observations of time spent by cells in the four phases of the division cycle. Also, these models do not incorporate cell size, so they cannot address the evident importance of cell growth in driving events of the cell cycle. Lastly, cyclins are treated as either absent or present (0 or 1), so the Boolean models cannot simulate the continuous accumulation and removal of cyclin molecules at different stages of the cell cycle. Whereas, for the continuous ODE models, there are always many kinetic parameters, these kinetic constants must be estimated from experimental data, and often there are insufficient kinetic data to determine their values.

To retain the elegance of the Boolean representation of the switching network, while introducing continuous variables for cell size, cell age, and cyclin composition, and in order to create a model that can be compared in quantitative detail to experimental measurements with a minimal number of kinetic parameters that must be estimated from the data, Singhania et al. [43] proposed a hybrid model for Mammalian cell cycle regulation, which combines a continuous representation of slowly changing protein concentrations with a discrete representation of components that switch rapidly between "on" and "off" states, and which combines the deterministic causality of network interactions with the stochastic uncertainty of random events. Specifically, they kept the cyclin regulators as Boolean variables but modeled the cyclins themselves as continuous concentrations that increase and decrease due to synthesis and degradation. Next, they replaced the Boolean model's metronome with real clock time to account for realistic rates of cyclin synthesis and degradation and for stochastic variability in the time spent in each Boolean state of the model. Finally, they introduced a cell size variable, $M(t)$, which affects progression through late G_1 phase. $M(t)$ increases exponentially with time as the cell grows and decreases by a factor of around 2 when the cell divides. (The assumption of exponential growth is not crucial; similar results are obtained by assuming linear growth between cell birth and division.) The hybrid approach can be easily tailored to the available knowledge of control systems, and it provides both qualitative and quantitative results that can be compared to experimental data to test the accuracy and predictive power of the model. In what follows, we briefly review the work reported by Singhania et al. [43].

5.5.2 *The Hybrid Model*

The hybrid model proposed by Singhania et al. [43] is hybrid in two senses. First, the authors employed both continuous and discrete variables, and second, they allowed for both deterministic and stochastic processes. Concerning the components of the control system, they tracked cyclin levels as continuous concentration variables, but they used discrete Boolean variables to represent the activities ("on" or "off") of the regulatory proteins (TFs and ubiquitinating enzymes) that control cyclin synthesis and degradation. This distinction is equivalent to a presumed "separation of time scales": the activities of the regulatory proteins change rapidly between 0 and 1, while the concentrations of cyclins change more slowly due to synthesis and degradation. The Boolean variables, they assumed, proceed from one state to the next according to a fixed sequence corresponding roughly to the super highway of Li et al. [11]. The time spent in each state, however, is not a "tick" of the metronome but rather the sum of a deterministic execution time (which may be 0) plus a random, exponentially distributed waiting time. In this sense, the model combines deterministic and stochastic processes.

The model (Fig. 5.13) tracks three cyclin species (A, B, and E), two TFs (TFE and TFB), and two different E3 ubiquitin-ligase complexes (APC-C and SCF). TFE drives the synthesis of cyclins E and A early in the cell cycle (comparable to the E2F family of TFs), and TFB drives the synthesis of cyclins B and A late in the cell cycle (comparable to FoxM1 and Myc). The Anaphase Promoting Complex–Cyclosome (APC-C) is active during M phase and early G_1, when it combines with Cdc20 and Cdh1 to label cyclins A and B for degradation by proteasomes. The authors made a further distinction between Cdc20 activity on cyclin A (Cdc20A, active throughout mitosis) and Cdc20 activity on cyclin B (Cdc20B, activated at anaphase). The SCF labels cyclin E for degradation via ubiquitination, but only when cyclin E is phosphorylated, which is assumed to be correlated primarily with cyclin A/Cdk2 activity. The proposed hybrid model is described as follows:

$$\begin{cases} \frac{[CycA]}{dt} = k_{sa} - k_{da}[CycA], \\ \frac{[CycB]}{dt} = k_{sb} - k_{db}[CycB], \\ \frac{[CycE]}{dt} = k_{se} - k_{de}[CycE]. \end{cases} \tag{5.7}$$

Here, $k_{sa}, k_{da}, k_{sb}, k_{db}, k_{se}$, and k_{de} are the rate constants for synthesis and degradation, as listed in Table 5.3, which depend on the Boolean variables, as shown in Table 5.4.

In the model (5.7), the two TFs and the four ubiquitination factors are each represented by a Boolean variable, B_{TFE}, etc. For each cyclin component, an ODE, $d[CycX]/dt = k_{sx} - k_{dx}[CycX]$ is used to describe its dynamics, where the rate constants for synthesis and degradation, k_{sx} and k_{dx}, depend on the Boolean variables (see Table 5.3). Hence, each cyclin concentration is governed by a piecewise linear ODE. The parameters in the model ($k_{sx}^{'}, k_{sx}^{''}$, etc.) are assigned numerical values (Table 5.3), chosen to fit observations of how fast cyclins accumulate and disappear during different phases of the cell cycle.

The rules for updating the Boolean variables in the model are assumed to follow a strict sequence of states (see Table 5.4) that corresponds roughly to the super highway discovered by Li et al. [11].

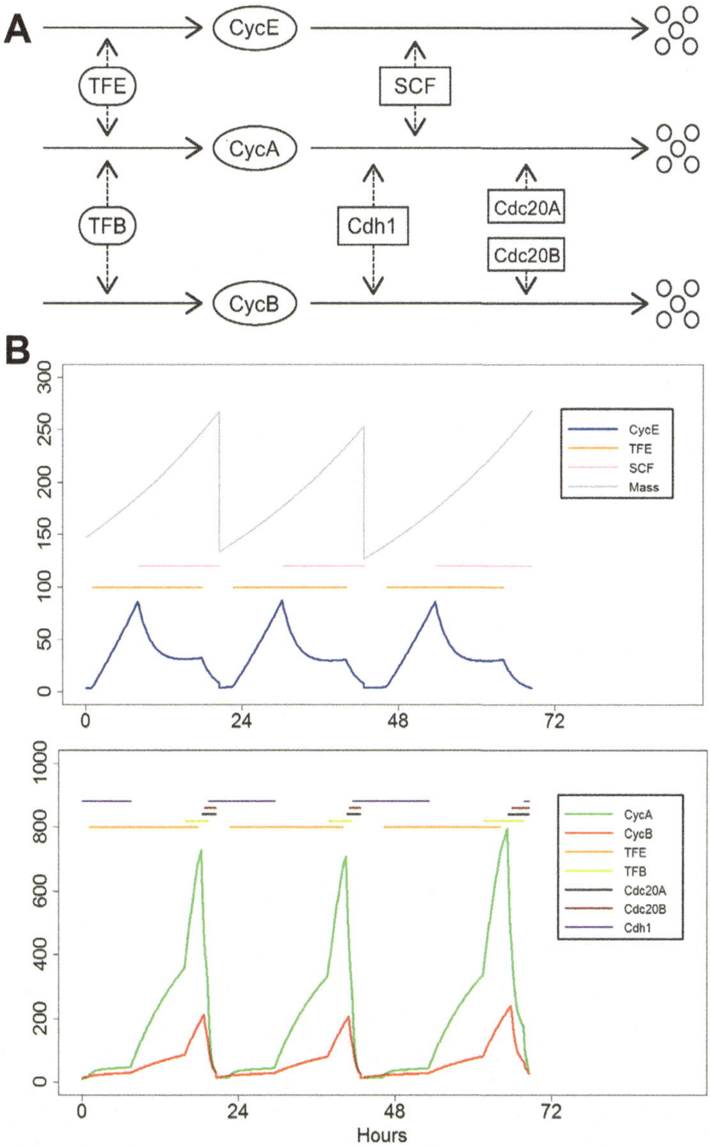

Fig. 5.13 The hybrid model network. (**a**) The synthesis and degradation of the cyclin proteins are regulated by TFs (TFE and TFB) and by ubiquitination machinery (SCF, Cdc20, and Cdh1). (**b**) Three successive cell cycles are simulated. Upper panel: gray curve, $30 \times M(t)$; blue curve, $[CycE] \times M(t)$; the gold line and the pink line indicate the time periods when $TFE = 1$ and $SCF = 1$, respectively. Lower panel: green curve, $[CycA] \times M(t)$; red curve, $[CycB] \times M(t)$; the colored bars indicate the time periods when the Boolean variables are active, according to the legend in the inset. Reprinted from Ref. [43]

Table 5.3 The rate constants and parameters for synthesis and degradation in Eq. (5.7) [43]

Rate constants	Parameter values	Parameter values	Parameter values
$k_{sa} = k'_{sa} + k''_{sa} B_{TFE} + k'''_{sa} B_{TFB}$	$k'_{sa} = 5$	$k''_{sa} = 6$	$k''_{sa} = 10$
$k_{da} = k'_{da} + k''_{da} B_{Cdc20A} + k'''_{da} B_{Cdh1}$	$k'_{da} = 0.2$	$k''_{da} = 1.2$	$k'''_{da} = 1.2$
$k_{sb} = k'_{sb} + k''_{sb} B_{TFB}$	$k'_{sb} = 2.5$	$k''_{sb} = 6$	
$k_{db} = k'_{db} + k''_{db} B_{Cdc20B} + k'''_{db} B_{Cdh1}$	$k'_{db} = 0.2$	$k''_{db} = 1.2$	$k'''_{db} = 0.3$
$k_{se} = k'_{se} + k''_{se} B_{TFE}$	$k'_{se} = 0.02$	$k''_{se} = 2$	
$k_{de} = k'_{de} + k''_{de} B_{SCF}$	$k'_{de} = 0.02$	$k''_{de} = 0.5$	
$M = \delta \times Matdivision$	$\gamma = 0.029hr^{-1}$	$\delta = 0.5 \times G$	
$G \sim N(\mu, \sigma^2)$	$\mu = 1$	$\sigma = 3.33$	

Table 5.4 Values of discrete variables for synthesis and degradation in Eq. (5.7) [43]

States	Phases	B_{TFE}	B_{SCF}	B_{TFB}	B_{Cdc20A}	B_{Cdc20B}	B_{Cdh1}	Condition for exit	$\lambda(h)$
1	G1a	0	0	0	0	0	1	none	2
2	Early G1b	1	0	0	0	0	1	$[CycE] * M = \theta_E$	0
3	Late G1b	1	0		0	0	0	$[CycA] > \theta_A$	0.01
4	S	1	1	0	0	0	0	$T_{\min} = 7h$	1
5	G2	1	1	1	0	0	0	$[CycB] > \theta'_B$	0.5
6	Prophae	0	1	1	0	0	0	none	0.75
7	Metaphase	0	1	1	1	0	0	none	1.5
8	Anaphase	0	1	1	1	1	0	none	0.5
9	Telophase	0	1	0	1	1	1	$[CycB] < \theta''_B$	0.025

5.5.3 Dynamical Analysis of the Hybrid Model

The first test for the hybrid model is to simulate flow cytometry measurements of the DNA content and cyclin levels in an asynchronous population of RKO (colon carcinoma) cells. In the dataset, a typical scatter plot has about 65,000 data points, each point displaying the measurements of two observables in a single cell chosen at random from the cell cycle (Fig. 5.14). When the data are plotted in this way, they form a cloudy tube of points through a projection of the state space (say, cyclin B versus cyclin A). Because there will be some cells from every phase of the cell cycle, the tube closes on itself. If the system was completely deterministic and the measurements were absolutely precise, the data points would be a simple closed curve (a "limit cycle") in the state space. The data actually present a fuzzy trajectory that snakes through state space before closing on itself. The indeterminacy of the points comes (presumably) from two sources: intrinsic noise in the molecular regulatory system (modeled by the random waiting times, T_i^r) and extrinsic measurement errors, which will be introduced momentarily. In Fig. 5.14, the authors compared the simulated flow-cytometry scatter plots with experimental results of Yan et al. [44]. The authors color-coded each cell in the simulated plot according to which Boolean state (Table 5.4) the cell is at the time of fixation. In Fig. 5.15, the authors plotted cyclin E fluctuations, as predicted by the hybrid

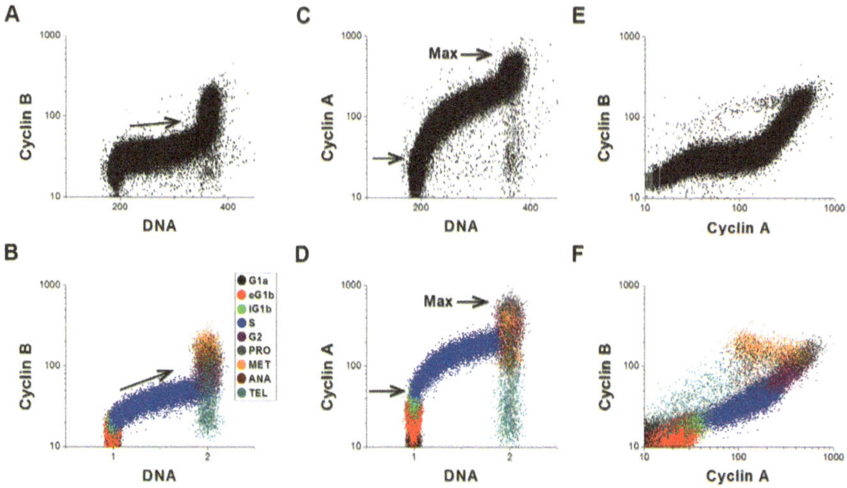

Fig. 5.14 Scatter plots. (**a**), (**c**), and (**e**). Flow cytometry data from Yan et al. [44]. $DNA = 190$ corresponds to G1 and $DNA = 380$ corresponds to G2/M. (**b**), (**d**), and (**f**). The total amounts of cyclin A and cyclin B per cell are plotted, i.e., $[CycA] * M(t)$ and $[CycB] * M(t)$. $DNA = 1$ in G0/G1 phase and $DNA = 2$ in G2/M phase. Some instrumental noise has been added to the calculated levels of cyclins and DNA. The arrows in A and B indicate the rate of cyclin B accumulation in S phase in the measurements and in the model. The arrows in C and D indicate the cyclin A level at the onset of DNA synthesis, compared to the maximum expression level of ~600 AU. Reprinted from Ref. [43]

model, along with a projection of the cell cycle trajectory in a subspace spanned by the three cyclin variables (A, B, and E). The model can capture the major features of cyclin fluctuations as measured by flow cytometry during the somatic division cycle of mammalian cells under appropriate parameters. However, there remain some inconsistencies between the mathematical simulations and the experimental observations that point out where future modifications to the model are needed. For example, in the model, DNA synthesis starts when cyclin A has accumulated to around 8% of its maximum level (see arrow in Fig. 5.14d; 50/600 $\approx$ 8%), whereas in the measurements, DNA synthesis starts when cyclin A is ~5% of its maximum level (see arrow in Fig. 5.14c).

The simulation (Fig. 5.14b) captures the observed accumulation of cyclin B in late G_1 (when Cdh1 turns off), but the simulated rise in cyclin B during S phase appears to be faster than the observed rise [45] (compare the arrows in Fig. 5.14a, b). The simulation does capture the rapid accumulation of cyclin B observed in G_2. Finally, while the authors did not calibrate the cyclin E expression parameters to any specific dataset, the pattern of expression in Fig. 5.15a is quite similar to the expected expression patterns for normal human somatic cells and some human tumor cell lines [46].

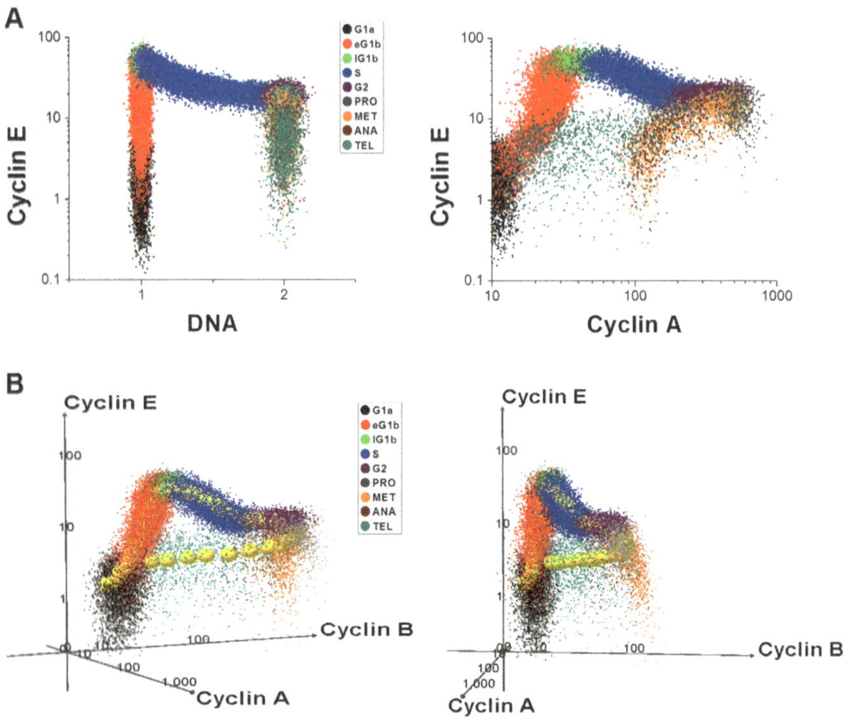

Fig. 5.15 Model predictions of cyclin E dynamics. (**a**) Scatter plots. (**b**) Stochastic limit cycle in the state space of cyclins A, B, and E. Two different perspectives of this three-dimensional figure are provided to help visualize how the cyclin levels go up and down. In addition, golden-colored balls have been added to help guide the eye along the cell cycle trajectory. Each ball represents the average of the cyclin levels of all the cells binned over a hundredth of the φ_i interval [0,1], where φ_i refers to the fraction of the cell cycle completed by cell i. Finally, it may help to recognize that Fig. 5.14e is a projection of the data on the CycA–CycB plane, and Fig. 5.15b is a projection on the CycA–CycE plane. Reprinted from Ref. [43]

5.5.4 *Summary*

Singhania et al. [43] have constructed a simple, effective model of the cyclin-dependent kinase control system in mammalian cells and used the model to simulate faithfully the accumulation and degradation of cyclin proteins during asynchronous proliferation of RKO (colon carcinoma) cells. The model is inspired by the work of Li et al. [11], who proposed a robust Boolean model of cell cycle regulation in the budding *yeast*. The goal was to retain the elegance of the Boolean representation of the switching network, while introducing continuous variables for cell size, cell age, and cyclin composition, in order to create a model that could be compared in quantitative detail to experimental measurements.

Singhania et al. [43] have shown that the established hybrid model can accurately simulate flow cytometric measurements of cyclin abundances in asynchronous populations of growing–dividing mammalian cells. The parameters in the model that allow for a quantitative description of the experimental measurements can be easily estimated from the data itself. The model is parameterized and validated for wild-type cells, it is currently extended to handle the behavior of cell populations perturbed by drugs and by genetic interference. In some cases, only the modest extensions of the model are required; in other cases, a more thorough overhaul of the way the discrete and continuous variables interact with each other is necessary.

However, there remain some inconsistencies between the mathematical simulations from the hybrid model and experimental observations that point out where future modifications to the model are needed. The simulation (Fig. 5.14b) captures the observed accumulation of cyclin B in late G_1 (when Cdh1 turns off), but the simulated rise in cyclin B during S phase appears to be faster than the observed rise (compare the arrows in Fig. 5.14a, b). The simulation does capture the rapid accumulation of cyclin B observed in G_2. Finally, while the authors did not calibrate the cyclin E expression parameters to any specific dataset, the pattern of expression in Fig. 5.15a is quite similar to expected expression patterns for normal human somatic cells and some human tumor cell lines.

The hybrid modeling approach will be generally useful for modeling macromolecular regulatory networks in cells, because it combines the qualitative appeal of Boolean models with the quantitative realism of reaction kinetic models.

5.6 General Hybrid Model for Large-Scale Bio-Molecular Networks

It is well known that theoretical analysis based on the mathematical predictive models of middle-sized GRNs is crucial to understand the life at system level for systems biology. Since there does not exist an efficient mathematical model for the general middle-sized or large-sized biological systems, it is still a challenge to theoretically investigate these networks. To overcome the limitations of the traditional continuous differential equations and discrete Boolean models, this section aims to introduce a general hybrid modeling approach for the general middle-sized GRNs. This general hybrid modeling approach is able to quantitatively investigate the middle-sized or large-sized biological networks by merging the advantages of both the continuous differential equation models and the Boolean network models.

5.6.1 Related Works and Motivations

It is well known that systems biology has received increasing attention in various disciplines over the last decade. GRNs, as an emerging research field, are composed of interactions among genes. Some of these networks have been quantitatively investigated from the mathematical point of view. Generally speaking, there are two kinds of representative mathematical descriptions: Boolean models and differential equation models. In 1969, Kauffman [10] firstly proposed the Boolean model to describe the metabolic stability and epigenesis in genetic circuits. After that, due to the relative simplicity and parameter free of the Boolean models, they had been extensively applied to investigate various biological systems. As we know now, the Boolean models are very useful for the modeling of middle-sized biological systems, where tens of nodes are involved in the network, such as the budding *yeast* cell cycle network [11, 31] and the fission *yeast* cell cycle network [32]. However, the Boolean models are not able to exactly characterize various biological processes. The differential equation models can elaborately reflect the evolution of molecular species in genetic circuits, especially for small-sized networks, such as the interlinked positive and negative feedback circuits [47, 48] and FFLs [49]. But for the middle-sized networks, for example, the *yeast* cell cycle network [18–21], differential equation models are very complex. Moreover, how to determine the associated extensive parameters is often beyond the complexity of the original differential equations. Therefore, the modeling of these middle-sized or large-sized biological circuits based on differential equations is also a challenging problem [12].

Beyond the continuous differential equation models and the discrete Boolean network models, there are some emerging modeling methods, such as the SDE models, partial differential equation models [50], since they have the similar problems as the ODE models; here, we omit the discussions. As it has been mentioned in the above sections, both the Boolean models and the ODE models have their defects in modeling the middle-sized or the large-sized genetic networks; therefore, it is crucial to develop general effective models for middle-sized or large-sized bio-molecular networks. The general hybrid modeling framework is a promising modeling approach.

Considering the advantages and disadvantages of the Boolean network models and the differential equation models, a natural idea is how to utilize the advantages of the above two modeling approaches and also simultaneously avoid the disadvantages of the above two modeling methods. The hybrid Boolean and differential equation models are a trade-off. Noted that the hybrid model was initiated by Glass and coworkers in 1973 and 1978 [51, 52], respectively. Until recently, Singhania et al. [43] have applied the above hybrid modeling approach to investigate the mammalian cell cycle network, where the cyclin-dependent kinase in the network was described by continuous linear differential equations, while the other nodes are modeled as Boolean dynamics, and the Boolean dynamic paths are determined by the attractive path in [11]. Compared with the differential equation models in [20], the hybrid model is quite simple. Furthermore, the hybrid model can much better describe the detailed network dynamics than the differential equation models.

As we know now, one of the ultimate goals of bottom-up systems biology is how to understand life at system level. Since the quantitative research of middle-sized or large-sized GRNs [12, 53] is an effective way to realize the above goal, the modeling of middle-sized or large-sized networks has significant scientific meaning. Since the pioneering work of Glass et al. [51, 52], hybrid (discrete–continuous) models have been employed by systems biologists in a variety of forms and contexts [43, 54–56]. Engineers have been modeling hybrid control systems for many years [56–58], and they have created some powerful simulation packages for such systems [59], such as SIMULINK [58], SHIFT [60, 61], and CHARON [62], to name a few.

Motivated by the above various reasons, we try to introduce a general hybrid modeling approach for the general GRNs [63], especially for the middle-sized or large-sized networks. Hereafter, the central components or network motifs in the middle-sized networks are treated as continuous variables. However, the other elements are treated as discrete Boolean variables. As an effective attempt, the general hybrid Boolean and continuous differential equation models are introduced to model the general middle-sized GRNs. Moreover, the dynamical behaviors of a toy hybrid system will also be discussed.

5.6.2 *The General Hybrid Model*

The main idea behind hybrid modeling for middle-sized or large-sized bio-molecular networks is that, nodes that involved in central modules or that are key nodes are taken as continuous variables, while the other not so important nodes are modeled as Boolean variables. A natural arising question is how to determine whether a node is important or not, and we note that there has been many methods to measure the relative importance of a node, such as node degree, betweenness centrality, and closeness centrality [64]. Except for these traditional measures, there are some emerging measures to evaluate the node importance in bio-molecular networks in recent years, such as the motif centrality measure [65–69], the integrative measure [70], and the SpectralRank[71]. For the topic of the identification of key nodes in bio-molecular networks, we will discuss it in the subsequent chapters.

The general hybrid model is described as the following Eq. (5.8):

$$\begin{cases} \frac{dx_j}{dt} = f(a_{ij}, \lambda, x), \quad j \in G_K, i \in G; \\ \\ x_i(t+1) = \begin{cases} 1, & \sum_j a_{ji}x_j(t) > \theta_i, \\ 0, & \sum_j a_{ji}x_j(t) < \theta_i, \\ x_i(t), & \sum_j a_{ji}x_j(t) = \theta_i. \end{cases} \quad i \in G_U, j \in G. \end{cases} \tag{5.8}$$

Here, x_k denotes the state of gene $k(k \in G)$; $f(.)$ reflects the rate of change for the gene states; G denotes the node set; G_K denotes the key node set that will be modeled as continuous differential equations; G_U denotes the unimportant node set that will be treated as discrete Boolean variables; $G = G_K \cup G_U$, $G_K \cap G_U = \emptyset$; $a_{ij}, i, j \in G$ is element of the adjacency matrix; a_{ij} reflects the interactions in the network; if there is an activate interaction from gene X_i to gene X_j, then $a_{ij} = 1$, and if the interaction is a repression, then $a_{ij} = -1$, otherwise $a_{ij} = 0$. λ represents the parameter vector for the system. θ_i denotes the activation threshold for protein i.

5.6.3 Hybrid Modeling and Analysis of a Toy Genetic Network

As an example, we consider a toy GRN as shown in Fig. 5.16. Figure 5.16a shows an abstract middle-sized GRN, and Fig. 5.16b shows part of regulatory interactions in Fig. 5.16a and also can be seen as a middle-sized network, which contains a repressilator module [72], namely $X_1 \dashv X_2 \dashv X_3 \dashv X_1$. Taking Fig. 5.16b as an example, we consider hybrid modeling of such middle-sized regulatory network. The repressilator module may resort to oscillation in the network, if we want to investigate oscillation problems in such systems, then the repressilator module becomes important, and we can model gene X_1, X_2, X_3 as continuous variables and other nodes treated as discrete Boolean variables. X_1, X_2, and X_3 can be modeled as

$$\frac{dx_j}{dt} = \frac{\Sigma_i \alpha_{ij} a_{ij}^2 (x_i/K_{ij})^{n_i}}{1 + \Sigma_i a_{ij}^2 (x_i/K_{ij})^{m_i}} - d_j x_j, \quad j = 1, 2, 3. \tag{5.9}$$

Here, x_j denotes the concentration of protein j for gene X_j and K_{ij} denotes the disassociation coefficient; α_{ij} is the maximum contribution of the expression of gene X_i to gene X_j; d_j represents the disassociation rate; n_i and m_i are the Hill coefficients, for an activator, one has $n_i = m_i$, while for a repressor, $n_i = 0, m_i > 0$. a_{ij} reflects the interactions in the network; if there is an activate interaction from gene X_i to gene X_j, then $a_{ij} = 1$, and if the interaction is a repression, then $a_{ij} = -1$; otherwise $a_{ij} = 0$. For example, in Fig. 5.16b, when $j = 1$, a_{ij} is nonzero only for $i \in \{3, 5, 6, 8\}$.

The next crucial question is how to determine the state update rules for other discrete variables. Just similar to Ref. [11], we set states update rules for $X_4, \cdots, X_{15}$ as follows:

$$x_i(t+1) = \begin{cases} 1, & \sum_j a_{ji} x_j(t) > \theta_i, \\ 0, & \sum_j a_{ji} x_j(t) < \theta_i, \\ x_i(t), & \sum_j a_{ji} x_j(t) = \theta_i. \end{cases} \quad i \in \{4, 5, \cdots, 15\} \tag{5.10}$$

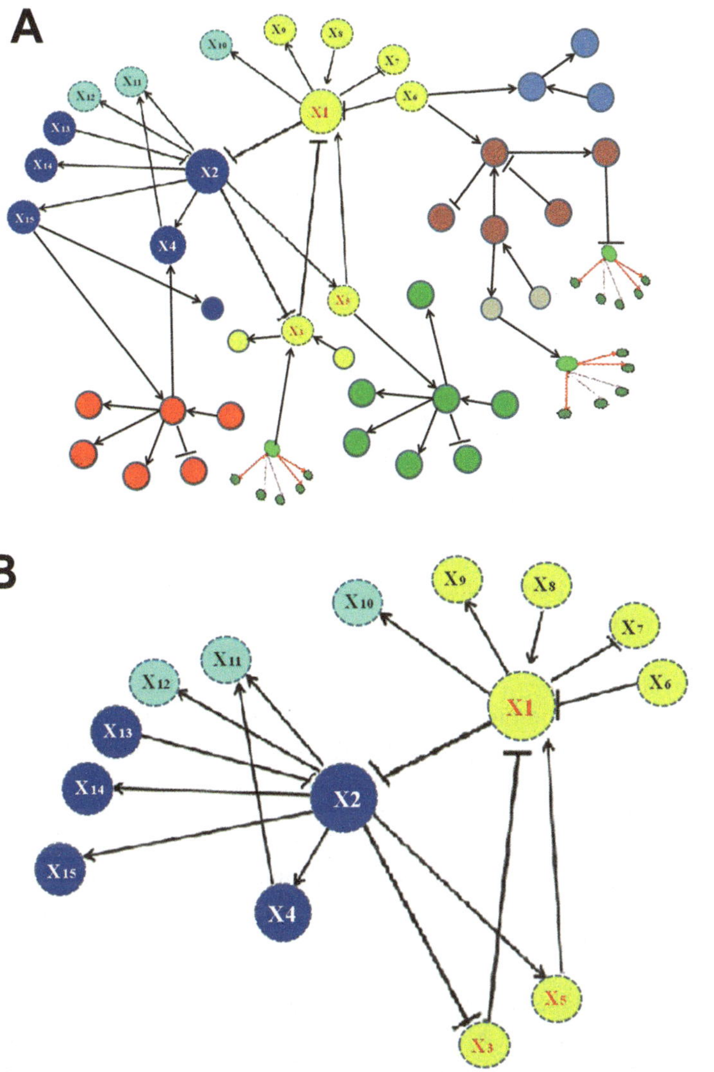

Fig. 5.16 A toy GRN and its subnetwork. (**a**) An abstract middle-sized GRN. (**b**) Part of regulatory modular in (**a**). Copyright (2013) IEEE. Reprinted, with permission, from Ref. [63]

Here, $x_i(t)$ denotes the state of protein i at time t, which can be 0 or 1 (represents inactivation or activation states, respectively). θ_i denotes the activation threshold for protein i.

After constructing the hybrid models for middle-sized networks, one needs to verify whether the constructed models can reveal real-world processes in the systems, while experimental data is needed to perform such verification. It is noted that, if one has enough data, one can tune the parameter values to facilitate the hybrid models with more biological sense.

5.6.3.1 Dynamical Analysis of the Hybrid Model

There are only 17 reaction rate coefficients, 7 Hill coefficients, and 12 threshold values θ_i; the Hill coefficients can be set as 1, 2 or 4, and the threshold value θ_i can be modulated by experience. Therefore, one can only determine 17 rate constants, compared with the network size $N = 15$, and the number of parameters have been greatly reduced.

Since there are no experimental data to estimate these 17 parameters, and the network in Fig. 5.16 is far from real biological systems, one sets $K_{ij} = 1, d_i = 0.5, n_i = 2, \alpha_{13} = 216, \alpha_{15} = 0.2, \alpha_{16} = 10, \alpha_{18} = 20, \alpha_{2,13} = 10, \alpha_{21} = 216$, and $\alpha_{32} = 216$. Following, we numerically simulate the system. For simplicity, we suppose that the update steps for the Boolean variables are the same as the discretized continuous variables. For the discrete Boolean variables, there are about $2^{12} = 4096$ possible initial values, and we randomly choose a set of initial values in each simulation. Initial values for the continuous variables are randomly taken as non-negative real values.

Chosen different activation threshold θ_i, for example, one chooses $\theta_i = 0, 0.5, 8$ for all $x_i, (i = 4, \cdots, 15)$, then time evolutions of the continuous variables x_1, x_2, x_3 are shown in Fig. 5.17, and states update for x_4 under $\theta_i = 8$ are shown in Fig. 5.17d. From Fig. 5.17, one sees that the oscillation behaviors in x_1, x_2, x_3 can be regulated by θ_i, with the increasing of θ_i, the system all can display damped oscillations, and the oscillation amplitudes become bigger and bigger. x_4 can realize switch between "on" and "off."

One can also choose different threshold value θ_i for different protein i, which may be more practical. For example, one takes $\theta_4 = 10, \theta_5 = 10, \theta_6 = 0.8, \theta_7 = 6, \theta_8 = 8, \theta_9 = 3, \theta_{10} = 4, \theta_{11} = 8, \theta_{12} = 5, \theta_{13} = 4, \theta_{14} = 6$, and $\theta_{15} = 2$, then the time evolutions of the continuous variables x_1, x_2, x_3 and the Boolean variable x_4 are shown in Fig. 5.18. The system can also display damped oscillation, and initially, the Boolean variable x_4 can switch between "off" and "on," but with the passage of time, x_4 can only rest on "off" state, which can be seen as damped switch.

The above example shows that the hybrid model can well reflect the dynamics in the network, and the hybrid model can overcome the defects of the continuous models and the Boolean models.

5.6.3.2 Statistical Analysis

In the above subsections, we mainly discuss the hybrid dynamical modeling and analysis of middle-sized or large-sized bio-molecular networks. In fact, real-world bio-molecular networks are always perturbed by noise; therefore, we can consider noise perturbed systems and describe the system with stochastic hybrid models.

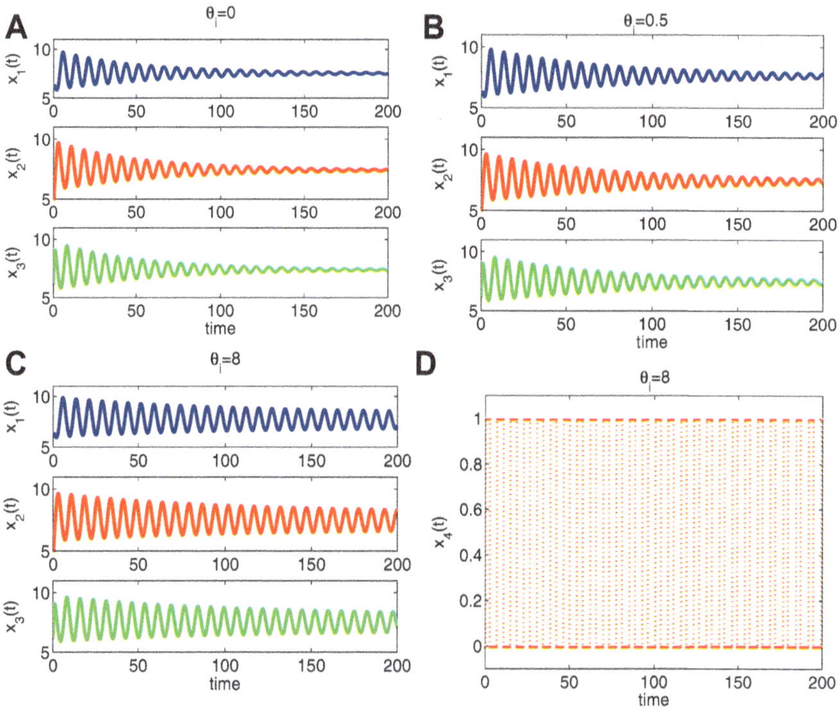

Fig. 5.17 Dynamics of the hybrid dynamical system described by Eqs. (5.9) and (5.10). (**a**)–(**c**) Time evolutions of the continuous variables $x_1 - x_3$ under different activation threshold θ_i. (**d**) The states update for protein x_4 under $\theta_i = 8$ for $i = 4, \cdots, 15$.

Assume that a bio-molecular network is perturbed by additive noise, then the stochastic hybrid model can be described as follows:

$$\begin{cases} \frac{dx_j}{dt} = f(a_{ij}, \lambda, x) + \xi_j(t), \quad j \in G_K, i \in G; \\ x_i(t+1) = \begin{cases} 1, & \sum_j a_{ji}x_j(t) > \theta_i, \\ 0, & \sum_j a_{ji}x_j(t) < \theta_i, \\ x_i(t), & \sum_j a_{ji}x_j(t) = \theta_i. \end{cases} \quad i \in G_U, j \in G. \end{cases} \tag{5.11}$$

The meanings of the symbols in Eq. (5.11) are the same as that in Eqs. (5.9) and (5.10), except that $\xi_j(t)$ represents the additive noise term. Traditionally, Gaussian white noise is often used to investigate such system, which satisfies

$$< \xi_j(t) >= 0, < \xi_j(t)\xi_j(s) >= D\delta(s-t). \tag{5.12}$$

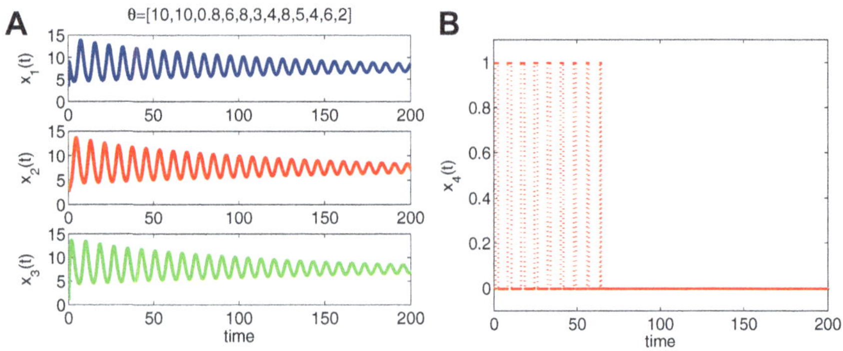

Fig. 5.18 Dynamics of protein x_1, x_2, x_3, x_4 under heterogeneous θ_i. (**a**) shows the cases for x_1, x_2 and x_3; (**b**) shows the case for x_4. Copyright (2013) IEEE. Reprinted, with permission, from Ref. [63]

Here, $\delta(s - t) = 1$ if $s = t$, and $\delta(s - t) = 0$ if $s \neq t$; D represents the noise strength.

The stochastic system (5.11) can be investigated via statistical methods, such as stochastic bifurcation analysis, stochastic dynamical analysis, and so on. We omit the detailed discussions here.

5.6.4 Summary

This section introduced a general hybrid modeling framework for the general GRNs [63], especially for the middle-sized or large-sized genetic networks. The central components or network motifs in the middle-sized networks are treated as continuous variables. However, the other elements are treated as discrete Boolean variables. The general hybrid Boolean and continuous differential equation models have been applied to model a toy GRN.

5.7 Discussions and Conclusions

This chapter has further discussed the modeling and analysis of middle-sized or large-sized bio-molecular networks. Especially, we discussed the hybrid modeling approach. The main idea behind hybrid modeling is that, nodes in central modules or key regulatory components are treated as continuous variables, while other nodes are treated as discrete Boolean variables. The hybrid models can greatly reduce parameters in comparison with the ODE models, and at the same time, they are more detailed than the Boolean models. A recent investigation on the cell cycle network, which modeled the mammalian cell cycle system as hybrid piecewise linear differential equations and Boolean models, revealed that the hybrid models can well reflect the evolution of biological processes in comparison with experimental data. We note that, though the investigations are about GRNs, the general modeling method can also be applied in PPI networks, metabolic networks, and so on. Moreover, the hybrid models can also be incorporated with external noise to facilitate the stochastic analysis of bio-molecular systems.

The bottom-up systems biology aims to understand life at the system level, mathematical models can provide predictive tools for quantitative investigations, and the hybrid modeling approach is promising for middle-sized or large-sized bio-molecular networks. The related investigations provide a possible way of hybrid modeling for general middle-sized or large-sized bio-molecular networks and may have its real-world applications in the near future.

References

1. Chen, B., Wu, W., Wang, Y., Li, W.: On the robust circuit design schemes of biochemical networks: steady-state approach. IEEE Trans. Biomed. Circ. Syst. **1**, 91–104 (2007)
2. Chen, B., Chen, P.: Robust engineered circuit design principles for stochastic biochemical networks with parameter uncertainties and disturbances. IEEE Trans. Biomed. Circ. Syst. **2**,114–132 (2008)
3. Wu, F.: Global and robust stability analysis of genetic regulatory networks with time-varying delays and parameter uncertainties. IEEE Trans. Biomed. Circ. Syst. **5**, 391–398 (2011)
4. Karlebach, G., Shamir, R.: Modelling and analysis of gene regulatory networks. Nat. Rev.: Mol. Cell Biol. **9**, 770–780 (2008)
5. McAdams, H.H., Arkin, A.: Stochastic mechanisms in gene expression. Proc. Natl. Acad. Sci. USA. **94**, 814–819 (1997)
6. Brandman, O., Ferrell, JE. Jr., Li, R., Meyer, T.: Interlinked fast and slow positive feedback loops drive reliable cell decisions. Science **310**, 496–498 (2005)
7. Mangan, S., Alon, U.: Structure and function of the feed-forward loop network motif, Proc. Natl. Acad. Sci. USA. **100**, 11980–11985 (2003)
8. Alon, U.: Network motifs: Theory and experimental approaches. Nat. Rev. Genetics **8**, 450–461 (2007)
9. Alon, U.: An introduction to systems biology: design principles of biological circuits. Chapman & Hall/CRC (2007)
10. Kauffman, S.A.: Metabolic stability and epigenesis in randomly connected nets. J. Theor. Biol. **22**, 437 (1969)
11. Li, F., Long,T., Lu,Y., Ouyang, Q., Tang, C.: The *yeast* cell-cycle network is robustly designed. Proc. Natl. Acad. Sci. USA. **101**, 4781–4786 (2004)
12. Bornholdt, S.: Less is more in modeling large genetic networks. Science **310**, 449–450 (2005)
13. Máayan, A., Jenkins, S.L., Neves, S., et al.: Formation of regulatory patterns during signal propagation in a mammalian cellular network. Science **309**,1078–1083 (2005)
14. Kazunari K., Ghosh, S., Matsuoka, Y., Moriya, H., Shimizu-Yoshida, Y., Kitano, H.: A comprehensive molecular interaction map of the budding *yeast* cell cycle. Mol. Syst. Biol. **6**, 415 (2010)
15. Tyson, J.J., Chen, K. C., Novak, B.: Network dynamics and cell physiology. Nat. Rev. Cell Biol. **2**, 908–916 (2001)

16. Wolkenhauer, O.: *Systems biology: dynamic pathway modelling*. in press (2012) http://ir.kib.ac.cn/handle/151853/17097.
17. Novak B., Tyson, J.J.: Modeling the control of DNA replication in fission *yeast*. Proc. Natl. Acad. Sci. USA. **94**, 9147–9152 (1997)
18. Tyson, J.J.: Size control of cell division. J. Theor. Biol. **126**, 381–391 (1987)
19. Chen, K.C., Csikasz-Nagy, A., Gyorffy, B., et al.: Kinetic analysis of a molecular model of the budding *yeast* cell cycle. Mol. Biol. Cell **11**, 369–391(2000)
20. Chen, K.C., Calzone, L., Csikasz-Nagy, A., et al.: Integrative analysis of cell cycle control in budding *yeast*. Mol. Biol. Cell **15**, 3841–3862 (2004)
21. Novak, B., Tyson, J.J.: A model for restriction point control of the mammalian cell cycle. J. Theor. Biol. **230**, 563–579 (2004)
22. Sveiczer, A., Tyson, J.J., Novak, B.: Modelling the fission *yeast* cell cycle. Brief. Funct. Genomics Proteom. **2**, 298–307 (2004)
23. Novak, B., Pataki, Z., Ciliberto, A., Tyson, J.J.: Mathematical model of the cell division cycle of fission *yeast*. Chaos **11**, 277–286 (2001)
24. Novak, B., Tyson, J.J.: Quantitative analysis of a molecular model of mitotic control in fission *yeast*. J. Theor. Biol. **173**, 283–305 (1995)
25. Cross, F.R., Schroeder, L., Kruse, M., Chen, K.C.: Quantitative characterization of a mitotic cyclin threshold regulating exit from mitosis. Mol. Biol. Cell **16**, 2129–2138 (2005)
26. Adames, N.R., Schuck, P.L., Chen, K.C., Murali, T.M., Tyson, J.J., Peccoud, J.: Experimental testing of a new integrated model of the budding *yeast* Start transition. Mol. Biol. Cell **26**, 3966–3984 (2015)
27. Novak, B., Chen, K.C., Tyson, J.J.: Systems biology of the *yeast* cell cycle engine. In Topics in Current Genetics, Vol. 13. Systems Biology: Definitions and Perspectives. Alberghina, L. and Westerhoff, H.V. eds. (Springer, Berlin /Heidelberg) 305–324 (2005)
28. Nurse, P., Fantes, P.A.: Cell cycle controls in fission *yeast*: a genetic analysis. In: The cell cycle, John, P.C.L. (ed.). Cambridge Univ. Press, Cambridge, 85–98 (1981)
29. Tyson, J.J., Novak, B., Odell, G.M., Chen, K., Thron, C.D.: Chemical kinetic theory: understanding cell-cycle regulation. Trends Biochem. Sci. **21**, 89–96 (1996)
30. Basi, G., Draetta, G.: p13suc1 of *Schizosaccharomyces pombe* regulates two distinct forms of the mitotic cdc2 kinase. Mol. Cell. Biol. **15**, 2028–2036 (1995)
31. Wang, G., Du, C., Chen, H., et al.: Process-based network decomposition reveals backbone motif structure. Proc. Natl. Acad. Sci. USA.**107**,10478–1048 (2010)
32. Davidich, M.I., Bornholdt, S.: Boolean network model predicts cell cycle sequence of fission *yeast*. PLoS One **3**, e1672 (2008)
33. Murray, A., Hunt, T.: The cell cycle: an introduction. Oxford Univ. Press, New York (1993)
34. Spellman, P.T., Sherlock, G., Zhang, M.Q., Iyer, V.R., Anders, K., Eisen, M.B., Brown, P.O., Botstein, D., Futcher, B.: Comprehensive identification of cell cycle-regulated genes of the *yeast Saccharomyces cerevisiae* by microarray hybridization. Mol. Biol. Cell **9**, 3273–3297 (1998)
35. Wolynes, P.G., Onuchic, J.N., Thirumalai, D.: Navigating the folding routes. Science **267**, 1619–1620 (1995)
36. Onuchic, J.N., Luthey-Schulten, Z., Wolynes, P.G.: Theory of protein folding: The energy landscape perspective. Annu. Rev. Phys. Chem. **48**, 545–600 (1997)
37. Dill, K.A., Chan, H.S.: From Levinthal to pathways to funnels. Nat. Struct. Biol. **4**, 10–19 (1997)
38. Kauffman, S.A.: The origins of order. Oxford Univ. Press, New York (1993)
39. Aldana, M.: Boolean dynamics of networks with scale-free topology. Physica D **185**, 45–66 (2003)
40. Sasai, M., Wolynes, P.G.: Stochastic gene expression as a many-body problem. Proc. Natl. Acad. Sci. USA. **100**, 2374–2379 (2003)
41. Kirschner, M., Gerhart, J.: Evolvability. Proc. Natl. Acad. Sci. USA. **95**, 8420–8427 (1998)
42. Bhalla, U.S., Iyengar, R.,: Emergent properties of networks of biological signaling pathways. Science **283**, 381–387 (1999)

43. Singhania, R., Sramkoski, R.M., Jacobberger, J.W., Tyson, J.J.: A hybrid model of mammalian cell cycle regulation. PLoS Comput. Biol. **7**, e1001077 (2011)
44. Yan, T., Desai, A.B., Jacobberger, J.W., Sramkoski, R.M., Loh, T., et al.: CHK1 and CHK2 are differentially involved in mismatch repair-mediated 6-thioguanine-induced cell cycle checkpoint responses. Mol. Cancer Ther. **3**, 1147–1157 (2004)
45. Jacobberger, J.W., Sramkoski, R.M., Wormsley, S.B., Bolton, W.E.: Estimation of kinetic cell-cycle-related gene expression in G1 and G2 phases from immunofluorescence flow cytometry data. Cytometry **35**, 284–289 (1999)
46. Darzynkiewicz, Z., Gong, J., Juan, G., Ardelt, B., Traganos, F.: Cytometry of cyclin proteins. Cytometry **25**, 1–13 (1996)
47. Tian, X., Zhang, X., Liu, F., Wang, W.: Interlinking positive and negative feedback loops creates a tunable motif in gene regulatory networks. Phys. Rev. E **80**, 011926 (2009)
48. Hasty, J., Dolnik, M., Rottschäfer, V., Collins, J.: Synthetic gene network for entraining and amplifying cellular oscillations. Phys. Rev. Lett. **88**, 148101 (2002)
49. Wang, P., Lü, J., Ogorzalek, M.J.: Global relative parameter sensitivities of the feed-forward loops in genetic networks. Neurocomput. **78**, 55–165 (2012)
50. Freund, J.A., Pöschel, T.: Stochastic process in physics, chemistry, and biology. Berlin, Heidelberg: Springer-Verlag (2000)
51. Glass, L., Kauffman, S.A.: The logical analysis of continuous non-linear biochemical control networks. J. Theor. Biol. **39**, 103–129 (1973)
52. Glass, L., Pasternack, J.: Stable oscillations in mathematical models of biological control systems. J. Math. Biol. **6**, 207–223 (1978)
53. Lü, J., Chen, G.: A time-varying complex dynamical network model and its controlled synchronization criteria. IEEE Trans. Automat. Contr., **50**, 841–846 (2005)
54. Bosl, W.J.: Systems biology by the rules: hybrid intelligent systems for pathway modeling and discovery. BMC Syst. Biol. **1**,13 (2007)
55. Li, C., Nagasaki, M., Ueno, K., Miyano, S.: Simulation-based model checking approach to cell fate specification during *Caenorhabditis elegans* vulval development by hybrid functional Petri net with extension. BMC Syst. Biol. **3**, 42 (2009)
56. Alur, R., Dang, T., Esposito, J.M., Fierro, R.B., Hur Y., et al.: Hierarchical hybrid modeling of embedded systems. In: Henzinger, T.A., Kirsch, C.M., eds. Embedded software: proceedings of the first international workshop. Berlin: Springer. 14–31 (2001)
57. Fishwick, P.A.: Handbook of dynamic system modeling. Boca Raton: Chapman & Hall/CRC (2001)
58. Klee, H., Allen, R.: Simulation of dynamic systems with MATLAB and Simulink. Boca Raton, FL: CRC Press, (2001)
59. Mosterman, P.: An overview of hybrid simulation phenomena and their support by simulation packages. In: Vaandrager F., van Schuppen J., eds. Hybrid systems: computation and control. Berlin: Springer (2001)
60. Deshpande, A., Gollu, A., Varaiya, P.: SHIFT: a formalism and programming language for dynamic networks of hybrid automata. In: Antsaklis, P., Kohn, W., Nerode, A., Sastry, S., eds. Hybrid systems IV. Berlin: Springer. 113–133 (1997)
61. Deshpande, A., Gollu, A., Semenzato, L.: The SHIFT programming language and run-time system for dynamic networks of hybrid systems. IEEE Trans. Automat. Contr. **43**, 584–587 (1997)
62. Alur, R., Grosu, R., Hur, Y., Kumar, V., Lee, I.: Modular specification of hybrid systems in CHARON. In: Lynch N, Krogh BH, eds. Hybrid systems: computation and control. Berlin: Springer. 6–19 (1997)
63. Wang, P., Lu, R., Chen, Y., Wu, X.: Hybrid modeling of the general middle-sized genetic regulatory networks. IEEE Int. Symp. Circ. Syst., Beijing, China, **May 19–22**: 2103–2106 (1997)
64. Kuhnert, M., Geier, C., Elger, C.E., Lehnertz, K.: Identifying important nodes in weighted functional brain networks: A comparison of different centrality approaches. Chaos **22**, 023142 (2012)

65. Koschützki, D., Schwöbbermeyer, H., Schreiber, F.: Ranking of network elements based on functional substructures. J. Theor. Biol. **248**, 471–479 (1997)
66. Koschützki, D., Schreiber, F.: Centrality analysis methods for biological networks and their application to gene regulatory networks. Gene Regulat. Syst. Biol. **2**,193–201 (1997)
67. Sporns O., Kötter, R.: Motifs in brain networks. PLoS Biol. **2**, e369 (1997)
68. Wang, P., Lü, J., Yu, X.: Identification of important nodes in directed biological networks: a network motif approach. PLoS One **9**, e106132 (2014)
69. Sporns, O., Honey, C.J., Kötter, R.: Identification and classification of hubs in brain networks. PLoS One **2**, e1049 (2014)
70. Wang, P., Yu, X., Lü, J.: Identification and evolution of structurally dominant nodes in protein-protein interaction networks. IEEE Trans. Biomed. Circ. Syst. **8**, 87–97 (2014)
71. Xu, S., Wang, P., Zhang, C., Lü, J.: Spectral learning algorithm reveals propagation capability of complex network. IEEE Trans. Cyber. **49**(12): 4253–4261 (2019).
72. Elowitz, M., Leibler, S.: A synthetic oscillatory network of transcriptional regulators. Nature **403**, 335–338 (2000)

Part II
Statistical Analysis of Biological Networks

This part discusses the statistical analysis of bio-molecular networks, which includes three chapters. Chapter 6 deals with the evolutionary mechanisms of network motifs in bio-molecular networks, where the duplication-divergence model with various duplication and divergence strategies will be considered to explore the evolution of network motifs. In Chap. 7, we present a focus topic in complex networks science and bio-molecular networks: the identification of important bio-molecules in biological systems. Three methods proposed by us will be introduced, including the motif-based methods, the integrative measure, and the SpectralRank method. Chapter 8 explores the statistical features of functional genes in the human protein interaction network. The ultimate goals of this part are intended to propose some statistical methods to explore bioinformatics from various bio-molecular networks.

Chapter 6
Evolutionary Mechanisms of Network Motifs in PPI Networks

Abstract Duplication and divergence are two basic evolutionary mechanisms of bio-molecular networks. Real-world bio-molecular networks and their statistical characteristics can be well mimicked by artificial algorithms based on the two mechanisms. Bio-molecular networks consist of network motifs, which act as building blocks of large-scale networks. A fundamental question is how network motifs are evolved from long time evolution and natural selection. By considering the effect of various duplication and divergence strategies, it is founded that the underlying duplication scheme of the real-world undirected bio-molecular networks would rather follow the anti-preference strategy than the random one. The anti-preference duplication mechanism and the dimerization can lead to the formation of various motifs and robustly conserve proper quantities of motifs in the artificial networks as that in the real-world ones. Furthermore, the anti-preference mechanism and edge deletion divergence can robustly preserve the sparsity of the networks. The investigations reveal the possible evolutionary mechanisms of network motifs and have potential implications in the design, synthesis, and reengineering of biological networks for biomedical purpose.

6.1 Backgrounds

Structures and functions of complex networks arising from various disciplines have been extensively investigated in the last several decades [1–3].

In 2002, Milo et al. [4–11] found that complex networks consist of network motifs. Network motifs are small subgraphs that more frequently appear in a network than in its random counterparts, where the random networks are permuted from the investigated networks, which keep the same node degrees as the concerned network. Milo et al. mainly considered various directed networks and found that the 3-node FFLs, the 4-node bi-fan, and bi-parallel subgraphs are typical motifs in various systems. Since then, structures and functions of network motifs have been extensively investigated both theoretically and experimentally [11–14]. It has been found that the FFLs, the bi-fan, and the bi-parallel subgraphs are with crucial

J. Lü, P. Wang, *Modeling and Analysis of Bio-molecular Networks*,
https://doi.org/10.1007/978-981-15-9144-0_6

functions in bio-systems [11–15]. For example, in 2003, Mangan et al. [12] found the relation between topological structures and functional characteristics of the FFLs. In 2012, through dynamical analysis, Wang et al. [13, 14] answered the question that why FFLs with certain structures can be selected by evolution.

PPI networks are typical undirected networks. Many works have been reported that PPI networks consist of modulars acting as basic functional building blocks [6–9]. In 2003, Wuchty et al. [8] investigated the evolutionary conservation of motif constituents within the *yeast* PPI network and found that the conservation of proteins within distinct topological motifs correlates with the interconnectedness and function of the motifs. Compared with the other subgraphs, the 3-node and 4-node fully connected subgraphs have the highest natural conservation rates [8]. It is noted that network motifs and subgraphs in their work are assumed to be interchangeable [8]. In 2004, Yeger–Lotem et al. [6] investigated network motifs in integrated *yeast* cellular networks of transcription-regulation and PPIs, where the TRN is directed and the PPI network is undirected. They found that the 3-node protein clique is the most abundant network motif, and 92% of the occurrences of this motif correspond to known protein complexes, such as RPN, PSA, PSB, and PRS [6, 8]. Furthermore, they found 63 four-node motifs, which include the undirected square, fully connected square. Some researchers have tried to explain why certain motifs are selected by organisms and conserved cross species [13–17]. However, few results have been reported on how network motifs came into being and evolved [18].

Statistical analysis of massive artificial networks generated by computer algorithms facilitates the explanation of the evolutionary mechanisms of network motifs. Duplication and divergence are two fundamental mechanisms for biological evolution. Based on different duplication and divergence strategies and real-world statistical characteristics of biological networks, such as sparsity, power law, small-worldness, modularity, and disassortativity, many DD algorithms have been proposed to mimic real-world bio-molecular networks [19–24]. Real-world bio-molecular networks consist of network motifs. Some naturally arising questions are (1) what about artificial networks? (2) what are the effects of different duplication and divergence strategies on the evolution of motifs? and (3) what is the underlying evolutionary mechanisms of motifs in bio-molecular networks?

Motivated by the above problems, we investigated the evolutionary mechanisms of network motifs in undirected bio-molecular networks [1]. The rest of the chapter is organized as follows. Section 6.2 presents the DD model. Section 6.3 discusses the evolving characteristics of motifs in real-world PPI networks by random sampling. The effects of duplication and divergence on the evolutionary mechanisms of motifs will be clarified in Sect. 6.4. Section 6.5 will present some theoretical results. Discussions and conclusions will be presented in Sect. 6.6.

6.2 Duplication-Divergence Model

In 1999, Barabási and Albert [25] proposed the well-known BA algorithm. The basic idea of the BA algorithm is the "rich-gets-richer" rule. As a result, the degree–degree correlation coefficients of the generated networks are positive. However, for bio-molecular networks, the degree–degree correlation coefficients are negative, that is, they are disassortative. Therefore, the possible growth strategy for bio-molecular networks will be different from the BA strategy. Based on the disassortativity, sparsity, SF, small-worldness, and modularity, many DD models have been proposed [19–24]. For example, in 2002 and 2003, Solé et al. [19] and Vázquez et al. [20] proposed two models to generate artificial PPI networks. In 2007, based on a random duplication model and an anti-preference duplication model, Zhao et al. [24] investigated the effects of duplication strategies on the disassortativity. In 2010, Xu et al. investigated several models with different ways of divergence [22], and they clarified that how divergence mechanisms influence the disassortativity of bio-molecular networks. Also in 2010, Wan et al. [23] proposed a simple DD model, which considered the anti-preference duplication, edge deletion, dimerization, and edge addition processes. They found that the DD model under proper parameters can well mimic the real-world PPI networks.

Algorithm 13 Algorithms for PPI network by considering different duplication and divergence strategies [1, 23]

1: Generate an initial connected network with n_0 nodes.
2: **repeat**
3: Duplication. Two duplication approaches can be separately considered (corresponding to two different models):
(i) Anti-preference strategy: At each time step, node i with degree k_i is chosen to duplicate with probability:

$$p_i = \frac{1/k_i}{\Sigma_j(1/k_j)}. \tag{6.1}$$

(ii) Random strategy: randomly chose a node to replicate.
4: Divergence. Four approaches of divergence are simultaneously considered.
(a) Edge deletion: For each node l linked to the newly selected target node i and its replica i', randomly choose one of the two links (i, l) or (i', l) and remove it with probability α.
(b) Dimerization: the target node i and its replica i' can be dimerized with probability β_0, that is, an edge between them will be added with probability β_0. According to [23], $\beta_0 = \beta k_i'$, where k_i' denotes the degree of the replica i', β is a constant, β_0 equals one if it is bigger than one.
(c) Edge addition: randomly choose a non-target node j, add a link between node j and i' with probability γ.
(d) After all the above processes, remove isolated nodes.
5: **until** Network size reaches the desired one.

Duplication can create new proteins, and divergence in the newly created proteins or interactions can lead to the emergence of novelty [26]. Artificially, duplication and divergence can be reflected in various ways. Different ways of duplication and divergence bring about different evolution models. For example, a pre-duplicated node can be selected randomly or by the anti-preference strategy [24]. Divergence can be reflected by node and edge deletion, addition, or edge rewiring [22]. By referring to the existing work [23], procedures of the DD algorithm considered in the following sections are the same as that in Algorithm 8 of Chap. 2, which are described as follows.

It is noted that the duplication and divergence processes correspond to real-world biological processes. For example, during the evolution of bio-molecular networks, a gene encoding an existing protein undergoes nucleotide substitutions, which leads to the creation of new links or deletion of existing links, and this process can be mimicked by the edge deletion and addition processes in the DD model [23]. The dimerization process mimics the probability that a duplicated node is self-interacting protein, or the links between the duplicated node and its replica are conserved during divergence [23, 27].

In the following, we suppose $n_0 = 2$, that is, all networks will be evolved from two ancestors. The node deletion divergence (d) is not controllable by parameters; therefore, we will not consider the effect of such process. Parameters α, β, γ can be tuned. Available data and works [19–24] suggested that α is far higher than γ, β is larger than γ, and γ is empirically very small. α, β, γ can be selected according to the following characteristics of real-world PPI networks. Firstly, researchers have reported that PPI networks are sparse. One result reported by Newman in 2002 was 2.12 [2], while Schwikowski et al. [28] and Yu et al. [29] showed that the average degrees of the *yeast* PPI networks were around 3. Secondly, PPI networks are SF, with $PLE = 2.5$ [19–24]. Thirdly, PPI networks are SW, with shorter APL and larger clustering coefficient than its random counterparts [2]. Moreover, biological networks are disassortative. Highly connected nodes tend to be neighbors of nodes with low degrees [2, 24].

Keeping the average degree and PLE in mind, under random duplication, from the theoretical mean-field analysis by Pastor-Satorras et al. [21], the best parameter α is $\alpha = 0.562$ and γ satisfies $\gamma = (\alpha - 0.5)\langle k \rangle / n$, where $\langle k \rangle$ is the average degree and n is the network size. But Pastor-Satorras et al. did not consider other duplication strategies and the dimerization processes. It has been reported that the anti-preference and random duplication mechanisms can both generate disassortativity, and the anti-preference scheme can strengthen the disassortativity. Therefore, it is also a possible network growth strategy. Dimerization is also a strategy of edge addition. However, since it can happen with a greater probability [23], it is better to separate the dimerization processes from the general edge addition processes. In the following, we set $\alpha = 0.562$ and $\beta = 0.12$, $\gamma = 0.000165$ as benchmark parameters for all simulation runs, unless otherwise noted.

6.3 Statistical Features of Network Motifs

To clarify the evolutionary mechanisms of network motifs, we introduce some measures about the significance of a subgraph in complex networks. For a m-node subgraph, the first index is the subgraph density or frequency, which is defined as the ratio of the number of subgraphs in the network to the total number of m-node subgraphs. The second index was proposed by Milo et al. [4], which is called Z_{score}, defined as

$$Z_{score} = \frac{N_{real} - N_{rand}}{SD}, \tag{6.2}$$

where N_{real} represents the occurrence of a subgraph in the investigated networks; N_{rand} denotes the average occurrence of the subgraph in randomized networks; and SD denotes the standard deviation of N_{rand}. Another index U is defined as the number of times that a subgraph appears in the network with distinct sets of nodes [4]; the higher U, the more frequent the appearance of such subgraph that is decentralized in the network and, thus, the more important the subgraph is. In the following, network motifs will be detected by mDraw (http://www.weizmann.ac.il/mcb/UriAlon). Subgraphs with $Z_{score} > 2$, $N_{real} - N_{rand} \geq 0.1N_{rand}$, and $U \geq 4$ are treated as motifs. The second restriction is to avoid detecting some common subgraphs as motifs, when these subgraphs have only a slight difference between N_{rand} and N_{real}, but also have a narrow distribution (small SD) in the randomized networks [4].

The undirected PPI networks also consist of network motifs. For example, for the *yeast* PPI network with 2238 interactions and 1825 proteins, constructed by Schwikowski et al. [28], we found that it contains 3-node protein clique motif with ID 238, 4-node motif with IDs 4958, 13260, 13278, and 31710 [9]. The subgraph density for the five motifs are $0.848\%, 0.127\%, 0.056\%, 0.065\%$, and 0.038%, respectively. The U values for the five motifs are 69, 63, 28, 32, and 19, respectively [9]. For the network with 2018 proteins and 2930 interactions reported by Yu et al. [29], the giant component contains 1647 proteins and 2518 interactions. It also consists of motifs 238, 13260, and 31710. The frequency and U of the motif 238 are 0.8% and 59. Compared with 1000 randomized networks, the Z_{score} is 5.13. The densities of subgraph 4958, 13260, 13278, and 31710 are $1.49\%, 0.537\%, 0.058\%$, and 0.0029%, respectively. The U values for the four subgraphs are 53, 32, 24, and 6, respectively. Figure 6.1a shows the network reported by Yu et al. [29]. Figure 6.1b shows some typical 3- and 4-node motifs in PPI networks.

Since subgraph 238 is a typical motif in the PPI networks, in the following, we mainly consider such subgraph. The 4-node subgraphs will be briefly discussed in the last section. To clarify the evolving characteristics of motifs in real-world networks, we take the *yeast* PPI network with 2018 proteins constructed by Yu et al. [29] as an example. We randomly sample subnetworks from the network and investigate the evolving characteristics of network motifs from the sampled networks. Researchers have reported that subnets of SF networks may be not SF

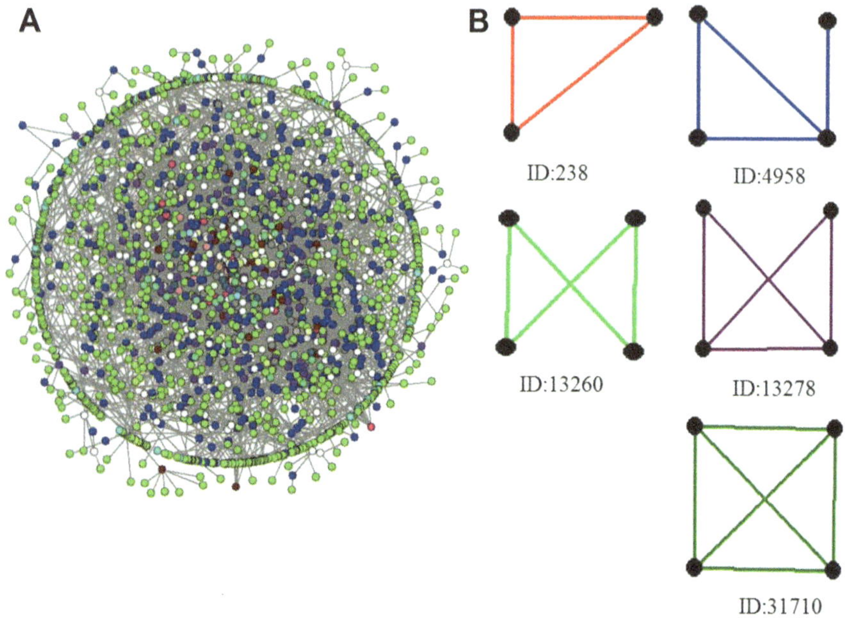

Fig. 6.1 A *yeast* PPI network and network motifs. (**a**) A *yeast* PPI network with 2018 proteins and 2930 interactions [29]. Nodes with different colors are with different degrees. (**b**) Some 3- and 4-node motifs in PPI networks. The ID number is transformed from its adjacent matrix. For details, one can refer to the mDraw user guide. (Color on line). ©[2015] IEEE. Reprinted, with permission, from Ref. [1]

[30], where the authors did not consider whether the sampled subnetworks are connected or not. Here, for the sampled connected subnetworks, we show their degree distributions in log–log scale in Fig. 6.2, where we can conclude that the subnetworks are SF, with the PLE around 2.5. Therefore, the sampled subnets may be used to explore the evolving characteristics of motifs in real-world networks. The green diamonds (Color on line) in Fig. 6.3 show the evolutions of the frequencies, Z_{score}, U of subgraph 238, and average degrees with the increasing of subnetwork sizes.

From Fig. 6.3, for the real-world networks, with the increasing of the sizes of sampled networks, averagely speaking, the frequency of subgraph 238 slightly decreases, but mostly around 0.6%. The Z_{score} for many networks are below 2. U increases linearly with the size of sampled networks, $U \geq 4$ for networks with sizes $n > 400$. The curves for Z_{score} and U indicate that in real-world PPI networks, whether subgraph 238 can be identified as motifs strongly depends on the network size. From Fig. 6.3d, the average degree slightly increases with subnetwork sizes and ranges from 2.1 to 3.1, which are consistent with the reported results for real-world networks. Keeping the characteristics of real-world networks in mind, in the following, we investigate how duplication and divergence affect the characteristics of network motifs, and clarify their evolutionary mechanisms.

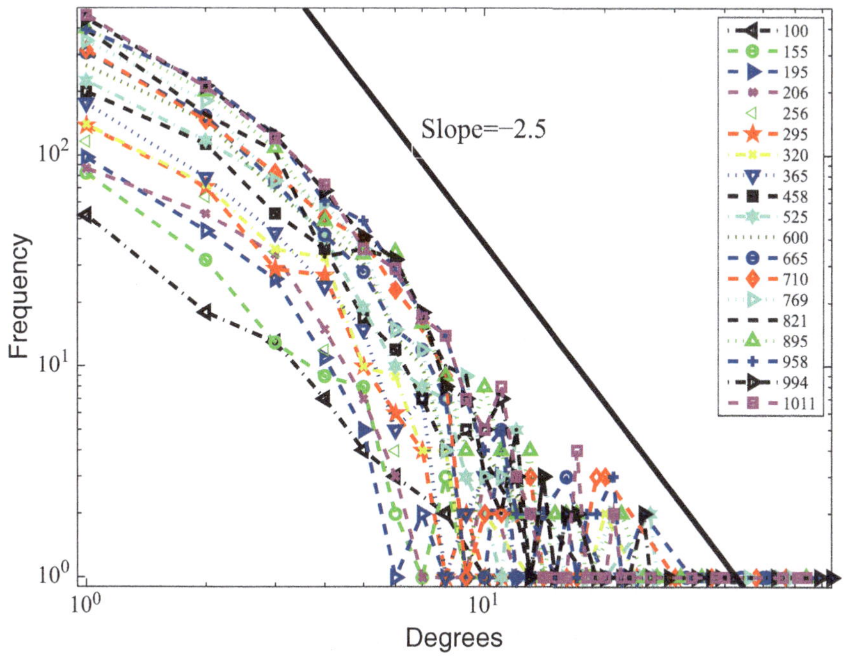

Fig. 6.2 Degree distributions in log–log scale for the sampled connected subnetworks. Numbers in the legend denote network sizes. The solid dark line with slope −2.5 acts as a reference line.

6.4 Evolutionary Mechanisms of Network Motifs

6.4.1 Effect of Duplication Strategies

Firstly, we consider the effect of various duplication strategies. It has been reported that highly connected proteins are more difficult to be duplicated than sparsely connected ones [22, 24, 25]. Therefore, the early versions of the DD models have assumed that proteins in the network can be selected randomly to duplicate [19, 20]. It has been reported that the anti-preference strategy can also well reflect the characteristics of real-world networks [24]. However, the effect of network growth strategy on the evolution of network motifs is still an open problem. To clarify this problem, we consider the random and anti-preference strategies. We set $\alpha = 0.562$, $\beta = 0.12$, and $\gamma = 0.000165$ for each duplication models and generate 10 sets of networks with sizes from 100 to 1000 under each strategy. Each set of networks contains 10 different samples. For each network, we generate 100 random networks to detect network motifs.

For subgraph 238, Fig. 6.3a draws the evolution of subgraph density, and Fig. 6.3b–d shows the evolutions of Z_{score}, U, as well as average degrees with

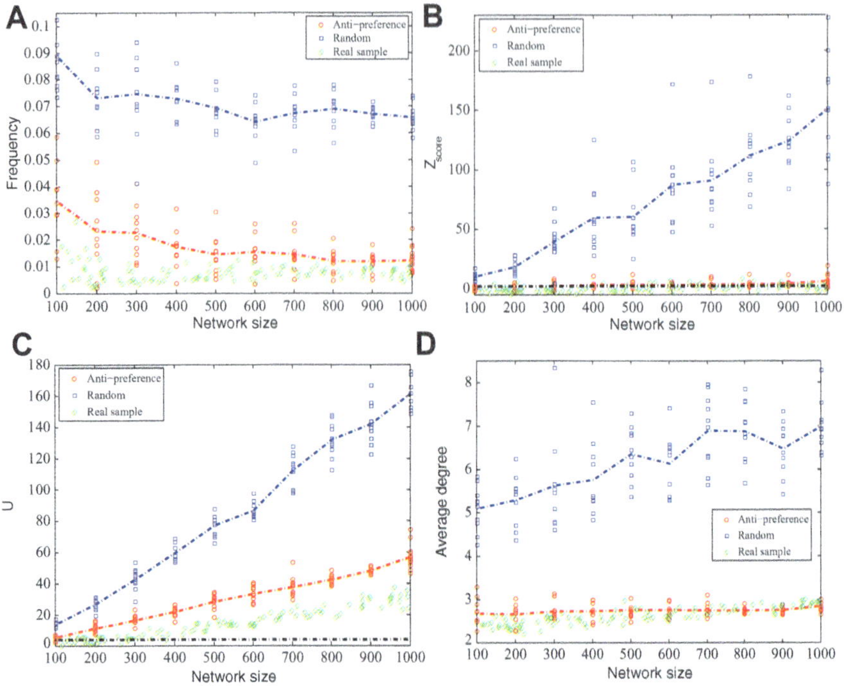

Fig. 6.3 Effect of different duplication strategies on motif 238. The evolutions of the frequency (**a**), Z_{score} (**b**), U (**c**), and average degrees (**d**) with the increasing of the network sizes. ©[2015] IEEE. Reprinted, with permission, from Ref. [1]

network sizes. From Fig. 6.3a, we can conclude that under the two duplication strategies, the density of subgraph 238 slightly decreases with network sizes. The density from the random strategy is around 7%, while it is around 2% for the anti-preference strategy. As we have reported in the above section, for the networks constructed by Schwikowski et al. [28] and Yu et al. [29], the density is below 1%. Therefore, the anti-preference strategy is more approximate to real-world networks. From Fig. 6.3b, c, the Z_{score} and U all increase with network sizes. The subgraph 238 under the random duplication strategy is obviously more significant than under the anti-preference strategy. More distinct sets of motifs can be produced by the random duplication model. Though the random duplication model can produce more and more subgraph 238 with the increasing of network sizes, its average degrees also increase quickly with network sizes, which range from around 5 to 7. While for the networks from the anti-preference strategy, the average degrees range from 2 to 3, which are in good agreement with the real-world ones. From Fig. 6.3b–d, we can conclude that the anti-preference strategy can better reflect the characteristics of real-world networks than the random ones. Therefore, the anti-preference duplication strategy is an optimal choice to simulate real-world biomolecular networks.

6.4.2 Effect of Divergence Strategies

Hereinafter, we analyze the effect of various divergence strategies, which will be investigated via tuning parameters α, β, γ. By setting the benchmark parameters as $\alpha = 0.562, \beta = 0.12, \gamma = 0.000165, n = 1000$, we consider three biologically relevant cases: (I) the effect of edge deletion: vary $\alpha \in [0.2, 0.8]$; (II) the effect of dimerization: vary $\beta \in [0, 0.2]$; and (III) the effect of edge addition: vary $\gamma \in [0, 0.00025]$.

Figure 6.4 shows the evolution of subgraph density and the evolutions of Z_{score}, U, and average degrees with α, β, and γ. From Fig. 6.4a–d, with the increasing of

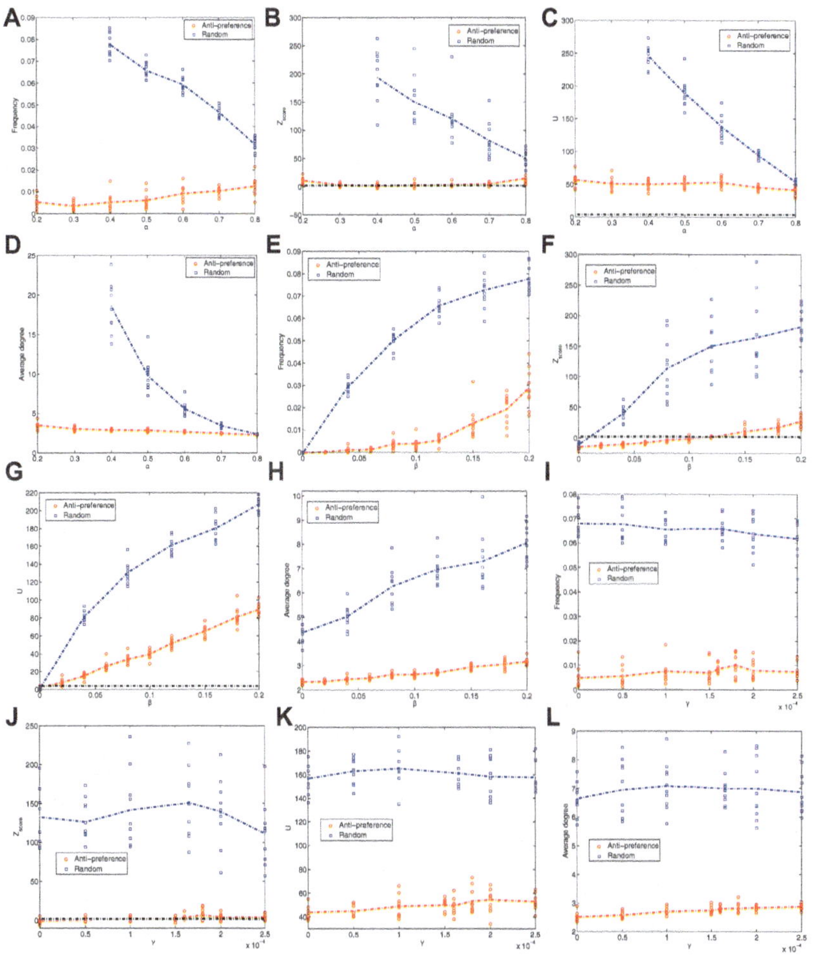

Fig. 6.4 Evolutions of the density, Z_{score}, and U of the subgraph 238 and the average degrees of networks with α **(a)**–**(d)**, β **(e)**–**(h)**, and γ **(i)**–**(l)**. ©[2015] IEEE. Reprinted, with permission, from Ref. [1]

α, the four indexes all decrease quickly under the random strategy, whereas they are very robust against α under the anti-preference strategy. The average degree under the anti-preference strategy slightly decreases with the increasing of α, while it decreases exponentially under the random strategy. The decreasing of the average degrees of the generated networks can be explained as follows. With the increasing of α, edges are prone to be deleted with higher probabilities in the network. Thus, for the fixed network size, the average degrees of the networks will naturally decrease. At each duplication step, under the anti-preference strategy, nodes with low degrees tend to be selected as targets, while under the random duplication, all nodes can be equally selected as targets. From the probabilistic perspective, under the random strategy, the available edges for the subsequent edge deletion tend to be more than those under the anti-preference strategy. Therefore, relatively more edges will be deleted with the increasing of α under the random strategy, and the average degrees will decrease more quickly than those under the anti-preference scheme.

From Fig. 6.4e–h, with the increasing of β, the four curves under each duplication strategy all increase. Under the anti-preference strategy, for $\beta < 0.12$, Z_{score} of subgraph 238 for most of the networks is below 2; therefore, it is not always a network motif under all β. For large β, subgraph 238 becomes a significant motif. The density, Z_{score}, and U of subgraph 238 under the random strategy increase exponentially, while they linearly increase with β under the anti-preference strategy. This indicates that the subgraph 238 under the anti-preference strategy can be more robustly conserved than that under the random model. The average degree increases linearly with β, and it increases more quickly under the random strategy. Finally, from Fig. 6.4i–l, with the increasing of γ, the four indexes are almost all parallel with the horizontal coordinates, which is mainly because γ is very small, and it has weak influence on the evolutions of subgraph 238.

6.4.3 Evolutionary Mechanisms of Network Motifs

From the above investigations, we have clarified the effect of different duplication and divergence strategies. Now, we discuss the evolutionary mechanisms of network motifs. Figure 6.5a shows a possible evolutionary route of a 5-node artificial network from two ancestral nodes. After two rounds of anti-preference duplication, a star network with three leaves can be generated, as shown in (III). We suppose v_4 is duplicated from v_3, then dimerization between v_4 and v_3 may happen with probability β, which results in network (IV). Further edge addition processes may create a link between v_1 and v_4, and thus we derive (V). Based on network (V) and the anti-preference strategy, node v_1, v_3 may be chosen as duplication target with higher probability than v_2, v_4. Suppose v_1 is chosen as a target, then one obtains (VI). By performing further edge deletion on the edges originated from v_1 or v_5, we derive (VII). Further dimerization between v_1 and v_5, edge addition between v_5 and v_3, one derives (VIII) and (IX). From Fig. 6.5a, we can see that all the subgraphs in Fig. 6.1b can be formed after several steps of duplication and divergence.

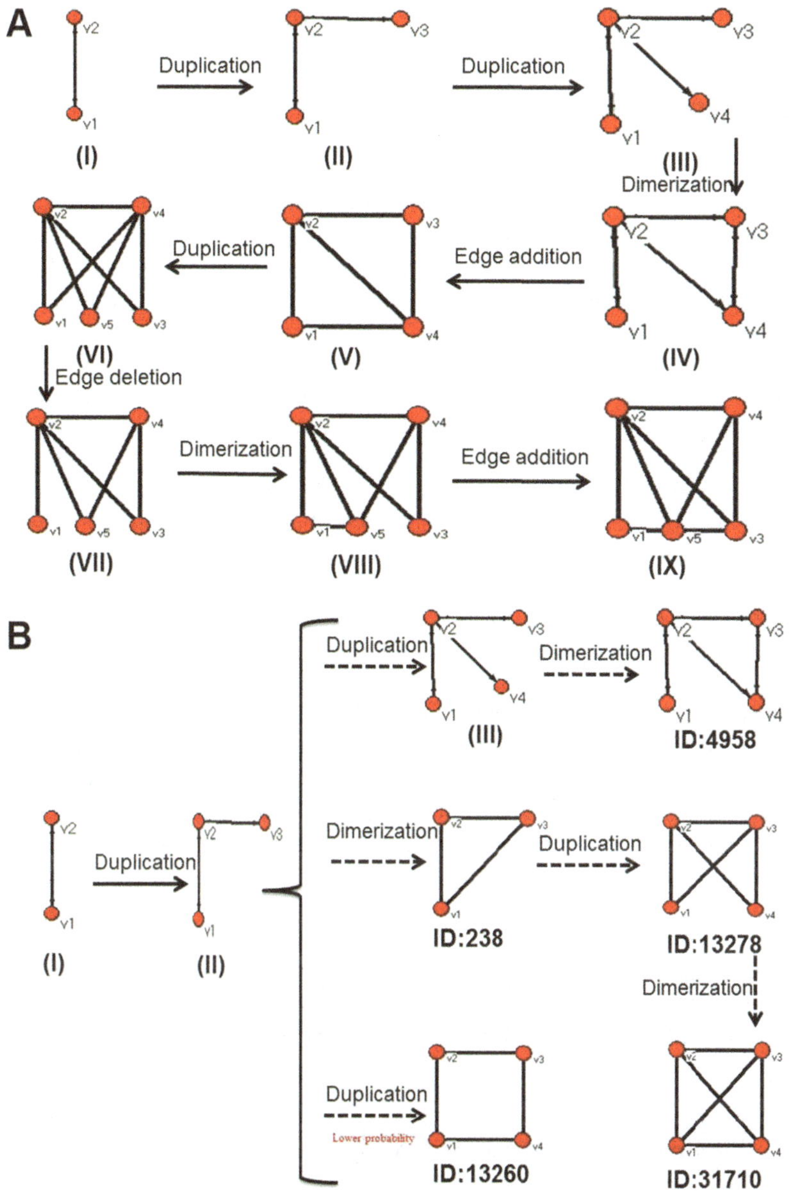

Fig. 6.5 Evolutionary mechanisms of network motifs under the DD model. (**a**) Under the anti-preference strategy, evolution processes of an artificial network with size 5. (**b**) Evolutionary mechanisms of the considered 3- and 4-node motifs. ©[2015] IEEE. Reprinted, with permission, from Ref. [1]

From Fig. 6.5a, the anti-preference strategy can effectively avoid highly connected nodes to be duplicated, thus inhibiting the rapid growth of average node degrees. Based on the simple example in Fig. 6.5a and our observations in the former section, we clarify the evolutionary mechanisms of network motifs, as shown in Fig. 6.5b. Suppose there are only two connected nodes in the initial network, and then after one round of duplication, a 3-node protein chain (II) can be formed. On one hand, based on this chain, further dimerization between the duplicated node v_1 and its replica v_3 can form the protein clique 238. This is a simple way to generate subgraph 238. Moreover, from Fig. 6.5a, we know that after each duplication event, any dimerization and edge addition events can produce a considerable quantity of subgraph 238. Therefore, it is easy to imagine that during continuous evolutions, the clique 238 can be conserved and selected as a significant building block. Based on subgraph 238, the second round of duplication creates subgraph 13278. Further dimerization between nodes in subgraph 13278 forms the 31710. On the other hand, based on (II), a 3-node star can be derived with very high probability after one duplication event. Dimerization between nodes in the star network can generate subgraph 4958. Finally, the subgraph 13260 can be easily generated by duplicating nodes with degree no less than 2. In small networks, the duplication probability of nodes with degree no less than 2 is very low. However, with the increasing of network size, this probability is considerable. Therefore, it is also easy to understand why subgraph 13260 can be a network motif. From Fig. 6.5, one can see that the dimerization and edge addition processes can promote the number of subgraphs constituted by triangles. Therefore, dimerization and edge addition are crucial for the formation of various motifs. The edge deletion process can lower the occurrences of the network motifs. However, since α is far larger than β and γ, in the divergence processes, averagely speaking, more edges would rather be deleted than added. Thus, the edge deletion processes are crucial for the preservation of the sparsity of bio-molecular networks. In summary, for bio-molecular networks, the potential evolutionary mechanisms of network motifs follow the anti-preference duplication and divergence. Duplication and dimerization are crucial for the evolutionary conservation and selection of various motifs, while the anti-preference mechanism and edge deletion processes are crucial to the preservation of the sparsity of bio-molecular networks.

6.5 Theoretical Analysis on Average Degrees

Theoretically, suppose at a time step, there are n nodes in the network, and the i'th node has degree k_i. For the next step, under the random duplication strategy, after one duplication event, the average increment of links will be

$$\langle \delta L_n \rangle_{rand} = \sum_{j=1}^{n} k_i p_i^{rand} = \frac{\sum_{j=1}^{n} k_i}{n}, \tag{6.3}$$

while under the anti-preference strategy, the links will be averagely increased by

$$\langle \delta L_n \rangle_{antipre} = \sum_{j=1}^{n} k_i p_i^{antipre} = \frac{n}{\sum_{j=1}^{n} \frac{1}{k_i}}. \tag{6.4}$$

Here, p_i^{rand} and $p_i^{antipre}$ represent the probabilities that node i is selected as a target node under the random and anti-preference duplication strategies, respectively. Since $1 \le k_i \le n-1$, suppose the degree distributions of the networks are the same at the former step under the two strategies. By the average inequalities, we have

$$\frac{n}{\sum_{j=1}^{n} \frac{1}{k_i}} \le \frac{\sum_{j=1}^{n} k_i}{n}. \tag{6.5}$$

Thus, the increment of links after one duplication event satisfies

$$\langle \delta L_n \rangle_{antipre} \le \langle \delta L_n \rangle_{rand}. \tag{6.6}$$

Therefore, under similar initial conditions, for the random strategy, the average degree of the network will be larger than that under the anti-preference strategy, and it increases relatively faster.

The above analysis is an initial result under ideal conditions. Generally, the degree distributions of the generated networks under the two duplication strategies are different. Moreover, isolated nodes will be deleted during the DD processes. However, it provides certain theoretical understanding of the numerical results. We denote L_N and K_N as total links and average degree for a network at step N, and when $N = 1$, the network has $n_0 = 2$ nodes. We denote $\langle \delta L_N \rangle$ as the increment of links by duplication from steps N to $N+1$. Based on the ideal condition and without considering the deletion of isolated nodes, after one round of duplication and divergence, we have

$$L_{N+1} = L_N + (1 - \alpha + \beta) \langle \delta L_N \rangle + \gamma (N - \langle \delta L_N \rangle), \tag{6.7}$$

where the term $(1 - \alpha + \beta) \langle \delta L_N \rangle$ corresponds to the contribution of duplication, edge deletion, and dimerization. The last term represents the effect of edge addition.

Generally, we have $K_N = 2L_N/N$. Therefore, Eq. (6.7) can be rewritten as

$$K_{N+1} = \frac{N K_N + 2\gamma N + 2(1 - \alpha + \beta - \gamma) \langle \delta L_N \rangle}{N+1}. \tag{6.8}$$

Under ideal conditions, Eq. (6.8) is suitable for different duplication strategies. For the random strategy, it is obvious that $\langle \delta L_N \rangle = K_N$. Therefore, one has

$$K_{N+1} = K_N + \frac{2\gamma N + (1 - 2\alpha + 2\beta - 2\gamma)K_N}{N+1}. \tag{6.9}$$

Similar to [21], using the continuous approximation

$$\frac{dK_N}{dN} \simeq K_{N+1} - K_N. \tag{6.10}$$

Equation (6.9) can be written as

$$\frac{dK_N}{dN} = \frac{2\gamma N + (1 - 2\alpha + 2\beta - 2\gamma)K_N}{N+1}, \tag{6.11}$$

whose solution is

$$K_N = \frac{2\gamma N}{1-\theta} + c_0(N+1)^\theta + \frac{2\gamma}{(1-\theta)\theta}, \tag{6.12}$$

where $\theta = 1 - 2\alpha + 2\beta - 2\gamma$.

$$c_0 = \frac{K_1 - \frac{2\gamma(1+\theta)}{(1-\theta)\theta}}{2^\theta}$$

is a constant, and K_1 is the average degree at step $N = 1$.

From Eq. (6.12), for $\theta \leq 1$ and $N \to \infty$, $K_N \propto N$. Therefore, K_N will linearly increase with N. For $\theta > 1$ and $N \to \infty$, $K_N \propto N^\theta$. Therefore, K_N will grow with N in power law, and the PLE is θ. For $\alpha = 0.562$, $\beta = 0.12$, $\gamma = 0.000165$, $\theta = 1-2\alpha+2\beta-2\gamma = 0.11567 < 1$. Therefore, under the random strategy, K_N will increase approximately linearly with N, as we can see from Fig. 6.3. When N, β, γ are fixed and α is varied, $\theta = 1.23967 - 2\alpha$, the effect of α on K_N can be analyzed. Similarly, one can analyze the effect of β and γ. Figure 6.6 shows numerical results of the average degree, where the first panel shows the evolution of K_N with N and the rest panels show the evolution of K_N with divergence parameters. It is noted that, in Fig. 6.6a, we have taken $K_{99} = 3.2$ as initial conditions, in the rest three panels, we have set $N = 999$, that is, the network size is $n = 1000$. From Fig. 6.6, for small N, the average degree increases with N in power law, but for $N > 500$, it almost linearly increases with N. From Fig. 6.6b, c, we can see that K_N decreases and increases exponentially with α and β, respectively, which are in accordance with the numerical results in the former section. From the last panel, K_N linearly increases with γ, since γ is small, the increment is very small, and this may be why the numerical results in Fig. 6.4 are almost parallel with the horizontal coordinates.

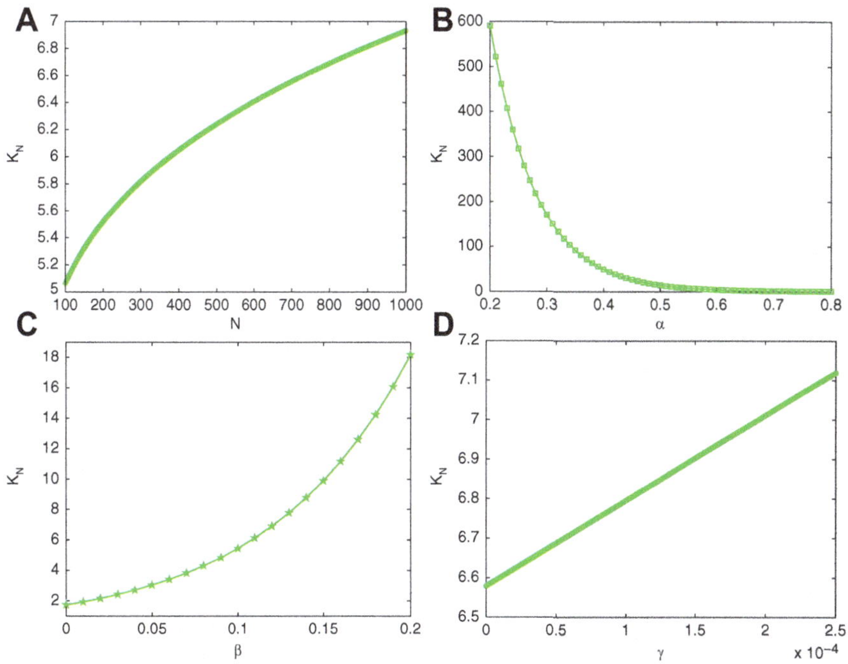

Fig. 6.6 Theoretical evolution of average degrees for the networks generated by the random duplication model. ©[2015] IEEE. Reprinted, with permission, from Ref. [1]

For the anti-preference model, it is difficult to deduce the relation between $\langle \delta L_N \rangle$ and K_N. Therefore, it is difficult to derive any theoretical results. However, since $\langle \delta L_N \rangle$ is smaller under the anti-preference strategy than that under the random strategy, with similar initial conditions, it is natural that K_N under the anti-preference strategy will be smaller. Finally, we note that the artificial networks under the anti-preference strategy are SW ones. Under the benchmark parameters and for the generated networks with $n = 1000$, the average degree is 2.87. The average clustering coefficient and APL are 0.0149 and 5.29 (averaged over 10 networks). For randomized networks with the same size and average degree, the clustering coefficient approximates to $2.87/1000 = 0.00287$ [2, 3]. Therefore, the generated networks under the anti-preference strategy are SW ones.

6.6 Discussions and Conclusions

Based on the DD model, we have considered the effect of two duplication strategies and four ways of divergence on the evolution of network motifs. We have investigated the evolution of subgraph density, Z_{score}, subgraph uniqueness U with network sizes, and divergence parameters. From numerical results and roughly theoretical analysis, we find that the anti-preference duplication with appropriate divergence parameters can well reflect the characteristics of motifs in real-world networks. Moreover, we find the underlying evolutionary mechanisms of network motifs in the real-world bio-molecular networks would be the anti-preference duplication, combined with divergences by edge deletion and addition, dimerization, and node deletion. Based on the findings, we have clarified how network motifs are evolved and conserved during long time evolution and natural selection.

Since subgraph 238 is a typical motif, we mainly consider the effects of different duplication and divergence strategies on subgraph 238. We note that the 4-node subgraphs can be similarly investigated. As to the effect of duplication strategies on the 4-node subgraphs, Fig. 6.7 shows the evolutions of subgraph density and U with network sizes, where the networks are the same as that in Fig. 6.3. To compare with the real-world cases, we show the subgraph density and U for the sampled real-world networks, just as Fig. 6.3. By taking subgraph 13260 as an example, Fig. 6.8 shows the effect of edge deletion on the evolution of density and U of subgraph 13260. The indexes under the anti-preference strategy can be more robustly conserved than that under the random strategy. From Figs. 6.7 and 6.8, we conclude that the 4-node subgraph features under the anti-preference strategy are also more approximate to the real-world cases, which further illustrates that real-world networks would rather follow the anti-preference strategy than the random one.

It is noted that although we have considered undirected bio-molecular networks, it can be easily extended to directed ones, provided that one can develop proper artificial algorithms [18, 31]. Another interesting issue is to investigate some other evolutionary features of bio-molecular networks, such as clarifying the effect of duplication and divergence on network entropy and modularity. Our future work will consider these topics. The evolutionary conservation of network motifs can help predict PPIs [32]. The other potential implications of the related investigations include the synthesis and design of artificial bio-molecular circuits or reengineering of real-world networks for the biomedical purpose [33–36].

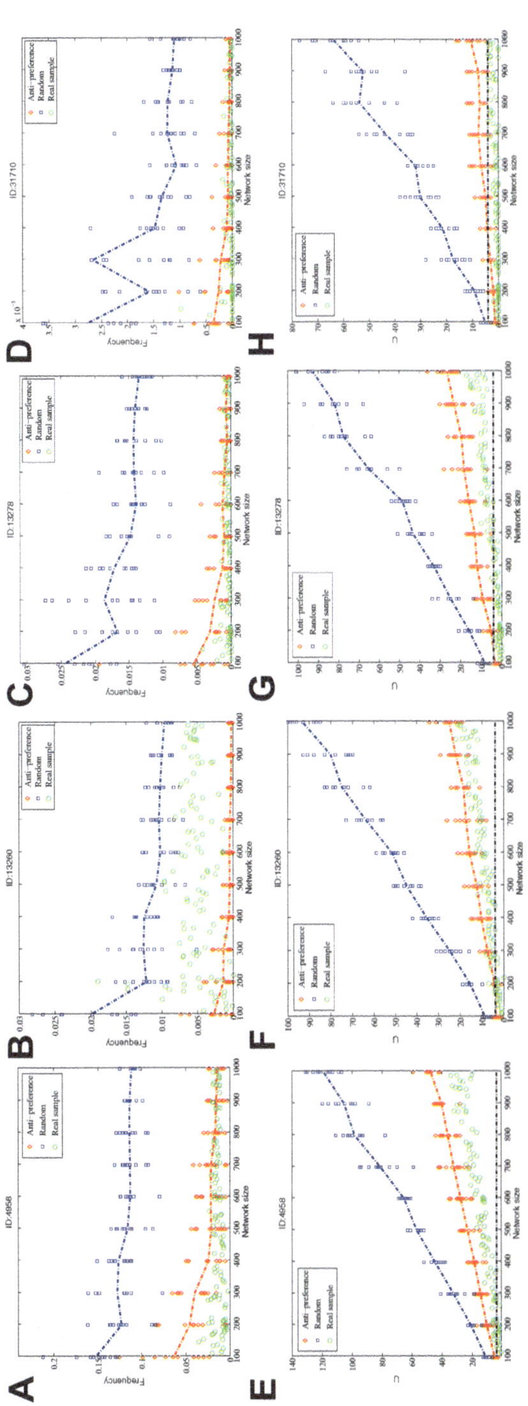

Fig. 6.7 Evolution of the 4-node subgraph densities **(a)**–**(d)** and U **(e)**–**(h)** with network size under different duplication strategies. ©[2015] IEEE. Reprinted, with permission, from Ref. [1]

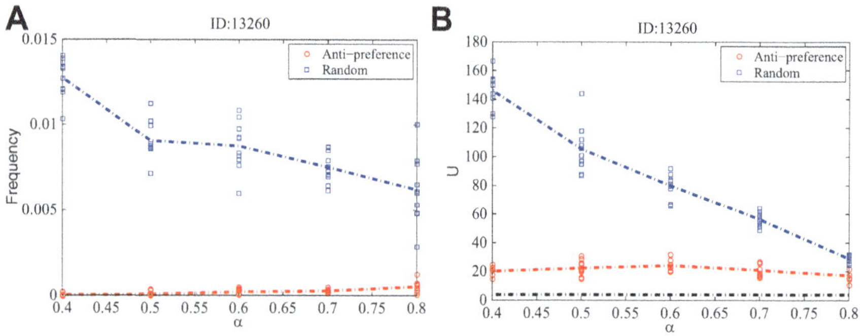

Fig. 6.8 Effect of edge deletion on density (**a**) and U (**b**) of subgraph 13260. ©[2015] IEEE. Reprinted, with permission, from Ref. [1]

References

1. Wang, P., Lü, J., Yu, X., Liu, Z.: Duplication and divergence effect on network motifs in undirected bio-molecular networks. IEEE Trans. Biomed. Circ. Syst. **9**, 312–320 (2015)
2. Newman, M.E.J.: The structure and function of complex networks. SIAM Rev. **45**,167–256 (2003)
3. Dorogovtsev, S.N., Mendes, J.F.F.: Evolution of networks. Adv. Phys. **51**,1079–1187 (2002)
4. Milo,R., Shen-Orr, S., Itzkovitz, S., et al.: Network motifs: simple building blocks of complex networks. Science **298**, 824–827 (2002)
5. Shen-Orr, S., Milo,R., Mangan, S., Alon, U.: Network motifs in the transcriptional regulation network of *Escherichia coli*. Nat. Genet. **31**, 64–68 (2002)
6. Yeger-Lotem, E., Sattath, S., Kashtan, N., et al.: Network motifs in integrated cellular networks of transcription-regulation and protein-protein interaction. Proc. Natl. Acad. Sci. USA. **101**, 5934–5939 (2004)
7. Spirin, V., Mirny, L.A.: Protein complexes and functional modules in molecular networks. Proc. Natl. Acad. Sci. USA. **100**,12123–12128 (2003)
8. Wuchty, S., Oltvai, Z.N., Barabási, A.L.: Evolutionary conservation of motif constituents in the *yeast* protein interaction network. Nat. Genet. **35**,176–179 (2003)
9. Wang, P., Yu, X., Lü, J.: Identification and evolution of structurally dominant nodes in protein-protein interaction networks. IEEE Trans. Biomed. Circ. Syst. **8**, 87–97 (2014)
10. Milo, R., Kashtan, N., Levitt, R., et al.: Superfamilies of evolved and designed networks. Science **303**,1538–1542 (2004)
11. Alon, U.: An introduction to systems biology: design principles of biological circuits. Chapman & Hall/CRC (2007)
12. Mangan, S., Alon, U.: Structure and function of the feed-forward loop network motif. Proc. Natl. Acad. Sci. USA. **100**,11980–11985 (2003)
13. Wang, P., Lü, J., Ogorzalek, M.J.: Global relative parameter sensitivities of the feed-forward loops in genetic networks. Neurocomput. **78**, 55–165 (2012)
14. Wang,P., Lü, J.: Control of genetic regulatory networks: opportunities and challenges. Acta Automatica Sin. **39**, 1969–1979 (2013) (In Chinese)
15. Lipshtat, A., Purushothaman, S.P., Iyengar, R., Maáyan, A.: Functions of bifans in context of multiple regulatory motifs in signaling networks. Biophys. J. **94**, 2566–2579 (2008)
16. Conant, G.C., Wagner, A.: Convergent evolution of genetic circuits. Nat. Genet. **34**, 264–266 (2003)

17. Camas, F.M., Poyatos, J.F.: What determines the assembly of transcriptional network motifs in *Escherichia coli*? PLoS One **3**, e3657 (2008)
18. Ward, J.J, Thomton, J.M.: Evolutionary models for formation of network motifs and modularity in the *Saccharomyces* transcription factor network. PLoS Comput. Biol. **3**, e198 (2007)
19. Solé, R.V., Pastor-Satorras, R., Smith, E., Kepler, T.B.: A model of large-scale proteome evolution. Adv. Complex Syst. **5**, 43–54 (2002)
20. Vázquez, A., Flammini, A., Maritan, A., Vespignani, A.: Modeling of protein interaction networks. Complexus **1**, 38–44 (2003)
21. Pastor-Satorras, R., Smith, E., Solé, R.V.: Evolving protein interaction networks through gene duplication. J Theor. Biol. **222**, 199–210 (2003)
22. Xu, C., Liu, Z., Wang, R.: How divergence mechanisms influence disassortative mixing property in biology. Physica A **389**, 643–650 (2010)
23. Wan, X., Cai, S., Zhou, J., Liu, Z.: Emergence of modularity and disassortativity in protein-protein interaction networks. Chaos **20**, 045113 (2010)
24. Zhao, D., Liu, Z., Wang, J.: Duplication: a mechanism producing disassortative mixing networks in biology. Chin. Phys. Lett. **24**, 2766–2768 (2007)
25. Barabási, A.L., Albert, R.: Emergence of scaling in random networks. Science **286**, 509–512 (1999)
26. Patthy, L.: Protein evolution. Blackwell, Oxford, (1999)
27. Wagner, A.: The *yeast* protein interaction network evolves rapidly and contains few redundant duplicate genes. Mol. Biol. Evol. **18**, 1283–1292 (2001)
28. Schwikowski, B., Uetz, P., Fields, S.: A network of protein-protein interactions in *yeast*. Nat. Biotechnol. **18**, 1257–1261 (2000)
29. Yu, H., Braun, P., Yildirim, M.A., et al.: High-quality binary protein interaction map of the *yeast* interactome network. Science **322**,104–110 (2008)
30. Stumpf, M.P.H., Wiuf, C., May, R.M.: Subnets of scale-free networks are not scale-free: sampling properties of networks. Proc. Natl. Acad. Sci. USA. **102**, 4221–4224 (2005)
31. Foster, D.V., Kauffman, S.A., Socolar, J.E.S.: Network growth models and genetic regulatory networks. Phys. Rev. E **73**, 031912 (2006)
32. Albert, I., Albert, R.: Conserved network motifs allow protein-protein interaction prediction. Bioinformat. **20**, 3346–3352 (2004)
33. Chen, B., Wu, W., Wang, Y., Li, W.: On the robust circuit design schemes of biochemical networks: steady-state approach. IEEE Trans. Biomed. Circ. Syst. **1**, 91–104 (2007)
34. Chen, B., Chen, P.: Robust engineered circuit design principles for stochastic biochemical networks with parameter uncertainties and disturbances. IEEE Trans. Biomed. Circ. Syst. **2**,114–132 (2008)
35. MacKay, S., Wishart, D., Xing, J.Z., Chen, J.: Developing trends in aptamer-based biosensor devices and their applications. IEEE Trans. Biomed. Circ. Syst. **8**, 4–14 (2014)
36. Wu, F.: Global and robust stability analysis of genetic regulatory networks with time-varying delays and parameter uncertainties. IEEE Trans. Biomed. Circ. Syst. **5**, 391–398 (2011)

Chapter 7
Identifying Important Nodes in Bio-Molecular Networks

Abstract Many issues in bio-molecular networks can be boiled down to the identification of important nodes or gene prioritization. Various measures have been proposed to characterize the importance of nodes in complex networks, such as the degree, betweenness, k-shell, clustering coefficient, closeness, semi-local centrality, PageRank, and LeaderRank. Different measures consider different aspects of complex networks. In this chapter, based on network motifs and principal component analysis, we introduced a new measure to characterize node importance in directed biological networks. Investigations on five real-world biological networks indicate that the proposed method can robustly identify actually important nodes in different networks. Further using the principal component analysis technique to integrate some existing centrality measures, we introduced a new integrative measure to find the structurally dominant proteins in protein interaction networks. Finally, the recently proposed SpectralRank and the weighted SpectralRank will be introduced, which can be used in various kinds of networks.

7.1 Backgrounds

Complex networks theory and its applications have been popular topics in recent years [1–8]. Many real-world systems can be described by complex networks and investigated through complex networks theory, such as social systems, biological systems. GRNs, signal transduction networks, neural networks, PPI networks, metabolic networks are typical biological networks, which have been extensively investigated [9–15].

Complex networks consist of nodes and edges. An edge denotes the interaction between two nodes, which can be directed or undirected. Many biological networks are directed ones. For example, in GRNs, nodes represent genes or TFs, edges represent the interactions between TFs and the regulated genes, or between TFs. Over the last decades, identification of important nodes in complex networks has been an intriguing topic [16–33]. For example, in social networks, provided that one knows which nodes are the most important ones, one can control these

J. Lü, P. Wang, *Modeling and Analysis of Bio-molecular Networks*,
https://doi.org/10.1007/978-981-15-9144-0_7

nodes in priority to prevent the spread of infectious diseases [16]. However, it is still a challenge to determine which nodes are important in a complex network. Traditionally, degree is frequently used to characterize the importance of a node [1, 2, 7, 8, 16, 34, 35]. The other indexes include the betweenness [19], closeness [1], k-shell [7], principal component centrality [17] based on adjacency matrix of the network, semi-local centrality [20], motif centrality [25, 27–31], PageRank (PR) [21], and others therein [36].

For undirected networks, some researchers believe that the most connected nodes are the most influential ones [1, 2, 35]. But recently, Kitsak et al. [32] investigated the spreading dynamics on four real-world complex networks. They found that for networks with a single initial spreader, k-shell can predict the outcome of spreading dynamics more reliably than degree and betweenness. Following, Chen et al. [20] proposed a semi-local centrality, which considers the degrees of both the nearest and next nearest neighbors of a node. The semi-local centrality can more effectively characterize influential spreaders in complex networks than the degree and betweenness. Recently, following the method in [32], we identified influential spreaders in artificial ER, SW, and SF networks. Some general conclusions have been obtained [33].

However, though there have been numerous results reported on undirected networks, few results have been reported on directed biological networks [25, 27–31]. In 2004, Sporns et al. [28] proposed a concept of motif fingerprint in brain networks, which counts the appearances of each node in network motifs with a given size as a measure. In 2007, based on the motif fingerprints and some of the other centrality measures, Sporns et al. [29] investigated the identification and classification of hubs in some brain networks. Also in 2007, based on the concept of network motif, Koschützki et al. [25, 27] proposed some new motif-based measures for GRNs. They took the occurrences of each node in the 3-node FFL as a measure, after further considering the direction of each edge, another two extended measures were proposed. Interesting results on finding the global regulators in the GRN of *E. coli* have been reported.

In the following sections, we firstly briefly discuss a motif centrality measure, and then based on the occurrences of each node in all 2-node, 3-node, and some 4-node network motifs and the principal component analysis (PCA), we develop a new method to characterize node importance in directed biological networks. To evaluate the performance of the new index, the in-degree, out-degree, total degree, PR, motif centrality, and betweenness are considered to compare with the proposed one. Investigations on five real-world biological networks will illustrate the performance of the proposed measure. Moreover, we develop an integrative measure to rank nodes in PPI networks, which is based on several topological indexes of the PPI networks, further using the artificial DD model, we investigate the evolutionary characteristics of important nodes in PPI networks. Finally, the recently developed efficient SpectralRank and weighted SpectralRank algorithms will be introduced.

7.2 Motif Centrality Measures

In the year 2007, Koschützki et al. [25] proposed a motif centrality measure and two extended measures to rank nodes in biological networks. In the following, we briefly discuss the motif centrality measures.

7.2.1 A Motif Centrality Measure

Denote a directed graph as $G = (V, E)$, consisting of a set of vertices V and a set of edges $E \subset (V \times V)$. V and E denote the node set and edge set of graph G. A graph $G' = (V', E')$ is a subgraph of graph $G = (V, E)$ if $V' \subset V$, $E' \subset E \bigcap (V' \times V')$. A graph $G_1 = (V_1, E_1)$ is isomorphic to a graph $G_2 = (V_2, E_2)$ if a bijective mapping $\phi : V_1 \mapsto V_2$ exists with $\forall v_a, v_b \in V_1 : (v_a, v_b) \in E_1 \Leftrightarrow (\phi(v_a), \phi(v_b)) \in E_2$. Such a mapping is called an isomorphism, and if $G_1 = G_2$, it is called an automorphism.

A centrality is a function $c : V \mapsto \mathbb{R}$ that assigns every vertex a real number. A vertex v_a is said to be more important (more central) than a vertex v_b if $c(v_a) > c(v_b)$. Based on the centrality values, vertices can be ordered or ranked. A recent review explains different concepts for centralities and describes more than 20 measures [26]. Small recurring subgraphs within a given graph are called motifs. A motif M is a directed graph according to the definition of graphs above. A match G_M of a motif M in a target graph G is a subgraph of $G (G_M \subset G)$ which is isomorphic to the motif $M (G_M \simeq M)$. See Fig. 7.1 for a graph, a motif, and a match of the motif in the graph. The motif match set $g_M = \{G_M | G_M \subset G\}$ of a motif M is the set of all matches of M in the graph G. It is a set of subgraphs of G and algorithms exist for its computation.

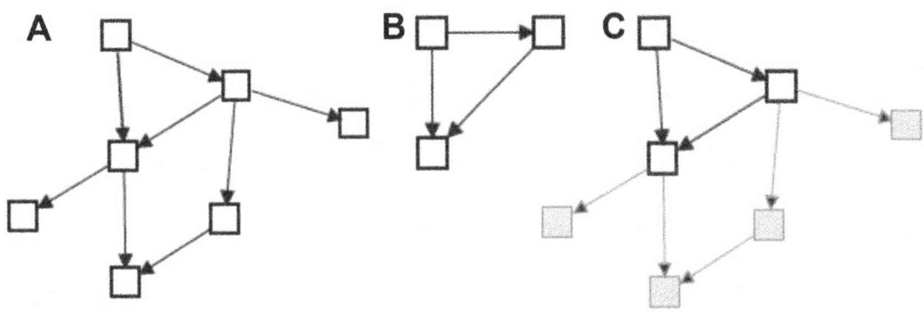

Fig. 7.1 A directed graph and illustration of network match. (**a**) A graph G; (**b**) a motif M; (**c**) a match G_M of M in G. The motif (**b**) occurs once in the graph (**a**). This occurrence is called a match and vertices and edges not participating in the match are marked in gray (**c**). Basically, motif-based centralities count matches of motifs in graphs. Reprinted from Ref. [25], with permission from Elsevier

Given a graph G, a motif M and the corresponding motif match set g_M, a centrality can be defined. The motif-based centrality C_m assigns to every vertex v the number of matches the vertex v occurs in.

The complexity of this algorithm is mainly determined by the function COMPUTEMOTIFMATCHSET, which is a NP-complete decision problem, similar to the SUBGRAPHISOMORPHISM problem [25]. The computation of motif-based centralities is therefore feasible for the same size of graphs and motifs that are currently investigated with existing motif analysis methods. The algorithm for the motif centrality is described in Algorithm 14.

Algorithm 14 Motif-based centrality [25]

1: **Input:** Graph G; Motif M;
2: **Output:** Centrality values $C_m(v)$ for the vertices $v \in V(G)$.
a. Initialise result vector;
For all $v \in V(G)$, do
$C_m(v) \leftarrow 0$;
b. Compute motif match set with existing algorithm;
$g_M \leftarrow COMPUTEMOTIFMATCHSET(G, M)$;
c. Compute motif-based centrality values;
for all $G_M \in g_M$, do
for all $v \in V(G_M)$, do;
$C_m(v) \leftarrow C_m(v) + 1$.

As an example, one considers the FFL motif (Fig. 7.2b), which matches three times in the target graph shown in Fig. 7.2a. Figure 7.2d shows the resulting centrality values for all vertices of this graph. Vertex v_2 is the most important vertex as it participates in all three matches of the motif.

7.2.2 *Extended Motif Centrality Measures*

Two extensions of this centrality exist: motif-based centrality with roles and motif-based centrality with classes. The two extended measures are described as follows.

7.2.2.1 Role-Based Motif-Based Centrality

Vertices of motifs may represent different functions. For example, in the GRN context, three different functions of the vertices of the FFL motif as shown in Fig. 7.2c can be identified: (1) the vertex at the top is the master regulator, this vertex regulates the other two vertices; (2) the vertex on the right side is the intermediate regulator, it is regulated by the master regulator and itself regulates together with the master regulator the vertex at the bottom; and (3) the vertex at the bottom of the drawing is regulated by both other vertices and is therefore called the regulated

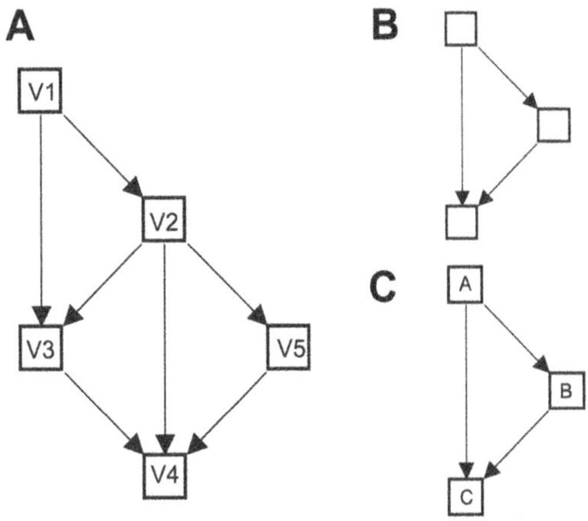

D

Vertex	Centrality c_{mc}
v_1	1
v_2	3
v_3	2
v_4	2
v_5	1

E

	Centrality c_{emc}		
Vertex	Role A	Role B	Role C
v_1	1	0	0
v_2	2	1	0
v_3	0	1	1
v_4	0	0	2
v_5	0	1	0

Fig. 7.2 (**a**) A target graph; (**b**) the FFL motif; (**c**) the FFL motif with three different roles A, B, and C; Tables (**d**) and (**e**) show the result of the centrality computation for the graph in (**a**); (**d**) The motif-based centrality given by the FFL motif without roles; (**e**) The extended motif-based centrality given by the FFL motif with roles. Reprinted from Ref. [25], with permission from Elsevier

vertex. Such different functions of vertices within motifs are called roles and three roles can be assigned to the vertices of the FFL motif.

Let R be a set of roles, G be a graph, M a motif, and g_M the corresponding motif match set. Koschützki et al. [25] defined a function role: $V \times g_M \mapsto \mathbb{R}$ which assigns a role to every vertex of G under a specific match. The role-based motif-based centrality $C_{emc}(v, r) : V \times g_M \mapsto \mathbb{R}$ assigns to every vertex $v \in V(G)$ the number of matches the vertex v occurs in and where it has the role r. It is defined as

$$C_{emc}(v, r) := |\{G_M | G_M \in g_M \wedge v \in V(G_M) \wedge role(v, G_M) = r\}|.$$

Considering a particular role r, the function C_{emc} is a centrality on the vertices of G.

The algorithm for the extended motif-based centrality is shown in Algorithm 15. The function GETROLEOFMATCHINGVERTEX returns the role of the vertex v within the motif M based on the match G_M. The result of the extended algorithm is not a single centrality vector for the vertices, as in the motif centrality algorithm, but a matrix consisting of rows and columns where the rows denote the vertices of the graph, the columns denote the roles, and the entries are the centrality values. The complexity of this algorithm is in the same class as the motif centrality algorithm as again the underlying decision problem of subgraph isomorphism, which is the most complex part of the algorithm.

Algorithm 15 Extended motif-based centrality [25]

1: **Input:** Graph G; Motif M with roles R;
2: **Output:** Centrality values $C_{emc}(v, r)$ for the vertices $v \in V(G)$ and roles $r \in R$.
  a. Initialise result table;
  For all $v \in V(G)$ do
  for all $r \in R$ do
  $C_{emc}(v, r) \leftarrow 0$;
  b. Compute motif match set with existing algorithm;
  $g_M \leftarrow COMPUTEMOTIFMATCHSET(G, M)$;
  c. Compute extended motif-based centralities;
  for all $G_M \in g_M$, do
  for all $v \in V(G_M)$ do
  $r \leftarrow GETROLEOFMATCHINGVERTEX(v, G_M)$;
  $C_{emc}(v, r) \leftarrow C_{emc}(v, r) + 1$.

Figure 7.2e shows the result of the extended algorithm for the FFL motif with roles shown in Fig. 7.2c in the same graph as before, see Fig. 7.2a. The column named Role A contains the number of matches for the vertices $v_1 - v_5$ for the role master regulator. Vertex v_2 is the most important vertex according to this role, followed by vertex v_1. A comparison with Fig. 7.2d shows that, based on the FFL motif, the vertex v_1 is a more important master regulator than the vertices $v_3 - v_5$. This is not obvious from the ranking based on the motif-based centralities without roles.

7.2.2.2 Motif-Class Centrality

Using the previously introduced concepts, Koschützki et al. [25] further extended the motif centrality measure. By assigning the same role to similar vertices of a group of similar motifs, a centrality based on a class (or a group) of motifs can be established. Consider, for example, a group of chains (see Fig. 7.3a), where all vertices at the start of such chains have a similar characteristic (no incoming edges) and all vertices at the end have another similar characteristic (no outgoing edges). For GRNs, several motif classes are known. For example, the regulatory chain motif class, as in the example above, consists of a set of chains of three or more regulators in which one regulator regulates another regulator, which in turn regulates a third one and so forth. In the single input motif (SIM) class, a set of vertices is exclusively regulated by a single vertex. Formally, motif classes can be described by graph grammars.

The role-based motif-based centrality for motif classes (motif-class centrality) is computed by using Algorithm 15 for the extended motif-based centrality. For each member of the motif class of interest, the matrix containing centrality values for vertices and roles is computed. If the motif class consists of l different motifs, then l matrices $c^1_{emc}, \cdots, c^l_{emc}$ with centrality values are obtained. The centrality value for a vertex $v \in V(G)$ and a role $r \in R$ for a motif class is defined as the sum of the

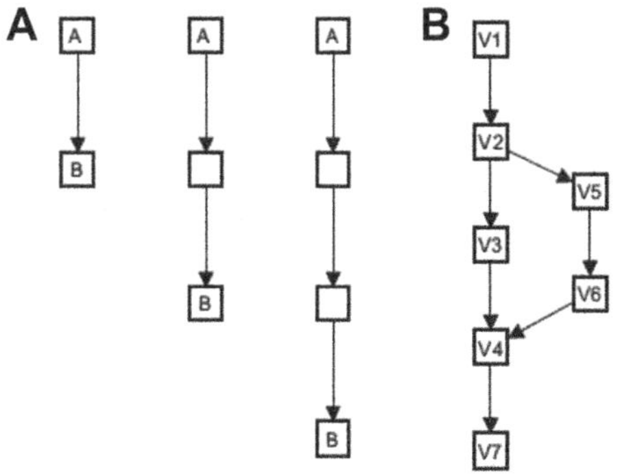

C

Vertex	Centrality c_{mcc}	Length of chains 2	3	4	5	6
v_1	8	1	2	2	2	1
v_2	7	2	2	2	1	0
v_3	2	1	1	0	0	0
v_4	1	1	0	0	0	0
v_5	3	1	1	1	0	0
v_6	2	1	1	0	0	0
v_7	0	0	0	0	0	0

Fig. 7.3 The motif-class centrality for the motif-class chains. (**a**) A sketch of the chain motif class with roles A and B; (**b**) An example graph; (**c**) The centrality values for the motif-class centrality for the motif-class chains. The length of the chain is defined as the number of vertices in the chain. The c_{mcc} values are the centrality values for role A for the chain motif class, that is, all different chains are considered and c_{mcc} is the sum of the centrality values of the different chains. Reprinted from Ref. [25], with permission from Elsevier

centrality values over all l matrices:

$$c_{mcc}(v, r) := \sum_{i=1}^{l} c^{i}_{emc}(v, r).$$

The demonstration of this method on the example graph is shown in Fig. 7.3b and the chain motif class (see Fig. 7.3a), where we are interested in the centrality value for vertices at the top of chains (role A in Fig. 7.3a). The size of the chain motif class is given by the number of chains of different lengths in the target graph. The centrality values for the vertices of the example graph in Fig. 7.3b are given in Fig. 7.3c. The vertex v_1 receives the highest centrality value, as it is the only vertex from which all other vertices are reachable.

7.2.3 *Motif Centralities for the GRN of* E. coli

The *E. coli* GRN network is based on the data of transcriptional regulatory interactions of genes from RegulonDB, Version 5.0. Genes are represented by vertices and transcriptional regulatory interactions between genes are modeled as edges, a common approach to model GRNs. The interactions between genes represent transcriptional control of TFs on the transcription of regulated genes. There are a few cases where TFs are formed by subunits of different gene products. They are here replaced by a common identifier which corresponds to the TF, e.g., ihfA or ihfB result in ihfAB. The regulatory interactions of such different subunits are assigned to this new identifier, and parallel edges which occurred due to the previous operation are replaced by a single edge. The resulting network consists of 1250 vertices and 2515 edges, of which 84 edges are self-loops representing autoregulation, i.e., the transcriptional control of a gene by its own gene product. It should be noted that autoregulation (self-loops) can be part of a motif.

In GRNs, genes at a high level within the hierarchy of regulatory control are of particular interest due to their far reaching influence on other genes within the network. These genes are commonly called global regulators. Some criteria for the characterization of global regulators have been proposed, such as the number of regulated genes, the number and type of coregulators, the number of other regulators they control, the size of their evolutionary family, and the variety of conditions where they exert their control.

For the motif-based centrality analysis, the FFL motif is used. It has been shown that depending on the type of interactions (activating or repressing), the FFL motif acts as an accelerator or delay element in the process of gene expression, and therefore has particular properties that control the expression of target genes. By using the three algorithms, the top 20 important nodes in the *E. coli* GRN are shown in Tables 7.1, 7.2, and 7.3.

The motif-based centrality c_m based on the FFL motif is computed for the GRN of *E. coli* and the resulting centrality values are used to rank the genes, see Table 7.1. The genes at the top position of this ranking have important functions in the regulation of cellular processes according to EcoCyc. The results of the motif-based centrality c_{mc} based on the FFL motif are consistent with current biological knowledge, as the genes which have been characterized as global regulators are assigned to top positions in the ranking. However, by consideration of the functional roles that genes adopt within the FFL motif, an extended motif-based centrality

Table 7.1 Top 20 out of 1250 genes of the *E. coli* GRN according to the motif-based centrality using the FFL motif with three vertices (see Fig. 7.2b) [25]

Rank	Gene	Centrality c_{mc}
1	crp	254
2	fnr	203
3	arcA	111
4	fis	110
5	narL	100
6	ihfAB	61
7	hns	53
8	fur	43
9	gadX	34
10	hyfR	33
11	marA	29
12	flhD	21
13	nagC, soxS	19
14	modE, tdcA, yiaJ	18
15	gutM, ompR, srlR	17

Table 7.2 Top 20 out of 1250 genes (see also Table 7.1) centrality values for different roles computed with the extended motif-based centrality using the FFL motif (see Fig. 7.2c) [25]

Gene	c_m	c_{emc}:A	c_{emc}: B	c_{emc}: C
crp	254	254	0	0
fnr	203	150	53	0
arcA	111	58	53	0
fis	110	40	70	0
narL	100	5	95	0
ihfAB	61	61	0	0
hns	53	14	39	0
fur	43	6	36	1
gadX	34	8	26	0
hyfR	33	0	33	0
marA	29	1	25	3
flhD	21	0	17	4
soxS	19	18	1	0
nagC	19	5	14	0
modE	18	18	0	0
tdcA, yiaJ	18	0	18	0
ompR	17	5	12	0
gutM, srlR	17	5	11	1

Table 7.3 Top ranking genes of the *E. coli* GRN shown for each role (numbers in brackets are the centrality values) [25]

Role A	Role B	Role C
crp (254)	narL (95)	marB (8)
fnr (150)	fis (70)	gadA (6)
ihfAB (61)	arcA, fnr (53)	fumB, gadC, gadB, lpdA, sodA (5)
arcA (58)	hns (39)	aceE, aceF, flhC and 27 further genes (4)
fis (40)	fur (36)	
modE, soxS (18)	hyfR (33)	
hns (14)	gadX (26)	
cpxR, fhlA, gadE (11)	marA (25)	
	tdcA, yiaJ (18)	

(studied in the following section) allows a further differentiation of the genes not covered by other centrality concepts.

The computation of the extended motif-based centrality c_{emc} based on the FFL motif with roles leads to three different rankings of the genes depending on the role under consideration, see Tables 7.2 and 7.3. The genes at top positions of these rankings allow an identification of important global regulators which receive a high rank for the master regulator role (role A in Table 7.3), important local regulators which receive a high rank for the intermediate regulator role (role B in Table 7.3) and important target genes which are controlled by at least two regulators as part of a functional FFL motif (role C in Table 7.3).

The genes at the top positions of these rankings can be clustered into four different groups: Genes that nearly exclusively adopt role A and therefore mainly act as global regulators without being controlled by many other genes (crp, ihfAB, soxS); Genes where both roles A and B are important and which selectively act as global and as local regulators (fnr, arcA, fis, hns); Genes that nearly exclusively adopt role B and therefore mainly act as local regulators, which are controlled by other genes (narL, fur, hyfR, gadX); Genes that nearly exclusively adopt role C and which are therefore regulated genes (marB, gadA, sodA, see Table 7.3). Consideration of roles of vertices allows a more detailed analysis of the functional properties of the elements within the network.

Comparing the ranking for the motif-class centrality based on the chain class and the motif-based centrality based on the FFL motif (see Tables 7.4 and 7.1, respectively) shows that crp, fnr, arcA, fis, and ihfAB are in both cases among the top six positions. The gene narL ranked at position 6 for the FFL-based motif-based centrality holds only position 20 for the motif-class centrality, even though it regulates a high number of genes directly, it influences only a low number of genes in total by indirect regulation. Furthermore, the composition of centrality values for individual motif chains shows some interesting characteristics. There are some

Table 7.4 Top 20 out of 1250 genes of the *E. coli* GRN according to the motif-class centrality based on the chain motif class for the role A (see Fig. 7.3) [25].

Gene	c_{mcc}	$l = 2$	$l = 3$	$l = 4$	$l = 5$	$l = 6$	$l = 7$
crp	1592	359	525	436	212	60	0
ihfAB	667	186	215	156	82	28	0
fnr	470	206	237	27	0	0	0
arcA	470	111	215	127	17	0	0
fis	387	156	121	82	28	0	0
evgA	325	4	27	90	125	51	28
ydeO	322	1	27	90	125	51	28
gadE	321	27	90	125	51	28	0
soxR	213	2	24	92	91	4	0
soxS	211	24	92	91	4	0	0
torR	191	10	15	87	51	28	0
gadW	185	4	15	87	51	28	0
cspE	184	1	2	88	65	28	0
cspA	183	2	88	65	28	0	0
gadX	181	15	87	51	28	0	0
hns	181	88	65	28	0	0	0
oxyR	166	15	73	74	4	0	0
fur	151	73	74	4	0	0	0
modE	141	32	94	15	0	0	0
narL	109	94	15	0	0	0	0

There are no chains with a length greater than 7. The number of chains of length 2 gives the number of genes that are directly regulated, excluding autoregulation genes among the top 20 that have a very low centrality value for motif chains of size 2: evgA (centrality value of 4 for chains of length 2 compared to 325 for centrality c_{mcc}), ydeO (1 vs. 322), soxR (2 vs. 213), gadW (4 vs. 185), cspE (1 vs. 184), and cspA (2 vs. 183). Therefore, these genes have a low range of direct control. However, all these genes indirectly control a large number of other genes.

These results show that the motif-class centrality with the chain class as motif family identifies genes that are important regulators within the GRN of *E. coli* which would be missed when only considering local approaches or other global approaches.

7.2.4 Summary

Koschützki et al. [25] presented a novel approach to rank vertices of networks based on network motifs, and discussed three particular methods. The first method (motif-based centrality) ranks vertices according to the number of motif matches such that the match contains the vertex of interest. The other two methods (extended motif-based centrality and motif-class centrality) are based on this method. The former additionally considers roles, and allows a more detailed analysis of the network of interest based on functions assigned to the vertices of the motif. The latter uses a whole group of similar motifs and therefore takes related functional network substructures into consideration. In contrast to existing centrality measures which consider either the local or the global network structure, the approach presented here deals with structural information between local and global information.

These methods can be applied to all kinds of networks by choosing appropriate motifs or motif classes. Koschützki et al. [25] applied them to the GRN of *E. coli*, where they yield interesting results in studying different functions of genes (by using the FFL motif) and in identifying key regulators which directly or indirectly regulate many genes (by using the chain motif class).

Table 7.5 shows the top ranked genes for the motif-based centrality c_m for the FFL motif with role A, the motif-class centrality c_{emc} for chains with role A, the out-degree centrality c_{odeg}, and the shortest-path betweenness centrality c_{spb}. For the *E. coli* GRN the rankings given by c_{odeg} and c_m are very similar for the top five positions. However, c_{odeg} only identifies key regulators which directly regulate a large number of genes, whereas c_m is able to identify important players for indirect regulation as well. The rankings given by c_{spb} and c_m are very different. c_{spb} finds some of the key regulators which are highly ranked by both c_{odeg} and c_m (hns, fur, gadE, fis, fnr, narL, arcA). It also identifies some genes important for indirect regulation such as gadX, soxS, and cspA, which obtain only a low ranking based on c_{odeg}, but a high ranking based on c_m. However, the gene crp, which is commonly regarded as the most important global regulator and which was ranked at the top position for c_{odeg} and c_m, is not under the top 20 positions for c_{spb}. For ihfAB, which has been characterized as a global regulator in previous studies, the same holds true: it receives positions 2 and 3 for c_m and c_{odeg}, respectively, but it is not

Table 7.5 Top 20 genes of the *E. coli* GRN according to $c_m, c_{emc}, c_{odeg}, c_{spb}$ (numbers in brackets are the centrality values) [25]

Order	c_m	c_{emc}	c_{odeg}	c_{spb}
1	crp (1592)	crp (254)	crp (360)	hns (1039.83)
2	ihfAB (667)	fnr (203)	fnr (207)	gadX (552)
3	arcA (470)	arcA (111)	ihfAB(187)	flhD (535)
4	fnr (470)	fis (110)	fis (157)	fur (488)
5	fis (387)	narL (100)	arcA (111)	gadE (418)
6	evgA (325)	ihfAB (61)	narL (94)	fis (342.5)
7	ydeO (322)	hns (53)	hns (89)	lrp (205.5)
8	gadE (321)	fur (43)	fur (74)	rcsAB (204)
9	soxR (213)	gadX (34)	lrp (63)	soxS (165)
10	soxS (211)	hyfR (33)	glnG (44)	fnr (159)
11	torR (191)	marA (29)	narP (40)	cspA (141)
12	gadW (185)	flhD (21)	cpxR (35)	caiF (135)
13	cspE (184)	nagC (19)	phoB (35)	purR (132)
14	cspA (183)	soxS (19)	fruR (32)	narL (99)
15	gadX (181)	modE (18)	modE (32)	marA (70.33)
16	hns (181)	tdcA (18)	fhlA (29)	metJ (61)
17	oxyR (166)	yiaJ (18)	lexA (29)	malT (51)
18	fur (151)	gutM (17)	flhD (28)	arcA (50.5)
19	modE (141)	ompR (17)	gadE (28)	glnG (50)
20	narL (109)	srlR (17)	purR (28)	ompR (23)

c_{odeg}: the out-degree centrality; c_{spb}: the shortest-path betweenness centrality. Both the motif-based centrality for the FFL motif with role A and the motif-class centrality for chains with role A are considered

under the top 20 ranked by c_{spb}. These results are not surprising, as c_{spb} assigns high centrality values to vertices that participate in many shortest-path communications. Since the GRN of *E. coli* has a hierarchical structure with global regulators on top, these regulators do not necessarily participate in shortest-path communications, but instead start the regulation and therefore are not captured by c_{spb}.

In conclusion, motif-based centrality is more effective in identifying important elements in biological networks (such as key regulators in GRNs) than previously used centrality measures. By using appropriate motifs or motif classes, it can be tailored to specific analysis tasks.

7.3 A Novel Network Motif Centrality and Its Performance

In this section, we introduce a new statistical centrality measure, which is mainly based on the PCA on network motifs. For details, one can refer to our work [37].

7.3.1 The New Motif Centrality Measure

Based on statistical analysis of network motifs, we introduce a new measure to characterize node importance in directed biological networks. Biological networks consist of some motifs, which act as functional units of the complex networks [37–42]. For example, it has been found that the FFLs play functional roles in GRNs, such as an incoherent FFL can act as a fold-change detector [9, 40]. Some other 3-node motifs and the 4-node bi-fan motif M_4204 (Fig. 7.4) are also found to play functional roles in biological systems [9, 14]. Therefore, nodes that frequently involved in network motifs may be more important. If a node involves in several different types of network motifs, then this node may potentially have multi-functional roles. Keeping the idea in mind, some related measures have been proposed to investigate the biological networks [25, 27–31]. We noted that in some works, network motifs are treated as subgraphs, such as the works of Rubinov et al. [31] and Wuchty et al. [43].

Hereinafter, different from the works in [25, 27–31], based on all 2, 3, and some 4-node motifs in directed networks and statistical analysis, we propose a new integrative measure. Specifically, suppose we have a directed network with n nodes, and there are totally m types of 2, 3, and 4-node motifs. Denote the occurrences of node i in the j-th type of motif as $u_{ij}, i = 1, \ldots, n, j = 1, \ldots, m$. Then, one can derive a matrix $A = (u_{ij})_{n\times m}$ for the network. In real-world networks, the importance of different types of motifs is varied. Therefore, one can endow each

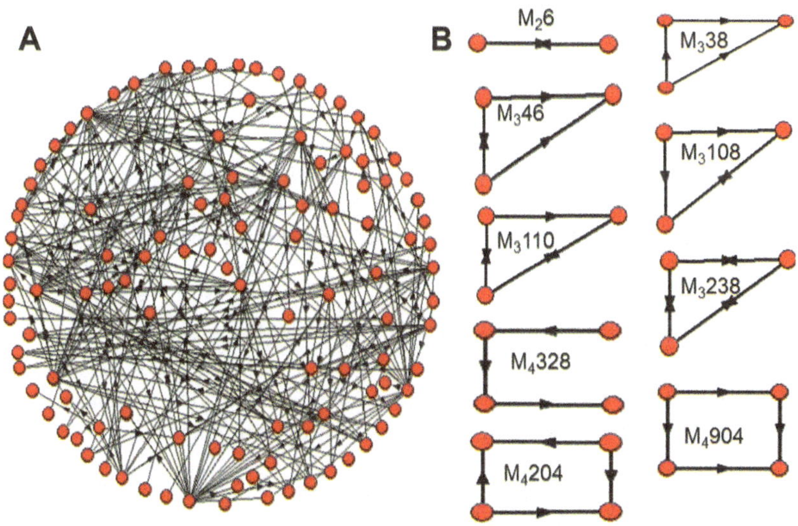

Fig. 7.4 A real-world biological network and some network motifs. (**a**) A *Drosophila* developmental transcriptional network with 119 nodes and 306 directed edges. (**b**) Some representative 2, 3, and 4-node motifs. Reprinted from Ref. [37]

motif with a weight w_j, $j = 1, 2, \ldots, m$, where

$$w_j = c_j / \sum_{k=1}^{m} c_k.$$

Here, $c_k (k = 1, 2, \ldots, m)$ denotes the number of the k-th type of motif. Subsequently, one derives a revised matrix:

$$B = (b_{ij})_{n \times m} = (b_1, b_2, \ldots, b_m) = (w_j u_{ij})_{n \times m}.$$

Based on B and the idea of the PCA [44–46], we construct the following index to obtain node importance score:

$$I^{\text{score}} = \sum_{j=1}^{m} \alpha_j b_j, \tag{7.1}$$

where $\alpha = (\alpha_1, \alpha_2, \ldots, \alpha_m)^\top$ are parameters to be determined. The best index vector I^{score} should have high distinguishability among different nodes. Therefore, the variance of I^{score} should be as large as possible. Take $\mathscr{B}_1, \ldots, \mathscr{B}_m$ as random variables, which represent the weighted counts of a node in the m types of motifs. For a certain network with size n, the $n \times m$ matrix $B = (b_1, b_2, \ldots, b_m)$ is an observation matrix of the m dimensional random vector $\mathscr{B} = (\mathscr{B}_1, \mathscr{B}_2, \ldots, \mathscr{B}_m)^\top$. The covariance matrix of $\mathscr{B}$ can be estimated by its observation matrix B. Denote the covariance matrix of B as Σ, then

$$\text{COV}(\mathscr{B}) \approx \text{COV}(B) = \Sigma = \frac{1}{n-1}\left(B^\top B - n\overline{B}\,\overline{B}^\top\right),$$

where $\overline{B}$ is the column mean vector of B, n is network size. It is noted that Σ is just the unbiased estimator of $\text{COV}(\mathscr{B})$ [45]. Based on the above notations, we have a stochastic form of I^{score} as $\mathscr{I}^{\text{score}} = \alpha^\top \mathscr{B}$. The variance of $\mathscr{I}^{\text{score}}$ can be estimated by

$$\text{Var}\left(\mathscr{I}^{\text{score}}\right) = \text{Var}(\alpha^\top \mathscr{B}) = \alpha^\top \text{Var}(\mathscr{B})\alpha \approx \alpha^\top \Sigma \alpha.$$

To determine the unique optimal vector α, we restrict $\alpha^\top \alpha = 1$. Thus, α can be determined through the following constrained extremal problem:

$$\begin{aligned} &\max\ \alpha^\top \Sigma \alpha, \\ &\text{s.t.}\ \ \alpha^\top \alpha = 1. \end{aligned} \tag{7.2}$$

To solve the optimization problem (7.2), by the Lagrangian multiplier method, we construct the following Lagrangian function:

$$L(\alpha, \lambda) = \alpha^\top \Sigma\alpha - \lambda(\alpha^\top\alpha - 1). \tag{7.3}$$

Let

$$\begin{cases} \frac{\partial L}{\partial \alpha} = 2(\Sigma - \lambda I)\alpha = 0, \\ \frac{\partial L}{\partial \lambda} = 1 - \alpha^\top\alpha = 0. \end{cases} \tag{7.4}$$

Here, I is the identity matrix. It follows from Eq. (7.4) that λ and α are just the eigenvalue and eigenvector of matrix Σ. Under Eq. (7.4), $\mathrm{Var}(\mathscr{I}^{\mathrm{score}}) \approx \alpha^\top\Sigma\alpha = \lambda\alpha^\top\alpha = \lambda$. Therefore, the optimal λ and α are just the biggest eigenvalue and the corresponding unit eigenvector of Σ. Denote the eigenvalues of Σ as $\lambda_1 \geq \lambda_2 \geq \cdots \geq \lambda_m \geq 0$, then the optimal $\lambda = \lambda_1$. From the theory of the PCA, the ratio $\lambda_1 / \sum_{i=1}^{m} \lambda_i$ can reflect the contribution of $\mathscr{I}^{\mathrm{score}}$, or how much information in $\mathscr{B}$ can be extracted by $\mathscr{I}^{\mathrm{score}}$.

So far we have determined α. For a concrete network, replacing b_j in Eq. (7.1) with concrete values, one determines the observation of $\mathscr{I}^{\mathrm{score}}$ as I^{score}. Finally, the nodes in the network can be ranked according to I^{score}. Nodes with larger I^{score} values are more important. Based on I^{score} and some well-defined distances, such as the well-known Euclidean distance, the n nodes can be classified into several clusters, where nodes in the same cluster are similarly important.

To sum up, for a network with n nodes, the procedures of the proposed measure are described in Algorithm 16.

Algorithm 16 The motif centrality based on PCA [37]

1: For a considered network, detect 2, 3 and 4-node network motifs in the network.
2: Count the occurrences of each node in m types of motifs, and derive a $n \times m$ matrix A.
3: Perform data processing on A, such as weighting and standardizing matrix A, then we obtain a matrix B. Compute the covariance matrix Σ of B.
4: For Σ, compute the biggest eigenvalue λ and the corresponding unit eigenvector α.
5: Compute I^{score} according to (7.1) and rank the n nodes accordingly.

7.3.2 An Illustrative Example

To illustrate the procedures of the proposed method, we give a simple example. The simple artificial network contains 6 nodes, and the topology of the network is shown in Fig. 7.5a. Suppose there are three motifs in the network, namely, M_338, M_3108, M_26, as shown in Fig. 7.5b. Figure 7.5c lists the members of the three motifs. Occurrences of nodes in each motif are summarized in Fig. 7.5d.

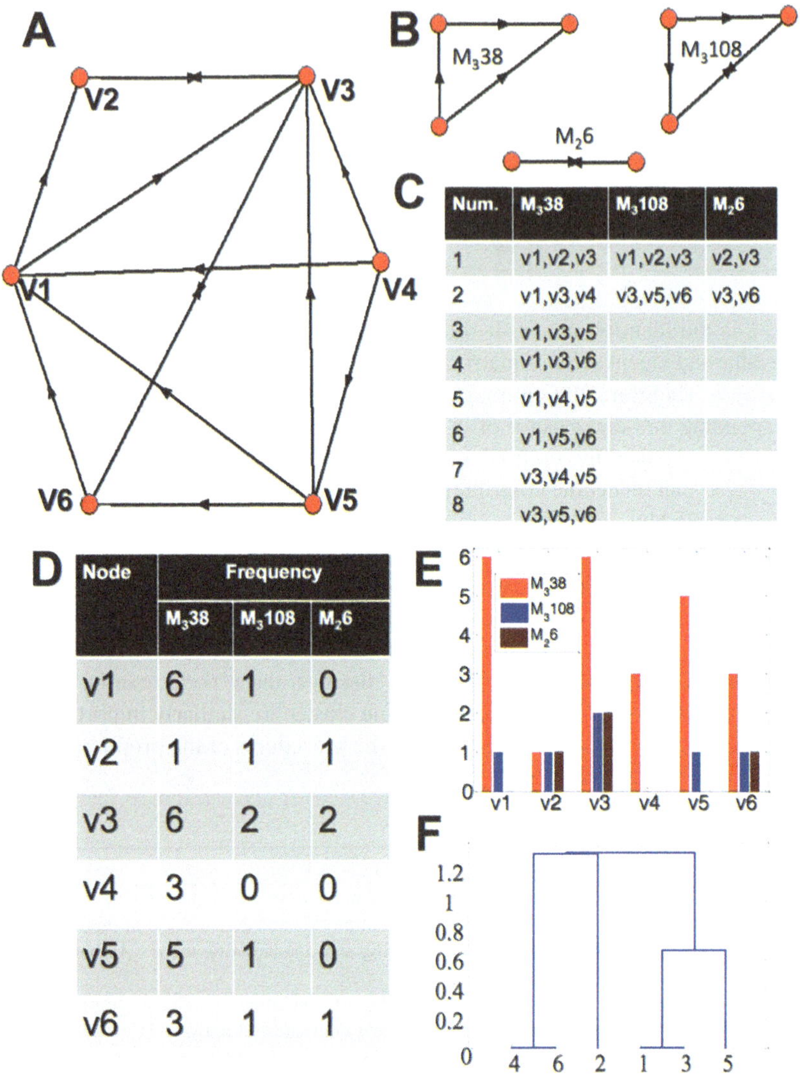

Fig. 7.5 An illustrative example. (**a**) A simple network with six nodes. (**b**) Subgraphs that are assumed to be motifs in network (**a**). (**c**) Members that compose the three types of motifs. (**d**) Appearances of nodes in each motif as shown in panel (**b**). (**e**) Frequency histograms for the six nodes. (**f**) Cluster analysis reveals that the six nodes can be remarkably classified into three classes. v_1, v_3, v_5 are the most important nodes, and v_2 forms the least important group, v_4, v_6 form another group, which is more important than v_2. Reprinted from Ref. [37]

As we see, the occurrences of $M_3 38, M_3 108, M_2 6$ are $8, 2$, and 2, respectively. Therefore, the weights of $M_3 38, M_3 108, M_2 6$ are $\omega_1 = 2/3, \omega_2 = 1/6, \omega_3 = 1/6$. Subsequently, we derive matrix B and its covariance matrix Σ:

$$B = (b_1, b_2, b_3) = \begin{bmatrix} 4 & 1/6 & 0 \\ 2/3 & 1/6 & 1/6 \\ 4 & 1/3 & 1/3 \\ 2 & 0 & 0 \\ 10/3 & 1/6 & 0 \\ 2 & 1/6 & 1/6 \end{bmatrix},$$

$$\Sigma = \begin{bmatrix} 1.7778 & 0.0667 & 0.0000 \\ 0.0667 & 0.0111 & 0.0111 \\ 0.0000 & 0.0111 & 0.0185 \end{bmatrix}.$$

The eigenvalues of Σ are $\lambda_1 = 1.7803, \lambda_2 = 0.0257, \lambda_3 = 0.0014$, and the unit eigenvector corresponding to λ_1 is

$$\alpha = (0.9993, 0.0377, 0.0002)^\top.$$

Thus, we have

$$I^{\text{score}} = 0.9993 b_1 + 0.0377 b_2 + 0.0002 b_3. \tag{7.5}$$

The contribution of I^{score} is $\lambda_1/(\lambda_1 + \lambda_2 + \lambda_3) = 98.50\%$. That is, 98.50% information that contained in b_1, b_2, b_3 can be extracted by I^{score}. Therefore, I^{score} can optimally rank the 6 nodes. Substitute b_1, b_2, b_3 in matrix B into Eq. (7.5), we have

$$I^{\text{score}} = (4.0034, 0.6725, 4.0098, 1.9986, 3.3372, 2.0049)^\top.$$

From I^{score}, the third value is the biggest. Therefore, we can judge that node v_3 is the most important one, and then v_1, the least important node is v_2. If one simply considers the total occurrences of a node in all the motifs, then v_2 and v_4 would be treated as equally important. Whereas, from the proposed method, v_4 is more important than v_2, which is reasonable in that the occurrences of $M_3 38$ are significantly more frequent than the other motifs. Based on I^{score} and through cluster analysis, the six nodes can be classified into three clusters, where v_1, v_3, v_5 are members of the most important cluster; v_4, v_6 are members of the less important cluster; while v_2 is the single member of the unimportant cluster.

7.3.3 Data Descriptions

The five real-world biological networks include the *C. Elegans* Neural (CEN) network [47, 48], the *E. Coli* Transcriptional (ECT) regulatory network from the RegulonDB database [49], the *Yeast* Transcriptional (YT) regulatory network [50], the *Drosophila* Developmental Transcriptional (DDT) network [13], and the Human Signal Transduction (HST) network [13]. We note that the investigated networks are with high quality and have been frequently used as models to detect network motifs [9, 11–13].

The five networks and their degree distributions are shown in Fig. 7.6. Simple statistical indexes for the five networks are summarized in Table 7.6. The numbers of nodes for these networks range from 119 to 1706. The numbers of edges range from 306 to 3870. The five networks are with abundant network motifs, such as the FFL $M_3 38$, $M_3 46$, the bi-fan $M_4 204$. It is noted that, we have considered all 2, 3-node motifs, but for simplicity, we have only considered three 4-node motifs: $M_4 204$, $M_4 328$, and $M_4 904$. There are totally 199 connected 4-node subgraphs, and there are many 4-node motifs in the five networks. For example, in the CEN and ECT, there are seven 4-node motifs. Since the bi-fan $M_4 204$ and the bi-parallel $M_4 904$ have been frequently investigated under various context [9], they are common motifs in many different real-world networks [11], and the 4-node chain $M_4 328$ may play crucial roles in signal transduction pathways, we will only consider these three 4-node motifs. From Table 7.6, the CEN has the most abundant of motifs. Subgraph $M_2 6$ is only a motif in the CEN and ECT, and the actual numbers are 233 and 10, respectively. The $M_4 328$ is only a motif in the HST, the actual number is 1570. There are no 3-node motifs in the HST. Whereas, for most of the networks, the FFL and bi-fan are motifs. The YT only consists of the FFL and the bi-fan.

7.3.4 Identifying Important Nodes in the Five Networks

Following the procedures as the illustrative example, one can obtain the order factor for each network. Noted that the occurrences of different motifs have different order of magnitude, we have performed standardized transformations to matrix B. Moreover, we denote the columns of matrix B as the vector $b_i j$, where i and j have the same meaning as that in $M_i j$. The I^{score} for the five networks are obtained as follows:

$$\begin{aligned} I_{\text{CEN}}^{\text{score}} &= 0.2654 b_2 6 + 0.3866 b_3 38 + 0.3924 b_3 46 + 0.3901 b_3 108 \\ &\quad + 0.3658 b_3 110 + 0.3285 b_3 238 + 0.3280 b_4 204 + 0.3529 b_4 904; \end{aligned} \tag{7.6}$$

$$I_{\text{ECT}}^{\text{score}} = 0.3778 b_2 6 + 0.0.5250 b_3 38 + 0.5352 b_3 46 + 0.5434 b_4 204; \tag{7.7}$$

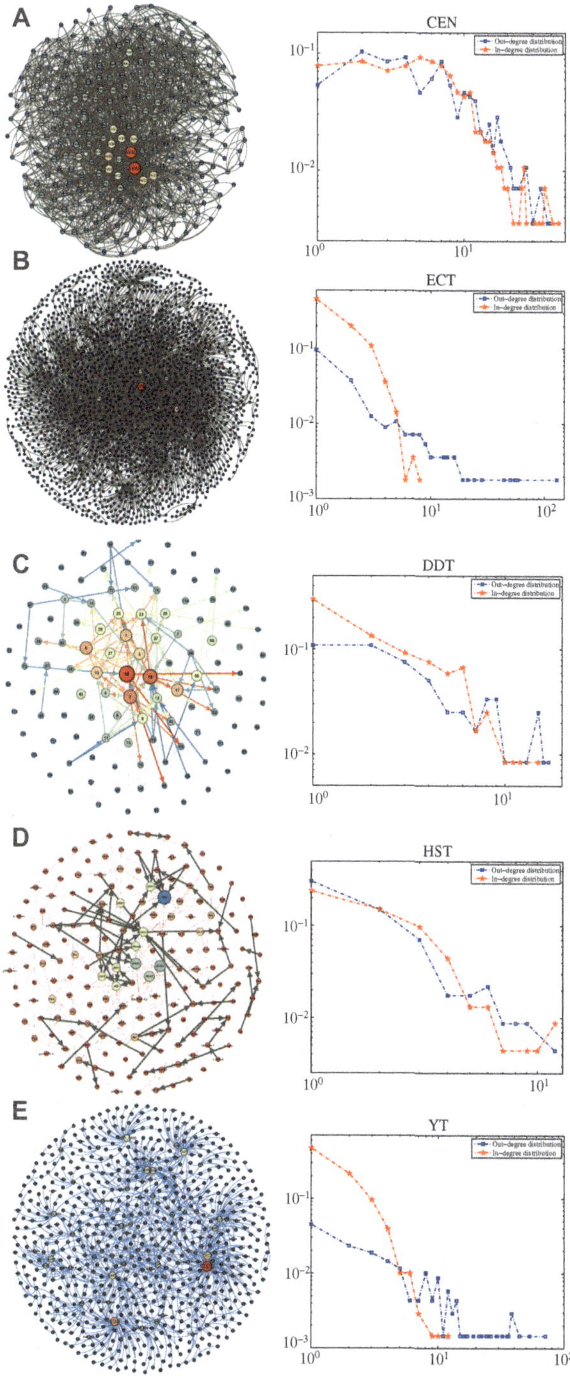

Fig. 7.6 The five real-world networks and its degree distributions. (**a**) CEN. (**b**) ECT. (**c**) DDT. (**d**) HST. (**e**) YT

Table 7.6 Statistical indexes for the five directed biological networks

Network	CEN	ECT	DDT	HST	YT
Node	280	1706	119	227	685
Edge	2194	3870	306	312	1052
Ave. in-degree	7.8357	2.2685	2.5714	1.3744	1.5358
Ave. out-degree	7.8357	2.2685	2.5714	1.3744	1.5358
Ave. total degree	15.6714	4.5369	5.1428	2.7489	3.0715
Ave. I^{score}	5.6753	35.9339	2.0367	12.3849	7.2407
M_26	233	10	–	–	–
M_338	1453	1196	174	–	62
M_346	552	226	26	–	–
M_3108	385	–	16	–	–
M_3110	175	–	–	–	–
M_3238	48	–	–	–	–
M_4204	2274	29,535	–	280	1812
M_4328	–	–	–	1570	–
M_4904	2253	–	–	275	–

"–" denotes no such motif

$$I_{\text{DDT}}^{\text{score}} = 0.5585b_338 + 0.5721b_346 + 0.6006b_3108; \tag{7.8}$$

$$I_{\text{HST}}^{\text{score}} = 0.5841b_4204 + 0.5696b_4328 + 0.5782b_4904; \tag{7.9}$$

$$I_{\text{YT}}^{\text{score}} = 0.7071b_338 + 0.7071b_4204. \tag{7.10}$$

Replacing b_ij with concrete values in the processed matrix B for each network, one obtains the importance score for each node. Average I^{score} values for the five networks are shown in Table 7.1. Based on I^{score}, we can characterize the node importance and classify the nodes for each network via cluster analysis. The basic idea of cluster analysis is as follows [45]. According to I^{score}, the Euclidean distance between any two nodes can be obtained. Firstly, two nodes with the shortest distance are merged as one group, each of the rest nodes forms a group. Then, one merges node groups via the single-linkage method, until all nodes are finally merged into one cluster. This cluster process can be mimicked by a dendrogram. From cluster analysis, one can classify nodes into groups, with similar important nodes in the same group. Furthermore, from the dendrogram, one can intuitively get some knowledge about the structural features of the network.

Figure 7.7 shows the dendrogram for the top-30 nodes of the five networks. We can see that these nodes can be roughly classified into three or four groups, detailed information of the top-30 nodes in the CEN, ECT, DDT, HST, YT and their corresponding rankings by the other methods are summarized in Tables 7.7, 7.8, 7.9, 7.10, and 7.11. In each table, we have shown the in and out-degree as well

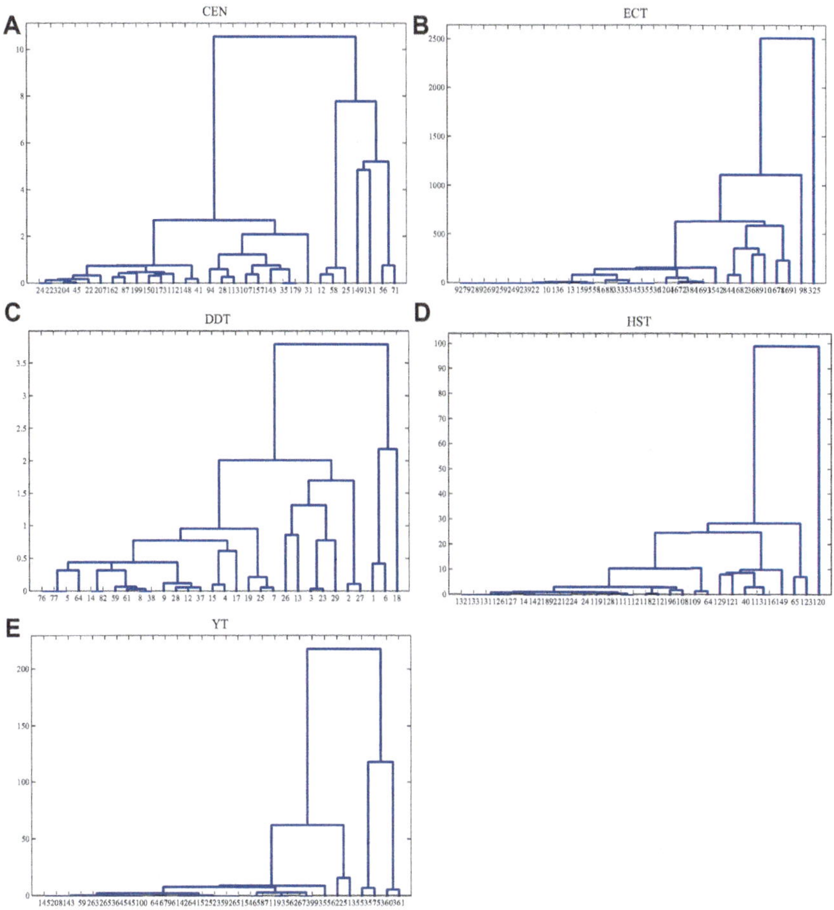

Fig. 7.7 Cluster analysis for the identified top-30 nodes in the five networks based on the I^{score}. **(a)**–**(e)** are corresponding to the CEN, ECT, DDT, HST and YT respectively. Reprinted from Ref. [37]

as their rankings by the other methods. Here, R_{total} is based on the total degree, R_{p} is based on the PR, R_{mc} is based on the motif centrality, and R_{bet} is based on the betweenness. The motif centrality only considers the FFL, since there is no such motif in the HST, it fails to work in the HST. For each network, the last group contains the largest amount of nodes, while the most important group G_1 contains only one to three nodes. From Fig. 7.7, for the five biological networks, only a few nodes are far more important than the others. There are clear hierarchical structures in these networks, which indicates that the proposed measure may also act as an effective hierarchical index.

Table 7.7 Clusters, members, rankings, and statistical characteristics of the identified top-30 nodes in the CEN

Group	Node	I^{score}	Out-deg.	R_{out}	In-deg.	R_{in}	R_{total}	R_p	R_{mc}	R_{bet}
G_1	12:AVER	54.90	18	11	33	6	10	126	6	22
	58:AVBR	54.50	15	14	38	4	8	119	4	28
	25:AVEL	53.81	16	13	36	5	9	129	5	21
G_2	149:AVDR	45.99	24	7	33	6	6	64	2	34
	131:AVDL	41.11	19	10	27	7	11	106	12	49
	56:AVBL	35.86	20	9	40	3	3	58	1	18
	71:AVAL	35.07	37	2	53	1	2	28	3	2
G_3	94:AVJR	24.48	12	17	18	13	21	114	7	96
	28:AVAR	23.85	49	1	49	2	1	13	8	1
	113:AIBR	23.55	11	18	25	9	16	143	11	42
	107:DVA	22.30	35	3	19	12	7	24	9	7
	157:PVCL	21.90	32	4	27	7	4	32	10	5
	143:PVCR	21.11	32	4	26	8	5	16	13	3
	35:RICR	20.48	8	21	14	17	29	173	36	39
	179:ADAL	20.46	14	15	8	23	29	97	15	102
	31:RICL	18.35	12	17	11	20	28	130	36	56
	148:AVL	15.63	12	17	15	16	24	83	19	58
	41:ADEL	15.42	26	5	5	26	20	6	28	54
	204:PVNL	14.63	19	10	8	23	24	12	18	70
	45:RIAL	14.58	15	14	27	7	13	117	20	19
	24:AIBL	14.43	13	16	24	10	15	66	15	11
	223:AVG	14.42	17	12	5	26	29	4	14	69
	22:RIAR	14.22	18	11	26	8	12	108	17	10
	207:ASHR	13.88	13	16	8	23	30	60	30	119
	162:AVJL	13.12	14	15	21	11	17	25	16	31
	87:RMGR	12.84	14	15	8	23	29	23	40	53
	199:ADLR	12.38	15	14	2	29	34	77	23	146
	150:HSNR	11.89	25	6	16	15	14	8	22	8
	173:PVNR	11.55	22	8	11	20	19	15	23	36
	112:RIML	11.13	12	17	16	15	23	139	21	73

R_{in} and R_{out} represent the rankings by the in and out-degree. R_{total} and R_p represent the results from the total degree and PR [21]. R_{mc} and R_{bet} denote the results from the motif centrality [25] and betweenness. Similarly hereinafter

7.3.5 *Functional Characteristics of the Top-Ranked Nodes*

In the following, for the CEN, ECT, and YT, we discuss whether the identified structurally top-ranked nodes are functionally important.

For the CEN, the identified top-30 nodes are shown in Table 7.7. The top-7 nodes are AVER, AVBR, AVEL, AVDR, AVDL, AVBL, and AVAL, which are

Table 7.8 Clusters, members, rankings, and statistical characteristics of the identified top-30 nodes in the ECT

Group	Node	I^{score}	Out-deg.	R_{out}	In-deg.	R_{in}	R_{total}	R_p	R_{mc}	R_{bet}
G_1	325:CRP	6643.21	496	1	1	13	1	2	2	6
G_2	98:FNR	4128.98	295	2	3	11	2	4	1	2
G_3	844:arcA	3014.10	173	6	1	13	6	6	18	9
	1682:IHF	2924.94	219	4	0	14	4	3	3	55
	368:fis	2564.89	226	3	2	12	3	1	7	8
	910:narL	2267.63	121	8	2	12	8	18	5	18
	1678:H-NS	1675.38	186	5	0	14	5	7	4	55
	1691:narP	1438.18	49	17	0	14	17	39	41	55
G_4	1542:cra	801.16	78	12	1	13	12	17	23	21
	1204:lrp	637.09	104	9	3	11	9	14	8	3
	1672:FlhDC	574.76	80	11	0	14	11	12	41	55
	384:fur	555.05	128	7	3	11	7	5	6	4
	1693:NsrR	538.26	83	10	0	14	10	11	26	55
	1688:ModE	386.47	46	18	0	14	18	20	25	55
	333:cysG	348.34	0	52	8	6	43	134	38	55
	534:nirB	348.34	0	52	8	6	43	134	38	55
	535:nirC	348.34	0	52	8	6	43	134	38	55
	536:nirD	348.34	0	52	8	6	43	134	38	55
	159:pflB	257.65	0	52	6	8	45	134	38	55
	558:pdhR	251.49	41	20	3	11	19	27	17	15
	13:gadX	234.85	27	29	13	1	22	30	12	5
	10:gadA	218.98	0	52	11	3	40	134	28	55
	922:nrfA	218.77	0	52	7	7	44	134	38	55
	923:nrfB	218.77	0	52	7	7	44	134	38	55
	924:nrfC	218.77	0	52	7	7	44	134	38	55
	925:nrfD	218.77	0	52	7	7	44	134	38	55
	926:nrfE	218.77	0	52	7	7	44	134	38	55
	927:nrfF	218.77	0	52	7	7	44	134	38	55
	928:nrfG	218.77	0	52	7	7	44	134	38	55
	136:lpd	207.41	0	52	7	7	44	134	36	55

all command interneurons. Additionally, the AVAR, PVCL, and PVCR are another three command interneurons, which are all top ranked. The AVAs, AVBs, AVDs, and PVCs are four bilaterally symmetric interneuron pairs with large diameter axons that run the entire length of the ventral nerve cord, and providing inputs to the ventral cord motor neurons. The AVAs are located at the lateral ganglia of head of the *C. elegans*, functioning as driver cell for backward locomotion [51]. The AVEs can drive backward movement of the animal along with AVAs, AVDs, and A-type motor neurons [51]. The AVDs function as touch modulator for backward locomotion induced by head-touch. The PVCs are ventral cord interneurons, a harsh

Table 7.9 Clusters, members, ranks, and statistical characteristics of the top-30 ranked nodes in the DDT

Group	Node	I^{score}	Out-deg.	R_{out}	In-deg.	R_{in}	R_{total}	R_p	R_{mc}	R_{bet}
G_1	1	20.01	15	3	4	9	5	7	1	23
	6	19.57	17	1	3	10	4	2	1	28
	18	17.37	9	6	15	1	2	14	2	2
G_2	26	13.57	8	7	7	6	8	16	3	15
	13	12.69	10	5	8	5	6	9	4	5
	3	11.36	9	6	8	5	7	10	6	10
	23	11.32	3	12	11	3	9	40	5	24
	29	10.53	15	3	0	13	8	3	7	49
	2	8.81	15	3	7	6	3	5	9	9
	27	8.69	8	7	5	8	10	17	8	22
G_3	19	6.66	16	2	12	2	1	6	10	1
	25	6.44	7	8	6	7	10	19	10	13
	7	6.37	5	10	3	10	14	26	10	43
	15	5.40	6	9	3	10	13	27	11	34
	4	5.28	4	11	6	7	12	18	12	14
	17	4.66	9	6	10	4	5	12	13	7
	9	3.87	6	9	8	5	9	11	14	4
	28	3.74	0	15	6	7	16	62	14	49
	12	3.67	7	8	5	8	11	22	14	18
	37	3.60	13	4	1	12	9	1	14	30
	14	3.15	3	12	5	8	14	46	15	25
	82	3.15	0	15	6	7	16	62	15	49
	59	2.82	9	6	6	7	8	13	17	3
	61	2.74	4	11	3	10	15	32	16	26
	8	2.70	5	10	3	10	14	24	16	11
	38	2.70	3	12	3	10	16	35	16	38
	5	2.25	8	7	0	13	14	8	17	49
	76	2.25	0	15	6	7	16	62	17	49
	77	2.25	0	15	6	7	16	62	17	49
	64	1.92	8	7	4	9	11	25	19	19

touch defect can be caused in the absence of PVC neurons [51]. From Table 7.7, the AVER has the largest I^{score} value 54.90, the in and out-degree of AVER are 33 and 18, which are not the largest. However, from our investigation, the AVER is the most important node in the CEN, which demonstrates that the I^{score} is different from the degree measures. The PR fails to identify most of the command interneurons as even among the top-50 level. The betweenness ranks many of the command interneurons out of the top-20 level. The results for the CEN indicate I^{score} can help to identify the actual important nodes.

Table 7.10 Clusters, members, rankings, and statistical characteristics of the identified top-30 ranked nodes in the HST

Group	Node	I^{score}	Out-deg.	R_{out}	In-deg.	R_{in}	R_{total}	R_{p}	R_{mc}	R_{bet}
G_1	120: MKK4	249.37	5	6	12	1	2	25	–	6
G_2	65:ERK	150.31	12	1	10	2	1	5	–	5
	123:MKK7	179.63	3	8	12	1	4	50	–	8
G_3	129:ASK1	114.55	6	5	5	6	6	21	–	3
	121:MKK5	106.32	9	2	7	4	3	10	–	11
	40:JNK	97.49	6	5	6	5	5	38	–	9
	113:JNK3	94.40	6	5	5	6	6	38	–	12
	116:JNK2	94.40	6	5	5	6	6	38	–	12
	149:p38beta	84.36	8	3	4	7	5	19	–	19
G_4	109:MSK2	59.39	3	8	4	7	9	69	–	28
	64:ERK1	57.87	6	5	6	5	5	26	–	17
	118:MKK2	47.15	2	9	4	7	10	62	–	15
	212:TAK1	46.36	4	7	2	9	10	32	–	21
	196:SHC	43.10	2	9	6	5	8	22	–	4
	108:MSK1	40.92	3	8	3	8	10	69	–	40
	24:B-Raf	37.60	3	8	2	9	11	44	–	7
	119:MKK3	36.33	3	8	2	9	11	39	–	22
	128:MEKK4	35.13	2	9	3	8	11	83	–	39
	111:MNK1	34.38	1	10	4	7	11	89	–	49
	112:MNK2	34.38	1	10	4	7	11	89	–	49
	14:ATF2	33.00	0	11	3	8	13	92	–	71
	142:NFAT4	33.00	0	11	3	8	13	92	–	71
	189:RNPK	33.00	0	11	3	8	13	92	–	71
	221: p53	33.00	0	11	3	8	13	92	–	71
	224: c-JUN	33.00	0	11	3	8	13	92	–	71
	126:MEKK2	31.87	3	8	1	10	12	61	–	38
	127:MEKK3	31.87	3	8	1	10	12	61	–	38
	131:MLK1	31.77	2	9	2	9	12	83	–	61
	132:MLK2	31.77	2	9	2	9	12	83	–	61
	133:MLK3	31.77	2	9	2	9	12	83	–	61

For the ECT, the identified top-30 nodes are shown in Table 7.8. In 2003, Martínez–Antonio et al. [52] identified global regulators in an ECT network. There are 18 global regulators in the network, namely, CRP, IHF, FNR, fis, arcA, lrp, hns, narL, ompR, fur, phoB, cpxR, soxR, soxS, mlc, cspA, rob, purR. Among which, the CRP, FNR, IHF, fis, arcA, narL, lrp are seven key regulators, which can regulate the expression of 51% of genes in *E. coli* [52]. From I^{score}, eight of the top-12 nodes (CRP, FNR, arcA, IHF, fis, narL, lrp, fur) are global regulators. The in-degree ranks most of the eight global regulators at the tail. The out-degree and total degree rank most of the eight global regulators at the top-10 level. According to the PR, motif

Table 7.11 Clusters, members, rankings, and statistical characteristics of the identified top-30 ranked nodes in the YT

Group	Node	I^{score}	Out-deg.	R_{out}	In-deg.	R_{in}	R_{total}	R_p	R_{mc}	R_{bet}
G_1	553:STE12	489.54	71	1	0	11	1	1	10	19
	575:TEC1	482.02	44	3	0	11	3	7	10	19
G_2	360:MSN2	363.05	35	6	0	11	6	14	10	19
	361:MSN4	356.90	32	7	0	11	7	16	10	19
G_3	622:YAP1	138.11	38	4	0	11	4	6	10	19
	513:SKN7	121.70	21	13	0	11	12	19	10	19
G_4	355:MIG1	58.87	26	9	0	11	9	4	7	19
	546:SSA4	49.23	0	32	4	7	28	100	10	19
	587:TKL2	49.23	0	32	4	7	28	100	10	19
	119:CTT1	45.81	0	32	6	5	26	100	10	19
	356:MIG2	45.79	12	20	0	11	20	31	9	19
	267:HSP78	42.39	0	32	4	7	28	100	10	19
	399:PGM2	42.39	0	32	6	5	26	100	10	19
	614:UME6	34.30	38	4	0	11	4	3	5	19
	264:HSP12	33.50	0	32	4	7	28	100	10	19
	152:DOG2	32.82	0	32	4	7	28	100	10	19
	523:SOD2	32.82	0	32	5	6	27	100	10	19
	592:TPS1	32.82	0	32	4	7	28	100	10	19
	651:YLR042C	32.82	0	32	4	7	28	100	10	19
	64:CAR2	30.08	0	32	6	5	26	100	10	19
	679:DAL81-DAL82	28.72	8	24	0	11	24	61	10	19
	263:HSP104	25.98	0	32	3	8	29	100	10	19
	265:HSP26	25.98	0	32	3	8	29	100	10	19
	364:MUC1	25.98	0	32	5	6	27	100	10	19
	545:SSA3	25.98	0	32	3	8	29	100	10	19
	100:CLN1	25.96	0	32	5	6	27	100	9	19
	59:BNI5	25.30	0	32	2	9	30	100	10	19
	143:DDR48	25.30	0	32	3	8	29	100	10	19
	145:DHH1	25.30	0	32	2	9	30	100	10	19
	208:GAT4	25.30	0	32	2	9	30	100	10	19

centrality, and betweenness, 2, 1, and 3 of the identified top-ranked global regulators are out of the top-10 level. The global regulator CRP is the most important node, which represents the cAMP receptor protein. The CRP can regulate cAMP, and genes regulated by the CRP are mostly involved in energy metabolism [53]. The CRP has the largest out-degree 496. But its in-degree is only 1. Though 280: csgE has the second largest in-degree 12, it is not top-30 ranked. From Table 7.8, the top-30 nodes can be classified into four clusters. The unimportant cluster contains the largest amount of nodes. The first three clusters are almost all global regulators.

The observations from the ECT indicate that the proposed measure can help to find global regulators.

For the YT, the top-30 nodes are shown in Table 7.11. STE12 and TEC1 are two most important nodes, with the I^{score} values 489.54 and 482.02, with the out-degree 71 and 44, and with the in-degree both 0. STE12 and TEC1 are two TFs. It has been reported that the STE12 controls two distinct developmental programs of mating and filamentation, therefore, it is a key regulator of cell fate determination [54]. Although the TEC1 gene has been reckoned as involving in the activation of expression of Tyl and the adjacent genes, it is not essential in the control of mating or sporulation processes [55]. It is intriguing to clarify why TEC1 is so frequently involved in network motifs and acts as building blocks of the YT network. From the results of the out-degree, total degree, PR, motif centrality, and betweenness, most of the nodes in G_4 are equally important, and thus have great differences from I^{score}.

7.3.6 Performance Evaluation Based on ROC Curves

To evaluate the performance of I^{score}, we perform Receiver Operating Characteristic (ROC) analysis. ROC curve is frequently used to evaluate the performance of a new test in the field of signal processing and medical diagnostic tests [56]. For a network with n nodes, the procedures of ROC analysis are as follows. Suppose the nodes can be classified into two groups: important and unimportant, and we know the actual classification. For a new index, the n nodes are with values in the interval $[a, b]$, for any threshold value $T \in [a, b]$, one can reclassify the n nodes into two classes. Comparing the actual classification with the new classification, several indexes can measure the accuracy of the new index, which are defined as follows [56]:

$$P_1 = \frac{n_2}{n_2 + n_4}, \tag{7.11}$$

$$P_2 = \frac{n_1}{n_1 + n_3}, \tag{7.12}$$

$$P_3 = \frac{n_1 + n_4}{n}, \tag{7.13}$$

where n_2 denotes the number of false positive nodes, which are considered important in the new classification but actually unimportant. n_4 gives the number of true negative nodes, where the nodes are both unimportant in the two classifications. Similarly, n_1 and n_3 denote the number of true positive and false negative nodes, respectively. P_1, P_2 are therefore called false and true positive rates, respectively. P_3 is called the accuracy of the new index. Given a T, one obtains a point (P_1, P_2). For $T \in [a, b]$, plotting the corresponding points in two-dimensional space, we derive the ROC curve. The area under the ROC curve (AUC) equals the probability that a classifier will rank a randomly chosen positive instance higher than a randomly chosen negative one [56], which can reflect the identification accuracy of the new index. The larger AUC, the more accurate of the index. Furthermore, the point in the upper left corner of a curve corresponds to the optimal threshold T, which gives the new classification of nodes with the highest P_3.

Hereinafter, based on the available information of some of the investigated networks and ROC curves, we evaluate the performance of I^{score} and the other indexes. In the following, for simplicity, we transform node ranks into fractional ones (range in $(0, 1]$). For nodes with rank k, its fractional ranks are the ratio of the number of nodes with ranks no more than k to n. Obviously, nodes with smaller fractional ranks are more important. For the CEN, on one hand, we have mentioned that the 10 command interneurons are known to be very important. If we take them as important nodes, one derives the ROC curves for each index, as shown in Fig. 7.8a. From Fig. 7.8a, the in-degree, total degree, I^{score}, and motif centrality all can well identify the command interneurons, the AUC (trapezoidal method) for these indexes are 0.9991, 0.9985, 0.9974, 0.9967, which are all above 0.99. The I^{score} is a little better than the motif centrality. The out-degree, PR, and betweenness are all worse than the other indexes. On the other hand, neurons in the *C. elegans* can be classified into interneurons, motor neurons, sensory neurons, where 117 neurons function as interneurons. If we take the 117 interneurons as important nodes, one obtains another ROC curve for each index, as shown in Fig. 7.8b, where all the seven measures have roughly similar performance. The I^{score} is a little better than the out-degree, in-degree, PR, and betweenness. For the ECT, there are 7 key and totally 18 global regulators, which are actually important in the network. If we take the 7 key global regulators and 18 global regulators as actually important nodes, we derive two ROC curves for each index, as shown in Fig. 7.8c, d. In Fig. 7.8c, the AUC for the seven indexes are 0.9996, 0.4385, 0.9996, 0.9997, 0.9983, 0.9987, and 0.9239. Except for the in-degree and betweenness, all the indexes can well identify the key global regulators. I^{score} is a little better than the other indexes. From Fig. 7.8d, the out-degree, total degree, PR, and motif centrality are with quite large AUC. The AUC for the I^{score} is 0.8628, which is only higher than that for the in-degree and betweenness, however, when $T = 0.0036$, the I^{score} can classify the nodes in the ECT with $P_3 = 99.30\%$.

For many biological networks, the actual classifications, known as gold standards, are not available. Fortunately, researchers have proposed several methods to evaluate the new test, such as constructing composite reference standards from available multiple tests [57, 58]. A single ranking from either the in, out, total degrees, PR, motif centrality, or betweenness is imperfect and cannot act as a gold standard. Subsequently, for each network, we construct a composite reference standard based on the six rankings (Five in the HST), and evaluate the accuracy of I^{score}. Specifically, in the composite reference standard, a node is defined as important if either one of the six rankings is among the top-T_0, where T_0 is a threshold, which can be taken as 10%, 20%, and so on. Thus, given a T_0, we derive a dichotomous reference classification of nodes in the network, either

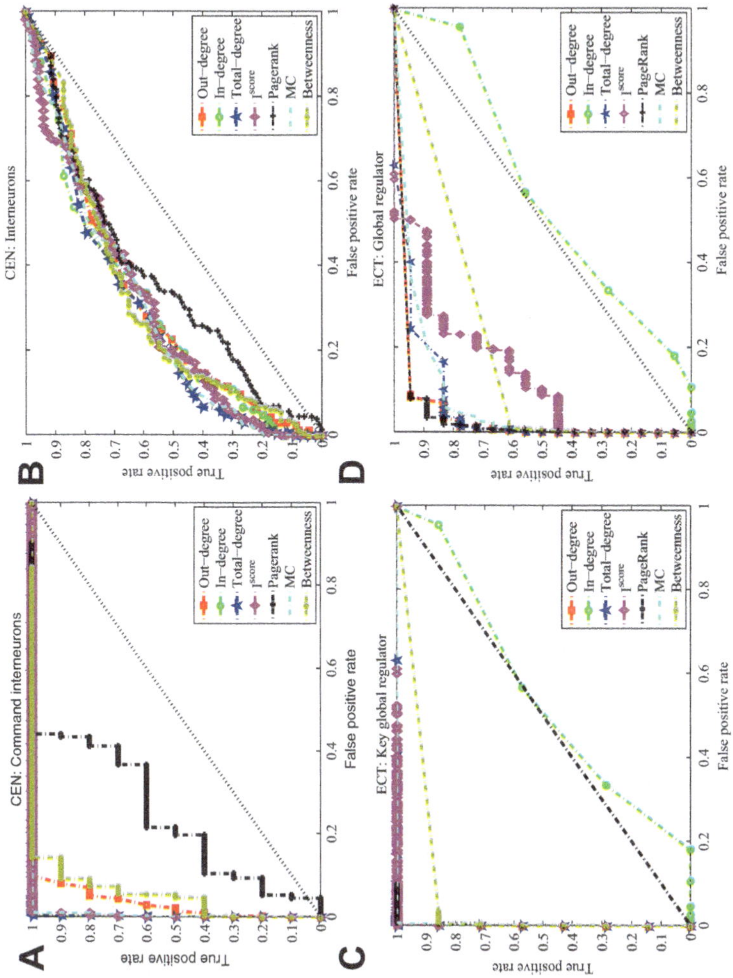

Fig. 7.8 ROC curves based on the available information in the CEN and ECT. **(a)** Performance of different indexes in identifying **(a)** the 10 command interneurons in the CEN, **(b)** the 117 interneurons in the CEN, **(c)** the 7 key global regulators in the ECT, and **(d)** the 18 global regulators in the ECT. Reprinted from Ref. [37]

positive (important) or negative (unimportant). According to the ranking from the I^{score}, we take several threshold values T to reclassify nodes, and finally derive the ROC curves for each network, as shown in Fig. 7.9. Figure 7.9a, b show the cases with $T_0 = 10\%$ and $T_0 = 20\%$, respectively. In Fig. 7.9a, the AUCs

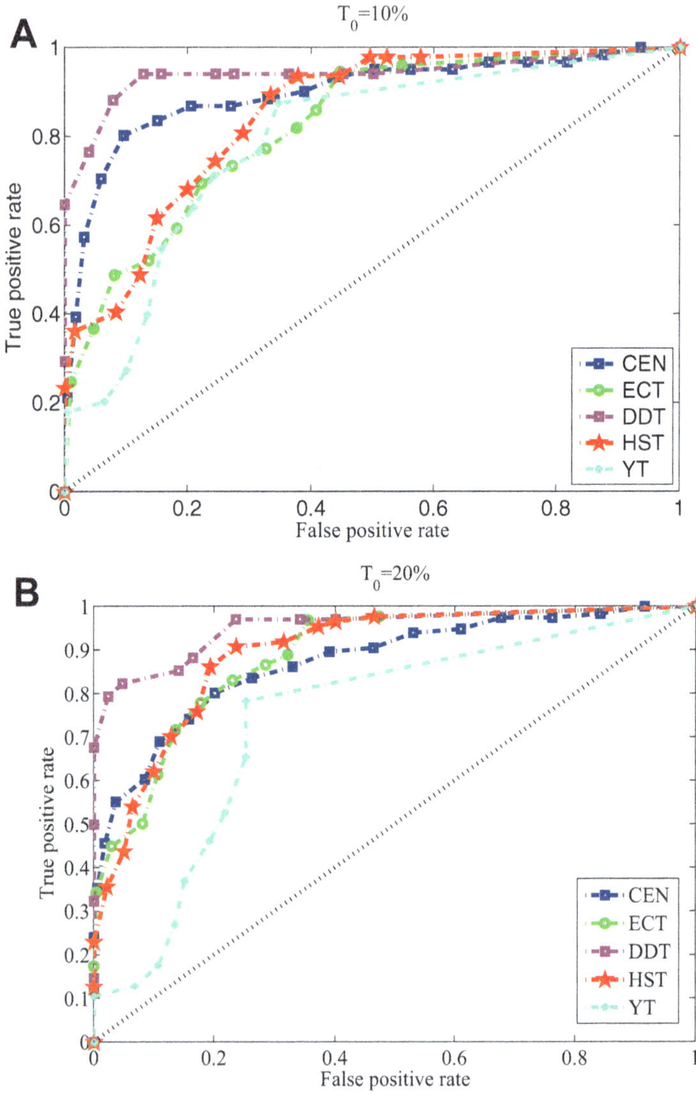

Fig. 7.9 Evaluation of I^{score} via ROC curves with composite reference standards for the five networks. (**a**) $T_0 = 10\%$. A node is defined as important if either its rankings by the in, out, total degree, PR, motif centrality, or the betweenness are at the top-T_0 level. (**b**) Similarly to (**a**), but with $T_0 = 20\%$. Reprinted from Ref. [37]

for the five networks are 0.8977, 0.8237, 0.9406, 0.8499, and 0.7878, respectively. The points in the upper left corner of the ROC curves in Fig. 7.9a correspond to $T = 20\%, 5\%, 10\%, 10\%, 5\%$, which lead to the highest P_3. For example, for the DDT, when the top-10% nodes are classified as important ones, the classification from the I^{score} has the best consistency with the reference classification, the P_3 can achieve 94.96%. For $T_0 = 20\%$, the AUCs for the five networks are 0.8740, 0.8884, 0.9521, 0.8955, and 0.7418, respectively. Under two different T_0 and for different networks, the AUCs are all above 0.74. Especially, in the DDT, the AUC is above 0.94, which indicates high identification accuracy of the proposed measure.

It is noted that, for the DDT, HST, and YT, since we still do not know how many nodes are actually important, it is difficult to compare among different measures via ROC curves. We also note that the ROC analysis without gold standards may subject to bias of the composite reference standard. However, since the composite reference standards for the five networks are based on six or five existing measures, it is trustworthy to treat them as reference standards. In conclusion, ROC analysis indicates the proposed measure is a remarkable alternative index to identify structurally important nodes in directed networks.

7.3.7 Topological Neighborhoods of Several Special Nodes

From the ROC analysis in the CEN and ECT, some measures are better than the I^{score} in identifying the command interneurons or global regulators. Hereinafter, through the specific analysis on topological neighborhoods of several nodes, we further illustrate the merits of the proposed measure. According to I^{score}, some hubs may be not important, whereas some non-hub nodes may be identified as very important ones. There are many highly connected but not highly ranked nodes, such as 946: soxs in the ECT; 22: b-catenin and 68: fak in the HST; 209: GCN4 and 332: MBP1-SWI6 in the YT. Examples of nodes with low degrees but ranked at top-20 include 333: cysG and 534-536:nirB-nirD in the ECT; 546: SSA4 and 587:TKL2 in the YT.

In the following, we take node 209 and 546 in the YT as two representative examples. Node 209 has out-degree 53 and in-degree 0, which is the second most important node according to the out and total degree, while its ranking is 62 according to I^{score}. Node 546 is with the in and total degree 4, the ranking is 28 according to the total degree, but it is ranked as the eighth most important node by the I^{score}. Figure 7.10a, b visualize the topological neighborhoods of the two nodes with their nearest and second nearest neighbors. From the topological neighborhoods of the two nodes, there are 81 nodes involved in the neighborhood of node 209, which are connected by 111 directed edges that centered at node 209, while 114 nodes and totally 182 directed edges consist of the neighborhood of node 546. The connection density of the neighborhood of node 209 is much lower than node 546. Moreover, from Fig. 7.10b, one can easily see that node 546 is directly regulated by four hub nodes and acts as a bridge or bottleneck of the

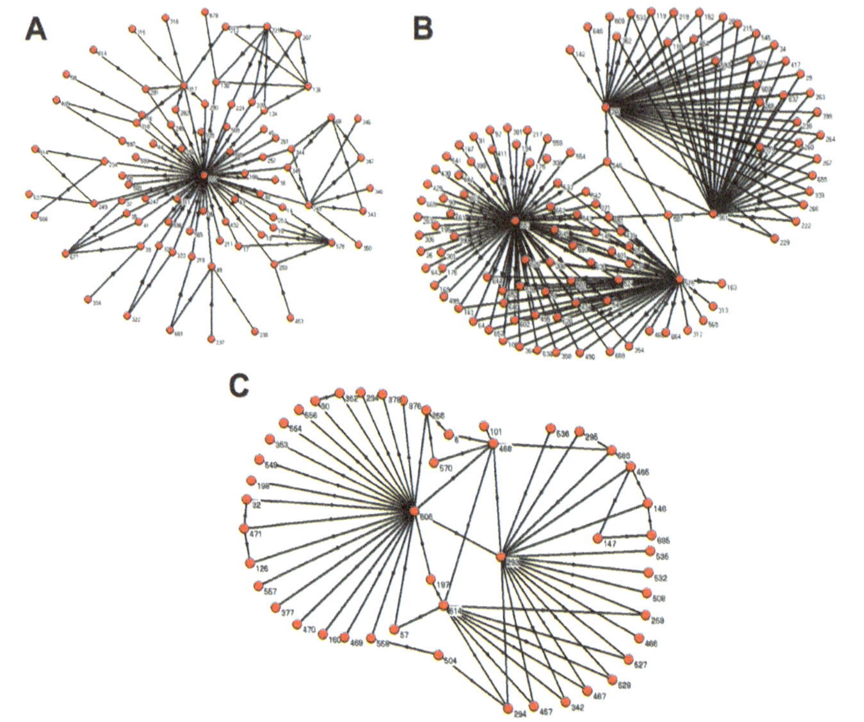

Fig. 7.10 Topological neighborhoods of several nodes. (**a**) Topological neighborhood of a hub but not top-ranked node: node 209 in the YT. (**b**) Topological neighborhood of a non-hub but top-ranked node: node 546 in the YT. (**c**) Topological neighborhood of a not top-ranked node but with the highest betweenness: node 293 in the YT. Reprinted from Ref. [37]

topological neighborhood. More importantly, the four hub neighbors of node 546 are just the top-4 nodes. Though node 209 can regulate 53 nodes, but its neighbors are neither hubs nor important nodes. Furthermore, node 546 involves in 1203 bi-fan subgraphs in its topological neighborhood, while there are only 39 such subgraphs for node 209, which indicate node 546 may play more functional roles in the system. Therefore, node 546 may be more important than 209. Finally, from the roles of biological functions, node 209 represents GCN4. It has been found that the GCN4 gene is conserved in *S. cerevisiae, K. lactis*, and *E. gossypii* [59]. SSA4 is widely conserved in human, *chimpanzee, Rhesus monkey, dog, cow, mouse, rat, chicken, zebrafish, fruit fly, C. elegans, S. cerevisiae*, and *A. thaliana* [59]. The cross species conservation of a gene indicates that it has been maintained by evolution despite speciation. It has been widely believed that mutation in a highly conserved gene can lead to a non-viable life form, or a form that is eliminated through natural selection [59, 60]. SSA4 is more widely cross-species conserved, which also indicates that SSA4 is more important than GCN4. Summing up, it is sufficient that the non-hub node 546 is actually more important than the hub node 209.

From the above analysis, it seems that node 546 similarly functions as nodes with high betweenness in undirected networks. However, we note that there are great differences between directed and undirected networks. In the YT, the node 209 has only 53 outgoing edges and the node 546 has only 4 ingoing edges, the betweenness [19] of the two nodes are both zeros, which are the least important nodes according to it. Therefore, the I^{score} is different from the classical betweenness. Furthermore, since the YT is a directed network, the betweenness of 96.06%(658/685) nodes are zeros, it fails to act as an effective ranking measure. It is noted that node 293: IME1 has the largest betweenness in the YT, whereas, it is not highly ranked according to I^{score}. Figure 7.10c shows the topological neighborhood of node 293. Node 293 is with 5 ingoing and 13 outgoing edges, but it is not frequently involved in network motifs. In conclusion, from the topological neighborhoods of several concrete nodes, we can further conclude that the proposed measure has its merits.

7.3.8 Some Further Issues of the New Measure

Biological networks are typical real-world complex networks. It has been reported that a single measure is insufficient to distinguish lethal nodes clearly from viable ones in some biological networks [27, 61]. Therefore, it is intriguing to find some more effective measures to characterize node differences in biological networks. In this subsection, based on the integration of the occurrences of each node in 2, 3, and some 4-node network motifs, we have proposed a new measure to characterize node importance in biological networks. Based on ROC curves and the analysis of the topological neighborhoods of several specific nodes, we have compared the obtained results with that from the degree, PR, motif centrality, and betweenness.

In the CEN and YT, when the command interneurons, interneurons, key global regulators, and global regulators are treated as actually important nodes, we compared the performance among different measures. The proposed measure has good performance in the two networks. The in-degree is good at identifying command interneurons in the CEN, but it is bad at finding global regulators in the ECT. The out-degree displays the contrary tendency as the in-degree. Though the PR can effectively identify the global regulators in the ECT, it is the worst measure in identifying command interneurons or interneurons in the CEN. Similarly, the betweenness is also not a good measure in the two networks. Therefore, the in-degree, out-degree, PR, and betweenness are not robust indicators of important nodes in different networks. The I^{score} provides an alternative robust measure for different types of biological networks.

Since the current knowledge on the five networks is limited, we note that it is still an open problem to further mining the advantages of the new measure. The number of command interneurons in the CEN and global regulators in the ECT are much fewer than the network sizes, the ROC analysis may suffer the effect of noise both in the interaction data and computation processes. We note that some other approaches may be used to further investigate the merits of the new measure, such as rich-cub

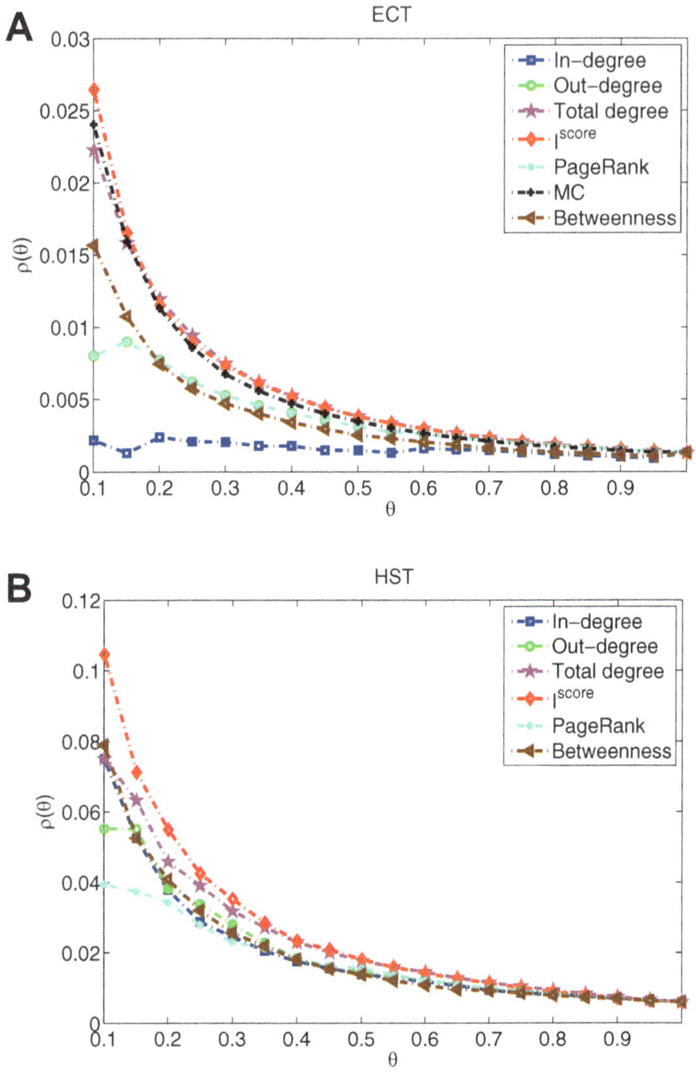

Fig. 7.11 The curves of connectivity density $\rho(\theta)$ against θ for different ranking measures in the ECT and HST. Reprinted from Ref. [37]

analysis [8, 62–66]. For simplicity, we simply examine the connectivity densities among the same amount of top-ranked nodes according to different measures in the ECT and HST, as shown in Fig. 7.11. Here, $\rho(\theta)$ is defined as the ratio of the total actual number of edges to the maximum possible number of edges among the top-$100\theta\%$ nodes. In Fig. 7.11, different curves correspond to different indexes. From Fig. 7.11, we can see that for many indexes, top-ranked nodes tend to be with higher connectivity densities than nodes ranked at the tail. The motif centrality fails to work in the HST, since the FFL is not a motif in such network. Moreover, comparing among different indexes, the I^{score} is very good at finding the cluster with high connectivity densities. That is, the connectivity density among a few motif-rich nodes is higher than the same number of top-ranked nodes by the other indexes. For example, in the HST, the connectivity density among the top-10% ($\theta = 0.1$) motif-rich nodes is above 0.10, while the top-10% large-degree nodes are with $\rho(\theta)$ below 0.08.

It has been reported that many bio-molecular networks are disassortative, which have negative PCCs [34]. For example, the PCCs of the CEN and YT are -0.0537 and -0.3496. The disassortativity indicates that large-degree nodes would connect with low-degree ones rather than with each other. Whereas, nodes with high I^{score} involve in many network motifs. Motif-rich nodes tend to form small connected subgraphs. Thus, the I^{score} may be helpful to find clusters with high connectivity density in disassortative networks.

Finally, we note that this subsection only considers five real-world biological networks, it is intriguing to further investigate the performance of the I^{score} in some artificial networks, such as artificial SF, SW networks, and networks with community structures. It is noted that for networks with large cliques at the periphery, nodes in the cliques may have very high I^{score} values, and therefore, these nodes may be highly ranked. Therefore, for such networks, the identified highly ranked clusters are probably just the large cliques. We will further investigate the related questions in our future works.

7.3.9 Summary

In this subsection, based on network motifs and multivariate statistical analysis, we have proposed a novel measure to characterize node importance in directed biological networks. The new measure enables us to further mining undiscovered characteristics of nodes in directed biological networks. Through the new measure, we have investigated five real-world biological networks, which include a neural network, three transcriptional regulatory networks, and one signal transduction network. These networks vary in sizes and link densities, and consist of various types of network motifs.

Based on the proposed measure, we have identified important nodes in the five networks. Our investigations suggest that the most important nodes in biological networks only take up a small fraction, but many of them are with important biological functions in real-world biological systems. Moreover, ROC analysis reveals that the proposed measure is a rather stable indicator of important nodes, and with very high prediction accuracy. Furthermore, the proposed measure can well characterize non-hub but very evolutionary conserved functional important nodes, and simultaneously, exclude hubs but not so functionally important nodes from the top rankings. Finally, we have discussed that the proposed measure may be used to reveal clusters with high connectivity density in disassortative networks. From these statistical analysis, we conclude that the proposed measure has some unique merits and it can be acted as an alternative network metric.

Although we have mainly investigated some directed biological networks, the proposed measure can be extended to some other networks, such as electrical networks, social networks. It is also noted that the proposed measure can be extended to involve more types of network motifs, but with the increasing of motifs, the computational complexity will be increased. Moreover, if the FFL is the unique network motif in a directed network, the proposed method will degenerate into the motif centrality [25]. Lastly, we note that this subsection provides an alternative way to characterize node features, it is still an open problem to find more effective ranking measures for nodes in directed biological networks, since it is generally difficult to obtain the actual rankings and a single measure is often insufficient to perfectly characterize all nodes. The related researches can help us to identify the actual key nodes in real-world systems. Real-world implications of identifying the key nodes include the finding of network control and regulation targets. For example, we can explore disease-associated or essential genes in cellular networks [67–70] for pharmacological or re-engineering purpose.

7.4 An Integrative Statistical Measure for Undirected Networks

7.4.1 Real-World PPI Networks

With the development of experimental technology for measuring interactions among proteins, such as the Y2H, the high-throughput affinity purification/mass spectrometry (AP/MS), and the yellow fluorescent protein complementation assay [71–73], a great quantity of data for PPIs of many organisms have been collected, and there have been many available online databases, such as BioGRID [74], YPD [75], MIPS [76], DIP [77]. Hereinafter, we consider the PPI networks for the *yeast Saccharomyces cerevisiae*. We construct four networks with network sizes $n = 907, 1640, 1825$, and 2675, respectively. The first three networks are constructed by following the work of Schwikowski et al. [71]. The fourth network is based on the third one, and further includes 850 new proteins and 1933 interactions, which was reported by Yu et al. [72]. It is noted that the first network is based on the MIPS database, the second network further considers 917 proteins and 808 interactions (some proteins and interactions are the same as the first network) reported by Uetz et al. [78]. The third network is reported by Schwikowski et al. [71], which is based on the second one and further includes 185 proteins and their interactions from Ito et al. [79] and the DIP Database [77]. The obtained four original networks are all disconnected, their LCCs contain 508, 1130, 1297, and 2307 nodes, and 796, 1619, 1862, and 3957 undirected edges, respectively (self-loops have been deleted). In this subsection, we mainly investigate the LCCs. The four connected networks consist

Table 7.12 Statistical characteristics of four real-world (above) and four artificial networks (below)

n	$\langle k \rangle$	$k_{\max}$	$k_{\min}$	APL	CC	PCC	PLE
508	3.1339	28	1	8.2249	0.1867	−0.0957	2.1100
1130	2.8655	29	1	7.7299	0.1254	−0.0538	2.6060
1297	2.8712	29	1	7.6270	0.1205	−0.0262	2.6720
2307	3.4304	89	1	5.6359	0.0935	−0.0785	2.5630
508	2.7795	45	1	6.1872	0.1688	−0.3615	2.2340
1130	2.9646	68	1	5.8609	0.1745	−0.2804	2.4580
1297	2.9915	72	1	5.7901	0.1680	−0.2677	2.4620
2307	3.0611	106	1	5.6218	0.1263	−0.2065	2.4470

of the evolving real-world PPI networks, with the former ones as subgraphs of the later ones.

Statistical characteristics of the four *yeast* networks are shown in the first four rows of Table 7.12, where $\langle k \rangle, k_{\max}, k_{\min}$ represent average degree, maximum and minimum node degree, respectively, APL denotes average path length [6], CC denotes average clustering coefficient [6] as defined in Chap. 1, PCC represents Pearson correlation coefficient, which can measure the disassortativity of the network [6, 80–85], PLE denotes the power-law exponent [6, 37, 86]. From Table 7.12, the average degree of the four networks are all around 3, which indicates the sparsity of the networks. The gaps between the largest degree and the smallest degree are very high, which provides clues for structural hierarchy. Moreover, the four networks have APL range from 5.6 to 8.2, and clustering coefficients range from 0.09 to 0.18. It is noted that, for random networks, clustering coefficient can be approximated by $\langle k \rangle/(n-1)$ [87], obviously, the cluster coefficients for real-world PPI networks are larger than their random counterparts. Therefore, APL and CC values indicate the SW property of these networks. Negative PCCs indicate the disassortativity [6, 85]. Finally, the degree distributions of the four PPI networks are all power-law, as shown in Fig. 7.12 (color online, the red circles, and downward sloping lines), the PLEs for the four networks are all around 2.5, which is consistent with the existing results in [6, 35, 78], where $PLEs$ for biological networks are reported to be between 2 and 3.

7.4.2 *Artificial PPI Networks*

Based the DD models, the artificial networks are generated according to [84], which have been described in Chaps. 2 and 6. We omit the details here. Hereinafter, we only consider the anti-preference duplication strategy.

We suppose $n_0 = 2$, that is, the network evolves from two ancestral nodes. For random duplication, from theoretical mean-field analysis by Pastor–Satorras et al.

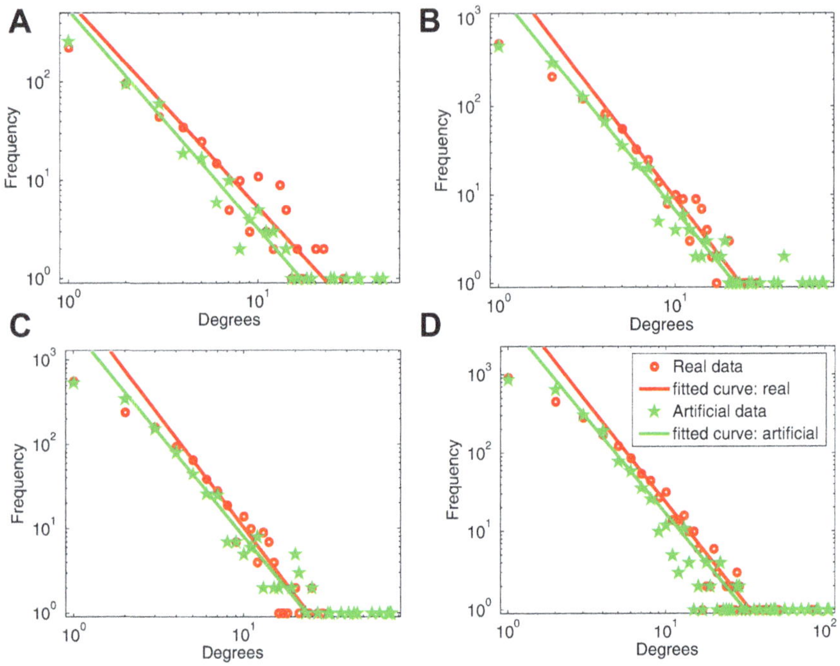

Fig. 7.12 Degree distribution of both real-world and artificial PPI networks. Network sizes are 508, 1130, 1297, and 2307. ©[2014] IEEE. Reprinted, with permission, from Ref. [46]

[35], the best parameter α is $\alpha = 0.562$, γ satisfies

$$\gamma = \frac{(\alpha - 0.5)\langle k \rangle}{n}, \tag{7.14}$$

where $\langle k \rangle$ is the average degree, n is the network size. But Pastor–Satorras et al. have not separated the dimerization processes from edge addition processes. Since dimerization between a duplicated node and its replica can happen with a greater probability [83], therefore, β should be greater than γ. Furthermore, we use the anti-preference duplication mechanism, which can strengthen the disassortativity [85]. In the following, we set $\alpha = 0.562$, $\beta = 0.1$, $\gamma = 0.000165$ for all the simulation runs. To keep the evolving structure of the artificial networks, when we generate networks with larger sizes, we take the newly generated smaller network as the initial network.

Statistical characteristics of four artificial PPI networks are presented in Table 7.12, with expected network size $n = 508, 1130, 1297, 2307$, respectively. From Table 7.12, the average node degree for the artificial networks ranges from 2.7795 to 3.0611, which are in good agreement with the real-world ones. The APLs are all around 6 and the clustering coefficients are all around 0.16, which are also well consistent with the real-world networks, and indicate the small world property

[37]. The PCC values are all negative, which indicates the disassortativity [85], the absolute value of PCC are a little larger than the real-world networks. The degree distributions of the artificial networks are also power-law, with exponents all approximate to 2.5. The degree distributions of the artificial networks with fitted curves are shown in Fig. 7.12 (Color online, green stars, and downward sloping lines). As one can see from Fig. 7.12, the degree distributions for the artificial networks are all well consistent with the real-world ones. To summarize, by comparing the statistical indexes between the real-world and artificial networks, we find that the artificial networks are in good agreement with the real-world ones. Furthermore, for the four artificial PPI networks, the average degrees and maximum degrees increase with the increasing of network sizes. The APL, the clustering coefficients, and the absolute values of PCC decrease with the increasing of network sizes. The evolutionary trends of these indexes are roughly the same as the real-world networks, which indicate that the parameters we have chosen in the DD model are biologically relevant, and one can use artificial networks to investigate the evolution characteristics of real-world ones.

7.4.3 Network Motifs in PPI Networks

mDraw (http://www.weizmann.ac.il/mcb/UriAlon) will be used for motif detection in the constructed real-world and artificial PPI networks. The detected motifs in the networks are summarized in Table 7.13, where, for each network, we have generated 100 random networks to compare with the investigated networks. In Table 7.13, the column n denotes network size; the column $Size$ denotes motif size, where we have considered 3-node and 4-node motifs; the column ID denotes the identification number of a motif, which is defined by the adjacency matrix of the motif (see the manual for mDraw); the column N_{real} represents the number of appearance of a subgraph in the investigated network; Z_{score} and U are defined as the former section. Following, subgraphs with $Z_{score} \geq 2$ and $U \geq 4$ are taken as network motifs.

Figure 7.13 shows the real-world *yeast* PPI networks with 1297 nodes, as well as the detected network motifs. From Table 7.13 and Fig. 7.13, we can see that both real-world and artificial PPI networks contain the 3-node motif 238 and the 4-node motif 13,278 and 31,710. Obviously, motifs in the real-world PPI networks are more abundant and more diverse than that in the artificial ones. In the artificial networks, subgraphs 4958 and 13,260 are not network motifs. However, the generated artificial networks can reflect most of the characteristics of the real-world ones.

Why are subgraphs 238 and 31,710 and others the building blocks of PPI networks? From evolutionary perspective, duplication and divergence are two basic mechanisms for biological evolution. During DD processes, on the one hand, many square structures can be produced by the duplication processes, since the target node and its replica share the same neighbors. On the other hand, the dimerization processes tend to bridge the target node and its replica with a probability, which can produce the triangular structure and 4-node fully connected structure. Therefore,

Table 7.13 Network motifs in the real-world (left) and artificial networks (right). $N_{r.}$, $Z_{s.}$ denote N_{real}, Z_{score}

			Real-world networks			Artificial networks		
n	Size	ID	N_{real}	Z_{score}	U	N_{real}	Z_{score}	U
508	3	238	667	105.99	39	141	5.81	30
	4	4958	2009	3.87	32			
		13,278	120	10.03	16			
		31,710	1381	2382.86	13	17	2.48	7
1130	3	238	776	141.39	62	359	9.14	62
	4	4958	3687	13.24	56			
		13,278	218	38.00	28	639	2.42	29
		31,710	1416	4922.79	18	47	4.04	13
1297	3	238	810	169.20	69	416	14.03	71
	4	4958	4200	20.17	63			
		13,260	151	5.73	28			
		13,278	274	78.84	32	722	2.55	34
		31,710	1425	10,127.41	19	58	4.60	15
2307	3	238	1072	66.62	116	560	11.19	118
	4	4958	18,007	4.18	107			
		13,260	3338	45.93	53			
		13,278	727	12.31	53			
		31,710	1457	921.66	27	53	3.30	18

one can conclude that the cooperation between duplication and dimerization bring about the motif 238 and 31,710, while random edge deletion and addition processes produce the other motifs.

7.4.4 The New Integrative Measure of Node Importance

By integrating some existing measures of node importance and based on the multivariate statistical analysis method [44, 45], we will propose a new integrative measure of node importance in PPI networks. Our main idea is as follows, firstly, we derive degree vector C_1, betweenness vector C_2, closeness vector C_3, clustering coefficient vector C_4, k-shell vector C_5, eigenvector centrality C_6, semi-local centrality vector C_7 for each network. Moreover, we count the appearances of each node in all network motifs and form another vector C_8, which can reflect the importance of each node in complex biological networks [15]. Secondly, based on the obtained $m = 8$ measures, similar to the proposed motif centrality in the above section, we perform PCA [44, 45, 88] and derive the first principal component

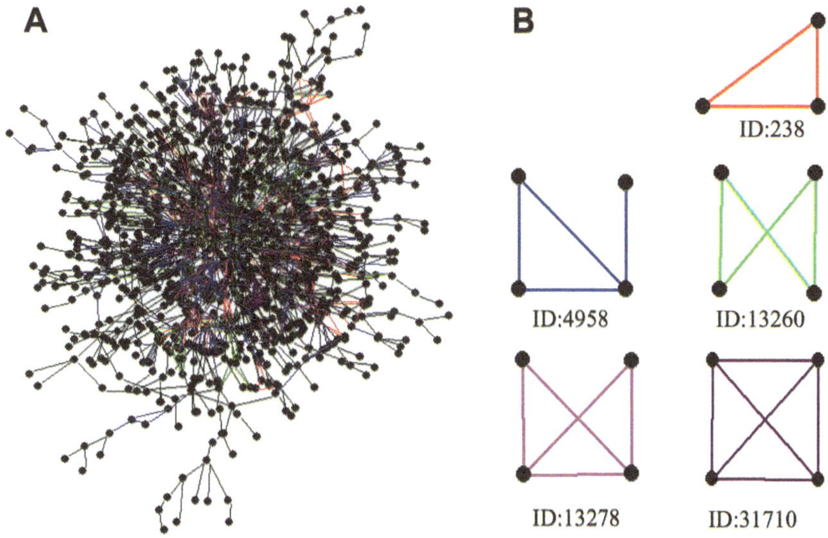

Fig. 7.13 The *yeast* PPI Network with 1297 nodes and motifs. Edges involved in network motifs in network (**a**) are marked out by colors as shown in (**b**) (Color online). ©[2014] IEEE. Reprinted, with permission, from Ref. [46]

(FPC) as a new integrative measure, which reads as follows:

$$FPC = \sum_{i=1}^{m} w_i C_i. \tag{7.15}$$

The FPC is actually a weighted sum of the m measures, the weight $w = (w_1, w_2, \ldots, w_m)^T$ is a unit vector, which is to be determined. It is noted that the FPC is the extension of the general summation operation. The weighted sum FPC is much better than the traditional direct sum, since different components often have different importance for determining the specific functions of networks. Existing works have reported that hubs are important nodes [89], some works suggest that nodes bridging different communities are important ones [34], while the other works have found that nodes with the highest semi-local [20] or k-shell [32] are more influential. From the PCA theory, the FPC can often be used to act as an ordering index. Therefore, we have extended the PCA theory to complex networks. The FPC integrates eight existing structural measures. The proteins with high FPC values are much more likely to be the structurally dominant nodes in the PPI networks. In this case, the proteins with high FPC values often play much more important roles than those of the proteins with low FPC values. Therefore, the proteins with high FPC values are dominant proteins for the functions of PPI networks. We call such important nodes as structurally dominant proteins (SDPs) or structurally dominant nodes (SDNs). The best index FPC should have the best discriminability among nodes, therefore, the variance of FPC should be as large as possible. Thus, w can be determined by following the rule of making the variance of FPC to realize maximization [44, 45, 88]. Take $\mathscr{C}_1, \ldots, \mathscr{C}_m$ as random variables, which represent the m measures of a node in complex networks, and the n dimensional values $C_1, \ldots, C_m$ from a concrete network as its n observations, denote $\mathscr{C} = (\mathscr{C}_1, \mathscr{C}_2, \ldots, \mathscr{C}_m)^T$ and its observation matrix as $C = (C_1, C_2, \ldots, C_m)_{n \times m}$. The covariance matrix of $\mathscr{C}$ can be estimated by its observation C. Denote the covariance matrix of C as Σ, then

$$COV(\mathscr{C}) \approx COV(C) = \Sigma = \frac{1}{n-1}\left(C^T C - n\overline{C}\,\overline{C}^T\right), \tag{7.16}$$

where $\overline{C}$ is the column mean vector of C, n is network size. Indeed, Σ is the unbiased estimator of $COV(\mathscr{C})$. Based on the above notations, we have the stochastic form of FPC as $\mathbf{FPC} = w^T \mathscr{C}$ and

$$Var(\mathbf{FPC}) = Var\left(w^T \mathscr{C}\right) = w^T COV(\mathscr{C}) w \approx w^T \Sigma w. \tag{7.17}$$

The unit vector w is therefore determined by the following optimization problem:

$$\begin{aligned} &\max\ w^T \Sigma w, \\ &s.t.\ \ w^T w = 1. \end{aligned} \tag{7.18}$$

By the Lagrangian multiplier method, let

$$L(w, \lambda) = w^T \Sigma w - \lambda\left(w^T w - 1\right), \tag{7.19}$$

the optimal w satisfies

$$\begin{cases} \frac{\partial L}{\partial w} = 2(\Sigma - \lambda I)w = 0, \\ \frac{\partial L}{\partial \lambda} = 1 - w^T w = 0. \end{cases} \tag{7.20}$$

Here, I is the identity matrix. The first sub-equation of Eq. (7.20) implies that $\Sigma w = \lambda w$, that is, λ and w are just the eigenvalue and eigenvector of Σ. Furthermore, we have $Var(\mathbf{FPC}) \approx w^T \Sigma w = \lambda w^T w = \lambda$. Therefore, the maximum value of $w^T \Sigma w$ equals the biggest eigenvalue of Σ, denoted as λ^*, then the optimal vector w^*, satisfying that $w^T \Sigma w$ reaches its maximum value, is just the corresponding unit eigenvector of λ^*.

The covariance matrix Σ is a real symmetric matrix, therefore, its eigenvalues are all non-negative. Denote $\lambda_1 \geq \lambda_2 \geq \cdots \geq \lambda_m \geq 0$ as the eigenvalues of Σ, then $\lambda^* = \lambda_1$. The index $\rho = \lambda_1 / \sum_{i=1}^{m} \lambda_i$ reflects the contribution ratio of the FPC, which can measure the amount of information in C extracted by the FPC [44, 45, 88].

Up to now, we have determined the integrative measure. Finally, for a concrete network, substituted the concrete values of C_i in Eq. (7.15), one derives the observation of **FPC** as FPC. Each node has a FPC value, based on the FPC value, one can identify SDNs in complex networks. Nodes with the largest FPC values are SDNs. It is noted that, if node v has larger $C_i(i = 1, 2, \ldots, m)$ than other nodes, then the FPC value must be also larger than other nodes. The integrative index FPC considers various structural characteristics of each node in a network, and obviously, it will be more reasonable than a single index.

7.4.5 Identifying Structurally Dominant Nodes in PPI Networks

For the eight networks in Sect. 7.4.2, the maximum eigenvalue of Σ, the contribution of the FPC, as well as the weight vector are obtained in Table 7.14. It is noted that values in matrix C have different orders of magnitudes, for example, the magnitude of the eigenvector centrality ranges from 10^{-1} to 10^{-13}, while the magnitude of the semi-local centrality ranges from 10 to 10^4. In order to avoid the malfunction of indexes with low order of magnitude, before our analysis, we have performed standardized transformation to each column of C.

From Table 7.14, we see that the maximum eigenvalues of Σ for each real-world network are around 4.5, while for artificial networks, the values are around 3.7. The contribution of the FPC for each real-world network is all above 55%, while the contribution is a little lower for artificial ones. This indicates that the variance among the eight indexes for the real-world networks is larger than that for the artificial ones, and the FPC index for the real-world networks can have better performance than for the artificial ones. For the real-world networks, the values for $w_i(i = 5, \ldots, 8, 1)$ are relatively larger than $w_j(j = 2, 3, 4)$. This indicates that the k-shell, the eigenvector centrality, the semi-local centrality, the network motif centrality, and the degree play more important roles than the other indexes in the

Table 7.14 Contribution of the FPC and weight vectors for the real-world and artificial PPI networks

Network	n	λ_1	ρ	w_1	w_2	w_3	w_4	w_5	w_6	w_7	w_8
Real.	508	4.5108	56.38%	0.3744	0.1313	0.1342	0.2488	0.4374	0.4126	0.4459	0.4498
	1130	4.4887	56.11%	0.3730	0.1986	0.2310	0.2115	0.4290	0.3876	0.4479	0.4341
	1297	4.4977	56.22%	0.3750	0.2120	0.2399	0.2041	0.4266	0.3844	0.4481	0.4297
	2307	4.4383	55.48%	0.4162	0.3558	0.3107	0.0910	0.3521	0.3596	0.4027	0.4242
Artificial	508	3.6011	45.01%	0.4502	0.3964	0.3052	0.1410	0.3727	0.2779	0.4121	0.3758
	1130	3.7939	47.42%	0.4217	0.3837	0.3574	0.0660	0.3228	0.3950	0.4345	0.3064
	1297	3.7103	46.38%	0.4285	0.3967	0.3590	0.0668	0.2955	0.3920	0.4333	0.3110
	2307	3.7494	46.87%	0.4304	0.4192	0.3558	0.0684	0.2778	0.3703	0.4119	0.3518

FPC. For the artificial networks, except w_4, there are no much differences among the other measures. Low value of w_4 indicates the clustering coefficient index has very weak contribution to the FPC.

For the real-world networks, by the FPC, the top-10 ranked nodes and their corresponding ranks by the other indexes are shown in the left column of Table 7.15, where we have listed the corresponding protein names and their ranks R_i by the measure $C_i (i = 1, \dots, 8)$. From Table 7.15, for each network, one can conclude that the rank by the integrative FPC index is different from all the eight single indexes. The $R_i (i = 5, \dots, 8, 1)$ have the best consistence with the R_F from the FPC measure, it indicates nodes with high $C_i (i = 5, \dots, 8, 1)$ are more prone to be SDNs. Indeed, this result just corresponds to the result in Table 7.14, where, the weight coefficients for $C_i (i = 5, \dots, 8, 1)$ are higher than the other indexes. By comparing the results among different real-world networks, more than 50% of the top-10 proteins are almost the same in different networks. With the evolution of network sizes, some nodes transit from not the top-10 to the top-10 ranked nodes, and even become the highest ranked nodes, such as GCD7, ATG17, SRP1, and SMT3. Whereas, a few nodes become less and less important, such as TRA1, TAF12, SPT7, and TAF5.

From the GO database [90], we can get more information about the functions of the proteins in the real-world networks. The twelve proteins: TAF10, ADA2, TRA1, NGG1, GCN5, SPT3, SPT20, SPT7, TAF5, TAF12, TAF9, TAF6 are all subunits of the SAGA (Spt-Ada-Gcn5-Acetyltransferase) complex, which perform similar functions or involved in the same biological processes. The SAGA complex is important for transcription in vivo and possesses histone acetylation function, mutations in SPT7 or SPT20 will disrupt the SAGA complex and cause severe phenotypes [91]. Furthermore, the SAGA complex also has important roles in transcript elongation, the regulation of protein stability, and telomere maintenance [92]. Bertolazzi et al. [89] reported that protein complexes are often evolutionary conserved. Here, with the evolution of networks, some of the SAGA complexes can keep to be the top-10 ranked proteins, therefore, they are functionally important nodes in the *yeast* PPI networks. For the rest proteins in Table 7.15, the reduction of GCD7 function will lead to decrease of competitive fitness and decrease of resistance to methyl methanesulfonate [90]. Over-expression of ATG17 inhibits filamentous growth. Mutation of ATG17 leads to bud morphology abnormal, the decrease of competitive Fitness, and so on [90]. Reduction of SRP1, SMT3, and MUK1 functions all lead to the decrease of competitive fitness, over-expression of SRP1 can resort to invasive growth increase and vegetative growth rate decrease [90], and over-expression of SMT3 can lead to decreased vegetative growth. Over-expression of MUK1 resorts to the abnormal vacuolar transport [90]. Obviously, most of the identified top-10 proteins are essential for the growth and reproduction of *yeast*.

For the artificial networks, the top-10 ranked nodes by the FPC and their corresponding ranks by the other indexes are obtained in the right column of Table 7.15, one can see that the ranks by the indexes with larger weight have better consistence with R_F. The clustering coefficient method is the most unusual, for

Table 7.15 The identified top-10 ranked nodes in the real-world (left) and artificial (right) PPI networks and the corresponding ranks by C_i

			Real-world networks									Artificial networks								
n	Nd	Prot.	R_F	R_1	R_2	R_3	R_4	R_5	R_6	R_7	R_8	Nd	R_1	R_2	R_3	R_4	R_5	R_6	R_7	R_8
	89	TAF10	1	6	53	149	84	4	1	2	1	15	1	2	31	152	16	1	2	5
	120	ADA2	2	8	118	150	69	7	2	3	3	2	2	1	1	156	9	54	1	14
	228	TRA1	3	13	18	25	65	10	7	1	5	14	3	4	143	150	15	10	20	11
	92	NGG1	4	7	86	153	83	5	6	7	2	22	4	5	2	155	20	82	7	27
508	214	GCN5	5	9	74	122	68	9	5	6	4	142	7	6	13	149	45	116	3	29
	114	SPT3	6	12	248	151	63	6	3	4	6	72	5	45	268	148	2	147	29	4
	451	SPT20	7	14	474	152	64	13	4	5	7	153	27	72	109	82	3	180	26	1
	27	SPT7	8	15	201	154	3	1	8	8	8	56	8	3	6	153	29	148	80	58
	41	TAF5	9	16	209	155	6	2	9	9	9	144	23	16	146	123	46	4	5	15
	87	TAF12	10	17	235	156	13	3	10	10	10	17	6	26	102	147	17	135	17	12
	121	TAF10	1	7	81	62	133	4	1	3	1	2	1	1	1	364	8	2	1	45
	158	ADA2	2	9	53	29	118	7	2	2	2	142	2	3	2	365	27	1	2	60
	125	NGG1	3	10	216	89	117	5	6	8	3	677	8	13	9	345	51	3	3	1
	518	TAF9	4	15	89	35	94	12	7	6	4	15	4	2	5	358	64	315	4	34
1130	304	TRA1	5	21	58	27	89	10	8	1	8	72	3	10	39	359	2	9	6	8
	287	GCN5	6	14	185	59	93	9	5	7	5	697	16	216	42	312	52	4	14	4
	233	TAF6	7	20	137	86	88	8	9	9	7	14	5	8	55	360	63	347	18	18
	151	SPT3	8	19	554	87	85	6	3	4	9	22	6	9	6	362	13	48	13	109
	591	SPT20	9	22	793	88	86	13	4	5	10	17	19	118	56	310	12	5	19	5
	119	TAF12	10	18	400	90	87	3	10	10	6	262	32	20	4	331	37	6	5	24

(continued)

Table 7.15 (continued)

			Real-world networks									Artificial networks								
n	Nd	Prot.	R_F	R_1	R_2	R_3	R_4	R_5	R_6	R_7	R_8	Nd	R_1	R_2	R_3	R_4	R_5	R_6	R_7	R_8
	127	TAF10	1	6	60	56	144	4	1	2	1	2	1	1	1	413	48	2	1	41
	164	ADA2	2	9	28	27	129	7	2	1	2	142	2	3	2	415	81	1	2	82
	131	NGG1	3	10	82	47	122	5	3	4	3	677	8	10	9	393	36	3	3	1
	542	TAF9	4	15	92	40	97	12	7	7	4	15	4	2	3	410	54	242	4	38
1297	315	TRA1	5	21	52	30	91	10	8	3	10	72	3	8	47	409	2	23	6	7
	298	GCN5	6	14	241	57	96	9	6	8	5	14	5	6	30	411	53	375	14	20
	157	SPT3	7	19	659	84	87	6	4	5	6	697	18	190	62	359	37	4	12	4
	619	SPT20	8	22	893	86	88	13	5	6	7	17	21	127	112	356	8	5	20	5
	243	TAF6	9	20	134	85	90	8	9	9	9	22	7	9	7	414	56	63	13	188
	125	TAF12	10	18	468	90	89	3	10	10	8	873	15	21	63	378	154	468	21	3
	1232	GCD7	1	1	1	1	540	61	15	1	1	2	1	1	1	636	42	1	1	6
	1252	ATG17	2	3	2	2	543	62	16	3	4	142	2	3	2	641	84	2	2	24
	662	SRP1	3	2	3	3	547	51	17	2	3	15	5	2	3	633	47	241	4	15
	189	ADA2	4	8	10	37	384	7	1	26	2	677	10	11	5	611	152	3	3	2
2307	151	NGG1	5	28	51	41	174	5	2	31	6	72	4	8	17	632	2	215	7	5
	147	TAF10	6	24	100	230	176	4	3	71	7	14	3	4	9	637	46	247	5	17
	200	SMT3	7	5	4	6	541	84	59	29	22	41	6	5	22	634	56	424	35	22
	622	TAF9	8	38	137	236	148	12	6	74	8	22	7	6	10	638	50	214	12	72
	1428	MUK1	9	17	34	8	470	67	20	6	5	150	13	13	20	623	87	8	11	39
	705	SPT20	10	51	270	246	122	13	4	69	9	13	16	7	13	619	8	290	19	46

Nd and Prot. are the abbreviation of node and protein, respectively each network, the clustering coefficients of the identified top-10 ranked nodes are all very small. However, by comparing the results among different artificial networks, we can conclude that with the evolution of network sizes, most of the top-10 ranked nodes can keep to be top-10.

Figure 7.14 shows the clustering dendrogram of top-100 ranked nodes for each networks. From Fig. 7.14, we can see that both the real-world and the artificial PPI networks have clear hierarchical structures. Nodes can be classified into clusters, with the most important cluster on the right. The most important cluster only contains a few nodes. A large number of nodes are not SDNs. The phenomenon just indicates the scale-free structural feature of these networks. Figure 7.15 displays the ranking-based rich-club phenomenon for the PPI networks with 2307 nodes, where $\rho(\theta)$ denotes the density of links among the top-$100\theta\%$ ranked nodes [20, 88]. In Fig. 7.15, we have shown the outcomes of the FPC ranking method and the other eight methods. The monotone decreasing of $\rho(\theta)$ with θ indicates rich-club ordering. By comparing between different methods, the clustering coefficient C_4 has the worst performance, the curves for the two networks are all not monotone decreasing, while the proposed FPC measure has very good performance, where top-ranked nodes are highly connected, and the curve from our method decreases faster than the other method.

Table 7.16 shows the correlation coefficients between FPC and C_i for all the networks. For real-world networks, the correlation coefficients between FPC and $C_i(i = 5, \ldots, 8, 1)$ are all very high, one can say that the FPC index is highly affected by these measures. Furthermore, the FPC index is the most relevant with the network motif centrality C_8, which results in the ranking similarity between the two measures. For artificial networks, the correlation coefficients between FPC and $C_i(i = 1, 2, 6, 7, 8)$ are larger than the others, which indicate nodes in artificial networks with larger degree, betweenness, eigenvector centrality, semi-local centrality, or network motif centrality are more prone to be SDNs. The correlation coefficient between FPC and C_4 is negative, which indicates the identified SDNs tend to be with low closeness values. As we can see from Table 7.15, the R_4 of the identified top-10 ranked nodes are all very large.

7.4.6 Evolution of Structurally Dominant Nodes in PPI Networks

Following, we discuss the evolutionary characteristics [93] of SDPs in PPI networks. For simplicity, we only consider the evolution of the top-10 nodes in networks with 508 nodes. Figure 7.16 shows the ranks of top-10 nodes in networks with 508 nodes and their ranks in larger networks. The line corresponding to $n = 508$ denotes the ranks in the network with size 508, and this line acts as a baseline. The ranks of the considered nodes in larger networks consist of another three curves. By comparing the cases between real-world and artificial networks, similar

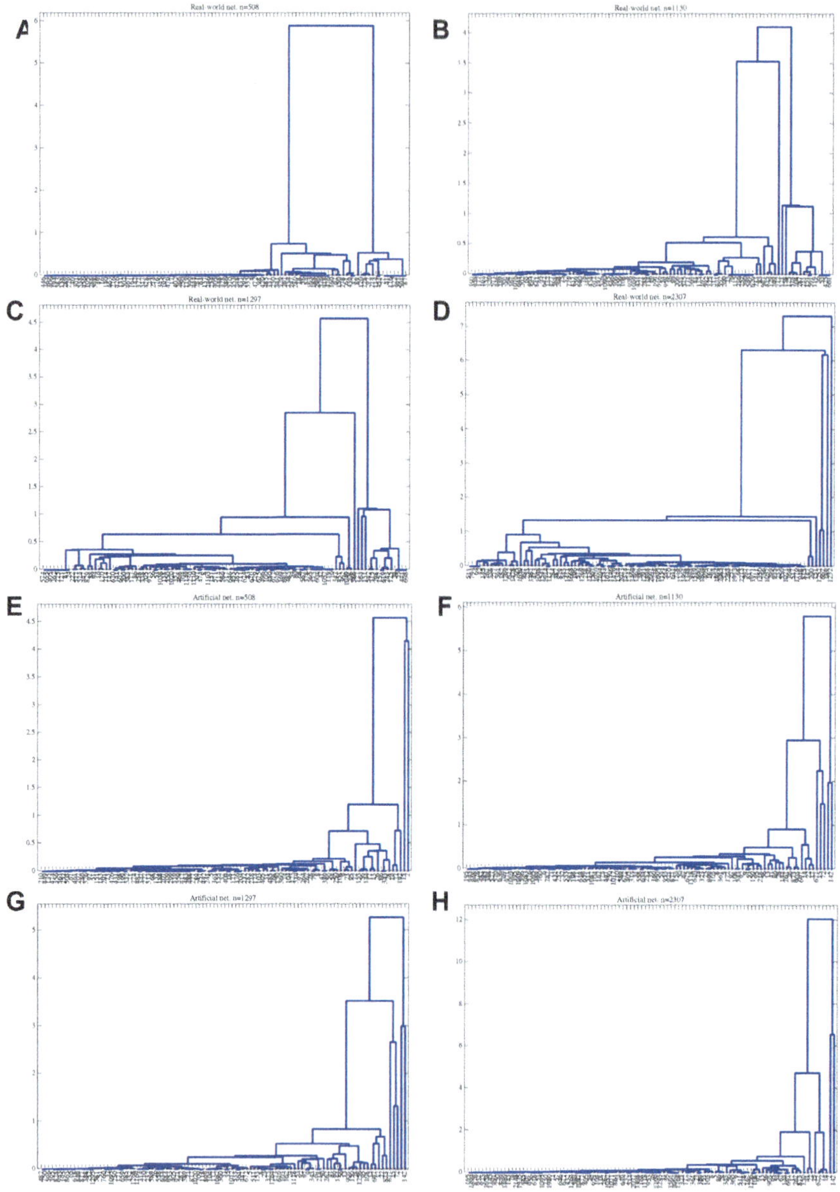

Fig. 7.14 Clustering dendrogram of the top-100 ranked nodes for the real-world and artificial networks. Panels (**a**)–(**d**) are corresponding to the results for real-world networks with 508, 1130, 1297 and 2307 nodes; (**e**)–(**h**) are corresponding to the cases in artificial networks. ©[2014] IEEE. Reprinted, with permission, from Ref. [46]

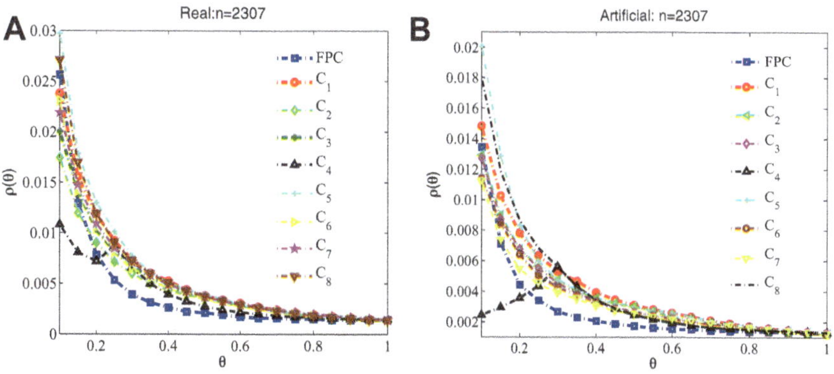

Fig. 7.15 Rich-club analysis by the FPC and the other ranking measures. (**a**) The case in the real-world network with 2307 nodes; (**b**) The case in the artificial networks with 2307 nodes. ©[2014] IEEE. Reprinted, with permission, from Ref. [46]

Table 7.16 Correlation coefficients between FPC and C_i

	Real-world networks				Artificial networks			
C_i	508	1130	1297	2307	508	1130	1297	2307
C_1	0.6364	0.6340	0.6314	0.7848	0.8219	0.8101	0.8130	0.8242
C_2	0.1315	0.2870	0.3193	0.7484	0.7634	0.7748	0.7914	0.8528
C_3	0.1069	0.1906	0.2019	0.2887	0.1795	0.2452	0.2392	0.2451
C_4	0.3188	0.2724	0.2559	0.0458	−0.048	−0.079	−0.069	−0.054
C_5	0.8410	0.8055	0.7882	0.5186	0.3796	0.3567	0.2752	0.2157
C_6	0.9209	0.8839	0.8810	0.7584	0.4457	0.6854	0.6798	0.6687
C_7	0.8885	0.8771	0.8834	0.7433	0.4983	0.5893	0.5746	0.5628
C_8	0.9516	0.9410	0.9355	0.9257	0.6111	0.5536	0.5622	0.6262

phenomenon can be observed. The curves for larger networks are mostly above the baseline, and therefore, we can say for PPI networks, roughly speaking, the ranks of most of the considered nodes drop with the increasing of network sizes. However, for the real-world networks, the ranks of proteins ADA2, NGG1, SPT20 have no much changes as compared with other proteins. The observations tell us that SDPs are evolving with the evolving of networks, some SDPs keep to be dominant ones, while a large amount of nodes become less and less dominant with the evolution of the PPI networks.

From the above section and the above investigations, we have concluded that the real-world and artificial networks have similar statistical characteristics, therefore, we can generate extensive artificial networks with different structures and sizes to infer the evolutionary characteristics of the real-world ones. To further investigate the evolution of SDPs, we randomly generate 5 sets of networks with sizes $n = 508, 800, 1130, 1297, 1700, 2307$, and further explore their evolutionary characteristics. By averaging over the simulation runs of these networks, Fig. 7.17 draws the evolution of ranks for the first 508 nodes in the first networks. The 508

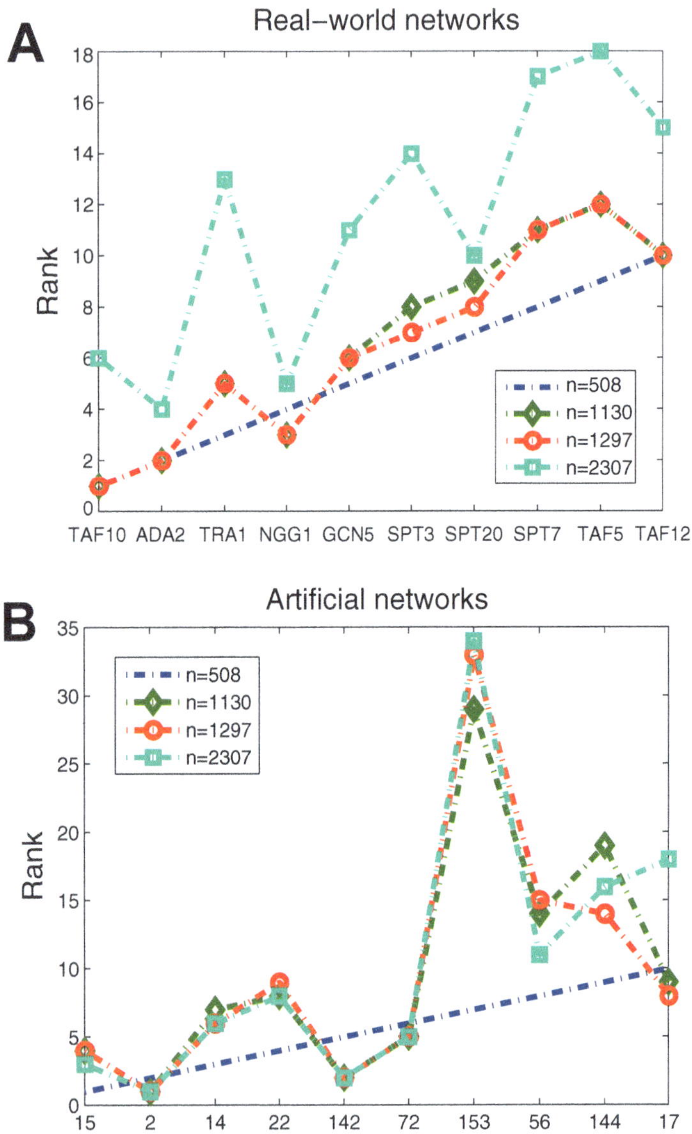

Fig. 7.16 Evolution of the top-10 SDPs in networks with 508 nodes. (**a**) and (**b**) show the cases in real-world networks and artificial networks respectively. ©[2014] IEEE. Reprinted, with permission, from Ref. [46]

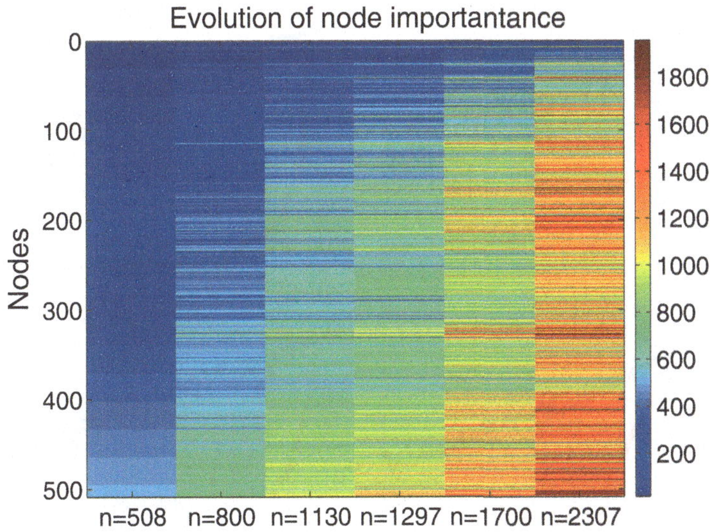

Fig. 7.17 Evolution of 508 nodes in artificial PPI networks. Averaged over 5 sets of networks.

nodes in the first network are sorted by their ranks in descending order, as described by the first column corresponding to $n = 508$. The rest columns show the ranks of the 508 nodes in larger networks. From Fig. 7.17, we can see that with the evolution of networks, only a few SDNs in smaller networks can keep to be dominant in larger networks, while for a large number of nodes, their ranks increase with network sizes, which indicate that they become less and less dominant. Furthermore, less dominant nodes in smaller networks can become less and less dominant in larger networks more rapidly, which indicates that SDNs evolve slower than non-dominant ones. From the review paper of Jancura and Marchiori [94], a generally accepted premise for the evolution of proteins in networks is that essential proteins should evolve at slower rates than non-essential ones. Interestingly, since we have found SDPs are prone to be essential ones, therefore our observations provide some evidences for the hypothesis.

7.4.7 Robustness Against Mutations

In 2000, Barabási et al. [2] found that SF complex networks have the property of random error tolerance and attack vulnerability. In 2001, it was found that PPI networks are lethal to the deletion of highly connected proteins [95]. Here, we investigate the robustness of PPI networks against targeted and random mutations. For targeted mutations, we consider both the removal of highly connected nodes and nodes with high FPC values. The curves of the fractions of the LCCs S versus the

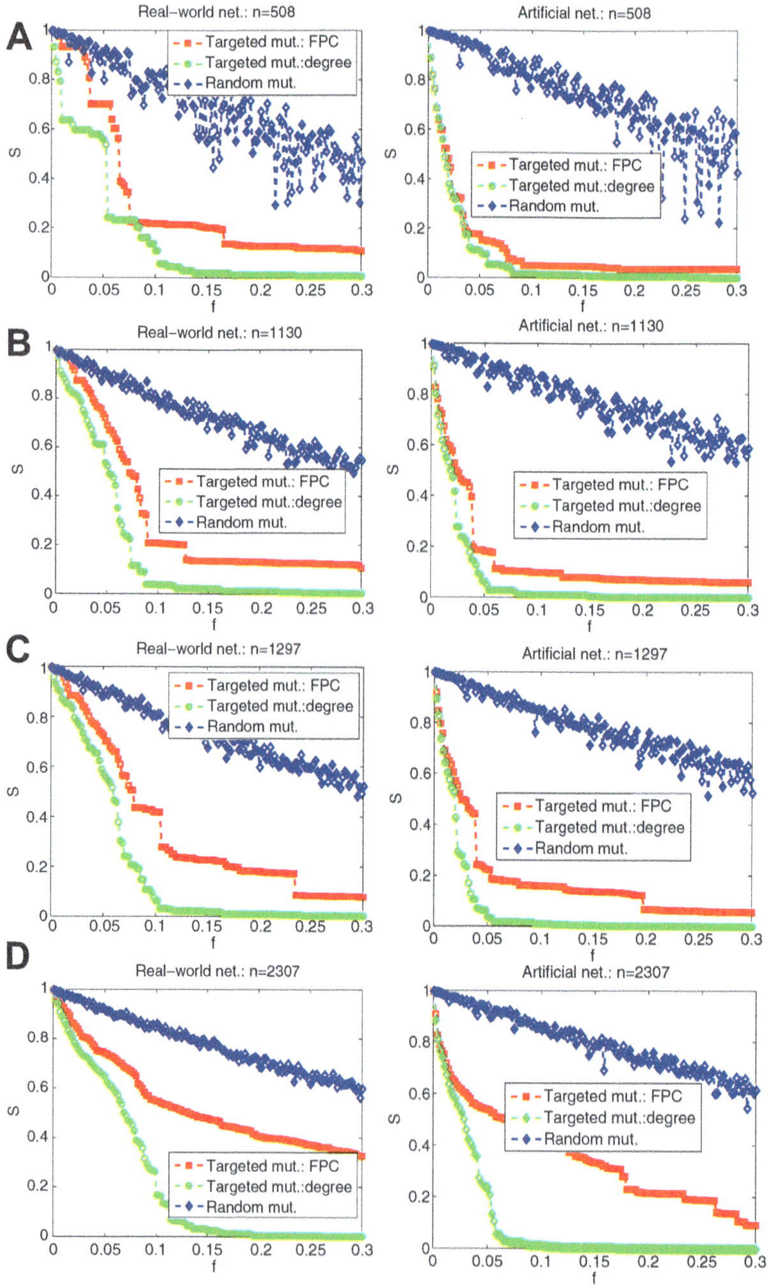

Fig. 7.18 Effect of targeted and random mutations on the PPI networks. The fractions of nodes in the LCCs S versus the fractions of the removed nodes f are plotted. Targeted mutations on proteins with large degree and with large FPC values have been considered. (**a**)–(**d**) show the cases with network sizes 508, 1130, 1297, and 2307, respectively. ©[2014] IEEE. Reprinted, with permission, from Ref. [46]

fractions of removed nodes f are frequently used to investigate the robustness of a network [2, 6]. In Fig. 7.18, we draw the curves for both the real-world and artificial networks as constructed in the above section.

From Fig. 7.18, we can conclude that the PPI networks are very robust to random mutations, while they are very fragile to targeted mutations. With the increasing of the fractions of randomly removed nodes, the curves S versus f decrease slowly, while with the increasing of the fractions of highly connected nodes or nodes with high FPC values, the sizes of the LCCs decrease quickly. An interesting finding is that for the real-world networks, by comparing the results between targeted mutations by FPC and degree, we find that there are certain robustness under targeted mutations on nodes with the largest FPC. Furthermore, targeted mutations on less than 2% nodes with the largest FPC values can produce roughly similar results as random mutations, which indicates that the structure of the PPI networks cannot be severely destroyed by removing a few SDNs. In real-world disease networks, it is crucial to keep network structures while curing disease, therefore, SDNs identified by our method can be taken as potential drug targets. It is also observed that with the growth of network sizes, the curves of S versus f obtained by the FPC decrease slower and slower, which indicates that targeted mutations on nodes with large FPC values cannot severely destroy the network as mutations on nodes with large degree, it also indicates network size may have certain effect on mutation robustness.

7.4.8 Summary

Identifying important nodes in complex networks has been a popular topic in recent years. In this subsection, based on some existing measures of node importance and the PCA method, we have proposed a new integrative node importance measure and identified SDNs in some evolving PPI networks. The evolving networks originate from the same small networks, but with increased network sizes and evolved topological structures. The investigated real-world *yeast* PPI networks are based on existing databases and publications, while the artificial networks are constructed by the DD model. Under properly chosen parameters, the generated artificial networks can well mimic the statistical characteristics of the real-world PPI networks, such as they have similar average node degrees, APLs, clustering coefficients, and PLEs. Most interestingly, both real-world and artificial PPI networks consist of similar network motifs as building blocks.

By the proposed new indicator, we have successfully identified the SDPs in real-world *yeast* and artificial PPI networks. Clustering analysis and rich-club analysis indicate that PPI networks have clear hierarchical structures, with only a few SDPs, and the few SDPs are highly connected. The proposed integrative measure FPC is closely correlated with the eigenvector centrality, the semi-local centrality, the network motif centrality, the degree, and betweenness measures, which reveals the relations between SDNs and important nodes identified by the other methods.

PPI networks evolve with time, which leads to the evolution of node importance. By performing numerical simulations on extensive artificial networks, we have also investigated the evolutionary characteristics of node importance. It is found that, after a long time evolution, most of the top-ranked nodes can keep to be dominant, while a few nodes may become less and less dominant. Moreover, the less dominant a node is, the more rapidly it evolves to be much less dominant. Furthermore, it is also found that PPI networks are robust to random mutations while fragile to targeted mutations. Targeted mutations on SDNs and high-degree nodes can bring about similar yet different consequences, the networks can keep certain robustness under the first mutation strategy.

As a representative example, we have only considered the *yeast* PPI networks. It is noted that the *yeast* PPI networks are the most frequently investigated bio-molecular networks, available data for the *yeast* PPIs are with very high quality [71–79]. We note that the related methods and investigations can be easily extended to other types of networks, such as social networks and directed gene networks, metabolic networks, electrical networks. They can also be easily extended to the other organisms, such as the PPI networks for Human, *E. Coli*, and *Xenopus laevis*. Our investigations on evolving PPI networks can shed some light on the future applications of the evolving characteristics of complex networks, such as re-engineering of particular bio-molecular networks for technological, synthetical, or pharmacological purposes [96–99]. For example, in real-world disease-related gene networks or PPI networks, one may detect SDNs and consider to take them as control targets during the treatment of diseases, since existing investigations have suggested that disease-related genes are characterized with certain structural features in the PPI networks [67–69], SDNs are more appropriate to be chosen as potential control targets [100], targeted mutations on these nodes will not severely destroy the networks.

7.5 Spectral Learning Algorithms Reveal Nodes' Spreading Ability

In network science and data mining field, a long-lasting and significant task is to predict propagation capability of nodes in a complex network. Recently, an increasing number of unsupervised learning algorithms, such as the prominent PR and LeaderRank (LR), have been developed to address this issue. However, in degree uncorrelated networks, we find that PR and LR are actually proportional to in-degree of nodes. As a result, the two algorithms fail to accurately predict nodes' propagation capability. To overcome the arising drawback, we propose a new iterative algorithm called SpectralRank (SR) [101], in which node's propagation capability is assumed to be proportional to the number of its neighbors after adding a ground node to the network. Moreover, a weighted SR algorithm is also proposed to further involve a priori information of a node itself. A probabilistic framework is established, which is provided as the theoretical foundation of the proposed algorithms. Simulations of the susceptible-infected-removed (SIR) model on 32 real-world networks, including directed, undirected, and binary ones, reveal the advantages of the SR-family methods (i.e., SR and weighted SR) over PR and LR. When compared with the other 11 well-known algorithms, the indexes in the SR-family always outperform the others. Investigations on biological networks reveal that the proposed methods also have good performance in identifying functional important nodes in bio-molecular networks. Therefore, the proposed measures provide new insights on the prediction of nodes' propagation capability and have great implications in the control of spreading behaviors in complex networks.

7.5.1 Related Works and Motivations

With the advent of the big data era [102–104], researchers are prone to focus on more complex data [105], where graph-based or system-based data has attracted much attention [6]. Graph-based data can be extracted from various fields, such as biology [106–111], social technology [112, 113], and industry [114]. The study of machine learning methods on graph-based data provides complex network practitioners with plenty of tools in data mining [115]. Identifying important node [116, 117] is one of the most prevailing applications. For instance, estimating node's propagation capability [33], finding vital proteins or genes [37, 46, 109, 111], mining network values of customers [118], and pinning control of multi-agent systems [119].

In complex networks, ranking nodes in the light of their importance is an unsupervised learning problem. Node importance is equivalent to its propagation capability under many circumstances. Evaluation of propagation capability of a node relies on spreading dynamics. Spreading dynamics on complex networks are so ubiquitous that the investigation of them might shed some light on controlling real world networks [120]. To approximate propagation capability, there are many traditional measures, including degree, k-core [32, 121], H-index [122], and many other Markov chain based methods. In particular, degree [33] is a basic measure but of little relevance in many cases. The k-core decomposing algorithm [121] identifies the core and periphery of network, and assigns each node with a layer (called coreness number) via removing low-degree nodes repeatedly. It has been reported that coreness outperforms degree to some extent [32]. The coreness can also be obtained by applying the H-index operation $\mathscr{H}$ [122] iteratively. The 0-order H-index is just the degree, while the steady state of the H-index operation is just the coreness. The degree, H-index, and coreness consist of the H-indexes family, which provide pretty prediction results of node importance.

Markov chain based methods are another class of learning algorithms, where the most representative one is PR [21]. It assumes that an Internet surfer walks randomly on the web and chooses one of links on this web stochastically. In the meantime, the surfer does not click on a hyperlink but jumps instead to a random web with a small probability. Even though PR is originally designed to rank webs, it has been applied to rank images, genes, and scientists [116]. Similar to the PR, the LR and its variants [22, 123, 124] are designed to mine the leaders in social networks. PR and LR have been widely used to evaluate node's spreading influence [116]. However, they are misused in network science, especially in learning propagation capability. According to the mean-field analysis and empirical analysis in degree uncorrelated and correlated networks (see Ref. [125] and Sect. 7.5.2.4), it shows that PR and LR are proportional to node's in-degree. This yields the conflict that in-degree actually can hardly extract the information of node's propagation capability [122, 126].

To learn nodes' propagation capability in complex networks, an effective algorithm should take the nature of spread dynamics into account. We study a class of node ranking spectral algorithms. The main contributions of this section are as follows:

1. The parameter-free learning algorithms called SR-family are proposed to elaborately measure node's propagation capability. It is shown that indexes in the SR-family are closely related to the dominant eigenvector of the augment network, i.e., the original network with a ground node.
2. In 32 representative networks (15 directed ones, 12 undirected ones, and 5 binary ones), we compare SR with 11 existing popular algorithms, including degree, H-index, coreness, mixed degree decomposition (MDD) [127], PR, LR, weighted LR (WLR), adaptive LR (ALR), ClusterRank (CluR) [126], eigenvector centrality (EC) [128], and cumulative nomination (CN) [129]. The obtained results reveal that the SR-family methods have advantages over the other measures in all of the considered networks.
3. We develop a probabilistic framework for a class of spectral-based algorithms, including EC, CN, and SR-family. We prove that the dominant eigenvector is the optimal solution under the probabilistic framework, which guides the application of spectral-based algorithms. Furthermore, this framework is able to explain why the addition of the ground node can enhance the performance of an algorithm.

7.5.2 Preliminaries

7.5.2.1 Descriptions of Real-World Networks

There are 32 real-world networks investigated in the following sections, including 15 directed ones, 12 undirected ones, and 5 binary ones. In fact, binary networks are special case of undirected ones, since there are two types of nodes and edges only exist across the two types of nodes. For detailed information of the associated networks, one can see Chap. 1 and Table 7.17.

Table 7.17 Basic topological features of the 32 real-world networks

Category	Network	N	M	$\langle k \rangle$	β	Type	Reference
Foodweb	RockLake	183	2494	13.6300	0.1500	D	[131]
HyperLink	PB	1222	16,714	27.3600	0.0190	U	[132]
Rating	SexEsc	16,730	35,051	4.6700	0.0560	B	[133]
Affiliation	AmeRev	141	160	2.2700	0.0680	B	[134]
	Leadership	44	99	4.5000	0.2940	B	[135]
Authorship	WikiBooks(fr)	30,616	67,613	4.4200	0.0250	B	[136]
	WikiNews(fr)	26,447	68,703	5.2000	0.0350	B	[136]
Collaboration	Jazz	198	2742	27.7000	0.0400	U	[137]
	NS	379	914	4.8200	0.2140	U	[138]
Contact	PrettyGood	10,680	24,316	4.5500	0.0840	U	[139]
	WikiVote	7118	103,675	14.5700	0.0070	D	[140]
Protein	Vidal	3133	6726	4.2900	0.1050	U	[141]
	Yeast	1870	2277	2.4400	0.2700	U	[142]
Communicate	UCsocial	1899	20,296	10.6900	0.0500	D	[143]
	Email	1133	5451	9.6200	0.0850	U	[144]
Lexical	SpaBook	12,643	57,453	4.5400	0.0500	D	[145]
	Bible	1773	9131	10.3000	0.0370	U	[146]
	David	112	425	7.5900	0.1180	U	[138]
Infrastructure	OpenFlights	2939	30,501	10.3800	0.0220	D	[147]
	USAirpot	1574	28,236	17.9400	0.0300	D	[148]
	Router	5022	6258	2.4900	0.1180	U	[149]
	USAir	332	2126	12.8100	0.0350	U	[150]
Citation	Cora	23,166	91,500	3.9500	0.0320	D	[151]
	DBLP	12,591	49,743	3.9500	0.0230	D	[152]
	Cite-th	27,770	352,807	12.7000	0.0320	D	[153]
	Cite-ph	34,546	421,578	12.2000	0.0330	D	[153]
Social	Advogato	5042	49,631	9.8400	0.0330	D	[154]
	Anybeat	12,645	67,053	5.3000	0.0290	D	[155]
	HighSchool	70	366	5.2300	0.1200	D	[156]
	JamesMoody	2539	12,969	5.1100	0.0300	D	[157]
	ResidenceHall	217	2672	12.3100	0.0370	D	[158]
	Hamster	1858	12,534	13.4900	0.0340	U	[159]

N and M denote the number of nodes and edges. D, U, and B mean directed, undirected, and binary networks, respectively. $\langle k \rangle$ is the average degree. β is the spreading rate and will be used in simulation of the SIR model. For detailed information of the associated networks, one can refer to Chap. 1

7.5.2.2 Propagation Capability

Overwhelming evidence has revealed that different nodes and edges in real-world networks play heterogeneous roles in dynamics, control, evolution, and function [119, 130]. Propagation capability measures a node's spreading impact and it is defined as the number of infected nodes if we set a node as a single infection source. A node leading to larger spreading scope has higher propagation capability. Nonetheless, this measure cannot be obtained unless disease breaks out. Hence, researchers have made great efforts to more efficiently learn node's propagation capability. A promising way is to predict and estimate it by unsupervised learning algorithms, such as PR, LR, EC, and so on. For each algorithm, a score that reflects the relative propagation capability is assigned to each node. A standard metric, Kendall $\tau(\tau_b)$ correlation coefficient (see Chap. 2), is used to quantify the accuracy of the prediction algorithms [160, 161]. We usually apply classical epidemic models to real networks so as to obtain a good approximation of node's propagation capability. Accordingly, we employ Kendall τ between propagation capability and algorithm score to estimate the accuracy. In the same network, an algorithm with higher τ indicates more accuracy and $\tau = 1$ means the perfect prediction. Hence, τ is called algorithmic accuracy or simply accuracy.

7.5.2.3 SIR Model and Parameter Settings

The SIR model [162] is employed to evaluate node's spread scope. In the SIR model, there are three states for all nodes, that is, susceptible, infected, and recovered. The susceptible nodes may be infected by its infected neighbors with probability β, and the infected ones may recover with probability μ. To obtain the spread scope of node i, we set i as a single infection seed and count the number of recovered nodes at the steady state of the SIR process. For each node, we average the spread scopes that are obtained over 100 independent simulation runs. The spread rate β is shown in Table 7.17, and the recover rate μ is set as 1. The parameters are chosen to guarantee the spreading of disease.

For undirected and binary networks, $\beta_c = \langle k \rangle / (\langle k^2 \rangle - \langle k \rangle)$ is an approximation of epidemic threshold via the degree-based mean-field approach [162]. Epidemic strength is defined as β/μ. If epidemic strength was higher than the epidemic threshold β_c, then the information or disease can be spread, while the infected numbers will be exponentially decreased if $\beta/\mu < \beta_c$ [163–165]. The chosen parameters β and μ guarantee that $\beta/\mu > \beta_c$, and information/disease can be spread in the networks. We select $\mu = 1$ and $\beta = 1.5\beta_c$ for the undirected and binary networks in the following sections. For directed networks, we set $\mu = 1$ and carefully select β to prevent the vanishing of propagation.

7.5.2.4 Drawbacks of PageRank and LeaderRank

Consider an unweighted network $G = (V, E)$, where V is the node set and E is the edge set. Meanwhile, $N = |V|$ and $M = |E|$ are the numbers of nodes and edges, respectively. The adjacency matrix $\mathbf{A}$ captures the wiring diagram, where the (i, j)'th entry $a_{ij} = 1$ if node i points to j and 0 otherwise. For PR, each node is initially assigned with an importance score $s_i(0) = 1(i = 1, 2, \ldots, N)$. Then, score for node i is updated according to the following iterative process [21]:

$$s_i(t+1) = q \sum_{j=1}^{N} a_{ji} \frac{s_j(t)}{k_j^{out}} + (1-q)\frac{1}{N}, \tag{7.21}$$

where k_j^{out} is the out-degree and q is a parameter, usually set as 0.85 [166]. When PR converges, the steady state $s(t_c)$ is employed to evaluate node importance.

However, PR is criticized for some drawbacks [166]. Firstly, the ranking result of PR is not unique if the considered network has disconnected components. As a result, some schemes attempt to overcome such a defect, and one prevalent algorithm is the LR [22]. In LR, a ground node that connects all other nodes via bidirectional edges is added. Consequently, an augmented strongly connected network with $N + 1$ nodes and $M + 2N$ edges is obtained. The updating rule of node importance score is designed as [22],

$$s_i(t+1) = \sum_{j=1}^{N+1} a_{ji} \frac{s_j(t)}{k_j^{out}}. \tag{7.22}$$

Remark that s_g or s_{N+1} denotes the score for the ground node and we set $s_g(0) = 0$ for the ground node, $s_i(0) = 1(i = 1, 2, \ldots, N)$ for the ordinary nodes. Different from PR, LR can ensure the uniqueness of node's importance score.

Secondly, although LR improves PR, a growing number of empirical analyses reveal that both PR and LR may fail under some circumstances [126]. Currently, it is still questionable about whether it is proper to use them to estimate node's propagation capability in all kinds of complex networks. Based on the mean-field analysis, it is reported that k^{in} is a good approximation of PR when network is degree uncorrelated [125]. Additionally, there is a similar conclusion for LR.

Theorem 7.1 *In a degree uncorrelated network, the average LR score for nodes within the node group with degree* $\mathbf{k} = (k^{out}, k^{in})$ *is proportional to* k^{in},

$$\bar{s}_{\mathbf{k}}(t+1) \approx \theta k^{in}. \tag{7.23}$$

Here, $\theta = N/[(N+1)\langle k^{in}\rangle]$.

Proof. We apply the mean-field analysis to LR. Nodes can be classified by its degrees, i.e., nodes in the same class $\mathbf{k}$ have the same out-degree and in-degree

(k^{out}, k^{in}). We consider the average LR score for a class of nodes,

$$\bar{s}_{\mathbf{k}}(t+1) \equiv \frac{1}{NP(\mathbf{k})} \sum_{i \in \mathbf{k}} s_i(t+1), \tag{7.24}$$

where $P(\mathbf{k})$ denotes the frequency of node with degree $\mathbf{k} = (k^{out}, k^{in})$. The updating rule of LR follows:

$$s_i(t+1) = \sum_{j=1}^{N+1} a_{ji} \frac{s_j(t)}{k_j^{out}}. \tag{7.25}$$

Hence, we have

$$\bar{s}_{\mathbf{k}}(t+1) = \frac{1}{NP(\mathbf{k})} \sum_{i \in \mathbf{k}} \sum_{j=1}^{N+1} a_{ji} \frac{s_j(t)}{k_j^{out}}. \tag{7.26}$$

Notice that nodes in the same class have the same out-degree. So in the right side of Eq. (7.26), we split the sum over j into two sums, one over all the degree classes $\mathbf{k}'$ and the other over all the nodes within each degree class $\mathbf{k}'$,

$$\bar{s}_{\mathbf{k}}(t+1) = \frac{1}{NP(\mathbf{k})} \sum_{\mathbf{k}'} \frac{1}{k'^{out}} \sum_{i \in \mathbf{k}} \sum_{j \in \mathbf{k}'} a_{ji} s_j(t). \tag{7.27}$$

We apply the mean-field theory, i.e., it is assumed that the LR scores of node i's predecessors that belong to class $\mathbf{k}'$ can be replaced by the mean value of LR scores for class $\mathbf{k}'$.

$$\sum_{i \in \mathbf{k}} \sum_{j \in \mathbf{k}'} a_{ji} s_j(t) \simeq \bar{s}_{\mathbf{k}'}(t) \sum_{i \in \mathbf{k}} \sum_{j \in \mathbf{k}'} a_{ji} = \bar{s}_{\mathbf{k}'}(t) E_{\mathbf{k}' \to \mathbf{k}}, \tag{7.28}$$

where $E_{\mathbf{k}' \to \mathbf{k}}$ denotes the number of edges from nodes within class $\mathbf{k}'$ to nodes within class $\mathbf{k}$.

$$E_{\mathbf{k}' \to \mathbf{k}} = k^{in} P(\mathbf{k}) N \frac{E_{\mathbf{k}' \to \mathbf{k}}}{k^{in} P(\mathbf{k}) N} = k^{in} P(\mathbf{k}) N P_{in}(\mathbf{k}'|\mathbf{k}), \tag{7.29}$$

where $P_{in}(\mathbf{k}'|\mathbf{k})$ is the frequency that the source node of an edge is with degree $\mathbf{k}'$, but the target node of the edge has degree $\mathbf{k}$. Therefore, we have

$$\bar{s}_{\mathbf{k}}(t+1) = \sum_{\mathbf{k}'} \frac{k^{in} P_{in}(\mathbf{k}'|\mathbf{k}) \bar{s}_{\mathbf{k}'}(t)}{k'^{out}}. \tag{7.30}$$

If network is uncorrelated (ground node does not affect the degree correlation of the original network), the conditional degree distribution $P_{in}(\mathbf{k}'|\mathbf{k})$ does not depend on $\mathbf{k}$ and we have

$$P_{in}(\mathbf{k}'|\mathbf{k}) = \frac{k'^{out} P(\mathbf{k}')}{\langle k^{in} \rangle}, \tag{7.31}$$

where $\langle k^{in} \rangle$ denotes the average in-degree. Furthermore, we have

$$\bar{s}_{\mathbf{k}}(t+1) = \frac{k^{in}}{\langle k^{in} \rangle} \sum_{\mathbf{k}'} \bar{s}_{\mathbf{k}'}(t) P(\mathbf{k}'). \tag{7.32}$$

According to the definition of statistical expectation, we have $\sum_{\mathbf{k}'} \bar{s}_{\mathbf{k}'}(t) P(\mathbf{k}') = E\left[\bar{s}_{\mathbf{k}'}(t)\right]$. Notice that $E\left[\bar{s}_{\mathbf{k}'}(t)\right] = N/(N+1)$. Thus, when N tends to infinity,

$$\bar{s}_{\mathbf{k}}(t+1) \approx \frac{k^{in}}{\langle k^{in} \rangle} \frac{N}{N+1} = \theta k^{in}. \tag{7.33}$$

Here, θ is a constant. Within finite steps, the iteration process converges, so the average LR score for nodes within class $\mathbf{k} = (k^{out}, k^{in})$ is proportional to k^{in}.

As for degree correlated networks, we conduct empirical analysis (See the supplementary material of our paper [101]), which demonstrates that correlations between k^{in} and PR/LR are statistically significant, as well. Nevertheless, k^{in} is not a good indicator for node's propagation capability. Such conclusion can be drawn by accuracy difference index $\Delta\tau = \tau_{k^{out}} - \tau_{k^{in}}$ ($\tau_{k^{out}}$ and $\tau_{k^{in}}$ are prediction accuracy of k^{out} and k^{in}, respectively) in the 15 directed networks, as shown in Fig. 7.19. Figure 7.19 indicates k^{in} is always inferior to k^{out} and this effect may be weakened only when k^{out} and k^{in} are strongly correlated. Thus, we conclude that k^{in} cannot effectively extract the information on propagation capability. At the same time, this phenomenon also implies that PR and LR fail to predict node's propagation capability.

The drawbacks of PR and LR urge us to propose new effective algorithms to evaluate node's propagation capability in any types of complex networks.

7.5.3 SpectralRank and Its Generlizations

7.5.3.1 SpectralRank

Node's propagation capability actually depends on their outgoing edges, which explains why PR and LR fail to learn node's propagation capability in networks where k^{in} and k^{out} sequences are not strongly correlated. In view of this point, methods that take outgoing edges into account have been proposed, such as the EC and CN. But similar to the PR, they also have some drawbacks, such as non-

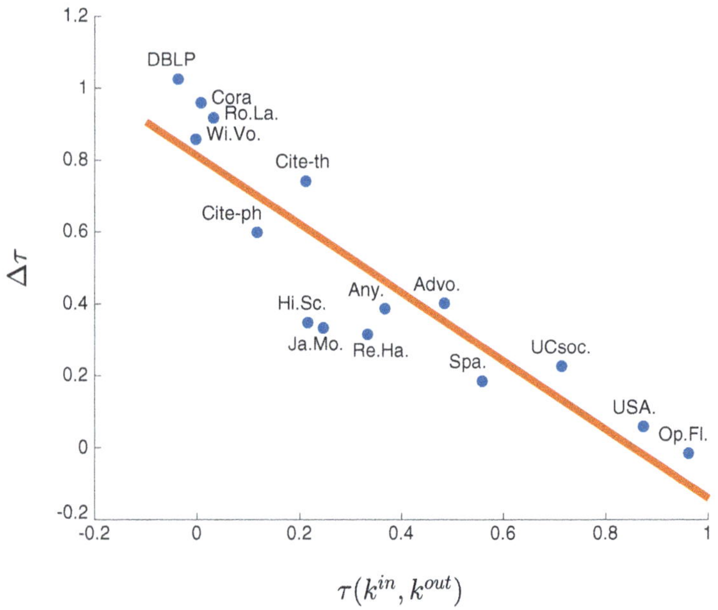

Fig. 7.19 k^{out} is a better indicator in the 15 directed networks. The fitted line is based on the OLS regression. Here $\tau(k^{in}, k^{out})$ is Kendall correlation coefficient between k^{in} and k^{out}. $\Delta\tau = \tau_{k^{out}} - \tau_{k^{in}}$ is the accuracy difference of k^{out} and k^{in}. ©[2019] IEEE. Reprinted, with permission, from Ref. [101]

unique rankings and dangling nodes. Consequently, it is urgent to develop a learning algorithm with universal applicability and high accuracy.

Hereinafter, we propose a new method called SR. In SR, each node is assigned with a score representing its propagation capability. To cope with the abovementioned problems, we insert a ground node (i.e., node connects to all other nodes via bidirectional edges) to obtain a strongly connected network. Nodes with larger k^{out} have more successive neighbors; however, these successive neighbors play heterogeneous roles in propagation process. Hence, node with more successive influential neighbors owns greater propagation capability. Specifically, we let the score of node i to be proportional to the sum of its successive neighbors' score, that is,

$$SR_i = s_i = c \sum_{j=1}^{N+1} a_{ij} s_j. \tag{7.34}$$

The matrix formation of Eq. (7.34) follows:

$$\mathbf{s} = c\widetilde{\mathbf{A}}\mathbf{s}, \tag{7.35}$$

where $\widetilde{\mathbf{A}}$ is the adjacency matrix of the augmented network,

$$\widetilde{\mathbf{A}} = \begin{pmatrix} \mathbf{A} & \mathbf{1} \\ \mathbf{1}^T & 0 \end{pmatrix}, \tag{7.36}$$

and c is a tuning parameter and is usually selected as the reciprocal of dominant eigenvalue of $\widetilde{\mathbf{A}}$, that is $c = 1/\lambda_1$. On the basis of this fact, the SR score reduces to the dominant eigenvector of $\widetilde{\mathbf{A}}$.

The eigenvector can be easily obtained by the iterative power method. Initially, we set the score $s_i(0) = 1 (i = 1, 2, \ldots, N)$ for the ordinary nodes and $s_g(0) = s_{N+1}(0) = 0$ for the ground node. Then the iteration includes two operations, i.e., linear transformation and normalization. The updating rule is,

$$\hat{s}_i(t+1) = \sum_{j=1}^{N+1} a_{ij} s_j(t),\ s_i(t+1) = \frac{\hat{s}_i(t+1)}{\max_k \hat{s}_k(t+1)}. \tag{7.37}$$

The matrix form can be written as,

$$\hat{\mathbf{s}}(t+1) = \widetilde{\mathbf{A}}\mathbf{s}(t),\ \mathbf{s}(t+1) = \frac{\hat{\mathbf{s}}(t+1)}{\max \hat{\mathbf{s}}(t+1)}. \tag{7.38}$$

The Perron–Frobenius theorem guarantees the iteration process (7.37) to be converged within finite steps.

Remark 7.1 In Eq. (7.35), we intuitively set the tuning parameter c to be $1/\lambda_1$ without any technical proof. In Sect. 7.5.4, we will build a probabilistic framework for EC and SR. It can be proved that $c = 1/\lambda_1$ is the optimal result under the probabilistic framework.

Remark 7.2 It is worth noticing that the SR procedures are similar to the eigenvector centrality whose updating rule follows $\mathbf{s}(t+1) = \mathbf{A}\mathbf{s}(t)$. The only difference is the iterative matrix in the former is augmented by a ground node. (As a summary, Fig. 7.20 demonstrates the relation and difference among PR, LR, EC, and SR.) But, there are two technical questions. Firstly, how to guarantee that the ground node improves our algorithm's performance? Secondly, the topology of the original network is changed by the ground node, so how to guarantee that the SR scores stand for the importance of nodes in the original network? These questions can be solved by our probabilistic framework in Sect. 7.5.4.

Remark 7.3 It is feasible to add a ground node even for a large-scale network (N is very large). Since, the sparsity of the augmented network is $M/N^2 + 2/N$, which is very close to the sparsity of the original network, i.e., M/N^2. The cost of adding a ground node is very low.

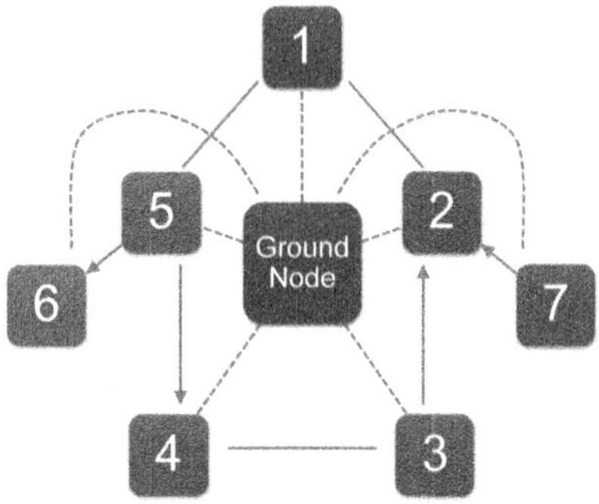

Consider node 2

- PR: $\hat{s}_2(t+1) = q\left(s_1(t) + s_2(t) + s_7(t)\right) + \frac{1-q}{7}$;
- LR: $\hat{s}_2(t+1) = s_1(t) + s_2(t) + s_7(t) + s_g(t)$;
- EC: $\hat{s}_2(t+1) = s_1(t)$;
- SR: $\hat{s}_2(t+1) = s_1(t) + s_g(t)$.

Consider node 5

- PR: $\hat{s}_5(t+1) = qs_1(t) + \frac{1-q}{7}$;
- LR: $\hat{s}_5(t+1) = s_1(t) + s_g(t)$;
- EC: $\hat{s}_5(t+1) = s_1(t) + s_4(t) + s_6(t)$;
- SR: $\hat{s}_5(t+1) = s_1(t) + s_4(t) + s_6(t) + s_g(t)$.

Fig. 7.20 A toy network with 7 nodes demonstrates the differences among PR, LR, EC, and SR. Directional edges are with arrows and bidirectional edges are without arrows. Solid lines represent the true topology edges and dashed lines represent the bidirectional edges between ground node and original ones. Specially, node 2 and 5 are focused on, which have large in-degree and out-degree, respectively. ©[2019] IEEE. Reprinted, with permission, from Ref. [101]

7.5.3.2 Weighted SpectralRank

Since SR assumes that a node's propagation capability is proportional to the sum of its successive neighbors, it suffers the *boundary effect* to some extent. In other words, the SR may (not always) underestimate a node as possibly the best spreader and overestimate it as the worst spreader. When learning propagation capability, it is appropriate to consider information from a node itself. To achieve this, we can add the prior knowledge into SR. The iteration matrix $\widetilde{\mathbf{A}}$ can be replaced by $\mathbf{W} = \widetilde{\mathbf{A}} + \mathbf{P}$, where $\mathbf{P}$ is a diagonal matrix and its (i, i)'th entry encodes our a prior knowledge of node i. In general, the weighted matrix $\mathbf{P}$ should be carefully selected. If we have no node information, $\mathbf{P}(i, i)$ can be set as 1 ($i = 1, 2, \ldots, N$). We can also consider results from some existing algorithms, such as degree k, H-index h, and coreness k_s. Notice that for the ground node, we always set $\mathbf{P}(N+1, N+1) = 0$. The new algorithm with weighted matrix $\mathbf{P}$ can be called as weighted SpectralRank (WSR). Since different kinds of a priori information correspond to different WSR algorithms, we denote the WSR with degree k, H-index h, and coreness k_s as diagonal elements of $\mathbf{P}$ as WSR-k, WSR-h, and WSR-k_s, respectively. In the subsequent discussions, we use the SR-family methods to refer to the SR and the WSR. Actually, SR can be viewed as a special case of the WSR, where all entities of $\mathbf{P}$ are zeros.

Algorithm 17 (Weighted) SpectralRank

1: Initialize importance score $\mathbf{s}(0)$ and set $t = 0$;
2: Construct the iteration matrix $\mathbf{W} = \widetilde{\mathbf{A}} + \mathbf{P}$;
3: **repeat**
4: Update importance score $\hat{\mathbf{s}}(t+1) = \mathbf{W}\mathbf{s}(t)$;
5: Normalize importance score $\mathbf{s}(t+1) = \frac{\hat{\mathbf{s}}(t+1)}{\max \hat{\mathbf{s}}(t+1)}$;
6: Set $t = t + 1$;
7: **until** change of $\mathbf{s}(t)$ is smaller than a predefined threshold

7.5.4 The Probabilistic Explanation

In the above subsection, we have proposed some heuristic algorithms to predict the node importance. Now, we discuss the physical mechanism of the new algorithms in undirected networks. At first, we build a data-driven framework for the node ranking problem, which bridges the gap between the heuristic algorithms (EC and SR-family) and statistical theory. Then, we prove that $c = 1/\lambda_1$ is an optimal parameter in our framework. At last, we explain the reason why we should add a ground node and weighted matrix $\mathbf{P}$ from the perspective of machine learning.

7.5.4.1 The Data-Driven Framework for Node Ranking

For decades, researchers have proposed numerous heuristic node ranking algorithms, including PR, LR, EC, and so on, and empirical experiments demonstrated that they are very effective in many applications. Nonetheless, to our best knowledge, there still lacks a literature to explain why these heuristic algorithms work.

Here, inspired by preferential attachment and statistical mechanics, we build a novel framework to provide a theoretical understanding of the EC and SR-family. Preferential attachment is a classical growth model in complex network theory [130]. Namely, newly added nodes tend to connect to the nodes with a specific property, such as large-degree nodes. This phenomenon has been proved by numerous evidences, one of which is called the Matthew effect or Gibrat's law, that is, the rich get richer. For example, the BA model [130] generates an undirected SF network, where the probability that a new node i connects to node j is proportional to the degree of node j. Here, we consider a so-called fitness model [167], where the link between i and j is created with a probability $p(i, j)$. Specifically, we assume that $p(i, j)$ is proportional to the product of their importance scores, $p(i, j) \propto s_i s_j$, where $s_i \geq 0, \forall i \in V$. Let $\mathscr{A}$ denote the network space which contains all possible complex networks constructed by nodes in V. Thus, the probability of observing $\mathbf{A}$ is given by the Boltzmann distribution [168]:

$$p(\mathbf{A}; \mathbf{s}) = \frac{e^{-H(\mathbf{A};\mathbf{s})}}{Z_{\mathscr{A}}}, \tag{7.39}$$

where the energy is given by the Hamiltonian function $H(\mathbf{A}; \mathbf{s}) = -\sum_{i,j \in V} a_{ij} s_i s_j$ and $Z_{\mathscr{A}} = \sum_{\mathbf{A} \in \mathscr{A}} p(\mathbf{A}; \mathbf{s})$ is the partition constant. Note that this distribution coincides with the Ising model without an external field [168].

In practice, our observed data is a network $\mathbf{A}$ and our task is to infer unknown parameters $\mathbf{s}$, that is, importance score. This issue can be solved by the maximum likelihood principle, $\mathbf{s}^* = \arg\max_{\mathbf{s}} \log p(\mathbf{A}; \mathbf{s})$.

Theorem 7.2 *The maximum likelihood estimate of importance score* $\mathbf{s}^*$ *in the fitness model (7.39) is exactly the eigenvector centrality of network* $\mathbf{A}$. *Furthermore,* $c = 1/\lambda_1$ *is the necessary condition of the maximum likelihood estimation.*

Proof It is able to rewrite the objective function as follows:

$$
\begin{aligned}
\mathbf{s}^* &= \arg\max_{\mathbf{s}} \log p(\mathbf{A}; \mathbf{s}) \\
&= \arg\max_{\mathbf{s}} \sum_{i,j \in V} a_{ij} s_i s_j \\
&= \arg\max_{\mathbf{s}} \mathbf{s}^T \mathbf{A} \mathbf{s}.
\end{aligned}
\tag{7.40}
$$

Now, without loss of generality, we must add the constraint, $\mathbf{s}^T \mathbf{s} = 1$, otherwise the quadratic form $\mathbf{s}^T \mathbf{A} \mathbf{s}$ may go to the infinite. Therefore, the Lagrange function of our problem can be constructed as follows:

$$
L = \mathbf{s}^T \mathbf{A} \mathbf{s} - \lambda(\mathbf{s}^T \mathbf{s} - 1), \tag{7.41}
$$

where λ is the Lagrange multiplier. Let $\partial L / \partial \mathbf{s} = 0$ and we know $\mathbf{s}$ ought to satisfy $\mathbf{A}\mathbf{s} = \lambda \mathbf{s}$, implying that λ must be the eigenvalue of $\mathbf{A}$. Thus, the original problem is equivalent to

$$
\mathbf{s}^* = \arg\max_{\mathbf{s}} \lambda \mathbf{s}^T \mathbf{s}, \quad \text{subject to } \mathbf{s}^T \mathbf{s} = 1. \tag{7.42}
$$

So, λ must be the largest eigenvalue of $\mathbf{A}$ and $\mathbf{s}^*$ is the dominant eigenvector. The proof is completed.

Theorem 7.2 provides a theoretical foundation of why we chose $c = 1/\lambda_1$ in (7.35). What is more, it gives new insightful understanding about the EC algorithm from the data-driven perspective. Besides, Theorem 7.2 also guides the application of the EC, i.e., the EC only works in the fitness model defined by Eq. (7.39).

7.5.4.2 The Theoretical Foundation of Weighted Matrix and Ground Node

As stated in Sect. 7.5.3, the main differences between EC and SR-family are the weighted matrix **P** and ground node. In spite of empirical studies showing these two differences may improve accuracy of ranking algorithms [22, 123, 124, 129], researchers still do not know the working mechanism. Here, from the viewpoint of Bayesian statistics, we study the roles that the weighted matrix **P** and ground node play.

In Bayesian statistics, we would encode a prior knowledge by a prior distribution of **s**. If we assume that **s** is governed by the conjugate prior of (7.39),

$$p(\mathbf{s}) = \frac{1}{Z_{\mathscr{S}}} e^{-\mathbf{s}^T \mathbf{P} \mathbf{s}}, \tag{7.43}$$

where $Z_{\mathscr{S}} = \int p(\mathbf{s})d\mathbf{s}$ is a partition constant, our task is to make the posterior estimate to reach its maximum,

$$\begin{aligned} \mathbf{s}^* &= \arg\max_{\mathbf{s}} \ \log p(\mathbf{s}|\mathbf{A}) \\ &= \arg\max_{\mathbf{s}} \ \log p(\mathbf{A}|\mathbf{s}) + \log p(\mathbf{s}) \\ &= \arg\max_{\mathbf{s}} \ \mathbf{s}^T \mathbf{A} \mathbf{s} + \sum_{i=1}^{N} s_i^2 \mathbf{P}(i,i). \end{aligned} \tag{7.44}$$

From the above equations, we know that **P** brings a weighted L_2 norm penalty (also called the ridge penalty), which is widely used in machine learning and pattern recognition [169]. The effect of the L_2 norm penalty (or say the prior $p(\mathbf{s})$) is to prevent over-fitting. In Eq. (7.44), $\mathbf{P}(i,i)$ is a penalty parameter. It is obvious that, based on optimization theory, larger/smaller $\mathbf{P}(i,i)$ leads to larger/smaller s_i.

As for the ground node, we have similar conclusions, which can be drawn if we replace **A** by $\tilde{\mathbf{A}}$. Thus, the problem turns into

$$\begin{aligned} (\mathbf{s}^*, s_g^*) &= \arg\max \mathbf{s}^T \mathbf{A} \mathbf{s} + 2 s_g \mathbf{s}^T \mathbf{1}, \\ &\text{subject to } \mathbf{s}^T \mathbf{s} + s_g^2 = 1, \end{aligned} \tag{7.45}$$

whose Lagrange function is

$$L = \mathbf{s}^T \mathbf{A} \mathbf{s} + 2 s_g \mathbf{s}^T \mathbf{1} - \lambda \left(\mathbf{s}^T \mathbf{s} + s_g^2 - 1 \right). \tag{7.46}$$

Let $\partial L/\partial s_g = 0$; then, we have $\lambda s_g = \mathbf{s}^T\mathbf{1}$. After plugging this equation into the objective function, we have

$$\mathbf{s}^* = \arg\max \ \mathbf{s}^T\mathbf{A}\mathbf{s} + \mathbf{s}^T\left(\frac{2}{\lambda}\mathbf{1}^T\mathbf{1}\right)\mathbf{s}. \quad (7.47)$$

Obviously, the ground node implies that we have a prior knowledge about importance score, which is encoded by $p(\mathbf{s}) = \frac{1}{Z}\exp(-2\mathbf{s}^T\mathbf{1}^T\mathbf{1}\mathbf{s}/\lambda)$ and, from the perspective of machine learning, the addition of the ground node can prevent over-fitting.

In summary, although **P** and ground node change the topology of the original network, the (W)SR scores still stand for the importance of nodes in the original network. But a natural question is whether the a prior information is correct, or in other words, whether the weighted matrix **P** and ground node can improve the accuracy of an algorithm. In the next section, the empirical studies show that the two over-fitting restriction strategies indeed enhance the performances of the algorithms.

7.5.5 Numerical Validations

7.5.5.1 Prediction of Propagation Capability

Thirty-two representative real-world networks, including 15 directed networks, 12 undirected ones, and 5 binary ones, are considered. The 32 networks cover social, biological, technological, transportation fields, with the numbers of nodes ranging from tens to tens of thousands. The detailed information of the 32 networks can be found in Chap. 1 or Sect. 7.5.2.1.

The SIR model is applied to the 32 representative real-world networks to obtain the propagation capability of nodes therein and the Kendall τ correlation coefficient is employed to evaluate algorithm accuracy. To show the superiority of the new learning algorithms, we selected 11 existing algorithms as benchmarks, including degree k, H-index h, coreness k_s, MDD, LR, WLR, ALR, CluR, PR, CN, and EC.

We introduce two measures to evaluate the performance of the proposed algorithms. The first one is median accuracy τ_m, it is the median of an algorithm's Kendall τ over all considered networks, measuring the average performance of algorithm x. That is, $\tau_m(x) = \text{median}_i \tau_i(x)$, where i and x are the indexes of the networks and the algorithms, respectively. $\tau_i(x)$ represents the accuracy of the algorithm x in network i. The second index is relative loss L, it is defined as $L(x) = \frac{1}{n}\sum_{i=1}^{n} \tau_i(x) - \tau_i^{opt}$, where n is the number of the considered networks, and $\tau_i^{opt} = \max_x \tau_i(x)$ is the best accuracy among the considered algorithms for network i. In this way, an algorithm with larger L indicates that it achieves better performance.

Tables 7.18, 7.19, and 7.20 list the accuracy of the considered algorithm on the 32 networks. Table 7.21 reports the overall performance as assessed by τ_m and L.

Table 7.18 Accuracies measured with τ for the 17 algorithms in the 15 directed networks

Algo.	Advo.	Any.	Hi.Sc.	Ja.Mo.	Re.Ha.	Op.Fl.	USA.	Spa.	Cora	DBLP	Cite-th	Cite-ph	UCsoc.	Wi.Vo.	Ro.La.
k	0.8180	0.6515	0.5421	0.5785	0.6362	0.7517	0.6667	0.5382	0.9648	0.9877	0.9485	**0.7354**	0.9226	0.8162	0.7695
h	0.8296	0.6925	0.5124	0.6100	0.6166	0.7638	0.6898	0.5693	0.8183	0.8818	0.8828	0.6899	0.9409	0.7848	0.8204
k_s	0.8276	0.6975	0.4845	0.6053	0.5515	0.7615	0.6992	0.5713	0.6322	0.8082	0.6276	0.4614	0.9393	0.7786	0.7317
MDD	0.8063	0.6772	0.5559	0.7445	0.7260	0.7877	0.6746	0.5692	0.7005	0.8081	0.7564	0.6050	0.9277	0.7643	0.6639
LR	0.2832	0.3903	0.1434	0.2201	0.2687	0.7140	0.5700	0.2692	−0.0489	−0.0550	0.0024	−0.0010	0.6822	−0.0507	−0.1629
WLR	0.3231	0.3922	0.1442	0.2444	0.2834	0.7911	0.6495	0.2735	−0.0098	−0.0408	0.1448	0.0762	0.6806	−0.0479	−0.1568
ALR	0.4554	0.4667	0.1600	0.3353	0.3752	0.8330	0.6959	0.3131	0.3714	0.4359	0.4702	0.3734	0.7031	0.2673	0.0639
CluR	0.8582	0.7721	0.6300	0.7663	0.7344	0.6467	0.7078	0.3116	0.7601	0.9462	0.8764	0.6636	0.9269	0.8195	0.8629
PR	0.6284	0.6483	0.2854	0.3802	0.5730	0.4716	0.4745	0.5026	0.5768	0.9324	0.5485	0.5038	0.8945	0.7685	0.7822
CohR	0.8978	0.8417	0.5917	0.8385	0.7369	0.7765	0.8645	0.5194	0.7152	0.8775	0.8183	0.6267	0.9400	0.8146	0.7919
CN	0.9105	0.8820	0.6613	0.5188	0.7369	0.7674	0.9077	0.5579	0.5481	0.9169	0.6602	0.4501	0.9598	0.8052	0.6455
EC	0.9083	0.8786	0.6604	0.5172	0.7369	0.7673	0.9073	0.5579	0.5265	0.7899	0.6207	0.4623	0.9576	0.7780	0.6455
SR	0.8852	0.8820	**0.8351**	**0.8426**	**0.8829**	**0.8708**	**0.9344**	0.5600	**0.9308**	0.9906	0.9526	0.7260	0.9679	0.8209	0.8378
WSR-k	0.8278	0.8571	0.5857	0.7186	0.6715	0.7935	0.8649	0.4385	0.6413	0.6592	0.7038	0.5740	0.8390	0.5356	0.8166
WSR-h	**0.9130**	0.8836	0.7020	0.7742	0.7555	0.8233	0.9205	0.5526	0.7581	0.6668	0.8122	0.6600	0.9617	0.5610	**0.8828**
WSR-k_s	0.8911	**0.8909**	0.6729	0.8211	0.8209	0.8457	0.9247	**0.5715**	0.7924	0.6854	0.8354	0.6899	**0.9735**	0.5188	0.8333
WSR-1	0.9107	0.8821	**0.8351**	**0.8426**	0.8815	0.8699	0.9343	0.5600	**0.9308**	**0.9907**	**0.9527**	0.7260	0.9679	**0.8229**	0.8563

The best prediction of each network is shown in bold face. The names of some networks are replaced by their abbreviations

Table 7.19 Accuracies measured with τ for the 17 algorithms in the 12 undirected networks

Network	Hamster	Vidal	Yeast	Router	USAir	Bible	David	Email	Jazz	NS	PrettyGood	PB
k	0.7032	0.5212	0.4423	0.3309	0.7256	0.6920	0.8374	0.7794	0.8021	0.5092	0.5255	0.8159
h	0.7359	0.5669	0.4948	0.2877	0.7540	0.7079	0.8525	0.8103	0.8431	0.5178	0.5235	0.8321
k_s	0.7445	0.5775	0.4991	0.2946	0.7529	0.7059	0.8158	0.8021	0.7958	0.4747	0.5178	0.8274
MDD	0.7127	0.4631	0.4227	0.3288	0.7308	0.6957	0.8444	0.7893	0.8233	0.5199	0.5309	0.8194
LR	0.6612	0.3284	0.2892	0.4198	0.6697	0.6420	0.7992	0.7440	0.7884	0.4541	0.5293	0.8001
WLR	0.7399	0.5882	0.6210	0.4549	0.7719	0.7277	0.8478	0.7959	0.8215	0.5707	0.6478	0.8125
ALR	0.7938	0.6583	0.6684	0.5351	0.8106	0.7980	0.8700	0.8306	0.8532	0.5672	0.6446	**0.8348**
CluR	0.7631	0.7224	0.6688	0.5675	0.6061	0.8015	0.7721	0.7347	0.7550	0.4644	0.5241	0.7591
PR	0.4875	0.1690	−0.0163	0.0591	0.5371	0.4492	0.7535	0.6828	0.6941	0.3357	0.2610	0.7579
CohR	0.8153	0.7879	0.7335	**0.5728**	0.7357	0.7906	0.8461	0.8369	0.6798	0.5309	0.6601	0.8123
CN	0.8203	**0.7992**	0.7724	0.5718	0.8220	0.8155	0.8729	0.8202	0.8387	0.5273	0.6038	0.8111
EC	0.8202	0.7980	**0.7732**	0.5707	0.8220	0.8154	0.8733	0.8202	0.8385	0.5203	0.6038	0.8110
SR	0.8402	0.6074	0.6031	0.4978	0.8285	0.8584	0.8987	0.8295	**0.8672**	0.6209	0.6195	0.8320
WSR-k	0.7303	0.7095	0.6298	0.5206	0.7655	0.7837	0.7738	0.7573	0.8073	**0.6867**	**0.6779**	0.6361
WSR-h	0.8349	0.6886	0.6459	0.5078	0.8222	0.8471	0.8858	**0.8476**	0.8383	0.6251	0.6257	0.7986
WSR-k_s	**0.8445**	0.6797	0.6454	0.5044	0.8252	**0.8585**	**0.9000**	0.8393	0.7896	0.6135	0.6203	0.8161
WSR-1	0.8405	0.7006	0.6410	0.4974	**0.8286**	**0.8585**	0.8977	0.8303	0.8670	0.6237	0.6194	0.8317

The best prediction of each network is shown in bold face

Table 7.20 Accuracies measured with τ for the 17 algorithms in the 5 binary networks

Network	AmeRev	Leadership	SexEsc	WikiBooks	WikiNews
k	0.5196	0.8049	0.5121	0.5431	0.5782
h	0.5111	0.7206	0.5439	0.5700	0.5985
k_s	0.5100	0.7008	0.5558	0.5719	0.6006
MDD	0.5196	0.8217	0.5208	0.5476	0.5823
LR	0.4536	0.7154	0.3620	0.3031	0.6206
WLR	0.2824	0.8299	0.6382	0.5402	0.5708
ALR	0.5651	0.8172	0.7060	0.6973	0.6600
CluR	0.7860	0.8961	**0.7790**	0.7908	0.7551
PR	0.3647	0.6391	0.1707	0.3448	0.3542
CohR	0.7860	0.8961	0.7776	0.7906	0.7551
CN	0.7931	0.8468	0.7521	0.8064	0.7559
EC	0.7042	0.8574	0.7357	0.7916	0.7529
SR	**0.8563**	0.9062	0.6730	0.8256	0.7680
WSR-k	0.7903	0.8702	0.7363	0.7178	0.7233
WSR-h	0.8311	0.8850	0.6738	**0.8263**	**0.7756**
WSR-k_s	0.8311	0.9062	0.6693	0.8250	0.7746
WSR-1	0.8560	**0.9083**	0.6720	0.8256	0.7692

The best prediction of each network is shown in bold face

Table 7.21 Two metrics of the 16 algorithms

Algorithm	τ_m-D	L-D	τ_m-U	L-U	τ_m-B	L-B
k	0.7517	−0.1158	0.6976	−0.1318	0.5431	−0.1380
h	0.7638	−0.1307	0.7219	−0.1132	0.5700	−0.1280
k_s	0.6975	−0.1924	0.7252	−0.1275	0.5719	−0.1446
MDD	0.7260	−0.1531	0.7042	−0.1422	0.5476	−0.1418
LR	0.2201	−0.6559	0.6516	−0.3051	0.4536	−0.3211
WLR	0.2444	−0.6211	0.7338	−0.2196	0.5708	−0.2624
ALR	0.3752	−0.4496	0.7959	−0.1514	0.6973	−0.1746
CluR	0.7663	−0.1188	0.7286	−0.0948	0.7860	−0.0713
PR	0.5730	−0.2729	0.4684	−0.3036	0.3542	−0.2830
CN	0.7369	−0.1424	0.8133	−0.0456	0.7931	−0.0562
EC	0.7369	−0.1566	0.8132	−0.0482	0.7529	−0.0661
SR	**0.8820**	−0.0096	0.8290	−0.0427	0.8256	−0.0210
WSR-k	0.7038	−0.1691	0.7199	−0.1000	0.7363	−0.0897
WSR-h	0.7742	−0.0958	0.8104	−0.0531	**0.8263**	−0.0421
WSR-k_s	0.8211	−0.0864	0.8029	−0.0605	0.8250	−0.0486
WSR-1	0.8815	**−0.0067**	**0.8295**	**−0.0324**	0.8256	**−0.0194**

The best and the second best predictions are shown in bold face and with underline, respectively

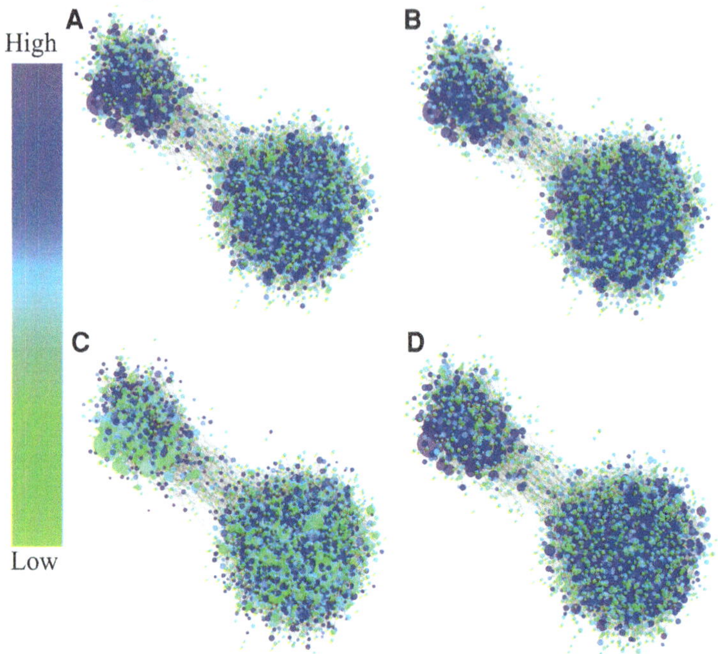

Fig. 7.21 Colormaps for the JamesMoody network. Node sizes are proportional to node degrees. From (**a**) to (**d**), node colors are proportional to the real spread ranges, SR values, PR values, and LR values, respectively. ©[2019] IEEE. Reprinted, with permission, from Ref. [101]

It can be observed that the SR-family methods show excellent predicting ability among all types of networks, where most of the optimal results are offered by the SR-family approaches. The SR-family methods also tend to have larger τ_m and L. The PR and LR-family always lead to the worst prediction results in all types of networks. This implies that the in-linkage based methods can poorly predict the propagation capability, especially in degree uncorrelated networks. As an example, Fig. 7.21 shows the colormaps for the JamesMoody network. It is observed that the colormap of SR can well match the actual spread range. Nevertheless, PR and LR failed. Furthermore, it can be seen that one of the WSRs may outperform the SR in a certain network, but the SR always adapts to all kinds of cases and offers outstanding results. For instance, WSR-h offers the best result in Advogato with $\tau = 0.9130$ and in Email with $\tau = 0.8476$; while for SR, whose accuracy for the two networks are $\tau = 0.8852$ and $\tau = 0.8295$, respectively, has similar performance with WSR-h. Nevertheless, the performances of WSR-h are far beyond that of SR in Cora and WikiVote (see Tables 7.18, 7.19, and 7.20). Furthermore, Table 7.21 reports that the performance of the SR is the best on average.

7.5.5.2 Application in Biological Networks

Recently, identification of key nodes (e.g., genes, proteins) in biological networks has attracted many attentions [37, 46]. Following, two real-world directed biological networks are considered, including the *C. Elegans* Neural [47] (CEN) network with 280 nodes and 2194 edges, the *E. Coli* Transcriptional [49] (ECT) regulatory network with 1706 nodes and 3870 edges. It has been known that 10 command interneurons (AVER, AVEL, AVAR, AVBL, AVBR, AVAL, AVDL, PVCR, AVDR, PVCL) in CEN, 18 global regulators (fnr, crp, fis, fur, mlc, ompR, cpxR, hns, arcA, narL, soxR, soxS, purR, lrp, rob, phoB, CspA, IHF), and 7 key global regulators (fnr, crp, fis, arcA, narL, lrp, IHF) in ECT are the key nodes, and they play profound biological roles in the normal life activities [37]. By taking the mentioned key nodes as gold standards, and in order to evaluate the accuracy of an algorithm, ROC analysis is employed. We can find that degree, PR, H-index, and MDD may outperform in a certain case, but SR gets the 4th, the 3rd, and the 1st positions in the three cases, respectively (See Fig. 7.22). Overall, SR achieves a better balance between precision and generalization.

The examples suggested that the proposed SR can also be applied to biological networks, which will help us to robustly find key nodes in bio-molecular networks.

7.5.6 Summary

Aided by the advanced data mining and machine learning techniques, the applications of complex networks have made great progress in various domains. For instance, the study of coarse-graining of complex networks [170], link prediction [171], recommender systems [172], community identification [173, 174], and vital node mining [116]. This section concentrates on identification of super spreaders, i.e., prediction of node's propagation capability. PR and LR are two popular algorithms. Nevertheless, some evidences have shown that they may fail in some situations. Taking both the merits and the drawbacks of PR and LR into account, the SR algorithm is proposed. The proposed SR reveals that node's propagation capability depends on the spectrum λ_1 of augmented network $\widetilde{\mathbf{A}}$. Simulations of the SIR model on the 32 real-world networks reveal that the SR is very competitive compared with some other 11 renowned algorithms.

We established a probabilistic framework for node ranking problem. Under this framework, we provide a theoretical foundation of the parameter chosen in the spectral-based algorithms. Our framework also tells us that spectral-based algorithms are the maximum likelihood or maximum a posteriori estimates in the fitness network. Note that researchers' empirical studies show that ground node can improve algorithm's performance. Nevertheless, no literature can explain the working mechanism of the ground node. Our framework gives a theoretical foundation on the addition of the ground node. Namely, it encodes a prior knowledge that helps the proposed algorithms to prevent over-fitting.

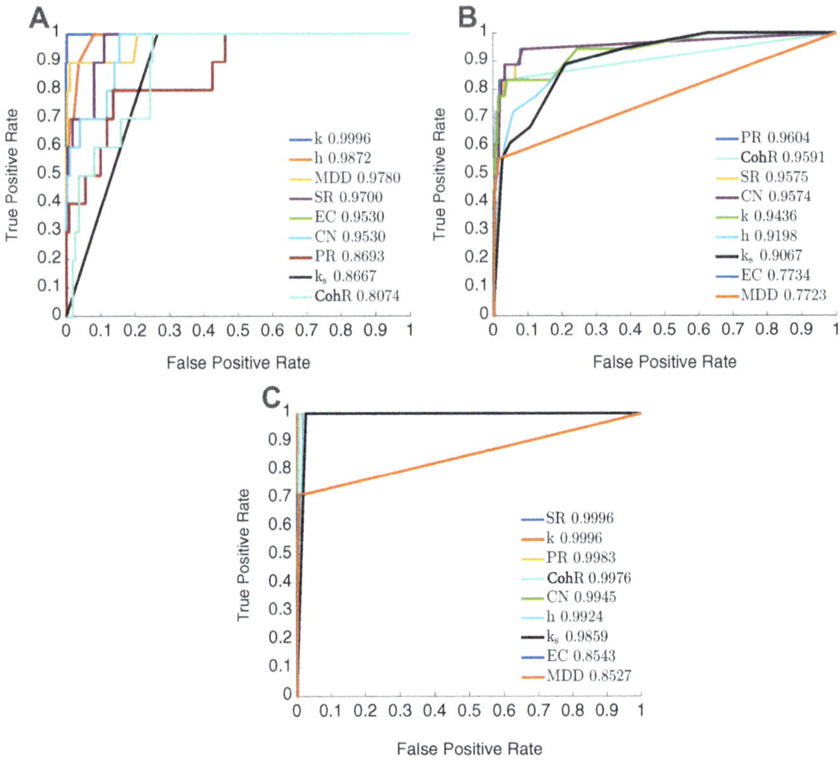

Fig. 7.22 ROCs and Areas under curves (AUCs) for the CEN and ECT. (**a**) Command interneurons in CEN as important nodes. (**b**). Global regulators in ECT as important nodes. (**c**). Key global regulators in ECT as important nodes. The numbers following the legends are AUCs, and algorithms are sorted by AUCs.

The proposed SR cannot only be applied to any type of complex networks, it also shows superiority in identification of key neurons and TFs in biological networks. In view of these facts, we conclude that the SR can help us to better understand the pattern of spread dynamics and to better identify important nodes in various complex networks.

7.6 Discussions and Conclusions

In this chapter, we mainly discuss some new measures to rank the nodes in biomolecular networks or general complex networks. Ranking nodes in complex systems is an important yet hot topic in recent years. Due to the complexity of real-world systems, it is generally difficult to establish a measure to rank nodes in different systems. On one hand, different measures consider different aspects of the nodes; on the other hand, the definition of node importance is different under different circumstances. For example, in bio-molecular systems, important nodes can be disease genes or proteins, they can also be housekeeping genes or proteins, conserved genes or proteins, which are determined by the concerns of researchers.

As to the scopes of human bio-molecular systems, especially human PPI networks, researchers have obtained some new insights through the ranking of nodes in PPI networks. For example, in the year 2006, Xu et al. [175] investigated the discovering of hereditary disease genes by topological features in PPIs network. Their investigations revealed that the hereditary disease genes ascertained from the OMIM database in the literature-curated PPIs network are characterized by a larger degree, tendency to interact with other disease genes, more common neighbors, and quick communication with each other. However, those properties could not be detected from the network identified from the high-throughput Y2H mapping approach and predicted PPI networks. The K-nearest neighbors classifier based on those features was created and on average gained overall prediction accuracy of 0.76 in cross-validation test. Then the classifier was applied to 5262 genes on human genome and predicted 178 novel disease genes. Some of the predictions have been validated by biological experiments. In 2008, Wu et al. [176] proposed a computational framework that integrates human PPIs, disease phenotype similarities, and known gene-phenotype associations to capture the complex relationships between phenotypes and genotypes. They developed a tool named CIPHER to predict and prioritize disease genes, and they showed that the global concordance between the human protein network and the phenotype network reliably predicts disease genes. Recently, based on several topological features of human PPI network and the statistical factorial analysis theory, we develop an integrative measure to find disease genes closely related to *esophageal squamous cell carcinoma*. Experiments validate the clinical relevance [177].

Up to now, new measures for node importance in complex networks are constantly proposed, such as the controllable centrality [178], the PhysarumSpreader [179], the H-index [122], just to name a few. The related investigations have potential implications in networked medicine and personalized medicine.

References

1. Freeman, L.C.: Centrality in social networks: conceptual clarification. Social Netw. **1**, 215–239 (1978)
2. Albert, R., Jeong, H., Barabási, A.L.: Error and attack tolerance of complex networks. Nature **406**, 378–482 (2000)
3. Pastor-Satorras, R., Vespignani, A.: Epidemic spreading in scale-free networks. Phys. Rev. Lett. **86**, 3200–3203 (2001)
4. Chen, Y., Lü, J., Yu, X., Lin, Z.: Consensus of discrete-time second order multi-agent systems based on infinite products of general stochastic matrices. SIAM J. Control Optim. **51**, 3274–3301 (2013)

5. Chen, Y., Lü, J., Lin, Z.: Consensus of discrete-time multi-agent systems with transmission nonlinearity. Automatica **49**,1768–1775 (2013)
6. Newman, M.E.J.: The structure and function of complex networks. SIAM Rev. **45**,167–256 (2003)
7. Carmi, S., Havlin, S., Kirkpatrick, S., Shavitt, Y., Shir, E.: A model of Internet topology using k-shell decomposition. Proc. Natl. Acad. Sci. USA. **104**,11150–11154 (2007)
8. Colizza, V., Flammini, A., Serrano, M.A., Vespignani, A.: Detecting rich-club ordering in complex networks. Nat. Phys. **2**,110–115 (2006)
9. Alon, U.: An introduction to systems biology: design principles of biological circuits. Chapman & Hall/CRC (2007)
10. Wang, P., Lü, J.: Control of genetic regulatory networks: opportunities and challenges. Acta Automat. Sin. **39**, 1969–1979 (In Chinese) (2013)
11. Milo, R., Shen-Orr, S., Itzkovitz, S., Kashtan, N., Chklovskii, D., Alon, U.: Network motifs: simple building blocks of complex networks. Science **298**, 824–827 (2002)
12. Shen-Orr, S., Milo, R., Mangan, S., Alon, U.: Network motifs in the transcriptional regulation network of *Escherichia coli*. Nat. Genet. **31**, 64–68 (2002)
13. Milo, R., Itzkovitz, S., Kashtan, N., Levitt, R., Shen-Orr, S., Ayzenshtat, I., Dheffer, M., Alon, U.: Superfamilies of evolved and designed networks. Science **303**,1538–1542 (2004)
14. Alon, U.: Network motifs: theory and experimental approaches. Nat. Rev. Genet. **8**, 450–461 (2007)
15. Wang, P., Lu, R., Chen, Y., Wu, X.: Hybrid modeling of the general middle-sized genetic regulatory networks. IEEE Int. Symp. Circ. Syst., Beijing, China, **May 19–22**, 2103–2106 (2013)
16. Goldenberg, J., Han, S., Lehmann, D., Hong, J.: The role of hubs in the adoption process. J. Market. **73**,1–13 (2009)
17. Canali, C., Lancellotti, R.: A quantitative methodology based on component analysis to identify key users in social networks. Int. J. Social Netw. Mining **1**, 27–50 (2012)
18. Probst, F., Grosswiele, L., Pfleger, R.: Who will lead and who will follow: identifying influential users in online social networks. Business and Informat. Syst. Eng. **3**, 179–193 (2013)
19. Kintali, S.: Betweenness centrality: algorithms and lower bounds. arXiv: 0809.1906v2 [cs.DS] (2008)
20. Chen, D., Lü, L., Shang, M.S., Zhou, T.: Identifying influential nodes in complex networks. Physica A **391**, 1777–1787 (2012)
21. Brin, S., Page, L.: Reprint of: The anatomy of a large-scale hypertextual web search engine. Comput. Netw. **56**(18), 3825–3833 (2012)
22. Lü, L., Zhang, Y., Yeung, C.H., Zhou, T.: Leaders in social networks, the delicious case. PLoS One **6**, e21202 (2011)
23. Gao, C., Lan, X., Zhang, X., Deng, Y.: A bio-inspired methodology of identifying influential nodes in complex networks. PLoS One **8**, e66732 (2013)
24. Salathé, M., Jones, J.H.: Dynamics and control of diseases in networks with community structure. PLoS Comput. Biol. **6**, e1000736 (2010)
25. Koschützki, D., Schwöbbermeyer, H., Schreiber, F.: Ranking of network elements based on functional substructures. J. Theor. Biol. **248**, 471–479 (2007)
26. Koschützki, D., Lehmann, K.A., Peeters, L., Richter, S., Tenfelde-Podehl, D., Zlotowski, O.: Network analysis: methodological foundations, Lect. Notes Comput. Sci., Tutorial. **3418**, Centrality Indices, Springer, Berlin, 16–61 (2005)
27. Koschützki, D., Schreiber, F.: Centrality analysis methods for biological networks and their application to gene regulatory networks. Gene Regulat. Syst. Biol. **2**,193–201 (2008)
28. Sporns, O., Kötter, R.: Motifs in brain networks. PLoS Biol. **2**, e369 (2004)
29. Sporns, O., Honey, C.J., Kötter, R.: Identification and classification of hubs in brain networks. PLoS One **2**, e1049 (2007)
30. Harriger, L., van den Heuvel, M.P., Sporns, O.: Rich club organization of macaque cerebral cortex and its role in network communication. PLoS One **7**, e46497 (2012)

31. Rubinov, M., Sporns, O.: Complex network measures of brain connectivity: Uses and interpretations. NeuroImage **52**, 1059–1069 (2010)
32. Kitsak, M., Gallos, L.K., Havlin, S., Liljeros, F., Muchnik, L., Stanley, H.E., Makse, H.A.: Identification of influential spreaders in complex networks. Nat. Phys. **6**, 888–893 (2010)
33. Wang, P., Tian, C., Lu, J.: Identifying influential spreaders in artificial complex networks. J. Syst. Sci. Complex. **27**, 650–665 (2014)
34. Newman, M.E.J.: A measure of betweenness centrality based on random walks. Social Netw. **27**, 39–54 (2005)
35. Pastor-Satorras, R., Smith, E., Solé, R.V.: Evolving protein interaction networks through gene duplication. J. Theor. Biol. **222**, 199–210 (2003)
36. Bonacich, P., Lloyd, P.: Eigenvector-like measures of centrality for asymmetric relations. Social Netw. **23**, 191–201 (2001)
37. Wang, P., Lü, J., Yu, X.: Identification of important nodes in directed biological networks: a network motif approach. PLoS One **9**, e106132 (2014)
38. Mangan, S., Alon, U.: Structure and function of the feed-forward loop network motif. Proc. Natl. Acad. Sci. USA. **100**, 11980–11985 (2003)
39. Mangan, S., Zaslaver, A., Alon, U.: The coherent feed-forward loop serves as a sign-sensitive delay element in transcription networks. J. Mol. Biol. **334**,197–204 (2003)
40. Goentoro, L., Shoval, O., Kirschner, M.W., Alon, U.: The incoherent feedforward loop can provide fold-change detection in gene regulation. Mol. Cell **36**, 894–899 (2009)
41. Wang, P., Lü, J., Ogorzalek, M.J.: Global relative parameter sensitivities of the feed-forward loops in genetic networks. Neurocomput. **78**,155–165 (2012)
42. Wang, P., Lü, J., Zhang, Y., Ogorzalek, M.J.: Global relative input-output sensitivities of the feed-forward loops in genetic networks. Proc. 31th Chin. Contr. Conf., Hefei, China, **July 25–27**, 7376–7381 (2012)
43. Wuchty, S., Oltvai, Z.N., Barabási, A.L.: Evolutionary conservation of motif constituents in the yeast protein interaction network. Nat Genet. **35**, 176–179 (2003)
44. Pearson K.: On lines and planes of closest fit to systems of points in space. Philos. Mag. **2**, 559–572 (1901)
45. Härdle, W.K., Simar, L.: Applied multivariate statistical analysis. Springer-Verlag, Berlin Heidelberg (2012)
46. Wang, P., Yu, X., Lü, J.: Identification and evolution of structurally dominant nodes in protein-protein interaction networks. IEEE Trans. Biomed. Circ. Syst. **8**, 87–97 (2014)
47. Chen, B.L., Hall, D.H., Chklovskii, D.B.: Wiring optimization can relate neuronal structure and function. Proc. Natl. Acad. Sci. USA. **103**, 4723–4728 (2006)
48. Varshney, L.R., Chen, B.L., Paniagua, E., Hall, D.H., Chklovskii, D.B.: Structural properties of the *Caenorhabditis elegans* neuronal network. PLoS Comput. Biol. **7**, e1001066 (2011)
49. Huerta, A.M., Salgado, H., Thieffry, D., Collado-Vides, J.: RegulonDB: a database on transcriptional regulation in *Escherichia coli*. Nucl. Acids Res. **26**, 55–59 (1998)
50. Costanzo, M.C., Crawford, M.E., Hirschman, J.E., et al.: YPD^{TM}, PombePD^{TM} and WormPD^{TM}: model organism volumes of the BioKnowledge^{TM} Library, an integrated resource for protein information. Nucl. Acids Res. **29**, 75–79 (2001)
51. Altun, Z.F., Herndon, L.A., Crocker, C., Lints, R., Hall, D.H. (eds) (2002–2012) WormAtlas. http://www.wormatlas.org/.
52. Martínez-Antonio, A., Collado-Vides, J.: Identifying global regulators in transcriptional regulatory networks in bacteria. Curr. Opin. Microbiol. **6**, 482–489 (2003)
53. Weickert, M.J., Adhya, S.: The galactose regulon of *Escherichia coli*. Mol. Microbiol. **10**, 245–251 (1993)
54. Chou, S., Lane, S., Liu, H.: Regulation of mating and filamentation genes by two distinct Ste12 complexes in *Saccharomyces cerevisiae*. Mol. Cell. Biol. **26**, 4794–4805 (2006)
55. Laloux, I., Dubois, E., Dewerchin, M., Jacobs, E.: TEC1, a gene involved in the activation of Tyl and Tyl-Mediated gene expression in *Saccharomyces cerevisiae*: cloning and molecular analysis. Mol. Cell. Biol. **10**, 3541–3550 (1990)
56. Fawcett T.: An introduction to ROC analysis. Pattern Recogn. Lett. **27**, 861–874 (2006)

57. Alonzo, T.A., Pepe, M.S.: Using a combination of reference tests to assess the accuracy of a new diagnostic test. Statist. Med. **18**, 2987–3003 (1999)
58. Rutjes, A.W.S., Reitsma, J.B., Coomarasamy, A., Khan, K.S., Bossuyt P.M.M.: Evaluation of diagnostic tests when there is no gold standard: a review of methods. Health Techn. Assess. **11**, iii, ix–51 (2007)
59. Wheeler, D.L., Barrett, T., Benson, D.A., et al.: Database resources of the National Center for Biotechnology Information. Nucl. Acids Res. **34**, D173–D180 (2006)
60. Gross, L.: Are "ultraconserved" genetic elements really indispensable? PLoS Biol. **5**, e253 (2007)
61. Wuchty, S.: Interaction and domain networks of *yeast*. Proteomics **2**, 1715–1723 (2002)
62. Zhou, S., Mondragon, R.J.: The rich-club phenomenon in the Internet topology. IEEE Commun. Lett. **8**,180–182 (2004)
63. van den, Heuvel, M.P., Sporns, O.: Rich-club organization of the human connectome. J. Neurosci. **31**,15775–15786 (2011)
64. de Reus, M.A., van den, Heuvel, M.P.: Rich club organization and intermodule communication in the cat connectome. J. Neurosci. **33**,12929–12939 (2013)
65. Towlson, E.K., Vértes, P.E., Ahnert, S.E., Schafer, W.R., Bullmore, E.T.: The rich club of the *C. elegans* neuronal connectome. J. Neurosci. **33**, 6380–6387 (2013)
66. Opsahl, T., Colizza, V., Panzarasa, P., Ramasco, J.J.: Prominence and control: the weighted rich-club effect. Phys. Rev. Lett. **101**, 168702 (2008)
67. Wang, E., Lenferink, A., O'Connor-McCourt, M.: Cancer systems biology: exploring cancer-associated genes on cellular networks. Cell Mol. Life Sci. **64**,1752–1762 (2007)
68. Wang, P., Lü, J., Yu, X., Liu, Z.: Duplication and divergence effect on network motifs in undirected bio-molecular networks, IEEE Trans. Biomed. Circ. Syst. **9(3)**, 312–320 (2015)
69. Wang, X., Gulbahce, N., Yu, H.: Network-based methods for human disease gene prediction. Brief Funct. Genomics. **10**, 280–293 (2011)
70. Östlund, G., Lindskog, M., Sonnhammer,E.L.: Network-based identification of novel cancer genes. Mol. Cell Proteom. **9**, 648–655 (2010)
71. Schwikowski, B., Uetz, P., Fields, S.: A network of protein-protein interactions in *yeast*. Nat. Biotechnol. **18**,1257–1261 (2000)
72. Yu, H., Braun, P., Yıldırım, M.A. et al.: High-quality binary protein interaction map of the *yeast* interactome network. Science **322**,104–110 (2008)
73. Jin, Y., Turaev, D., Weinmaier, T., Rattei, T., Makse, H.A.: The evolutionary dynamics of protein-protein interaction networks inferred from the reconstruction of ancient networks. PLoS One **8**, e58134 (2013)
74. Stark, C., Breitkreutz, B.J., Reguly, T., Boucher, L.,Breitkreutz, A., Tyers, M.: Biogrid: a general repository for interaction datasets. Nucl. Acids Res. **34**,D535–D539 (2006)
75. Payne, W.E., Garrels, J.I.: *Yeast* Protein database (YPD): a database for the complete proteome of *Saccharomyces cerevisiae*. Nucl. Acids Res. **25**, 57–62 (1997)
76. Mewes, H.W., Frishman, D., Mayer, K.F. et al.: MIPS: analysis and annotation of proteins from whole genomes in 2005. Nucl. Acids Res. **34**, D169–D172 (2006)
77. Xenarios, I., Rice, D.W., Salwinski, L., Baron, M.K., Marcotte, E.M., Eisenberg, D.: DIP: the database of interacting proteins. Nucl. Acids Res.**28**, 289–291 (2000)
78. Uetz, P., Giot, L., Cagney, G. et al.: A comprehensive analysis of protein-protein interactions in *Saccharomyces cerevisiae*. Nature **403**, 623–627 (2000)
79. Ito,T.,Tashiro,K., Muta,S. et al.: Toward a protein-protein interaction map of the budding *yeast*: a comprehensive system to examine two-hybrid interactions in all possible combinations between the *yeast* proteins. Proc. Natl. Acad. Sci. USA. **97**, 1143–1147 (2000)
80. Solé, R.V., Pastor-Satorras, R., Smith, E., Kepler, T.B.: A model of large-scale proteome evolution. Adv. Complex Syst. **5**, 43–54 (2002)
81. Vázquez, A., Flammini, A.,Maritan, A., Vespignani, A.: Modeling of protein interaction networks. Complexus **1**, 38–44 (2003)
82. Ispolatov, I., Krapivsky, P. L., Yuryev, A.: Duplication-divergence model of protein interaction network. Phys. Rev. E **71**, 061911 (2005)

83. Wan, X., Cai, S., Zhou, J., Liu, Z.: Emergence of modularity and disassortativity in protein-protein interaction networks. Chaos **20**, 045113 (2010)
84. Xu, C., Liu, Z., Wang, R.: How divergence mechanisms influence disassortative mixing property in biology. Physica A **389**, 643–650 (2010)
85. Zhao,D., Liu, Z., Wang, J.: Duplication: a mechanism producing disassortative mixing networks in biology. Chin. Phys. Lett. **24**, 2766–2768 (2007)
86. Watts, D.J., Strogatz, S.H.: Collective dynamics of 'small-world' networks. Nature **393**, 440–442 (1998)
87. Dorogovtsev, S.N., Mendes, J.F.F.: Evolution of networks. Adv. Phys. **51**, 1079–1187 (2002)
88. Wang, P., Yu, X., Lü, J.: Identification of important nodes in artificial bio-molecular networks. IEEE Int. Symp. Circuits Syst. *June 1–5*, 1267–1270 (2014)
89. Bertolazzi, P., Bock, M.E., Guerra, C.: On the functional and structural characterization of hubs in protein-protein interaction networks. Biotechnol. Adv. **31**, 274–286 (2013)
90. Ashburner, M., Ball, C.A., Blake, J.A., et al.: Gene ontology: tool for the unification of biology. The Gene Ontology Consortium. Nat. Genet. **25**, 25–29 (2000)
91. Sterner, D.E., Grant, P.A., Roberts, S.M., et al.: Functional organization of the *yeast* SAGA complex: distinct components involved in structural integrity, nucleosome acetylation, and TATA-binding protein interaction. Mol. Cell Biol. **19**, 86–98 (1999)
92. Koutelou, E., Hirsch, C.L., Dent, S.Y.R.: Multiple faces of the SAGA complex. Curr. Opin. Cell Biol. **22**, 374–382 (2010)
93. Wagner, A.: The *yeast* protein interaction network evolves rapidly and contains few redundant duplicate genes. Mol. Biol. Evol. **18**,1283–1292 (2001)
94. Jancura, P., Marchiori, E.: A survey on evolutionary analysis in PPI networks, Protein interaction/book 2. InTech (2011)
95. Jeong, H., Mason, S.P., Barabási, A.L., Oltvai, Z.N.: Lethality and centrality in protein networks. Nature **411**, 41–42 (2001)
96. Chen, B., Wu, W., Wang, Y., Li, W.: On the robust circuit design schemes of biochemical networks: steady-state approach. IEEE Trans. Biomed. Circ. Syst. **1**, 91–104 (2007)
97. Chen, B., Chen, P.: Robust engineered circuit design principles for stochastic biochemical networks with parameter uncertainties and disturbances. IEEE Trans. Biomed. Circ. Syst. **2**,114–132 (2008)
98. Gu, M., Chakrabartty, S.: FAST: a framework for simulation and analysis of large-scale protein-silicon biosensor circuits. IEEE Trans. Biomed. Circ. Syst. **7**, 451–459 (2013)
99. Wu, F.: Global and robust stability analysis of genetic regulatory networks with time-varying delays and parameter uncertainties. IEEE Trans. Biomed. Circ. Syst. **5**, 391–398 (2011)
100. Roy, S.: Systems biology beyond degree, hubs and scale-free networks: the case for multiple metrics in complex networks. Syst. Synth. Biol. **6**, 31–34 (2012)
101. Xu, S., Wang, P., Zhang, C., Lü, J.: Spectral learning algorithm reveals propagation capability of complex network. IEEE Trans. Cyber. **49**(12), 4253–4261 (2019)
102. Fan, J., Han, F., Liu, H.: Challenges of big data analysis. Natl. Sci. Rev. **1**(2), 293–314 (2014)
103. Wen, G., Yu, W., Li, Z., et al.: Neuro-adaptive consensus tracking of multiagent systems with a high-dimensional leader. IEEE Trans. Cyber. **47**(7), 1730–1742 (2017)
104. Wen,G., Huang, T., Yu, W., et al.: Cooperative tracking of networked agents with a high-dimensional leader: qualitative analysis and performance evaluation. IEEE Trans. Cyber. **48**(7), 2060–2073 (2018)
105. Yoon, B., Park, Y.: A text-mining-based patent network: analytical tool for high-technology trend. J. High Tech. Manage. Res. **15**(1), 37–50 (2004)
106. Bullmore, E., Sporns, O.: Complex brain networks: graph theoretical analysis of structural and functional systems. Nat. Rev. Neurosci. **10**(3),186–198 (2009)
107. Dunne, J.A., Williams, R.J., Martine, N.D.: Food-web structure and network theory: the role of connectance and size. Proc. Natl. Acad. Sci. USA. **99**(20), 12917–12922 (2002)
108. You, Z. H., Zhou, M., Luo, X., Li, S.: Highly efficient framework for predicting interactions between proteins. IEEE Trans. Cyber. **47**(3), 731–743 (2017)

109. Wang, Z., Yang, C., Chen, H., et al.: Multi-gene co-transformation can improve comprehensive resistance to abiotic stresses in *B napus L.*. Plant Sci. **274**, 410–419 (2018)
110. Wang, P., Wang, D., Lü, J.: Controllability analysis of a gene network for *Arabidopsis thaliana* reveals characteristics of functional gene families. IEEE/ACM Trans. Comput. Biol. Bioinform. **16**(3), 912–924 (2019)
111. Wang, P., Yang, C., Chen, H., et al.: Transcriptomic basis for drought-resistance in *Brassica napus L.*. Sci. Rep. **7**, 40532 (2017)
112. Kossinets, G., Watts, D.J.: Empirical analysis of an evolving social network. Science **311**(5757), 88–90 (2006)
113. Mei, G., Wu, X., Wang, Y., et al.: Compressive-sensing-based structure identification for multilayer networks. IEEE Trans. Cyber. **48**(2), 754–764 (2018)
114. Pagani, G. A., Aiello, M.: The power grid as a complex network: a survey. Physica A **392**(11), 2688–2700 (2013)
115. Zanin, M., Papo, D., Sousa, P.A., et al.: Combining complex networks and data mining: why and how,. Phys. Rep. **635**, 1–44 (2016)
116. Lü, L., Chen, D., Ren, X., et al.: Vital nodes identification in complex networks. Phys. Rep. **650**,1–63 (2016)
117. Liu, W., Deng, Z., Cao, L.: Mining top K spread sources for a specific topic and a given node. IEEE Trans. Cyber. **45**(11), 2472–2483 (2015)
118. Domingos, P., Richardson, M.: Mining the network value of customers. Proc 7th ACM SIGKDD Inter. Conf. Knowledge Discovery and Data Mining, San Francisco, California, USA. 57–66 (2001)
119. Xu, W., Ho, D. W. C., Li, L., Cao, J.: Event-triggered schemes on leader-following consensus of general linear multiagent systems under different topologies. IEEE Trans. Cyber. **47**(1), 212–223 (2017)
120. Zhang, Z., Liu, C., Zhan, X.: Dynamics of information diffusion and its applications on complex networks. Phys. Rep. **651**, 1–34 (2016)
121. Seidman, S.B.: Network structure and minimum degree. Social Netw. **5**(3), 269–287 (1983)
122. Lü, L, Zhou, T., Zhang, Q, Stanley, H.E.: The H-index of a network node and its relation to degree and coreness. Nat. Commun. **7**, 10168 (2016)
123. Li, Q., Zhou, T., Lü, L., Chen, D. : Identifying influential spreaders by weighted leaderrank. Physica A **404**, 47–55 (2014)
124. Xu, S., Wang, P.: Identifying important nodes by adaptive LeaderRank. Physica A **469**, 654–664 (2017)
125. Fortunato, S., Boguñá, M., Flammini, A., Menczer, F.: Approximating PageRank from in-degree. Int. Workshop on Algorithms and Models for the Web-Graph. Springer Berlin Heidelberg, 59–71 (2006)
126. Chen, D., Gao, H., Lü, L., Zhou, T.: Identifying influential nodes in large-scale directed networks: the role of clustering. PLoS One **8**, e77455 (2013)
127. Zeng, A., Zhang, C.: Ranking spreaders by decomposing complex networks. Phys. Lett. A **377**(14), 1031–1035 (2013)
128. Bonacich, P.: Factoring and weighting approaches to status scores and clique identification. J. Math. Sociol. **2**(1), 113–120 (1972)
129. Poulin, R., Boily, M.C., Mâsse, B.R.: Dynamical systems to define centrality in social networks. Social Netw. **22**(3), 187–220 (2000)
130. Barabási, A.L., Albert, R.: Emergence of scaling in random networks. Science **286**(5439), 509–512 (1999)
131. Martinez, N.D., Magnuson, J.J., Kratz, T., Sierszen, M.: Artifacts or attributes? Effects of resolution on the Little Rock Lake food web. Ecological Monographs **61**, 367–392 (1991)
132. Adamic, L.A., Glance, N.: The political blogosphere and the 2004 U.S. election: Divided they blog. Proc. 3rd Int. Workshop on Link Discovery, LinkKDD'05. (ACM, New York, NY, USA), 36–43 (2005)
133. Rocha, L.E.C., Liljeros, F., Holme, P.: Information dynamics shape the sexual networks of internet-mediated prostitution. Proc. Natl. Acad. Sci. USA. **107**(13), 5706–5711 (2010)

134. Kunegis, J.: American revolution network dataset, KONECT (2016)
135. Barnes, R., Burkett, T.: Structural redundancy and multiplicity in corporate networks. Int. Network for Social Netw. Anal. **30**(2), (2010)
136. Wikimedia Foundation (2010) Wikimedia downloads (http://dumps.wikimedia.org/).
137. Gleiser, P.M., Danon, L.: Community structure in jazz. Adv. Complex Syst. **6**(4), 565–573 (2003)
138. Newman, M.E.J.: Finding community structure in networks using the eigenvectors of matrices. Phys. Rev. E **74**(3), 036104 (2006)
139. Boguna, M., Pastor-Satorras, R., Díaz-Guilera, A., Arenas, A.: Models of social networks based on social distance attachment. Phys. Rev. E **70**(5), 056122 (2004)
140. Leskovec, J., Huttenlocher, D., Kleinberg, J.: Governance in social media: a case study of the Wikipedia promotion process in Proc. Int. Conf. on Weblogs and Social Media (2010)
141. Rual, J.F., Venkatesan, K., Hao, T., et al.: Towards a proteome-scale map of the human protein-protein interaction network. Nature **437**(7062), 1173–1178 (2005)
142. Stumpf, M.P.H., Wiuf, C., May, R.M.: Subnets of scale-free networks are not scale-free: sampling properties of networks. Proc. Natl. Acad. Sci. USA.**102**(12), 4221–4224 (2005)
143. Opsahl, T., Panzarasa. P.: Clustering in weighted networks. Social Netw. **31**(2),155–163 (2009)
144. Guimerá, R., Danon, L., Díaz-Guilera, A., Giralt, F., Arenas, A.: Self-similar community structure in a network of human interactions. Phys. Rev. E **68**(6), 065103 (2003)
145. Kunegis, J.: Spanish book network dataset, KONECT, (2016)(accessed on 2016.08.06)
146. Harrison, C.: Bible cross-references (http://chrisharrison.net/projects/bibleviz/index.html(accessed on 2014.08.22)) (2014)
147. Opsahl, T., Agneessens, F., Skvoretz, J.: Node centrality in weighted networks: generalizing degree and shortest paths. Soc. Netw. **3**(32), 245–251 (2010)
148. Opsahl, T.: Why anchorage is not (that) important: binary ties and sample selection (2011) (accessed on 2016.08.06)
149. Spring, N., Mahajan, R., Wetherall, D., Anderson, T.: Measuring ISP topologies with rocketfuel. IEEE/ACM Trans. Networking **12**(1), 2–16 (2004)
150. Batagelj, V., Mrvar, A.: Pajek datasets. (2006) (accessed on 2016.08.06)
151. Subelj, L., Bajec, M.: Model of complex networks based on citation dynamics. Proc. WWW Workshop on Large Scale Netw. Anal. 527–530 (2013)
152. Ley, M.: The DBLP computer science bibliography: evolution, research issues, perspectives. Proc. Int. Symp. String Processing and Information Retrieval. 1–10 (2002)
153. Leskovec, J., Kleinberg, J., Faloutsos, C.: Graph evolution: densification and shrinking diameters. ACM Trans. Knowledge Discovery from Data **1**(1),1–40 (2007)
154. Massa, P., Salvetti, M., Tomasoni, D.: Bowling alone and trust decline in social network sites. The Eighth IEEE Int. Conf. Dependable, Autonomic and Secure Comput. 658–663 (2009)
155. Fire, M., Puzis, R., Elovici, Y.: Link prediction in highly fractional data sets, Subrahmanian V. (ed.). Springer New York, New York, NY, 283–300 (2013)
156. Coleman, J.S.: Introduction to mathematical sociology. London Free Press Glencoe, (1964)
157. Moody, J.: Peer influence groups: identifying dense clusters in large networks. Social Netw. **23**(4), 261–283 (2001)
158. Freeman, L.C., Webster, C.M., Kirke, D.M.: Exploring social structure using dynamic three-dimensional color images. Social Netw. **20**(2),109–118 (1998)
159. Kunegis, J., Hamsterster friendships network dataset, KONECT, (2016)
160. Kendall, M.G.: A new measure of rank correlation. Biometrika **30**(1/2), 81–93 (1938)
161. Xu, S., Wang, P., Lü, J.: Iterative neighbour-information gathering for ranking nodes in complex networks. Sci. Rep. **7**, 41321 (2017)
162. Pastor-Satorras, R., Castellano, C., Van Mieghem, P., Vespignani, A.: Epidemic processes in complex networks. Rev. Mod. Phys. **87**(3), 925–979 (2015)
163. Newman, M.E.J.: Spread of epidemic disease on networks. Phys. Rev. E **66**(1), 016128 (2002)
164. Cohen, R., Erez, K., ben Avraham, D., Havlin S.: Resilience of the Internet to random breakdowns. Phys. Rev. Lett. **85**(21), 4626–4628 (2000)

165. Castellano, C., Pastor-Satorras, R.: Thresholds for epidemic spreading in networks. Phys. Rev. Lett. **105**(21), 218701 (2010)
166. Gleich, D. F.: PageRank beyond the Web. SIAM Rev. **57**(3), 321–363 (2015)
167. Garlaschelli,D., Loffredo, M.I.: Fitness-dependent topological properties of the World Trade Web. Phys. Rev. Lett. **93**(18), 188701 (2004)
168. Metzner, R.: Fundamental of statistical and thermal physics. Phys. Today **20**(12), 85–87 (1967)
169. Bishop, C.M.: Pattern recognition and machine learning. Springer-Verlag, New York (2006)
170. Xu, S., Wang, P.: Coarse graining of complex networks: a k-means clustering approach. Chin. Contr. Deci. Confer. (CCDC), Yinchuan, China, 4113–4118 (2016)
171. Lü, L., Pan, L., Zhou, T., et al.: Toward link predictability of complex networks. Proc. Natl. Acad. Sci. USA. **112** (8), 2325–2330 (2015)
172. Zhou, T., Kuscsik, Z., Liu, J., et al.: Solving the apparent diversity-accuracy dilemma of recommender systems. Proc. Natl. Acad. Sci. USA. **107**(10), 4511–4515 (2010)
173. Yang, L., Cao, X., Jin, D., et al.: A unified semi-supervised community detection framework using latent space graph regularization. IEEE Trans. Cyber. **45**(11), 2585–2598 (2015)
174. He, T., Chan,K. C.C.: MISAGA: an algorithm for mining interesting subgraphs in attributed graphs. IEEE Trans. Cyber. **48**(5), 1369–1382 (2018)
175. Xu, J., Li, Y.: Discovering disease-genes by topological features in human protein-protein interaction network. Bioinformat. **22**, 2800–2805 (2006)
176. Wu, X., Jiang, R., Zhang, M.Q., Li, S.: Network-based global inference of human disease genes. Mol. Syst. Biol. **4**, 189 (2008)
177. Liu, R., Gao, S., Zhao, Y., Wang, P., et al.: Integrative topological analysis of mass spectrometry data reveals molecular features with clinical relevance in esophageal squamous cell carcinoma. Sci. Rep. **6**, 21586 (2016)
178. Liu, Y., Slotine, J.J., Barabási, A.L.: Control centrality and hierarchical structure in complex networks. PLoS One **7**, e44459 (2012)
179. Wang, H., Zhang, Y., Zhang, Z., Mahadevan, S., Deng, Y.: PhysarumSpreader: A new bio-inspired methodology for identifying influential spreaders in complex networks. PLoS One **10**, e0145028 (2015)

Chapter 8
Statistical Analysis of Functional Genes in Human PPI Networks

Abstract In this chapter, based on the up-to-date data from various databases or literature, two large-scale human protein interaction networks and six functional subnetworks have been constructed. The six functional subnetworks consist of essential genes, viable genes, disease genes, conserved genes, housekeeping genes, and tissue-enriched genes, respectively. We illustrate that the human protein interaction networks and most of the subnetworks are sparse, small-world, scale-free, disassortative, and with hierarchical modular structures. The essential, the disease and the housekeeping subnetworks are more densely connected than the others. Statistical analysis reveals that the lethal genes, the conserved genes, the housekeeping genes, and the tissue-enriched genes are with hallmark topological features. Receiver operating characteristic curves indicate that the essential genes can be distinguished from the viable ones with accuracy as high as almost 70%. Closeness, semi-local and eigenvector centralities can distinguish the housekeeping genes from the tissue-enriched ones with accuracy around 82%. Furthermore, statistical analysis of disease genes reveals that some classes of disease genes are with hallmark topological features, especially for the cancer genes, the housekeeping disease genes, and the tissue-enriched disease genes. The findings facilitate the identification of some functional genes via their topological structures in protein interaction networks.

8.1 Backgrounds

With the development of high-throughput technologies, such as the Y2H and the mass spectrometry technique, various interactome resources for species ranging from model organisms to human have been available [1]. Data of protein interactions are especially rich in amount. Many databases have been established to provide the binary protein interactions data for various organisms, such as the online predicted human interaction database (OPHID) [2], the human protein reference database (HPRD) [3], the biological general repository for interaction datasets (BioGRID) [4], the Münich information center for protein sequence (MIPS) [5], the bio-

J. Lü, P. Wang, *Modeling and Analysis of Bio-molecular Networks*,
https://doi.org/10.1007/978-981-15-9144-0_8

molecular interaction network database (BIND) [6], the database of interacting proteins (DIP) [7], the molecular interaction database (MINT) [8], and the protein interaction database (IntAct) [9].

Numerous researches on bio-molecular networks focused on the *E. coli* and *yeast S. cerevisiae* [10–14], which cover hundreds or thousands of nodes. It is estimated that the complete human protein interactome contains about 25,000 protein-coding genes and more than 375,000 interactions among them [15–17]. In 2005, based on the Y2H technology and literature curation, Rual et al. [18] constructed a human protein interaction network with 2784 nodes and 6438 interactions. Stelzl et al. [19] identified a high-confidence HPIN with 401 proteins and 911 interactions. As of January 2014, the BioGRID database has collected 153,379 interactions among 16,287 proteins for the HPIN. The HPRD database has included 39,008 interactions. The OPHID database integrates some of the data from the BioGRID, HPRD, MINT, BIND, IntAct, and the existing literature, which covers 14,601 proteins and 208,763 interactions. The increasingly accumulated datasets facilitate the exploration of the structural characteristics of large-scale HPINs.

Nodes in the HPIN are proteins, which are encoded by the corresponding human genes. Essential genes, viable genes, disease genes, conserved genes, housekeeping (HK) genes, and tissue-enriched (TE) genes are six groups of functional genes. The definitions, data sources, and abbreviations for these genes are summarized in Table 8.1. The essential genes are essential or lethal for the survival of an organism; They can cause lethal phenotypes in gene knockout or mutation experiments [20]. It is reported that only a small fraction of genes are essential or lethal for the survival of an organism [20]. In the following, we treat "essential" and "lethal" as interchangeably. Contrary to the lethal genes, the viable genes cannot lead to lethal effect in gene knockout or mutation experiments. Human diseases are determined by disease genes. Now it is well known that a human disease is rarely a consequence

Table 8.1 The six groups of functional genes

Genes	Definitions or descriptions	References	Subnetwork abbreviations
Essential/Lethal	Essential or lethal for the survival of an organism. They can cause lethal phenotypes in gene knockout or mutation experiments.	[20]	EGS or LGS
Viable	Not lethal for the survival of an organism. They have no lethal effect in gene knockout or mutation experiments.	[20]	VGS
Disease	Related to disease phenotypes.	[22]	DGS
Conserved	Evolutionary conserved in various species.	[21]	CGS
Housekeeping	Required for maintaining basic cellular function. Express in various tissues of an organism under normal and patho-physiological conditions.	[32]	HKGS
Tissue-enriched	Only express in some specific tissues.	[32]	TEGS

of an abnormality in a single gene but caused by the malfunctions of the underlying complex bio-molecular interaction networks [16, 21]. Some works reveal that hubs tend to be essential genes, and they are highly susceptible disease genes [20–23], whereas some works on the HPIN and human disease network obtained conflicting conclusions [15, 20, 24–29]. For example, in 2007, Goh et al. [27] found that the vast majority of disease genes are viable and show no tendency to encode hubs. The conserved genes are cross-species conserved. In 2005 and 2006, Sharan et al. [30] and Gandhi et al. [24] compared the HPIN with that for the other species and found that among more than 70,000 binary protein interactions of human, *yeast, worm*, and *fly*, 42 are common to human, *worm*, and *fly*, and only 16 are common for all species. The cross-species comparisons of protein interactions facilitate the identification of conserved genes. The HK genes are required for maintaining basic cellular function, which express in various tissues of an organism under normal and patho-physiological conditions [31, 32], whereas the TE genes only express in some specific tissues. Among the six groups of functional genes, neither genes are both lethal and viable, nor genes are both HK and TE. Beyond that, there may be overlaps among any other two groups of genes. With the available data, it is crucial to construct functional gene subnetworks in the HPIN and clarify their statistical characteristics. For the identification purpose, it is also crucial to reveal the hallmark graphical features of each group of functional genes.

Motivated by the above problems, we construct two large-scale HPINs and six functional subnetworks. The six subnetworks correspond to the six groups of functional genes. Our aim is to clarify the statistical characteristics of the HPINs and hallmark graphical features of the functional genes, especially for the disease-related genes. We find that the HPINs and most of the functional gene subnetworks are sparse [33], SW [33], SF [34], disassortative [33], and with hierarchical modularity [35]. Based on nine statistical indexes of the HPIN, we reveal hallmark graphical features of some functional genes, which facilitate the identification of them via topological interaction networks.

8.2 Construction of Human PPI Networks and Functional Subnetworks

8.2.1 The Human PPI Networks

Based on the up-to-date data, we construct two undirected HPINs from different sources for mutual corroboration. The first network $HPIN_1$ is based on the BioGRID, HPRD, and literature [18]. The raw network of $HPIN_1$ contains 17,423 proteins and 178,469 undirected interactions. The LCC contains 17,311 nodes and 151,412 undirected interactions (without self-interactions). The second network $HPIN_2$ is constructed from the OPHID. The raw data includes 14,601 unique proteins and 208,763 undirected interactions. The LCC of $HPIN_2$ encompasses

14,423 proteins and 152,146 undirected interactions (without self-interactions). We note that the constructed HPINs are all unweighted. In practice, the interactions among proteins may evolve with time. However, due to the limitations of the current technology, most of the investigated protein interactions are unweighted and static [4–13]. In the following, we use $HPIN_1$ to construct six functional gene subnetworks, since this network covers almost 70% of human genes. Nodes in these subnetworks correspond to proteins encoded by the corresponding functional genes, and edges represent the interactions among functional proteins. It is noted that, though the considered networks may still suffer from the sampling effect [36, 37], they cover more than half of the whole human interactome, and the investigations on such large-scale networks provide hints for the understanding of the whole interactome.

8.2.2 The Lethal and the Viable Subnetworks

It is reported that mutations on essential genes tend to cause developmental abnormality, rather than adult disease [20]. Therefore, some researchers concluded that essential genes are different from disease ones [27, 38], whereas some other researchers assumed essential genes as severe disease genes, since mutations on these genes may hinder human survival and reproduction [26]. Furthermore, it is reported that the most connected genes tend to be essential and are topologically yet functionally central [23, 27].

Dickerson et al. [20] collected 1308 lethal and 697 viable human genes. They investigated the characteristics of these genes, and the relations with disease genes. In this chapter, we use these data to construct the essential and the viable gene subnetworks. Among these data, 1243 lethal and 593 viable genes correspond to the nodes in the $HPIN_1$. There are 7846 and 481 interactions among the 1243 lethal and the 593 viable genes. These subnetworks are shown in Fig. 8.1a and b. The LCCs of the lethal and the viable subnetworks encompass 1098 and 291 nodes and 7846 and 465 interactions, respectively. Obviously, the lethal genes subnetwork is more densely connected than the viable one.

8.2.3 The Disease Subnetwork

Human disease networks have attracted an increasing attention over the last decades [15, 16, 20, 24–29]. In 2007, Goh et al. [27] constructed the first human disease network by linking diseases that share disease genes. They concluded that most of disease genes are viable ones and encode proteins at the periphery of the network. A well-known assertion is that disease genes tend to connect with each other [15]. Therefore, based on the graphical features of disease genes, in 2006, Xu et al. [15]

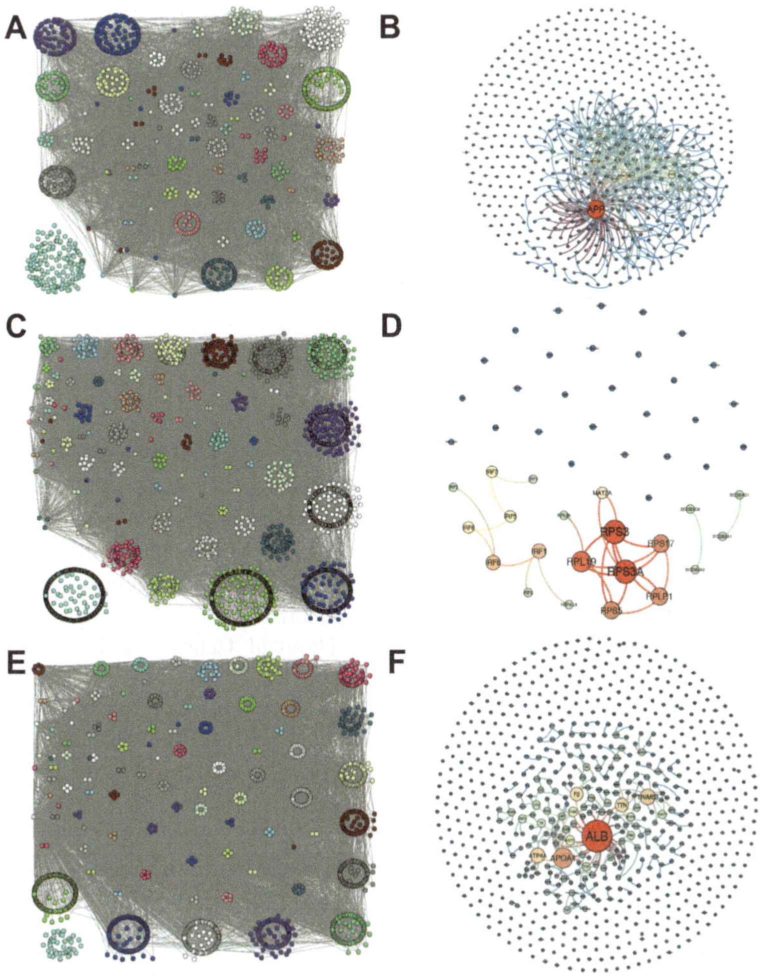

Fig. 8.1 The six subnetworks. (**a**) The essential subnetwork with 1243 nodes, where the LCC contains 1098 nodes and 7846 interactions. (**b**) The viable subnetwork with 593 nodes, where 291 nodes and 465 links consist of the LCC. (**c**) The disease subnetwork with 3068 nodes, where the LCC contains 2549 nodes and 11,438 interactions. (**d**) The conserved subnetwork with 49 nodes, where the LCC contains 9 nodes and 8 interactions. (**e**) The HK gene subnetwork with 1389 nodes, where the LCC includes 1346 nodes and 10,306 interactions. (**f**) The TE subnetwork with 697 nodes, where 138 nodes and 179 links consist of the LCC. Different clusters of colored nodes are with different degrees (Color online). ©[2016] IEEE. Reprinted, with permission, from Ref. [1]

proposed a machine learning algorithm to identify disease genes in the HPIN, where the HPIN contains only 5955 proteins and 17,183 interactions.

In this chapter, disease genes are obtained from the Online Mendelian Inheritance in Man (OMIM) database [22]. We collect 3068 disease genes from 6545 diseases, which are nodes in the $HPIN_1$. Different from Goh et al. [27], we construct a disease gene subnetwork from the interactions among the 3068 disease genes in the $HPIN_1$. The constructed subnetwork is shown in Fig. 8.1c, where 2549 genes and 11,438 interactions consist of the LCC. The rest genes are isolated or loosely connected with two or three of the other nodes.

8.2.4 The Conserved Subnetwork

Conserved genes are conserved in various species [39], which subject to evolutionary pressure of natural selection. The cross-species conservation of genes allows researchers to predict new genes and interactions [16, 24, 29]. For example, if one knows conserved genes in *yeast* and human, then interactions among the conserved genes in human can be predicted by those in *yeast*, since experimental detection of the interactions in *yeast* is more executable [24, 25].

From the homologene database of the National Center for Biotechnology Information database (NCBI) [21], we collect 58 conserved genes among Human, *C. elegans, S. cerevisiae, Drosophila melanogaster, Pan troglodytes, Bos taurus*, and other species. 49 out of the 58 genes are nodes of the $HPIN_1$. The 49 conserved genes and 26 interactions among them consist of the conserved subnetwork, which is shown in Fig. 8.1d. The conserved genes are loosely connected, and many of them are isolated. The LCC only contains 9 nodes and 8 interactions. The second LCC contains 8 nodes. The LCC mainly consists of IRFs, which are a large family of master TFs involved in host immune response, hematopoietic differentiation, and immunomodulation [40]. The second LCC mainly consists of RPLs and RPSs, which are all ribosomal proteins.

8.2.5 The Housekeeping and the Tissue-Enriched Subnetworks

HK genes broadly express in various tissues, which involve in some processes necessary for cell survival. Some HK genes may involve in sustaining cell function, while others may involve in cell maintenance. HK genes tend to produce proteins at steady rates. Errors in their expression can lead to cell death. Like a housekeeper, they keep a cell running smoothly so that it can continue to function [32–44]. The TE genes only express in a few specific tissues. It is reported that the HK genes are more conserved than the other genes and evolve more slowly [43]. Therefore, they have been widely used as experimental control and normalization references in gene expression experiments [32]. Since the TE genes express in one or a few specific tissues or cell types, they can serve as bio-markers of particular tissues or biological processes, and some TE genes may act as drug targets [32].

Identification of the HK genes has attracted an increasing attention over the last decade, which is mainly based on microarray gene expression profiling analysis. For example, in 2008, Zhu et al. [41] reported 1206 HK genes from microarray data, which widely expressed in 18 human tissues. In 2009, She et al. [32] found 1522 HK genes and 975 TE genes from 18,149 genes, where the HK genes highly express in 42 human tissues. In this chapter, genes predicted by She et al. [32] will be used to construct the two subnetworks. We extract 10,306 interactions among 1389 HK genes, where 1346 nodes are connected in the $HPIN_1$. Of the 975 TE genes, 697 are nodes in the $HPIN_1$, but the largest part only contains 138 nodes and 179 links. The constructed networks are shown in Fig. 8.1e, f. Obviously, the HK genes tend to be more densely connected with each other than the TE genes.

8.3 Network Metrics and Connection Ratio

8.3.1 Network Metrics

To clarify the structural characteristics of each network, we consider the LCC of each network and their average degree $\langle k \rangle$, maximum and minimum degrees $k_{\max}, k_{\min}$, network diameter D, average path length (APL), average clustering coefficient $\langle cc \rangle$ and global clustering coefficient C^*, small-world index SW, Pearson correlation coefficient (PCC), and power-law exponent (PLE).

The definitions of average degree, maximum and minimum degrees, and APL are very simple and follow from reference [33]. Network diameter D is defined as the longest shortest path length between any two nodes in the network. There are two definitions of clustering coefficient. From Watts and Strogatz [45], the clustering coefficient of node i is defined as

$$C_i = \frac{2n_i}{k_i(k_i - 1)}, \tag{8.1}$$

where k_i denotes the degree of node i and n_i represents the number of links among the k_i neighbors. The average clustering coefficient $\langle cc \rangle$ of the overall network is obtained by averaging C_i over all the nodes [45]. The definition of the global clustering coefficient [33] is as follows:

$$C^* = \frac{3 \times number\ of\ triangles}{number\ of\ paths\ of\ length\ 2}. \tag{8.2}$$

The SW index [46] is defined as

$$SW = \frac{C^*/C^*_{rand}}{APL/APL_{rand}}. \tag{8.3}$$

$SW > 1$ indicates that the test network is SW. Here, C^* and C^*_{rand} are defined in Eq. (8.2) for the test network and the average of that for randomized networks. APL_{rand} is the averaged APL for randomized networks. For ER random networks with n nodes and average degree $\langle k \rangle$, the clustering coefficient C^*_{rand} can be approximated by Newman [33]:

$$C^*_{rand} = \frac{\langle k \rangle}{n}. \tag{8.4}$$

APL_{rand} can be approximated by Newman [47]:

$$APL_{rand} = \frac{ln(n)}{ln(\langle k \rangle)}. \tag{8.5}$$

In this chapter, we use the approximation of C^*_{rand} and APL_{rand} as defined in Eqs. (8.4) and (8.5) to obtain SW.

The PCC [33] can act as an indicator of assortativity and disassortativity, which is defined as

$$PCC = \frac{M^{-1}\sum_i j_i k_i - \left[M^{-1}\sum_i \frac{1}{2}(j_i + k_i)\right]^2}{M^{-1}\sum_i \frac{1}{2}\left(j_i^2 + k_i^2\right) - \left[M^{-1}\sum_i \frac{1}{2}(j_i + k_i)\right]^2}.$$

Here, M is the total number of edges, k_i, j_i are the degrees of the nodes at the ends of the i'th$(i = 1, 2, \ldots, M)$ edge. $PCC < 0$ indicates the disassortativity and $PCC > 0$ indicates the assortativity, while $PCC = 0$ indicates no degree correlations.

For hierarchical modularity [35], we use the index introduced by Ravasz and Barabási [35] to investigate the constructed networks. If the average clustering coefficient [45] $C(k) \sim k^{\theta}$ for nodes with the degree k $(k = k_{\min}, \cdots, k_{\max})$, and $\theta \approx -1$, then the network is hierarchical modularity.

Power-law degree distribution is an important attribute of SF networks. The degree distribution of a SF network follows $p(k) \sim k^{-PLE}$, and PLE is called the power-law exponent [34].

To clarify hallmark graphical features of different functional genes, we obtain the degree, k-shell [48], clustering coefficient [45], betweenness [49], semi-local centrality [50], eigenvector centrality [33], PR [51], closeness [50], and motif centrality [13, 14, 52] for each node in the HPIN. These indexes for large-scale HPIN will be computed by using the complex networks package for Matlab, which is developed by Lev Muchnik [53].

8.3.2 Connection Ratio

For a group of functional genes, to verify whether they tend to connect with each others, we define connection ratio k_r as

$$k_r = \frac{E_{in}}{E_{among}}. \tag{8.6}$$

Here, E_{in} denotes the number of edges within the group of functional genes, E_{among} represents the number of edges that connect the functional genes and the other genes. For a group of functional genes, large k_r indicates the functional genes tend to connect with each other.

8.4 Statistical Characteristics of the HPINs and the Subnetworks

Based on the network metrics introduced in Sect. 8.3, we obtain the structural characteristics of the HPINs and the six subnetworks, as shown in Table 8.2. For the two HPINs, we show the degree distributions and the curves $C(k)$ versus k in Fig. 8.2. For the six subnetworks, we show their degree distributions and clustering coefficient $C(k)$ versus degree k in Fig. 8.3.

Based on the obtained indexes and figures, firstly, we show the constructed networks are sparse. From Table 8.2, the average degree $\langle k \rangle$ of the largest components of the $HPIN_1$ and $HPIN_2$ are 17.4932 and 21.0977, respectively. The connection densities of them are 0.1011% and 0.1463%, respectively. Therefore, they are sparse. Similarly, from Table 8.2, the six subnetworks are all sparse. The HK, the lethal, and the disease subnetworks are more densely connected than the other networks, and there are many isolated nodes in the other three networks. In the $HPIN_1$, the first two largest degrees are 9638 and 2475, which correspond to proteins UBC and NRF1, respectively. The UBC is encoded by the gene ubiquitin C, which is a HK gene and has wide interactions with the other genes in various human tissues [21]. The NRF1 can activate the expression of some key metabolic genes, regulating cellular growth and nuclear genes required for respiration, heme biosynthesis, mitochondrial DNA transcription, and replication [21]. The gene UBC in the HK gene subnetwork is with the maximum degree 1216, which indicates that the UBC connects with almost 90% of the other HK genes. The degree of the most connected TE protein ALB is 27. The ALB gene can encode albumin, which is a soluble, monomeric protein that comprises about one-half of the blood serum protein. The ALB is also a disease gene, connected with 70 disease genes in the disease subnetwork, which can trigger hepatorenal syndrome and dysalbuminemic hyperthyroxinemia [21].

Table 8.2 Statistical characteristics of the HPINs and the six subnetworks (Data collected: Jan. 2014)

	Networks	$HPIN_1$	$HPIN_2$	EGS	VGS	DGS	CGS	HKGS	TEGS
	Source	BioGrid [4], HPRD [3], Ref. [18]	OPHID [2]	Ref. [20]	Ref. [20]	OMIM [22]	NCBI [21]	Ref. [32]	Ref. [32]
Raw	Nodes	17,423	14,601	1243 (1308)	593 (697)	3068	49 (58)	1389 (1522)	697 (975)
	Edges	178,469	208,763	7846	481	11,458	26	10,306	246
LCC	Node	17,311	14,423	1098	291	2549	9	1346	138
	Edge	151,412	152,146	7846	465	11,438	8	10,306	179
	$\langle k \rangle$	17.4932	21.0977	14.2914	3.1959	8.9745	1.7778	15.3135	2.5942
	k_{max}	9638	1138	165	76	336	3	1216	27
	k_{min}	1	1	1	1	1	1	1	1
	D	11	10	10	10	9	6	6	11
	APL	2.7736	3.5316	3.0623	3.9846	3.5550	2.8333	2.1520	4.6149
	CC	0.2281	0.1412	0.2022	0.0813	0.1457	0.0000	0.4163	0.1548
	C^*	0.0070	0.0381	0.0494	0.0168	0.0251	0.0000	0.0366	0.0334
	SW	8.5168	23.1635	3.2625	1.8746	7.1678	0.0000	3.9471	1.9900
	PCC	−0.0637	−0.0375	−0.0892	−0.1946	−0.1222	−0.4359	−0.1338	−0.1751
	PLE	1.8300	1.5900	1.6340	2.1290	1.8100	–	1.4900	1.6810

LCC: the largest component; the other abbreviations are noted in the text or Table 8.1

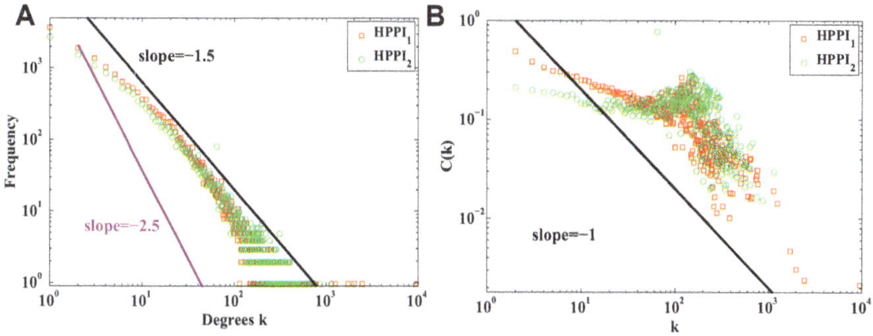

Fig. 8.2 Degree distributions and hierarchical modularity of the HPINs. (**a**) Degree distributions of the two HPINs. (**b**) The average clustering coefficient $C(k)$ as a function of degree k indicates hierarchical modularity. ©[2016] IEEE. Reprinted, with permission, from Ref. [1]

Secondly, we show the SW property of the considered networks. The $APLs$ of the two HPINs are 2.7736 and 3.5316, which indicate any two nodes in the HPINs can have an interaction with each other through a few nodes. Except the conserved subnetwork, the SW indexes for the other subnetworks are all above one, which indicate the small-worldness. Especially, the disease subnetwork is with the strongest SW property ($SW = 7.1678$), which supports the assertion that disease genes tend to be connected [15]. The clustering coefficient and SW for the LCC of the conserved subnetwork are all 0, since there are no links among the neighbors of any node. The diameters of all the networks are between 6 and 11. Compared with network sizes, the diameters of most of the networks are very short, which further support their SW properties.

Thirdly, we discuss whether the constructed networks are SF. From Table 8.2 and Figs. 8.2 and 8.3, except the conserved subnetwork, the LCCs of the HPINs and the constructed subnetworks are all power-law. The PLE of the HPIN falls in the interval [1.5, 2.5]. The approximated $PLEs$ for the two HPINs are 1.8300 and 1.5900, respectively, which are a little bigger than that for the *yeast* [10–13]. The viable and the disease subnetworks have very high $PLEs$, which are 2.1290 and 1.8100, respectively. For the conserved subnetwork, the 9 nodes have only 3 different degree values, and the PLE is not statistically meaningful.

Finally, from the PCC values in Table 8.2, we can conclude that the constructed networks are all disassortative. The disassortativity of these networks indicate that highly connected proteins or genes would connect with low degree ones rather than among themselves. From Figs. 8.2 and 8.3, for the LCCs of the considered networks, the distributions of $C(k)$ versus k roughly disperse along the line with slop -1. Therefore, these networks are hierarchical modularity. The hierarchical modularity of the TE genes can be attributed to the tissue-specific feature of these genes. It is reported that disease genes tend to show specific expression in human tissues where the diseases originated [32, 54, 55]; therefore, the hierarchical modularity of the disease subnetwork may also relate to their tissue-specific feature.

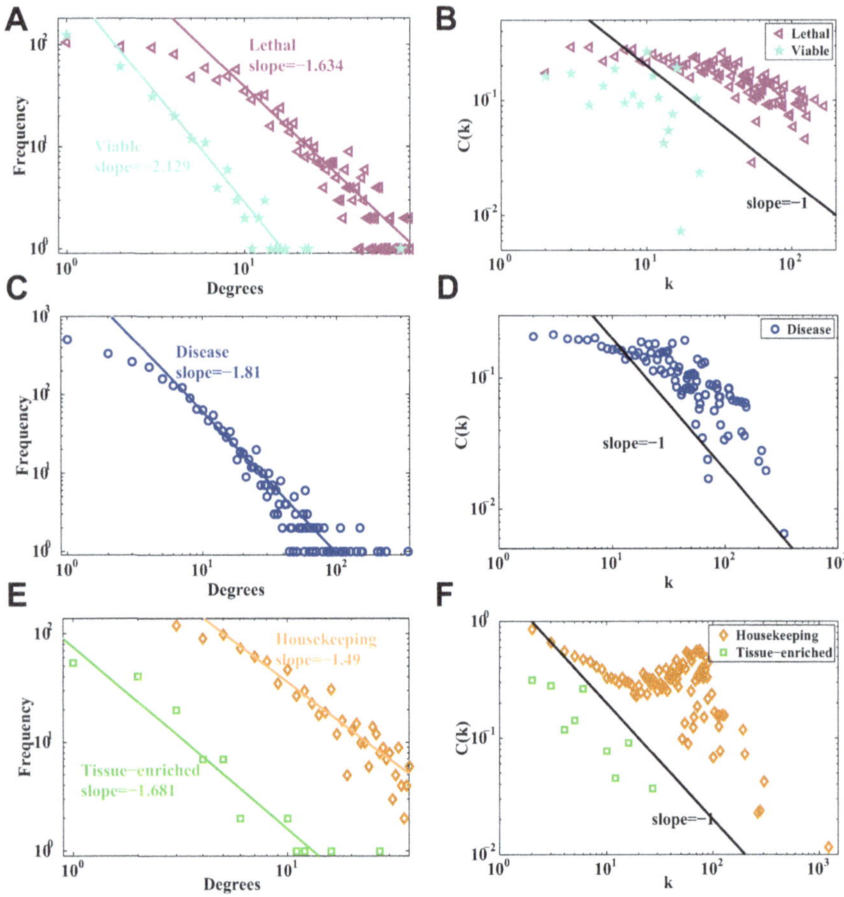

Fig. 8.3 Degree distributions and hierarchical modularity of the five subnetworks. (**a**) Degree distributions. (**b**) The $C(k)$ as a function of degree k in the essential and the viable subnetworks. (**c**) and (**d**) are for the disease subnetwork. (**e**) and (**f**) are for the HK and the TE subnetworks. ©[2016] IEEE. Reprinted, with permission, from Ref. [1]

8.5 Statistical Analysis of Functional Genes in the HPIN

An increasingly popular topic in the field of networked systems biology is to identify functional genes for specific purpose, such as identifying disease genes for drug targets and drug design [16, 27]; identifying HK genes for experimental controls and normalization references in gene expression experiments [32]; and identifying TE genes for bio-markers or drug targets [32]. Graphical features of different genes in the HPIN can provide clues for their identification. Therefore, it is crucial to clarify hallmark graphical features of functional genes.

Table 8.3 shows the statistics of the nine indexes for the HPIN_1, the six groups of functional genes and the unclassified genes, where we have shown the average, standard deviation (Std), coefficient of variation (CV), and median for each index. Moreover, for the six groups of functional genes, we have shown the connection ratio k_r as defined in Eq. (8.6).

Since different indexes have different ranges, to facilitate the comparison among different indexes, we perform range normalized transformation. The range normalized vector for $x = (x_1, \ldots, x_n)'$ is defined as $\tilde{x} = (\tilde{x}_1, \ldots, \tilde{x}_n)'$, where

$$\tilde{x}_i = \frac{x_i - \min(x)}{\max(x) - \min(x)}.$$

Thus, $0 \leq \tilde{x}_i \leq 1$. Fig. 8.4 shows the error bars of the nine normalized average indexes for the seven groups of genes. Since some groups of genes are overlapped, to exclude the effect of such overlap, we draw Venn diagrams [56] for the six groups of functional genes, as shown in Fig. 8.5. The six groups of functional genes totally contain 5693 unique genes and can be classified into 27 nonempty subgroups. For simplicity, we mainly consider the non-overlapped functional genes and the disease-related genes.

Based on Fig. 8.5a, Fig. 8.5b shows the error bars of the nine range normalized indexes for the 27 groups of functional genes. To investigate the distributions of different groups of functional genes according to each index, we divide the nodes in the HPIN_1 into five groups of roughly equal sizes according to each index (top 20% ranked, (20%, 40%] ranked, ..., (80%, 100%] ranked) and show the percentages of each subgroup of functional genes in each interval in Figs. 8.6 and 8.7; where Fig. 8.6 shows the cases for the six groups of non-overlapped functional genes, Fig. 8.7 shows the cases for the overlapped subgroups. In the following, based on Figs. 8.4, 8.5, 8.6, and 8.7 and Table 8.3, we explore the hallmark features of the lethal, the conserved, the disease, the HK, and the TE genes, as well as the disease-related overlapped genes.

8.5.1 *The Lethal Genes*

Of the 1243 lethal genes, 614 (49.40%) are non-overlapped with the other groups, while 420 (33.79%) are also disease ones (Fig. 8.5). From Table 8.3, the lethal genes are with remarkable graphical features; compared with the other groups, the 1243 lethal genes are featured with very large average degree, k-shell, semi-local and motif centralities, the second largest average betweenness, and PR, but with very low average clustering coefficient. The indexes for the 593 viable genes are almost all lower than those for the lethal ones. Moreover, the CV of the motif centrality, the semi-local centrality, and the PR for the lethal genes are smaller than that for the viable ones, which indicate the lethal genes tend to cluster according to these indexes. From Table 8.3, $k_r = 0.3414$ for the lethal genes, which is 4.8 times larger

Table 8.3 Statistics of the nine structural indexes for functional genes

Index	Statistics	$HPIN_1$	Lethal	Viable	Disease	Conserved	HK	TE	Unclassified
Degree	Average	17.4932	49.6018	24.5396	25.0763	36.5306	56.0382	9.6399	10.9222
	Std	88.0187	86.4643	90.3825	65.6941	53.9278	277.0540	18.7116	33.6815
	CV	5.0316	1.7432	3.6831	2.6198	1.4762	4.9440	1.9411	3.0838
	Median	5	22	9	9	9	22	4	4
	k_r	–	0.3414	0.0708	0.4190	0.0299	0.3602	0.0790	–
k-shell	Average	8.9300	17.9517	11.0270	11.5532	16.7347	19.5961	6.4548	6.9297
	Std	10.0170	12.5114	10.0692	10.8406	16.7826	13.7832	6.9950	8.2009
	CV	1.1217	0.6969	0.9131	0.9383	1.0029	0.7034	1.0837	1.1834
	Median	5	16	8	8	9	17	4	3
Betweeness	Average	3.1548e4	7.4532e4	5.9773e4	3.9935e4	1.9733e4	2.1377e5	1.1835e4	1.0195e4
	Std	1.4793e6	2.2815e5	8.0634e5	3.8335e5	4.2373e4	5.1870e6	4.6928e4	1.8610e5
	CV	46.8893	3.0611	13.4901	9.5995	2.1473	24.2640	3.9652	18.2547
	Median	407.2920	1.1226e4	2.0744e3	2.0291e3	3.9371e3	7.8555e3	246.7625	111.8366
Clustering coefficient	Average	0.2281	0.1841	0.1975	0.2184	0.2649	0.2341	0.1673	0.2338
	Std	0.2783	0.1842	0.2264	0.2420	0.2406	0.2010	0.2530	0.2972
	CV	1.2203	1.0005	1.1466	1.1080	0.9086	0.8587	1.5117	1.2709
	Median	0.1429	0.1421	0.1309	0.1538	0.2088	0.1820	0.0635	0.1291
Semi-local	Average	7.9727e7	1.2726e8	7.4940e7	8.9679e7	1.4313e8	1.5951e8	4.2218e7	6.9256e7
	Std	7.7820e7	1.0046e8	7.7198e7	8.2958e7	1.3235e8	1.1797e8	5.5144e7	6.4524e7
	CV	0.9761	0.7894	1.0301	0.9251	0.9246	0.7396	1.3062	0.9317
	Median	9.8138e7	1.2205e8	7.3892e7	1.0180e8	1.0894e8	1.3611e8	1.0733e7	9.8138e7

Eigenvector	Average	0.0039	0.0076	0.0039	0.0047	0.0095	0.0099	0.0020	0.0031
	Std	0.0065	0.0093	0.0059	0.0067	0.0125	0.0157	0.0028	0.0039
	CV	1.6596	1.2225	1.5051	1.4119	1.3068	1.5875	1.4121	1.2665
	Median	0.0033	0.0052	0.0032	0.0037	0.0045	0.0060	7.3393e−4	0.0032
PR	Average	1.0000	2.4758	1.4377	1.3737	1.6310	2.9127	0.6583	0.6786
	Std	6.0722	4.0405	6.3279	3.7288	2.0271	20.0326	0.9193	2.0468
	CV	6.0722	1.6320	4.4013	2.7144	1.2429	6.8776	1.3964	3.0164
	Median	0.4244	1.1897	0.6531	0.6269	0.6406	1.1697	0.4004	0.3527
Closeness	Average	2.0988e−5	2.2309e−5	2.0672e−5	2.1306e−5	2.2137e−5	2.3250e−5	1.9370e−5	2.0704e−5
	Std	3.1473e−6	2.6846e−6	3.2595e−6	3.0535e−6	2.8775e−6	1.9457e−6	2.9639e−6	3.1627e−6
	CV	0.1500	0.1203	0.1577	0.1433	0.1300	0.0837	0.1530	0.1528
	Median	2.3300e−5	2.3600e−5	2.0600e−5	2.3300e−5	2.3400e−5	2.3800e−5	1.8400e−5	2.3300e−5
Motif centrality	Average	82.6217	266.8584	78.9545	112.6320	528.5306	437.2167	16.4634	35.6780
	Std	865.9390	803.5000	352.0152	516.7670	1.2456e3	2.8633e3	77.6538	213.1180
	CV	10.4808	3.0110	4.4585	4.5881	2.3567	6.5490	4.7167	5.9734
	Median	3	38	7	8	12	44	1	1

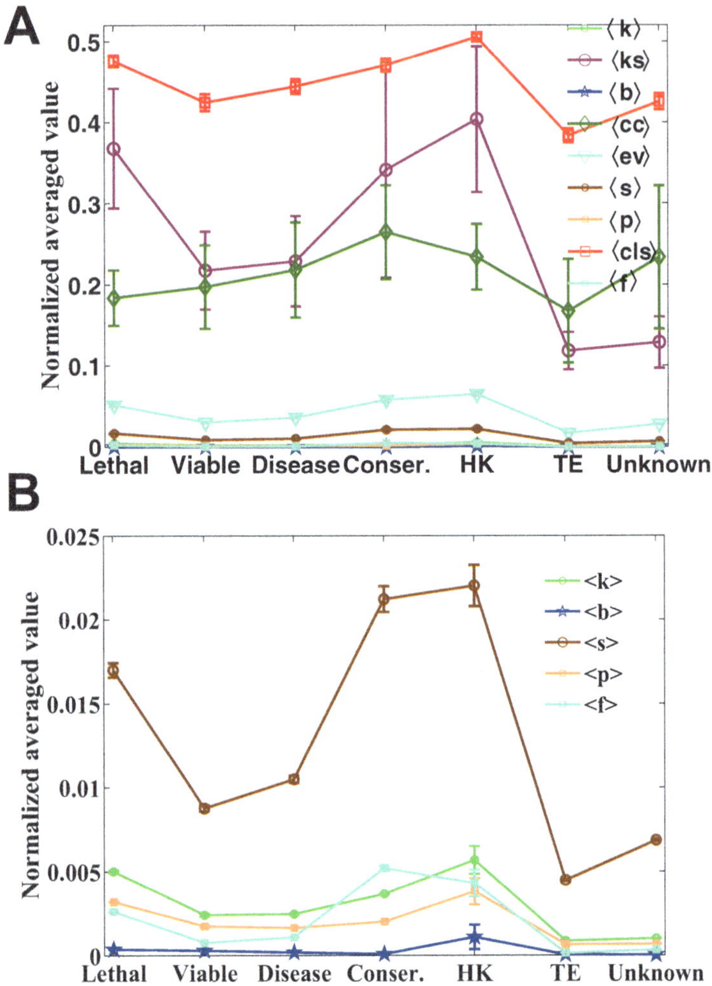

Fig. 8.4 Error bars of the nine range normalized average statistical indexes for different groups of genes in the HPIN_1. $\langle k\rangle$, $\langle ks\rangle$, $\langle cc\rangle$, $\langle b\rangle$, $\langle s\rangle$, $\langle ev\rangle$, $\langle p\rangle$, $\langle cls\rangle$, and $\langle f\rangle$ represent the range normalized average degree, k-shell, clustering coefficient, betweenness, semi-local centrality, eigenvector centrality, PR, closeness, and motif centrality, respectively. Panel (B) is part of (A), which is enlarged from (A) to guide the eyes. ©[2016] IEEE. Reprinted, with permission, from Ref. [1]

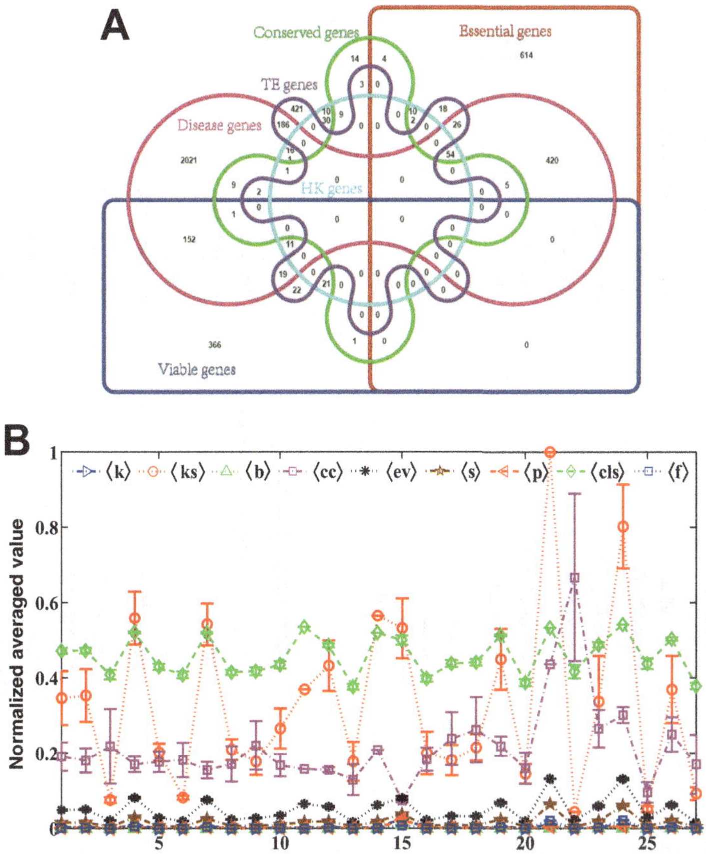

Fig. 8.5 Venn diagrams for the six sets of functional genes (**a**) and error bars of the nine range normalized indexes for the 27 nonempty subgroups of functional genes (**b**). The 27 nodes at the horizontal ordinate denote functional gene subgroups, which are abbreviated as L, LD, LC, LHK, LTE, LCD, LHKD, LTED, V, VD, VC, VHK, VTE, VCD, VHKD, VTED, D, CD, HKD, TED, CHKD, CTED, C, CHK, CTE, HK, and TE, respectively. Here, L, V, D, and C denote the lethal, viable, disease, and conserved genes, respectively. ©[2016] IEEE. Reprinted, with permission, from Ref. [1]

than that for the viable ones. Therefore, compared with the viable genes, the lethal genes tend to be connected with each other.

By taking the 1243 lethal genes and the 593 viable genes as a binary classification, we draw ROC curves for each index, as shown in Fig. 8.8a. Here, for each index, the thresholds are taken as the fractional ranks 0%, 5%, . . . , 100%. In Fig. 8.8a, the AUCs for the nine indexes are 0.6675, 0.6679, 0.6386, 0.5239, 0.6770, 0.6781, 0.6606, 0.6659, and 0.6700, respectively. The accuracy for most of the indexes is around 66%, while the clustering coefficient is with the lowest accuracy, which is not much better than random classifications. The highest *acc*s of the nine

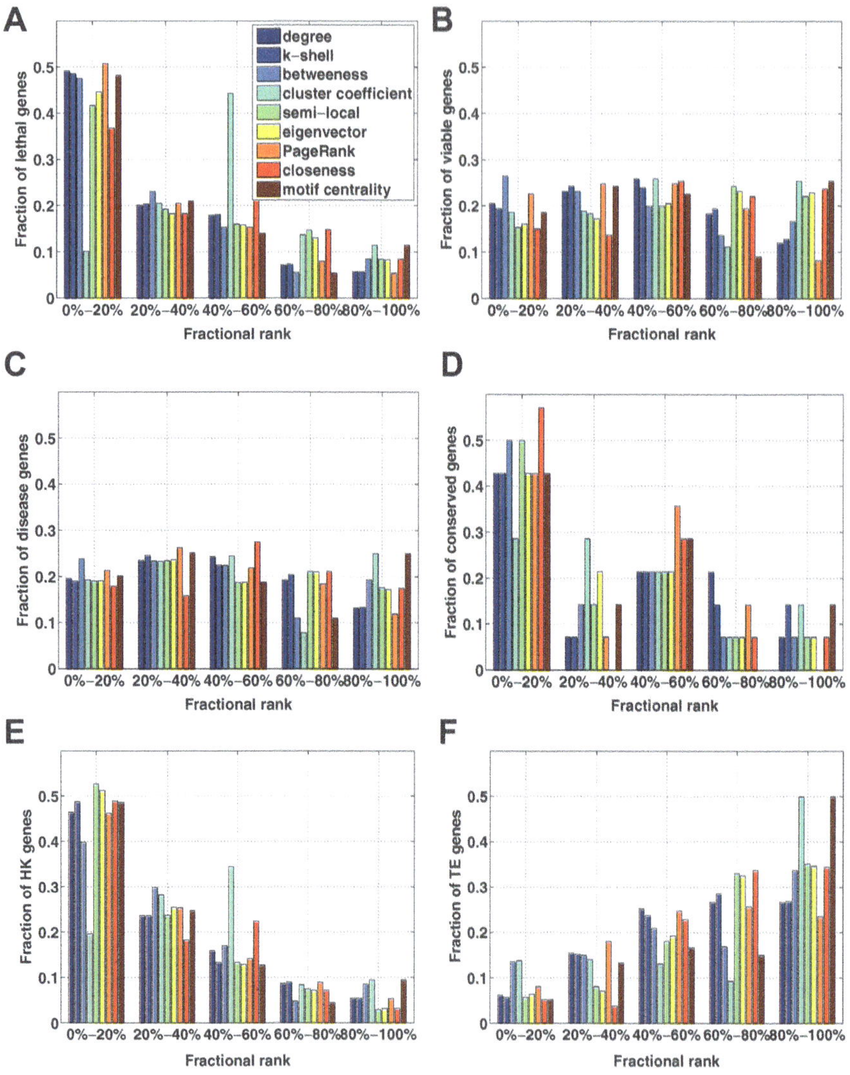

Fig. 8.6 Percentages of non-overlapped functional genes in five intervals according to the nine indexes. For each index, nodes in the $HPIN_1$ are divided into five groups with roughly equal numbers of nodes. The vertical axis shows the percentages of the functional genes. ©[2016] IEEE. Reprinted, with permission, from Ref. [1]

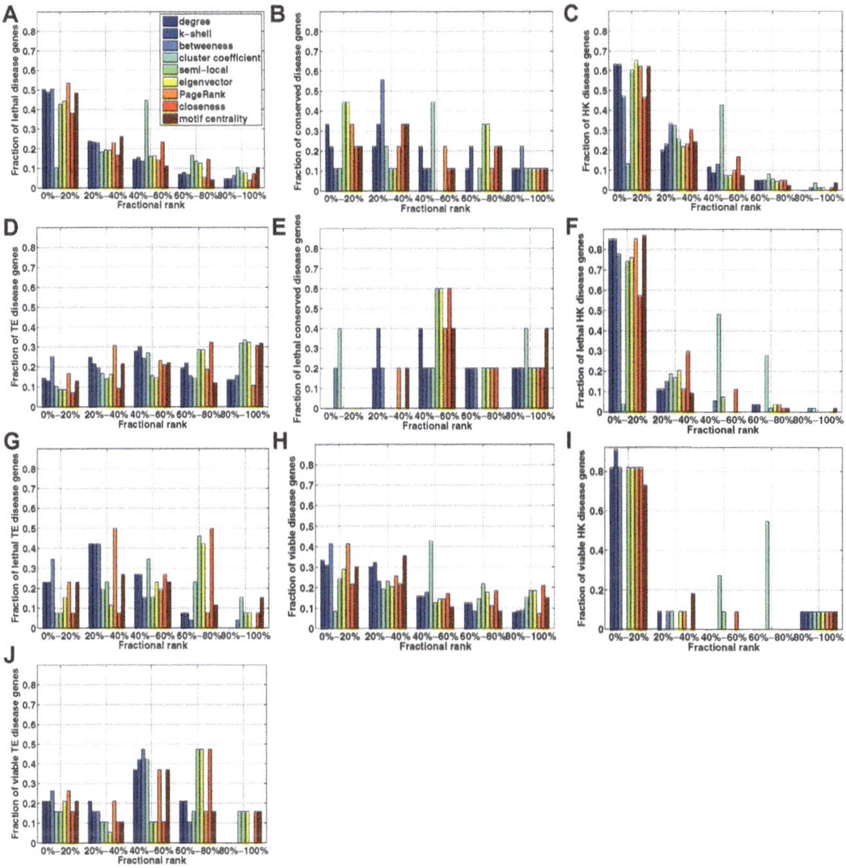

Fig. 8.7 Percentages of various overlapped disease-related subgroups of functional genes in five intervals for each of the nine indexes. ©[2016] IEEE. Reprinted, with permission, from Ref. [1]

indexes are 0.6852, 0.6868, 0.6776, 0.6786, 0.6923, 0.6944, 0.6808, 0.6993, and 0.6786. Obviously, under appropriate threshold, the prediction accuracy for most of the indexes can achieve as high as almost 70%. Therefore, ROC curves reveal that the lethal genes are structurally distinguished from the viable genes. We note that it is an interesting future topic to further find some combined indexes with higher prediction accuracy [13, 57].

For the non-overlapped 614 lethal and 366 viable genes, from Fig. 8.6a, b, more than 50% of the lethal genes are top 20% ranked according to PR, more than 45% are top 20% ranked according to degree, k-shell, betweenness, and motif centrality, whereas according to the clustering coefficient, only 10% are top 20% ranked; however, about 45% of the lethal genes are ranked among (40%, 60%]. The 366 viable genes tend to be uniformly distributed among the five intervals according

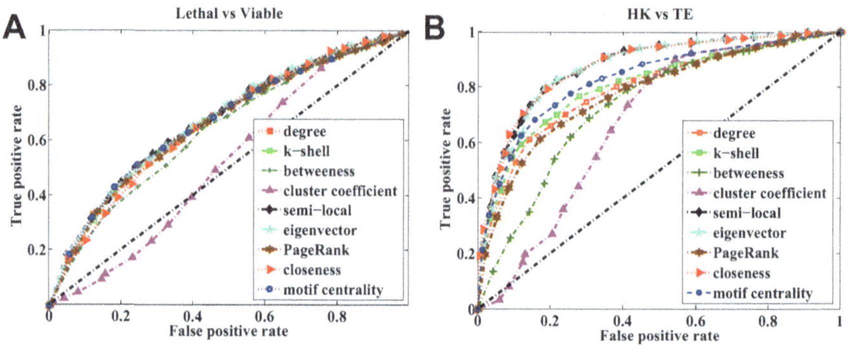

Fig. 8.8 ROC curves reveal the prediction accuracy of the nine indexes. (**a**) The 1243 lethal and the 593 viable genes act as a binary classification. (**b**) The 1389 HK and the 697 TE genes act as a binary classification. ©[2016] IEEE. Reprinted, with permission, from Ref. [1]

to each of the nine indexes. Therefore, compared with the viable genes, the lethal genes tend to be structurally dominant [13] (except for the clustering coefficient).

8.5.2 *The Conserved Genes*

Based on Table 8.3 and Fig. 8.4, the conserved genes are characterized with the largest average motif centrality, which is more than 6 times larger than that for the $HPIN_1$, and the CV of the motif centrality is the smallest. Among the six groups of functional genes, the conserved genes are also with the largest average clustering coefficient. The average betweenness of the conserved genes is lower than that for the $HPIN_1$. The average degree is more than two times larger than that for the $HPIN_1$, which indicates that conserved genes tend to encode proteins with large degrees. However, the connection ratio for the conserved genes is 0.0299, which is the smallest among the six groups. This indicates that the conserved genes are loosely connected with each other but densely connected with the other groups and frequently involve in network motifs, consisting of building blocks [13] of the HPIN.

The Venn diagram indicates that only 14 conserved genes are non-overlapped with the other groups and only take up to 28.57%, which indicates the conserved genes may highly overlap with the other groups and act as multi-functional genes. Figure 8.6d indicates the distributions of the conserved genes are somewhat bimodal under most of the indexes, more than 40% of the conserved genes are top 20% ranked according to eight of the nine indexes; and more than 20% of them are ranked among (40%, 60%] according to all the indexes. However, according to the clustering coefficient, the 14 conserved genes are roughly uniformly distributed.

8.5.3 The Housekeeping and the Tissue-Enriched Genes

Graphical features of the HK genes are significantly different from the TE genes. From Table 8.3 and Fig. 8.4, most of the indexes for the HK genes are the largest. For example, the average betweenness is almost 7 times, and the average degree is more than 3 times larger than that for the $HPIN_1$. However, the Std and CV of the degree, betweenness, PR, and motif centrality for the HK genes are also the largest, which indicate the HK genes are extensively dispersed according to these indexes, whereas the Std and CV of the closeness, semi-local centrality and clustering coefficient for the HK genes are the smallest, which indicate that these genes tend to be clustered according to these indexes. More importantly, the medians of the closeness and semi-local centrality for the HK genes are far larger than that for $HPIN_1$, which indicates they may be efficient in identifying HK genes. TE genes are featured with the lowest of most of the indexes, which indicates they tend to be at the periphery of the network. The unclassified group contains 11,618 genes. The averaged indexes for these genes are similar to the TE ones, which may because most of the unclassified genes are probably enriched in specific tissues.

From the Venn diagram (Fig. 8.5), 1030 of the 1389 HK genes (74.15%) and 421 of the 697 TE genes (60.40%) are non-overlapped with the others. Among the rest 359 HK genes, 102, 21, 161, and 9 are overlapped with the lethal, the viable, the disease, and the conserved ones, respectively. Thus, about 11.59% of the HK genes are disease ones. For the rest 276 TE genes, 18, 22, 186, and 3 TE genes are also the lethal, the viable, the disease, and the conserved ones, respectively. Therefore, about 26.69% of the TE genes are disease ones. Obviously, compared with the HK genes, TE genes are more likely to overlap with the other groups and are more closely related to disease.

Distributions of the non-overlapped 1030 HK and 421 TE genes in Fig. 8.6e, f indicate the significant differences between them. The distributions of the two groups are opposite. Except the clustering coefficient, more than 40% of the HK genes are ranked at the top 20% in the $HPIN_1$ according to the other eight indexes, whereas almost 50% of the TE genes are ranked at the tail 20% according to the clustering coefficient and the motif centrality. By considering the 1389 HK and the 697 TE genes as a binary classification, Fig. 8.8b shows the ROC curves for the nine indexes. The AUC for the closeness, the semi-local, and the eigenvector centralities are 0.8763, 0.8741, and 0.8706, respectively. The $accs$ for the three indexes are 0.8236, 0.8226, and 0.8198, respectively. Therefore, high closeness, semi-local, or eigenvector centralities are hallmark features of the HK genes; while on the contrary, the TE genes tend to be with small centrality scores.

8.5.4 *The Disease Genes*

8.5.4.1 Statistical Features of the Disease Genes

From Table 8.3 and Fig. 8.4, it seems the disease genes have no notable graphical features. However, the connection ratio of the disease genes is 0.4190, which is the largest among the six groups and indicates that they tend to be connected, which further verifies the assertion in Ref. [15, 16].

From Fig. 8.5, 2021 of the 3068 disease genes are non-overlapped with the other five groups. For the left 1047 disease genes, 420, 152, 9, 161, and 186 are overlapped with the lethal, the viable, the conserved, the HK, and the TE genes, and they take up to 33.79%, 25.63%, 18.37%, 11.59%, and 26.69% in these groups, respectively. This indicates that the lethal genes and the TE genes may act as suspicious disease genes candidates.

From Fig. 8.6, the 2021 non-overlapped disease genes tend to be uniformly distributed according to the nine indexes. The distributions of the disease and the viable genes are very similar (Fig. 8.6b, c). Therefore, it is difficult to identify non-overlapped disease genes through any one of the nine indexes. Thus, it is interesting to develop integrative structural indexes or some new indexes to effectively characterize these disease genes [13, 15].

8.5.4.2 Classification of the Disease Genes via Functional Overlaps

Hereinafter, based on Figs. 8.6 and 8.7, we further investigate the classification and graphical features of disease-related genes, especially for the overlapped disease genes. 1047 out of the 3068 disease genes are overlapped with the other groups. The 14 overlapped subgroups correspond to a classification of the 1047 disease genes. In fact, 420 genes are also lethal, 152 genes are viable, 9 genes are conserved, 161 genes are HK, 186 genes are TE, 5 genes are lethal conserved, and 54 genes and 26 genes are lethal HK and lethal TE, respectively; while 11 and 19 genes are viable HK and viable TE disease ones. Two genes are conserved TE disease ones, only one gene is viable conserved disease, and only one gene is conserved HK disease.

From the error bars of the nine range normalized indexes as shown in Fig. 8.5, one can conclude that the HK related genes have relatively high average values. The conserved HK disease genes encompass the largest average normalized k-shell and semi-local centrality values. The conserved TE disease genes have the largest clustering coefficient, but the variance of this index is also very large. Genes that belong to the intersections of multiple sets encompass relatively higher mean values, which indicates they may have more identifiable graphical features.

From the distributions of various disease genes as shown in Fig. 8.7, one can also conclude that the HK-related disease genes have more distinguishable graphical features. For examples, most of the HK disease genes, the lethal HK disease genes, and the viable HK disease genes are top 20% ranked according to most of the graphical indexes. Furthermore, the distributions of the lethal disease genes and the HK disease genes are very similar to that for the lethal and the HK genes, respectively. The TE disease, the lethal TE disease, and the viable TE disease genes are all without remarkable graphical features. Almost 50% of the lethal disease genes are top 20% ranked according to most of the nine indexes, while few lethal conserved disease genes and viable disease genes are top ranked.

8.5.4.3 Classification of the Disease Genes Through Disease Categories

In 2007, Goh et al. [27] manually classified some of the OMIM diseases into 21 categories, which include bone disease, cancer, cardiovascular disease, connective tissue disease, dermatological disease, developmental disease, ear–nose–throat disease, endocrine disease, gastrointestinal disease, hematological disease, immunological disease, metabolic disease, multiple disease, muscular disease, neurological disease, nutritional disease, ophthamological disease, psychiatric disease, renal disease, respiratory disease, and skeletal disease. For the 3068 disease genes investigated in this chapter, 1543 different genes are labeled with known categories, and 7 genes are related to two different disease categories; for example, gene MIF is related to both developmental disease and connective tissue disease, and PRPH is related to both ophthamological disease and neurological disease. In the following, for simplicity, we treat the 7 multiple-category genes in the 1543 genes as different genes, and we have totally 1550 disease genes.

The numbers of genes that involve in the 21 categories are shown in Fig. 8.9a. 209, 201, and 168 genes are related to the neurological disease, metabolic disease, and cancer, respectively. The numbers of disease genes in these three categories are relatively more than those in the other categories. On the contrary, only 15 genes are related to nutritional disease and 20 genes are related to psychiatric disease. The nine average graphical indexes for the 21 categories are shown in Fig. 8.9b. From Fig. 8.9b, it is observed that all the 21 categories have the highest average closeness. Among the 21 categories, on one hand, the 168 cancer associated genes have the highest average degree, k-shell, betweenness, semi-local centrality, eigenvector centrality, PR, closeness, and motif centrality; the developmental disease genes are with the highest average clustering coefficient. On the other hand, the metabolic disease genes are with the lowest average degree, betweenness, and PR; the ophthamological disease genes have the lowest average k-shell; the gastrointestinal disease genes are with the lowest average clustering coefficient; the nutritional disease genes have the lowest average semi-local, eigenvector, closeness, and motif centralities. These results indicate that the cancer genes are graphical dominant in disease genes, while the metabolic and the nutritional disease genes tend to encompass some trivial graphical features. The related findings facilitate the future identification of some special disease genes via graphical features.

Hereinafter, by considering the classification from the overlapped functional genes and the above 21 disease categories, Fig. 8.10 shows the cluster dendgrams [58] based on the nine normalized average indexes. Here, Fig. 8.10a shows the case

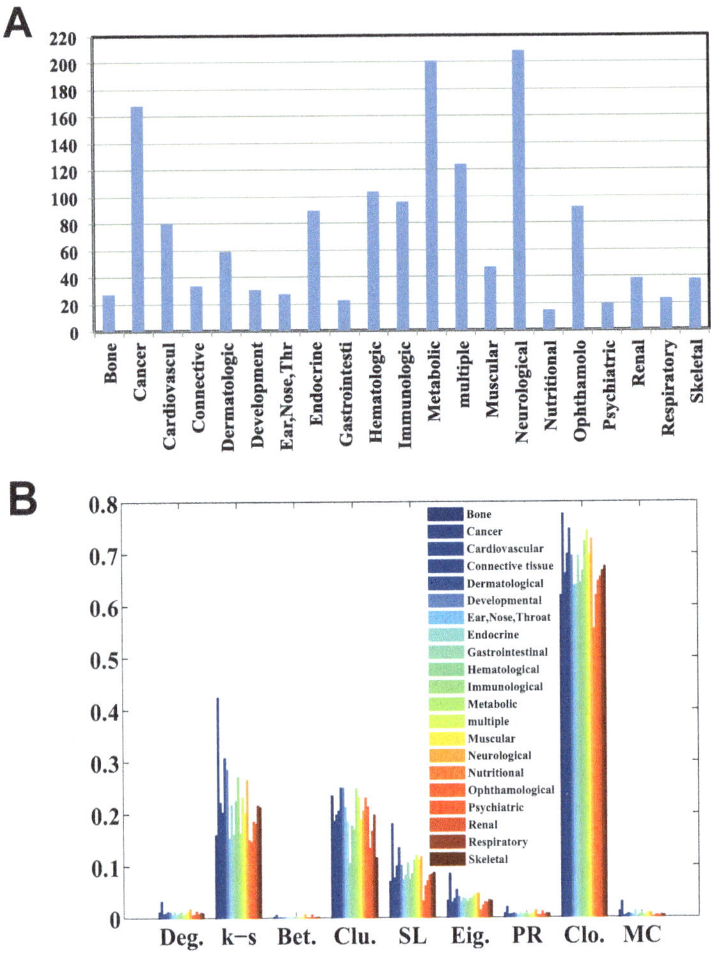

Fig. 8.9 Disease genes classification and their categorical average graphical indexes. (**a**) The numbers of disease genes that involve in the 21 disease categories. (**b**) The nine average graphical indexes for the 21 disease categories.

for the 14 groups of the overlapped disease genes; Fig. 8.10b shows the case for the 21 categories.

From Fig. 8.10a, the 14 groups of disease-related genes can be classified into four groups, the first group includes the disease, conserved disease, viable disease, TE disease, viable TE disease, lethal TE disease, and lethal conserved disease genes, and there are three sets of TE specific genes in this group. The second group includes the lethal disease, HK disease, lethal HK disease, viable conserved disease, and viable HK disease genes, where three sets are related to the HK genes. The last two groups are the conserved HK disease genes and the conserved TE disease

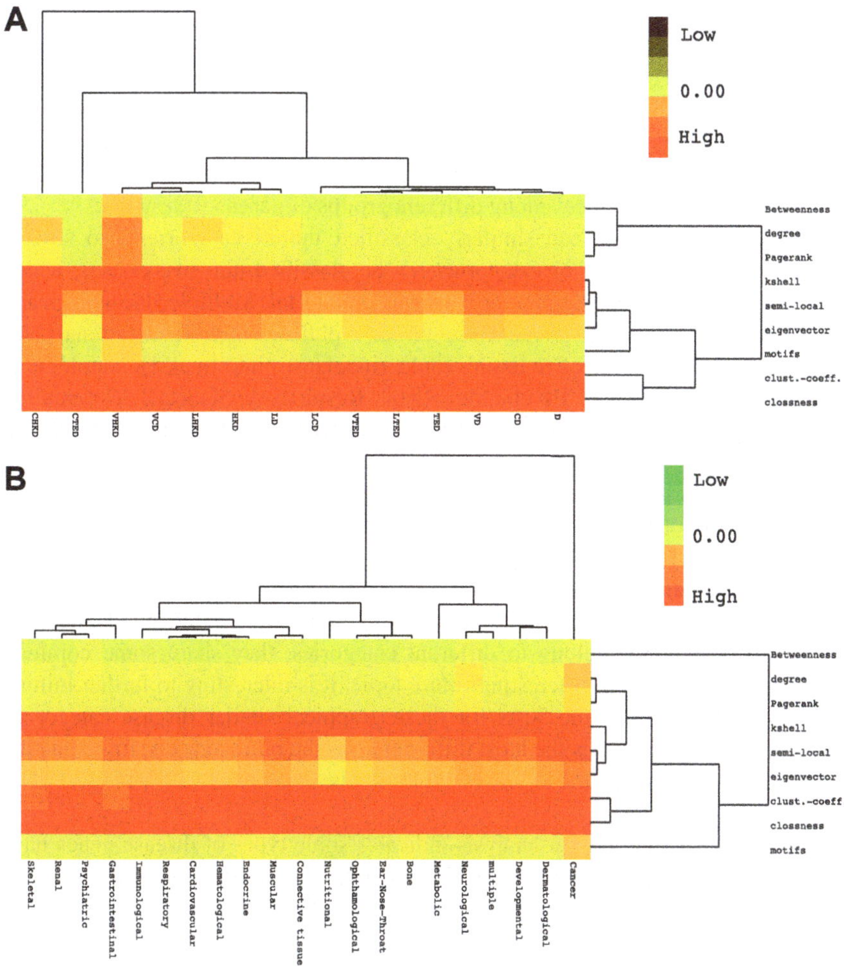

Fig. 8.10 Cluster analysis for the disease genes. The two dendgrams are based on the two different ways of classification. (**a**) The case for the functional overlapped classification, where the nine average normalized indexes are used. (**b**) The case for the 21 categories of diseases. In both panel, the distances between any two groups of disease genes are defined as the Euclidean distance. The pairwise distances among the nine indexes are defined from their correlation coefficients. During the clustering processes, the average linkage method is used [58]. ©[2016] IEEE. Reprinted, with permission, from Ref. [1]

genes, respectively. This further indicates that the HK genes and the TE genes can be well distinguished from each other, especially when these genes are also both conserved. The nine indexes can be classified into three classes. Class one includes the betweenness, the degree, and the PR; class two includes the k-shell, the eigenvector, the semi-local centrality, and the motif centrality; the last class includes the clustering coefficient and the closeness. Indexes belonging to the same class tend to have similar performance among different groups of genes.

From Fig. 8.10b, one can similarly classified the 21 categories into several classes. For example, the 21 categories can be classified into five classes, which are C_1={Cancer}, C_2={Dermatological, Developmental, Multiple, Neurological}, C_3={Metabolic}, C_4={Bone, Ear–nose–throat, Ophthamological, Nutritional}, C_5 = {Cardiovascular, Connective tissue, Endocrine, Gastrointestinal, Hematological, Immunological, Muscular, Psychiatric, Renal, Respiratory, Skeletal}. Hereinafter, based on this figure, we discuss some novel phenomenon. Firstly, it is notable that the cancer genes are the most different from the other categories. In fact, this result just corresponds to the findings in Fig. 8.9b, where the cancer genes are with the largest of most of the normalized indexes. Secondly, some disease categories are with very similar average graphical indexes, for example, the connective tissue and the muscular diseases; the respiratory and the cardiovascular diseases; the endocrine and the hematological diseases. This graphical similarity indicates that though these diseases belong to different categories, they share some common graphical features. As an interesting future topic, it is interesting to further mining the differences on genomic basis for these graphical similar disease categories. Thirdly, the betweenness is the most different from the other indexes, following the motif centrality, whereas the degree and the PR, the semi-local, and the eigenvector centralities are similar.

As a summary, the cluster analysis indicates some types of disease genes have distinguishable graphical features, which facilitates the further classification and identification of them via graphical characteristics of the HPIN.

8.6 Discussions and Conclusions

Systems biology is an emerging interdisciplinary field, highlighting the study of nonlinear behaviors in both small functional bio-molecular circuits [59–62] and complex bio-molecular networks [10–18]. PPI networks are bio-molecular networks. The *yeast* PPI networks have attracted an increasing attention from various aspects over the last decades [10–18]. Recently, the HPINs begin to attract the attention of more and more researchers. Based on the up-to-date data, we construct two large-scale HPINs. Our investigations reveal that the HPINs are sparse, SW, SF, disassortative, and hierarchical modularity. The average degrees of the HPINs are more than 17, and the PLEs are among [1.5, 2.5], which are a little larger than that for the *yeast*. Furthermore, we construct six subnetworks of the $HPIN_1$, where nodes in the six subnetworks correspond to the essential, the viable, the disease, the conserved, the HK, and the TE genes, respectively. Investigations on the statistical characteristics of the six subnetworks show that most of them are sparse, SW, SF, disassortative, and many of them are hierarchical modularity. Moreover, the essential, the disease, and the HK subnetworks are more densely connected than the other subnetworks.

We have obtained nine topological indexes for the HPIN, which include the degree, k-shell, betweenness, clustering coefficient, semi-local centrality, eigenvector centrality, PR, closeness, and motif centrality. The nine indexes are used to investigate hallmark graphical features of functional genes. We find the lethal genes can be distinguished from the viable ones via many indexes. Under appropriate thresholds, the closeness, eigenvector, and semi-local centralities can predict the lethal genes with accuracy around 70%. The disease genes are featured with indexes that are almost all alike but higher than the average level, while the TE genes tend to be at the periphery of the network. The conserved genes are characterized with the most frequently involvement in 3-node motifs and high average clustering coefficient. The HK genes encompass the highest average degree, betweenness, PR, and semi-local centrality, but the four indexes have very large CV. Among the nine indexes for the HK genes, ROC analysis indicates that the closeness, eigenvector, and semi-local centralities can all predict the HK genes with accuracy around 82%. Thus, the investigations facilitate the identification of the lethal or HK genes. Furthermore, we analyze the overlaps among different groups of genes and find that such overlaps allow us to classify disease genes into several groups, where some groups are with distinguishable topological features. Finally, by classifying the disease-related genes into 21 categories, we can reclassify them into several classes via graphical similarity. We find that some classes of functional genes encompass hallmark graphical features, especially for cancer genes and HK and TE related disease genes, while some classes of genes share similar graphical features. It is intriguing to explore the relations between such disease gene classification and disease phenotypes [27].

The HPIN covers almost 70% of the human genes. However, for the six subnetworks, due to the limitation of inadequate data, we have constructed the viable and the TE subnetworks with hundreds of nodes, and the conserved subnetwork has only 49 nodes. It is interesting to further verify our conclusions based on large-scale subnetworks in the future. It is also noted that, for the motif centrality, due to computational complexity, we have only considered the 3-node fully connected network motif. Investigations on the *yeast* protein networks indicate that there are plenty of 4-node network motifs, and it is interesting to consider more motifs in the motif centrality [13, 62]. Furthermore, it is also interesting to consider more topological indexes of the HPIN, which may be helpful to find hallmark graphical features of functional genes that are currently without highlighted statistical features. For the classification of disease genes, another way is based on the OMIM database, where diseases are classified into autosomal dominant inherited, autosomal recessive inherited, X-linkage, Y-linkage, mitochondrial inherited, and chromosome hereditary. It is also interesting to investigate the features of such classification of disease genes.

The thorough investigations on large-scale HPIN and its functional subnetworks help us to understand the emerging rules of the complex organization of human body. The findings provide clues for the future identification of various functional genes. Most importantly, it sheds some light on the characterization, classification, and identification of disease genes and thus has potential implications in networked medicine and biological network control.

References

1. Wang, P., Chen, Y., Lü, J., Wang, Q., Yu, X.: Graphical features of functional genes in human protein interaction network. IEEE Trans. Biomed. Circ. Syst. **10(3)**, 707–720 (2016)
2. Brown, K.R., Jurisica, I.: Online predicted human interaction database. Bioinformat. **21**, 2076–2082 (2005)
3. Peri, S., Navarro, J.D., Amanchy, R., et al.: Development of human protein reference database as an initial platform for approaching systems biology in humans. Genome Res. **13**, 2363–2371 (2003)
4. Stark, C., Breitkreutz, B.J., Reguly, T., et al.: BioGRID: a general repository for interaction datasets. Nucl. Acids Res. **34**, D535–D539 (2006)
5. Güldener, U., Münsterkötter, M., Oesterheld, M., et al.: MPact: the MIPS protein interaction resource on *yeast*. Nucl. Acids Res. **34**, D436–D441 (2006)
6. Bader, G.D., Hogue, C.W.: BIND—a data specification for storing and describing biomolecular interactions, molecular complexes and pathways. Bioinformat. **16**, 465–477 (2000)
7. Xenarios, I., Rice, D.W., Salwinski, L., et al.: DIP: the database of interacting proteins. Nucl. Acids Res. **28**, 289–291 (2000)
8. Zanzoni, A., Montecchi-Palazzi, L., Quondam, M., et al.: MINT: a molecular interaction database. FEBS Lett. **513**, 135–140 (2002)
9. Aranda, B., Achuthan, P., Alam-Faruque, Y., et al.: The IntAct molecular interaction database in 2010. Nucl. Acids Res. **38**, D525–D531 (2010)
10. Uetz, P., Giot, L., Cagney, G., et al.: A comprehensive analysis of protein-protein interactions in *Saccharomyces cerevisiae*. Nature **403**, 623–627 (2000)
11. Yu, H., Braun, P., Yıldırım, M.A., et al.: High-quality binary protein interaction map of the *yeast* interactome network. Science **322**, 104–110 (2008)
12. Wan, X., Cai, S., Zhou, J., Liu, Z.: Emergence of modularity and disassortativity in protein-protein interaction networks. Chaos **20**, 045113 (2010)
13. Wang, P., Yu, X., Lü, J.: Identification and evolution of structurally dominant nodes in protein-protein interaction networks. IEEE Trans. Biomed. Circ. Syst. **8**, 87–97 (2014)
14. Koschützki, D., Schwöbbermeyer, H., Schreiber, F.: Ranking of network elements based on functional substructures. J. Theor. Biol. **248**, 471–479 (2007)
15. Xu, J., Li, Y.: Discovering disease-genes by topological features in human protein-protein interaction network. Bioinformat. **22**, 2800–2805 (2006)
16. Barabási, A.L., Gulbahce, N., Loscalzo, J.: Network medicine: a network-based approach to human disease. Nat. Rev. **12**, 56–68 (2011)
17. Ramani, A.K., Bunescu, R.C., Mooney, R.J., Marcotte, E.M.: Consolidating the set of known human protein-protein interactions in preparation for large-scale mapping of the human interactome. Genome Biol. **6**, R40 (2005)
18. Rual, J., Venkatesan, K., Hao, T., et al.: Towards a proteome-scale map of the human protein-protein interaction network. Nature **437**, 1173–1178 (2005)
19. Stelzl, U., Worm, U., Lalowski, M., et al.: A human protein-protein interaction network: a resource for annotating the proteome. Cell **122**, 957–968 (2005)

20. Dickerson, J.E., Zhu, A., Robertson, D.L., Hentges, K.E.: Defining the role of essential genes in human disease. PLoS One **6**, e27368 (2011)
21. NCBI HomoloGene database: http://www.ncbi.nlm.nih.gov/homologene
22. Hamosh, A., Scott, A.F., Amberger, J.S., et al.: Online Mendelian Inheritance in Man (OMIM), a knowledge base of human genes and genetic disorders. Nucl. Acids Res. **33**, D514-D517 (2005)
23. Jeong, H., Mason, S.P., Barabási, A.L., Oltvai, Z.N.: Lethality and centrality in protein networks. Nature **411**, 41–42 (2001)
24. Gandhi, T.K. B., Zhong, J., Mathivanan, S., et al.: Analysis of the human protein interactome and comparison with *yeast, worm* and *fly* interaction datasets. Nat. Genet. **38**, 285–293 (2006)
25. Gonzalez, M.W., Kann, M.G.: Protein interactions and disease. PLoS Comput. Biol. **8**, e1002819 (2012)
26. Tu, Z., Wang, L., Xu, M., et al.: Further understanding human disease genes by comparing with housekeeping genes and other genes. BMC Genom. **7**, 31 (2006)
27. Goh, K., Cusick, M.E., Valle, D., et al.: The human disease network. Proc. Natl. Acad. Sci. USA. **104**, 8685–8690 (2007)
28. Yıldırım, M.A., Goh, K., Cusick, M.E., Barabási, A.L.: Drug-target network. Nat. Biotech. **25**, 1119–1126 (2007)
29. Ideker, T., Sharan, R.: Protein networks in disease. Genome Res. **18**, 644–652 (2008)
30. Sharan, R., Suthram, S., Kelley, R.M., et al.: Conserved patterns of protein interaction in multiple species. Proc. Natl. Acad. Sci. USA. **102**, 1974–1979 (2005)
31. Eisenberg, E., Levanon, E.Y.: Human housekeeping genes are compact. Trends Genet. **19**, 362–365 (2003)
32. She, X., Rohl, C.A., Castle, J.C., et al.: Definition, conservation and epigenetics of housekeeping and tissue-enriched genes. BMC Genomics **10**, 269 (2009)
33. Newman, M.E.: The structure and function of complex networks. SIAM Rev. **45**, 167–256 (2003)
34. Barabási, A.L., Albert, R.: Emergence of scaling in random networks. Science **286**, 509–512 (1999)
35. Ravasz, E., Barabási, A.L.: Hierarchical organization in complex networks. Phys. Rev. E **67**, 026112 (2003)
36. Stumpf, M.P.H., Wiuf, C., May, R.M.: Subnets of scale-free networks are not scale-free: sampling properties of networks. Proc. Natl. Acad. Sci. USA. **102**, 4221–4224 (2005)
37. Han, J.J., Dupuy, D., Bertin, N., et al.: Effect of sampling on topology predictions of protein-protein interaction networks. Nat. Biotechnol. **23**, 839–844 (2005)
38. Park, D., Park, J., Park, S.G., et al.: Analysis of human disease genes in the context of gene essentiality. Genomics **92**, 414–418 (2008)
39. Jordan, I.K., Rogozin, I. B., Wolf, Y.I., Koonin, E.V.: Essential genes are more evolutionarily conserved than are nonessential genes in bacteria. Genome Res. **12**, 962–968 (2002)
40. Huang, B., Qi, Z., Xu, Z., Nie, P.: Global characterization of interferon regulatory factor (IRF) genes in vertebrates: glimpse of the diversification in evolution. BMC Immunol. **11**, 22 (2010)
41. Zhu, J., He, F., Song, S., Wang, J., Yu, J.: How many human genes can be defined as housekeeping with current expression data? BMC Genomics **9**, 172 (2008)
42. Butte, A.J., Dzau, V.J., Glueck, S. B.: Further defining housekeeping, or "maintenance," genes focus on "a compendium of gene expression in normal human tissues". Physiol. Genomics **7**, 95–96 (2001)
43. Zhang, L., Li, W.: Mammalian housekeeping genes evolve more slowly than tissue-specific genes. Mol. Biol. Evol. **21**, 236–239 (2004)
44. De Jonge, H.J., Fehrmann, R.S., De Bont, E.S., et al.: Evidence based selection of housekeeping genes. PLoS One **2**, e898. (2007)
45. Watts, D.J., Strogatz, S.H.: Collective dynamics of 'small-world' networks. Nature **393**, 440–442 (1998)
46. Humphries, M.D., Gurney, K.: Network 'small-world-ness': a quantitative method for determining canonical network equivalence. PLoS One **3**, e0002051 (2008)

47. Newman, M.E., Strogatz, S.H., Watts, D.J.: Random graphs with arbitrary degree distributions and their applications. Phys. Rev. E **64**, 026118 (2001)
48. Carmi, S., Havlin, S., Kirkpatrick, S., et al.: A model of Internet topology using k-shell decomposition. Proc. Natl. Acad. Sci. USA. **104**, 11150–11154 (2007)
49. Brandes, U.: A faster algorithm for betweenness centrality. J. Math. Sociol. **25**, 163–177 (2001)
50. Chen, D., Lü, L., Shang, M., Zhou, T.: Identifying influential nodes in complex networks. Physica A **391**, 1777–1787 (2012)
51. Brin, S., Page, L.: The anatomy of a large scale hypertextual web search engine. Comput. Networks ISDN Syst. **30**, 107–117 (1998)
52. Wang, P., Lü, J., Yu, X.: Identification of important nodes in directed biological networks: a network motif approach. PLoS One **9**, e106132 (2014)
53. Muchnik, L.: Complex networks package for MatLab (Version 1.6).(2013) http://www.levmuchnik.net/Content/Networks/ComplexNetworksPackage.html
54. Lage, K., Hansen, N.T., Karlberg, E.O., et al: A large-scale analysis of tissue-specific pathology and gene expression of human disease genes and complexes. Proc. Natl. Acad. Sci. USA. **105**, 20870–20875 (2008)
55. Reverter, A., Ingham, A., Dalrymple, B.: Mining tissue specificity, gene connectivity and disease association to reveal a set of genes that modify the action of disease causing genes. BioData. Min. **1**, 8 (2008)
56. Martin, B., Chadwick, W., Yi, T., Park, S.S., et al. VENNTURE—a novel Venn diagram investigational tool for multiple pharmacological dataset analysis. PLoS One **7**, e36911 (2012)
57. Wang, P., Zhang, Y., Lü, J., Yu, X.: Topological characterization of housekeeping genes in human protein-protein interaction network. The 8th Int. Conf. Syst. Biol. (ISB). **Oct. 25–27**, 1–6 (2014)
58. De Hoon, M. J., Imoto, S., Nolan, J., et al.: Open source clustering software. Bioinformat. **20**, 1453–1454 (2004)
59. Wang, P., Lü, J., Yu, X.: Colored noise induced bistable switch in the genetic toggle switch systems. IEEE/ACM Trans. Comput. Biol. Bioinformat. **12**, 579–589 (2015)
60. Chen, B., Chen, P.: Robust engineered circuit design principles for stochastic biochemical networks with parameter uncertainties and disturbances. IEEE Trans. Biomed. Circ. Syst. **2**, 114–132 (2008)
61. Wu, F.: Global and robust stability analysis of genetic regulatory networks with time-varying delays and parameter uncertainties. IEEE Trans. Biomed. Circ. Syst. **5**, 391–398 (2011)
62. Wang, P., Lü, J., Yu, X., Liu, Z.: Duplication and divergence effect on network motifs in undirected bio-molecular networks. IEEE Trans. Biomed. Circ. Syst. **9**, 312–320 (2015)

Part III
Data-Driven Statistical Approaches for Omics Data Analysis

This part mainly considers the data mining of omics data, where some state-of-the-art statistical methods will be introduced.

Chapter 9
Data-Driven Statistical Approaches for Omics Data Analysis

Abstract With the rapid development of high-throughput technology, various omics data for biological systems increases exponentially. A challenge problem for biologists is how to explore useful bioinformatics from high-dimensional or ultrahigh-dimensional omics data. In this chapter, we introduce some recent progresses on the topic of omics data analysis, paying special attention on the related data-driven statistical approaches. Especially, the weighted gene co-expression network analysis, the genome-wide association study, the general linear models, and the hidden Markov random field model will be introduced.

9.1 Backgrounds

9.1.1 Various High-Throughput Sequencing Technologies

With the rapid development of high-throughput technology, various omics data for biological systems increases exponentially [1–17]. The human genome sequence was completed in draft form in 2001 [18, 19]. Shortly thereafter, the genome sequences of several model organisms were determined [20–22]. These feats were accomplished with Sanger DNA sequencing, which was limited in throughput and high cost. Commercially available high-throughput sequencing (HTS) platforms (Figs. 9.1 and 9.2) that have improved the traditional Sanger sequencing include the following: (1) The Illumina Genome Analyzer II that was released by Illumina/Solexa in 2006. Illumina currently has produced a suite of sequencers (MiSeq, NextSeq 500, and the HiSeq series) optimized for a variety of throughputs and turnaround times. In early 2014, Illumina introduced the NextSeq 500 as well as the HiSeq X Ten. The NextSeq 500 is designed as a fast benchtop sequencer for individual labs, while the HiSeq X Ten is a population-scale whole-genome sequencing (WGS) system. (2) Life Technologies commercialized Ion Torrent's semiconductor sequencing technology in 2010 in the form of the benchtop Ion PGM sequencer. The template preparation and sequencing steps are conceptually similar to the Roche/454 pyrosequencing platform [23]. (3) Single-molecule real-time (SMRT) sequencing

J. Lü, P. Wang, *Modeling and Analysis of Bio-molecular Networks*,
https://doi.org/10.1007/978-981-15-9144-0_9

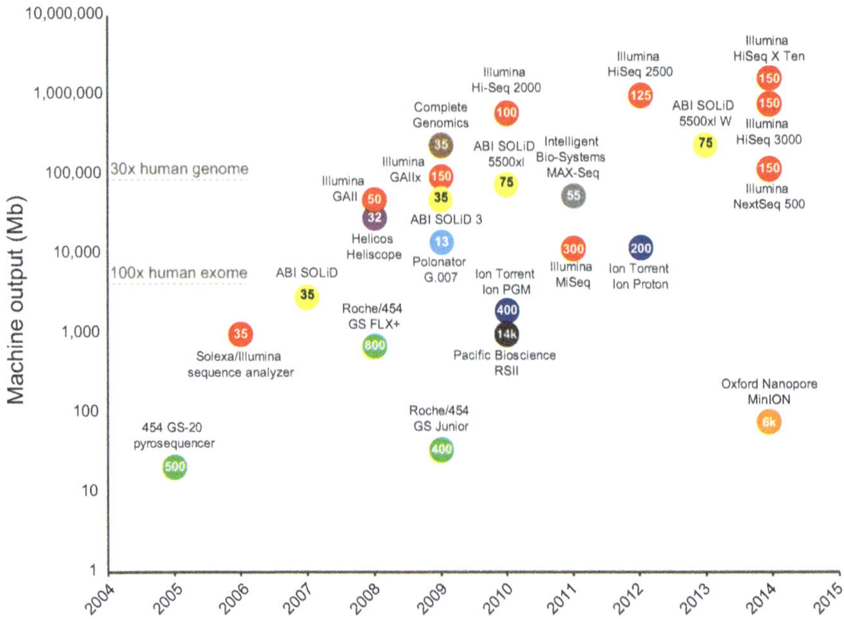

Fig. 9.1 Timeline and comparison of commercial HTS instruments. Plot of commercial release dates versus machine outputs per run is shown. The numbers inside the data points denote the current read lengths. Sequencing platforms are color coded. Reprinted from Ref. [24], with permission from Elsevier

was pioneered by Nanofluidics, Inc. and commercialized by Pacific Biosciences. (4) Oxford Nanopore Technologies. Nanopore-based sequencing is an emerging single-molecule strategy that has made significant progress in recent years, with Oxford Nanopore Technologies leading the development and commercialization of this method.

9.1.2 Applications of High-Throughput Sequencing Technologies

Since the invention of the mentioned sequencing technologies, decreasing costs and increased accessibility have enabled researchers to develop a rich catalog of HTS applications [24] (Table 9.1 and Fig. 9.2), and they have become a powerful tool to study the presence and quantity of bio-molecules in biological samples and have revolutionized omics studies. HTS machines have become widely present in university core facilities and even in individual labs. Some of these technologies were initially developed using DNA microarrays, but many are enabled only by using sequencing. HTS offers many advantages over DNA microarrays. In partic-

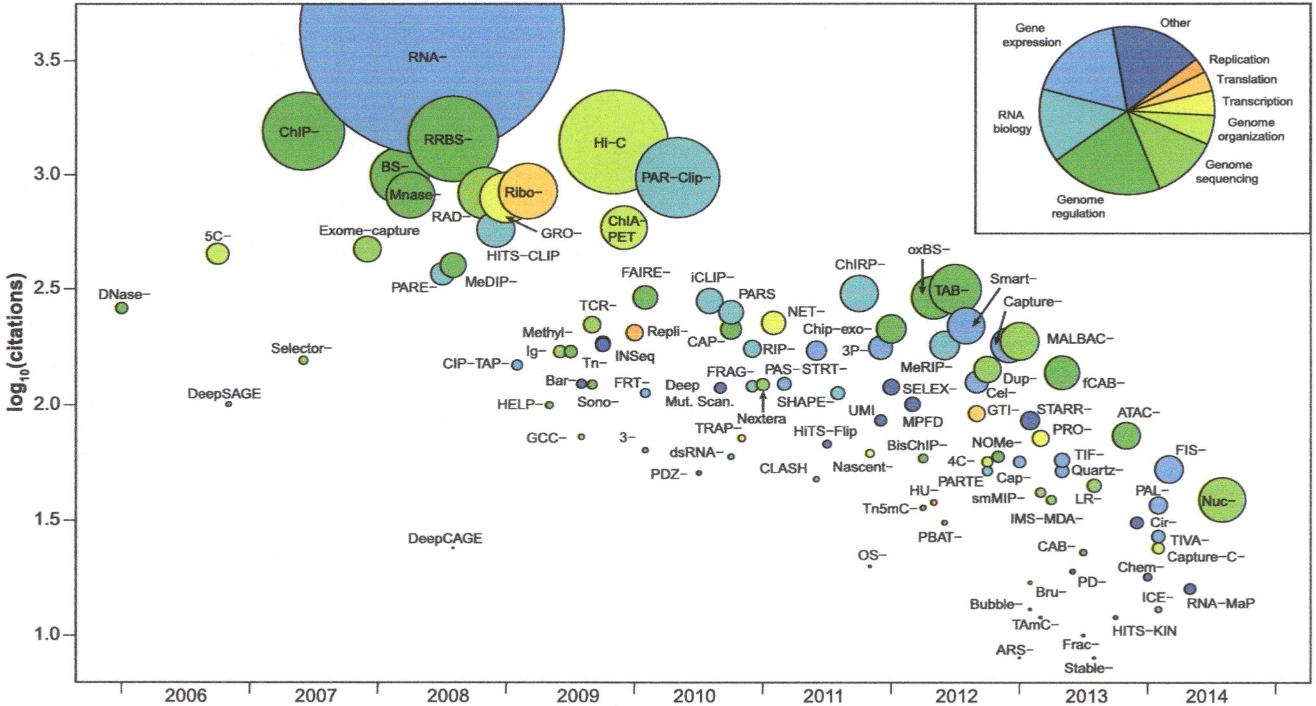

Fig. 9.2 Overview of selected HTS applications. Publication date of a representative article describing a method versus the number of citations that the article received. Methods are colored by category, and the size of the data point is proportional to publication rate (citations/months). The inset indicates the color key as well the proportion of methods in each group. For clarity, the word "seq" has been omitted from the labels. Reprinted from Ref. [24], with permission from Elsevier

Table 9.1 Selected HTS methods [24]

Methods	Purposes	References
RNA-seq	Transcript analysis	[26]
Global run-on sequencing (GRO-seq)	Transcription	[27]
Nascent-seq	Transcription	[28]
Native elongating transcript sequencing (NET-seq)	Transcription	[29]
Ribo-seq	Translation	[30]
Replication sequencing (Repli-seq)	Replication	[31]
Hi-C	Chromatin conformation	[32]
Chromatin interaction analysis by paired-end tag sequencing (ChIA-PET)	Chromatin conformation	[33]
Chromosome conformation capture carbon copy (5-C)	Chromatin conformation	[34]
Chromatin isolation by RNA purification sequencing (ChIRP-seq)	Genome localization	[35]
Reduced representation bisulfite sequencing (RRBS-seq)	Genome methylation	[36]
Bisulfite sequencing (BS-seq)	Genome methylation	[37]
DNAse-seq	Open chromatin	[38]
Assay for transposase-accessible chromatin using sequencing (ATAC-seq)	Open chromatin	[39]
Parallel analysis of RNA structure (PARS)	RNA structure	[40]
Structure-seq	RNA structure	[41]
RNA on a massively parallel array (RNA-MaP)	RNA–protein interactions	[42]
RNA immunoprecipitation sequencing (RIP-seq)	RNA–protein interactions	[43]
Parallel analysis of RNA ends sequencing (PARE-seq)	microRNA target discovery	[44]
Massively parallel functional dissection sequencing (MPFD)	Enhancer assay	[45]

ular, it is more precise and not subject to cross-hybridization, thereby providing a higher accuracy and a larger dynamic range. Similar to microarrays, however, HTS-based applications can be biased by a number of variables, such as sequencing platform and library preparation method. The sequencing quality control consortium and similar initiatives are designed to study these biases and develop approaches to control for them, as has been recently demonstrated for RNA-seq [25]. Use of HTS applications by both individual laboratories and the large consortia has enabled researchers to illuminate previously intractable topics in biology, some of which include: (1) genome sequencing and variation; (2) mapping regulatory information of the genome; (3) mapping the 3D organization of the genome; (4) characterizing the transcriptome; (5) microbiome sequencing; (6) genome sequencing of rare diseases; (7) cancer genome sequencing, and so on.

It is becoming increasingly clear that while the technologies of today may be capable of providing population-level sequencing to both researchers and clinicians, key limitations remain [24]. From a technological perspective, accuracy and coverage across the genome are still problematic, particularly for GC-rich regions and long homopolymer stretches. In addition, the short read lengths produced by most current platforms severely limit our ability to accurately characterize large repeat regions, many indels, and structural variations, leaving significant portions of the genome opaque or inaccurate. In addition to genomes, quantitative analysis of complete transcriptomes, with individual allelic and spliced isoforms, is hindered by short reads. Improvements in the throughput and accuracy of current long-read technologies, such as Pacific Biosciences and Oxford Nanopore Technologies, as well as the use of "synthetic long-read methods" in which longer fragments can be sequenced and assembled from short reads will help overcome these limitations [24].

9.1.3 RNA-seq Analysis at Four Different Levels

Various HTS methods generated massive omics data. How to explore bioinformatics from omics data is an important yet challenge scientific problem. With the decreasing of experimental costs, RNA-seq becomes one of the most widely used technologies by biologists for different purpose. The analysis of RNA-seq data at four different levels (samples, genes, transcripts, and exons) involve multiple statistical and computational questions (Fig. 9.3), some of which remain challenging up to date [2].

In the sample-level analysis, the availability of numerous public RNA-seq datasets has created an unprecedented opportunity for researchers to compare multi-species transcriptomes under various biological conditions. Comparing transcriptomes of the same or different species can reveal molecular mechanisms behind important biological processes and help us understand the conservation and differentiation of these molecular mechanisms in evolution. Researchers need similarity measures to directly evaluate the similarities among different samples (i.e., transcriptomes) based on their genome-wide gene expression data summarized from RNA-seq experiments. Such similarity measures are useful for outlier sample detection, sample classification, and sample clustering analysis. When samples represent individual cells, similarity measures may be used to identify rare or novel cell types. In addition to gene expression, it is also possible to evaluate transcriptome similarity based on alternative splicing events. Correlation analysis is a classical approach to measure transcriptome similarity of biological samples. The most commonly used measures are the Pearson and Spearman correlation coefficients. The analysis starts with calculating pairwise correlation coefficients of normalized gene expression between any two biological samples, resulting in a correlation matrix. Users can visualize the correlation matrix (usually as a heatmap) to interpret the pairwise transcriptome similarity of biological samples, or they may use the correlation matrix in downstream analysis such as sample clustering.

For the gene-level analysis, a common and important question in a large cohort of biological studies is how to compare gene expression levels across different

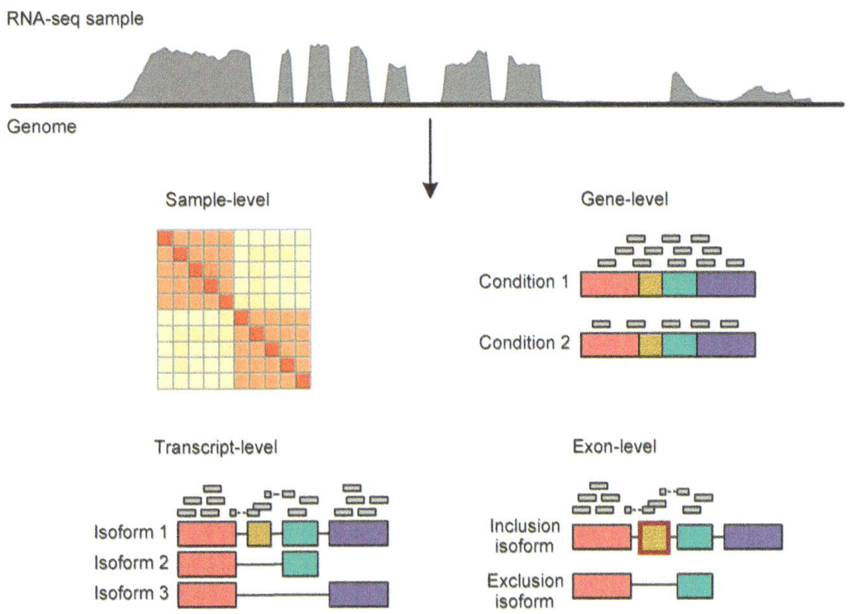

Fig. 9.3 RNA-seq analyses at four different levels: sample level, gene level, transcript level, and exon level. Reprinted by permission from Springer, Ref. [2]

experimental conditions, time points, tissues and cell types, or even species. When a biological study concerns two different biological conditions, differential gene expression (DGE) analysis is useful for comparing RNA-seq samples of the two conditions. When the number of biological conditions far exceeds two, though DGE analysis can still be used to compare samples in a pairwise manner, a more useful way is to simultaneously measure the transcriptome similarity of multiple samples. A gene is defined as "differentially expressed" (DE) if it is transcribed into different amounts of mRNA molecules per cell under the two conditions [46]. However, since we do not observe the true amounts of mRNA molecules, statistical tests are principled approaches that help biologists understand to what extent a gene is DE. It is commonly acknowledged that normalization is a crucial step prior to DGE analysis due to the existence of batch effects, which could arise from different sequencing depths or various protocol-specific biases in different experiments [47]. The reads per kilobase per million mapped reads (RPKM) [48], the fragments per kilobase per million mapped reads (FPKM) [49], and the transcripts per million mapped reads (TPM) [50] are the three most frequently used units for gene expression measurements from RNA-seq data, and they remove the effects of total sequencing depths and gene lengths. The RPKM is defined as

$$RPKM = \frac{total \quad exon \quad reads}{mapped \quad reads(Millions) \times exon \quad length(KB)}. \tag{9.1}$$

The FPKM is defined as

$$FPKM = \frac{total \quad exon \quad fragment}{mapped \quad fragment(Millions) \times exon \quad length(KB)}. \tag{9.2}$$

The TPM for gene i in a sample is defined as

$$TPM_i = \frac{(N_i/L_i) \times 10^6}{\sum_{j=1}^{S}(N_j/L_j)}. \tag{9.3}$$

Here, N_i denotes the number of reads that are mapped to gene i; L_i denotes the total length of the exon of gene i; S represents the total reads in the considered sample. The main difference between RPKM and FPKM is that the former is a unit based on single-end reads, while the latter is based on paired-end reads and counts the two reads from the same RNA fragment as one instead of two. The difference between RPKM/FPKM and TPM is that the former calculates sample-scaling factors before dividing read counts by gene lengths, while the latter divides read counts by gene lengths first and calculates sample-scaling factors based on the length-normalized read counts.

The transcript-level analysis focuses on reads mapped to different isoforms. An important use of RNA-seq data is to recover full-length mRNA transcript structures and expression levels based on short RNA-seq reads. This application involves two major tasks. The first task, identification of novel transcripts in RNA-seq samples, is commonly referred to as transcript/isoform reconstruction, discovery, assembly, or identification. This is one of the most challenging problems in this area due to the large search space of candidate isoforms (especially for complex genes) and inadequate information contained in short reads. The second task, estimation of the expression of known or newly discovered transcripts, is usually referred to as transcript/isoform quantification or abundance estimation. In recent years, it is a common practice to combine the two tasks into one step, and many popular computational tools simultaneously perform transcript reconstruction and quantification [51]. This is usually achieved by estimating the expression levels of all the candidate isoforms with penalty or regularity constraints, and the resulting isoforms with nonzero estimated expression are treated as reconstructed isoforms.

The exon-level analysis mostly considers the reads mapped to or skipping the exon of interest [2]. Since transcript-level analysis of complex genes in eukaryotic organisms remains a great challenge, there are approaches focusing on exon-level signals, seeking to study alternative splicing based on exons and exon–exon junctions instead of full-length transcripts. For the thorough reviews on the related questions and challenges of RNA-seq data analysis, one can refer to the review work by Li et al. [2].

9.2 Weighted Gene Co-Expression Network Analysis

We have introduced the topic of how to reconstruct bio-molecular networks in Chap. 2. However, as an important application of gene co-expression network (GCN) in omics data analysis, we will further introduce gene co-expression network analysis in this subsection. A GCN is an undirected graph, where nodes correspond to genes and edges connecting the nodes denote the co-expression relationships between genes. GCNs can help people learn the functional relationships between genes and infer and annotate the functions of unknown genes. It is reported that [2] the first GCN analysis on a genome-wide scale across multiple organisms was completed in 2003, enabled by the availability of high-throughput microarray data [52]. One of the most commonly used GCN analysis methods, weighted gene co-expression network analysis (WGCNA) [53, 54], was initially developed for microarray data but can also be used on normalized RNA-seq data [53]. It is widely applied to gene expression datasets to detect gene clusters and modules and to investigate gene connectivity by analyzing correlation networks (Fig. 9.4). Here we introduce the GCN methods based on the framework proposed in Refs. [55] and [53].

We denote the gene expression matrix as $X_{n\times p}$, where the p columns represent genes and the n rows represent samples. The p genes are considered as p nodes in the co-expression network. The first step is to construct a symmetric adjacency matrix $A_{p\times p}$, where A_{ij} is a similarity score in the range from 0 to 1 between genes i and j. A_{ij} measures the level of concordance between gene expression vectors X_i and X_j, the i-th and j-th rows of X. Transcriptome similarity measures can be calculated based on the correlation coefficients, the transcriptome overlap measure (TROM), or the mutual information measures (as discussed in Chap. 2), depending on the type of gene co-expression relationships of interest in the analysis. The elements in the adjacency matrix only consider each pair of genes when evaluating their similarity in expression profiles. However, it is important to consider the relative connectedness of gene pairs with respect to the entire network in order to detect co-expression gene modules. Therefore, one needs to calculate the topological overlap matrix $T = (T_{ij})_{p\times p}$, where T_{ij} is the topological overlap between nodes i and j. One such example used in previous studies is [56]

$$T_{ij} = \frac{\sum_{k=1}^{p} A_{ik}A_{kj} + A_{ij}}{\min\left\{\sum_{k=1}^{p} A_{ik}, \sum_{k=1}^{p} A_{jk}\right\} + 1 - A_{ij}}. \tag{9.4}$$

The final distance between nodes i and j is defined as $d_{ij} = 1 - T_{ij}$. Clustering methods can then be applied to search for gene modules based on the resulting distance matrix. The identified gene modules are of great biological interest in many applications. For example, the modules can serve as a prioritizer to evaluate functional relationships between known disease genes and candidate genes [57]. Gene modules can also be used to detect regulatory genes and study the regulatory mechanisms in various organisms [58].

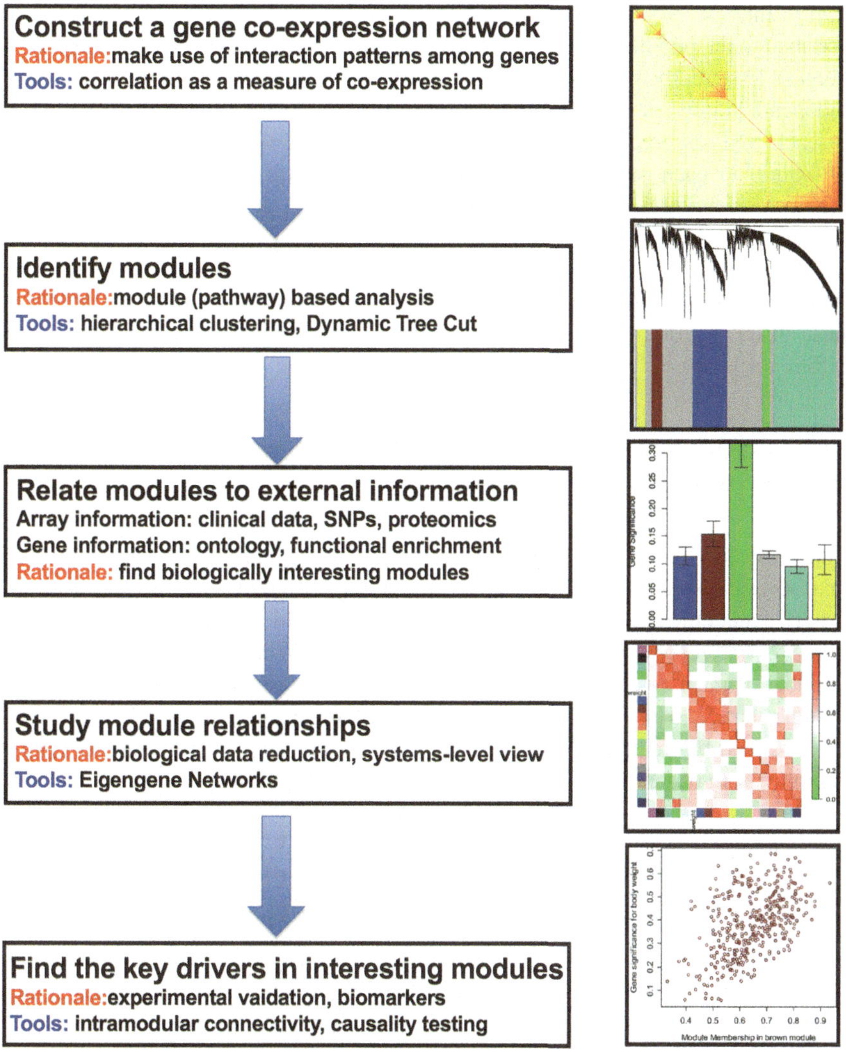

Fig. 9.4 Flowchart and illustration of the WGCNA methodology. This flowchart presents a brief overview of the main steps of WGCNA. Reprinted from Ref. [53]

9.3 Genome-Wide Association Study for Omics Data

Genome-wide association study (GWAS) [59–63] for omics data can be used to explore the link between phenotype and genotype. Specifically, proper statistics model was applied to fastly identify the genes' underlying trait with single-nucleotide polymorphism (SNP) and accurate phenotype as input data. GWASs scan an entire species genome for association between up to millions of SNPs and a given trait of interest. Notably, the trait of interest can be virtually any sort of phenotype ascribed to the population, be it qualitative (e.g., disease status) or quantitative (e.g., height). Essentially, given p SNPs and n samples or individuals, a GWAS will fit p independent univariate linear models, each based on n samples, using the genotype of each SNP as predictor of the trait of interest. The significance of association (P-value) in each of the p tests is determined from the coefficient estimate of the corresponding SNP. Note that because these tests are independent and quite numerous, there is a great computational advantage in setting up a parallelized GWAS. Quite reasonably, it is necessary to adjust the resulting P-values using multiple hypothesis testing methods such as Bonferroni, Benjamini–Hochberg, or false discovery rate (FDR). GWASs are now commonplace in genetics of many different species.

9.4 General Linear Models

Supposed that we have a response variable **y**, if **y** is a categorical variable, it can take several discrete values that distinguish among different samples. For example, it may equal either 1 or 0. $y = 1$ corresponds to treated samples or diseased ones, while $y = 0$ represents controls or normal samples. Whereas if **y** is a continuous variable, it must be experimentally measurable, and $\mathbf{y} \in R$ in theory. We further assume that $\mathbf{X} = (\mathbf{X}_1, \mathbf{X}_2, \cdots, \mathbf{X}_p)'$ represents features or independent variables that are possibly related to experimental phenotypes ($y = 1$). Suppose we have n samples, with observations taking the following matrix form:

$$Y = \begin{pmatrix} y_1 \\ y_2 \\ \vdots \\ y_n \end{pmatrix}_{n\times 1}, \quad X = \begin{pmatrix} x_{11} & x_{12} & \cdots & x_{1p} \\ x_{21} & x_{22} & \cdots & x_{2p} \\ \vdots & \vdots & \ddots & \vdots \\ x_{n1} & x_{n2} & \cdots & x_{np} \end{pmatrix}_{n\times p} = \begin{pmatrix} X_{(1)}^T \\ X_{(2)}^T \\ \vdots \\ X_{(n)}^T \end{pmatrix} = (X_1, X_2, \cdots, X_p). \tag{9.5}$$

Here, $X_{(i)} = (x_{i1}, x_{i2}, \cdots, x_{ip})^T$ denotes the ith sample vector ($i = 1, 2, \cdots, n$).

Generally, one assumes that the n samples are independent. Based on the given data as shown above, the aims for data analysis include the following three points. **(1) How to select crucial genes that are closely related to phenotype variation among samples under different experimental groups (treatment versus control)?** This question is also well known as gene selection, gene prioritization, and gene ranking in the area of bioinformatics, and it was widely called as variable selection or feature selection in the fields of statistics and machine learning. **(2) How to classify samples as treatments and controls?** This question is well known as a supervised machine learning problem. And it is called as discriminant analysis in statistical theory. **(3) How to infer the interactions among genes?** This is also known as the network reconstruction problem, which has been introduced in Chap. 2. A big challenge to solve the abovementioned problem is that the number of genes (variables) p is far larger than the amount of samples n. Hereinafter, we introduce some of the state-of-the-art statistical approaches to solve these problems.

9.4.1 Penalized Linear Regression

If the response variable $\mathbf{y}$ is continuous, and suppose genes are independent and identical distributed (i.i.d.), researchers have constructed the following linear regression model [64] to answer the question of gene selection and sample classification:

$$\mathbf{y} = \beta_0 + \beta_1\mathbf{X}_1 + \beta_2\mathbf{X}_2 + \cdots + \beta_p\mathbf{X}_p + \varepsilon. \tag{9.6}$$

Here, $\mathbf{X}_i (i = 1, 2, \ldots, p)$ denotes a random variable that corresponds to the expression value of gene i and ε is a random error term. For the observed data as shown in Eq. (9.6), we have

$$y_j = \beta_0 + \beta_1 X_{j1} + \beta_2 X_{j2} + \cdots + \beta_p X_{jp},\ j = 1, 2, \ldots, n. \tag{9.7}$$

If we denote $\beta = (\beta_0, \beta_1, \ldots, \beta_p)^T$, denote the n samples in matrix form as that in Eq. (9.7), and rewrite X as

$$X^* = \begin{pmatrix} 1 & x_{11} & x_{12} & \cdots & x_{1p} \\ 1 & x_{21} & x_{22} & \cdots & x_{2p} \\ \vdots & \vdots & \vdots & \ddots & \vdots \\ 1 & x_{n1} & x_{n2} & \cdots & x_{np} \end{pmatrix}. \tag{9.8}$$

Then, Eq. (9.7) can be written in the following matrix form:

$$Y = X^*\beta = f(X^*, \beta). \tag{9.9}$$

For $n << p$, Eq. (9.9) is ill-conditioned, the solution or parameter β of Eq. (9.9) is not unique. Many methods have been proposed to obtain the conditional optimal solution of Eq. (9.9). The OLS estimation is a famous solution. The OLS estimation of the parameter vector β can be easily deduced as

$$\hat{\beta} = (X^{*T}X^*)^{-1}X^{*T}Y. \tag{9.10}$$

Equation (9.10) requires that $X^{*T}X^*$ is invertible. If it is uninvertible, one can use the following formulation to correct it:

$$\hat{\beta} = \left(X^{*T}X^* + \lambda I\right)^{-1} X^{*T}Y. \tag{9.11}$$

Here, λ is a tunable parameter and I is the identity matrix. In fact, the solution obtained from Eq. (9.11) is equivalent to the following constraint optimization problem:

$$\begin{aligned}\hat{\beta} &= arg\min\left\{||Y - X^*\beta||_2^2 + \lambda||\beta||_2^2\right\} \\ &= arg\min\left\{\left(Y - X^*\beta\right)^T\left(Y - X^*\beta\right) + \lambda\beta^T\beta\right\}.\end{aligned} \tag{9.12}$$

It is noted that the objective function in Eq. (9.12) is also known as the ridge regression. In fact, to obtain the optimal β, we compute the first-order derivative of $F(\lambda, \beta) = (Y - X^*\beta)^T(Y - X^*\beta) + \lambda\beta^T\beta$ with respect to β and set

$$\frac{\partial F}{\partial \beta} = -2X^{*T}Y + 2\left(X^{*T}X^*\right)\beta + 2\lambda\beta = 0. \tag{9.13}$$

Solving Eq. (9.13), one obtains $\hat{\beta} = (X^{*T}X^* + \lambda I)^{-1}X^{*T}Y$. Thus, under this meaning, $\hat{\beta}$ is just the optimal solution of the L_2 penalized optimization problem (9.12).

Generally, to obtain the optimal β, one can consider the following optimization problem:

$$\hat{\beta} = arg\min\left\{||Y - X^*\beta||_2^2 + g(\beta, \lambda)\right\}. \tag{9.14}$$

Here $g(\beta, \lambda)$ is the penalized term or regularization term. Different kinds of $g(\beta, \lambda)$ correspond to different existing algorithms, which are described as follows [64]:

(1) OLS estimation [64]: $g(\beta, \lambda) = 0$;
(2) Bridge regression [65–67]: $g(\beta, \lambda) = \lambda||\beta||^q$ with $0 < q < 1$;
(3) LASSO [68]: $g(\beta, \lambda) = \lambda||\beta||_1$;
(4) Tikhonov regularization, L_2 penalization, or ridge regression [69]: $g(\beta, \lambda) = \lambda||\beta||_2^2$;
(5) Elastic net [70]: $g(\beta, \lambda) = \lambda\alpha||\beta||_1 + \lambda(1 - \alpha)||\beta||_2^2$;
(6) Fused LASSO [71]: $g(\beta, \lambda) = \lambda_1||\beta||_1 + \lambda_2\sum_{j=1}^{p}|\beta_j - \beta_{j-1}|$;
(7) Network-constrained regularization [72]: $g(\beta, \lambda) = \lambda_1||\beta||_1 + \lambda_2\beta^T L\beta$, where L denotes the graph Laplacian matrix;
(8) Group LASSO [73]: $g(\beta, \lambda) = \lambda\sum_{g=1}^{|G|}\sqrt{d_g}||\beta_{u_g}||_2$, here, $|G|$ represents the total groups of the p features, u_g denotes the gth group of parameters, and d_g represents the numbers of features in group g;

(9) Pairwise structured penalization [74]:

$$g(\beta, \lambda) = \lambda\alpha \sum_{j=1}^{p} \theta_j |\beta_j + \lambda(1-\alpha) \sum_{i=1}^{p}\sum_{j=1}^{p} w_{ij}(\mu_i\beta_i - \mu_j\beta_j)^2,$$

with $\theta_j = 1/\rho(Y, X_j)^2$, $\mu_j = \rho(Y, X_j)$, and w_{ij} denotes the weights between features X_i, X_j, which measures the similarity between the two features;

(10) ...

After we obtain the estimated parameter β, we can use it to realize feature selection. If $\beta_i \neq 0$, then the ith gene is the feature that should be selected. Moreover, the importance of genes can be evaluated by the absolute value of its coefficients. Higher β_i indicates more importance of gene i. As the linear regression method can only be used to realize feature selection, it is difficult to be used to classify samples and reconstruct networks. Moreover, this model implicitly assumed that the p explanatory variable $\mathbf{X}_i (i = 1, 2, \ldots, p)$ were independent with each other, which was not the actual case in real-world biological systems.

9.4.2 Penalized Logistic Regression

The logistic regression can simultaneously realize feature selection and sample classification. In the logistic regression, the response variable $\mathbf{y}$ can be a discrete categorical variable or a probability. Under this condition, we cannot use Eq. (9.9) to model the relationships between $\mathbf{y}$ and $\mathbf{X}$, since $\mathbf{y}$ is discrete and with limited categorical values, but $\mathbf{X}$ is continuous. To model the relationship among $\mathbf{y}$ and $\mathbf{X}$, we can use the logit transformation:

$$logit(\mathbf{y}) = ln\left(\frac{\mathbf{y}}{1-\mathbf{y}}\right). \tag{9.15}$$

Obviously, if $\mathbf{y} \in [0, 1]$, $logit(\mathbf{y}) \in (-\infty, +\infty)$. In the following, we mainly consider the binary classification problem. The binary function $y = 1(x > 0)$ or $0(x < 0)$ can be fitted by the Sigmoid function $y = 1/(1 + e^{-x})$. We suppose the observation of the response variable $y \sim Bernoulli(\pi)$, here $\pi = p\{y = 1|X^*\}$. The logistic regression model can be established as follows:

$$logit(\pi) = f(X^*, \beta) = X^*\beta. \tag{9.16}$$

This model also implicitly assumed that the p explanatory variable $\mathbf{X}_i (i = 1, 2, \ldots, p)$ were independent with each other. To estimate parameter β, we can construct and maximize the following objective function:

$$\Pi_{i=1}^{n} \pi_i^{y_i} (1 - \pi_i)^{1-y_i}. \tag{9.17}$$

It is equivalent to minimize the negative log likelihood of the above function:

$$-\sum_{i=1}^{n} [y_i ln\pi_i + (1 - y_i) ln(1 - \pi_i)]. \tag{9.18}$$

Similar to the linear regression, one can consider various penalized terms to realize sparse or smooth estimation of β as

$$\hat{\beta} = arg \min \left\{ -\sum_{i=1}^{n} [y_i ln\pi_i + (1 - y_i) ln(1 - \pi_i)] + g(\beta, \lambda) \right\}, \tag{9.19}$$

where $g(\beta, \lambda)$ is the same as those for the linear regression model.

The optimization problem (9.19) can be solved by various optimization algorithms, such as stochastic gradient descent (SGD), alternating direction method of multipliers (ADMM) [75], and so on. After the parameter was estimated, one can realize gene selection and samples classification. Specifically, if $\beta_i \neq 0$, then the ith gene should be selected, and the importance of genes can be ranked according to the absolute values of their corresponding coefficients. Furthermore, denoting the jth row of X^* as $X_{(j)}^{*T}$, if

$$\hat{\pi}_j = \frac{e^{X_{(j)}^{*T} \hat{\beta}}}{1 + e^{X(j)^{*T} \hat{\beta}}} > 0.5,$$

then the jth sample should be classified as class 1; otherwise, it should be classified as class 0.

9.4.3 *Optimization Methods for Parameter Estimation*

9.4.3.1 The Ordinary Least Square Estimation

The OLS estimation is based on the minimization of the sum of square for error. Based on the observation matrix X^* for covariates $\mathbf{X_1} - \mathbf{X_p}$ and the observed vector Y for response variable $\mathbf{y}$, the OLS estimation for the parameter vector β can be obtained as

$$\hat{\beta} = \left(X^{*T} X^*\right)^{-1} X^{*T} Y. \tag{9.20}$$

The OLS estimation needs that $X^{*T}X^*$ is reversible, and it is inappropriate for high-dimensional data, since high-dimensional matrix computation is time costly and easy to produce calculation error.

9.4.3.2 The Least Angle Regression

The least angle regression (LAR) [76] is a fast iteration algorithm for feature selection and parameter estimation of linear regression problems. The LAR is similar to the forward stepwise method, and it can be seen as a high efficiency algorithm for the LASSO.

The LAR procedure works roughly as follows. As with classic forward selection, one starts with all coefficients equal to zero and finds the predictor most correlated with the response, say x_{j_1}. One takes the largest step possible in the direction of this predictor until some other predictor, say x_{j_2}, has as much correlation with the current residual. At this point, LAR parts company with forward selection. Instead of continuing along x_{j_1}, LAR proceeds in a direction equiangular between the two predictors until a third variable x_{j_3} earns its way into the "most correlated" set. LAR then proceeds equiangularly between x_{j_1}, x_{j_2}, and x_{j_3}, that is, along the "least angle direction," until a fourth variable enters, and so on.

There are three main properties for the LAR [76]: (1) A simple modification of the LAR algorithm implements the LASSO, an attractive version of the OLS that constrains the sum of the absolute regression coefficients; the LAR modification calculates all possible LASSO estimates for a given problem, using an order of magnitude less computer time than previous methods. (2) A different LAR modification efficiently implements forward stagewise linear regression, another promising new model selection method; this connection explains the similar numerical results previously observed for the LASSO and stagewise and helps us understand the properties of both methods, which are seen as constrained versions of the simpler LAR algorithm. (3) A simple approximation for the degrees of freedom of a LAR estimate is available, from which one derives a C_p estimate of prediction error; this allows a principled choice among the range of possible LAR estimates. LAR and its variants are computationally efficient: the algorithm requires only the same order of magnitude of computational effort as OLS applied to the full set of covariates.

9.4.3.3 Stepwise Regression

The basic idea of stepwise regression is to introduce variables into the model one by one, a statistical F test will be performed after introducing each variable, and at the same time, statistical t test will be performed for each variable that has been introduced into the model. If a variable is not significant after the introducing of the new variable, then it should be deleted from the model. Before we introduce new variables to the model, existing variables that have been introduced into the model should be significant. The procedure stops when no more significant variables can be introduced into the model, and no more non-significant variables can be deleted from the model.

Stepwise regression can be used to reject variables that may cause multicollinearity. The procedure is as follows. Initially, one establishes the simple linear regression model for the response variable with each of the covariates, selects a covariate that best explained the response variable, takes the associate regression model as foundation, and introduces the other variables into the model one by one. Finally, covariates that selected in the model are important, and multicollinearity can be avoided.

9.4.3.4 Newton's Method

Newton's method was also called as the Newton–Raphson method, it was firstly proposed by Newton in the seventeenth century, and it can be applied to obtain the approximate solutions of an equation. To obtain the root of equation $f(x) = 0$, the Newton's iteration formula can be written as

$$x_{n+1} = x_n - \frac{f(x_n)}{f'(x_n)}. \tag{9.21}$$

Here, we need $f'(x_0) \neq 0$ and $f(x) \in C^2[a, b]$.

The Newton's method has the following advantages: (1) With relative high computational speed. (2) If $f(x) = 0$ has a single root, then the Newton's method is with square convergence. (3) The Newton's method can be also used to find multiple roots and complex roots. (4) The Newton's method needs the initial value of the iteration to be approximate to the real solution of the equation. (5) One should consider the continuity of the function $f'(x)$.

9.4.3.5 Gradient Descent Method/Steepest Descent Method

Gradient descent (GD) is an optimization method to find a local (preferably global) minimum of a function. In backpropagation, it is used to iteratively update the weights in order to minimize the error function.

9.4.3.6 Stochastic Gradient Descent Method

In a GD optimization, all the training samples are used for each update of the weights, whereas in stochastic gradient descent (SGD) [77], only one or a small batch of training samples are used for each step. With large training sets, the computation time of the GD optimization can become extremely long as one must compute the outputs, errors, and gradients of all the samples at each iteration. SGD is therefore almost always preferred to GD in neural networks.

The GD method cannot guarantee to find global optimal solution of a loss function, and the obtained solution may be its local minima. However, if the loss function is a convex function, then the obtained optimal solution will be a global one.

9.4.3.7 Coordinate Descent Method

The coordinate descent method [78] is a derivative-free optimization method. In each iteration step, one should search the local minimum along one coordinate. Different coordinate directions are iteratively used during the searching of the global minimum point. The coordinate descent method needs the objective function to be continuous. However, if the objective function is not smooth, the solution of the coordinate descent method may fall into a non-stationary point.

9.4.3.8 Alternating Direction Method of Multipliers

The alternating direction method of multipliers (ADMM) [75] is an algorithm that solves convex optimization problems by breaking them into smaller pieces, each of which are then easier to be handled. It has recently found wide applications in a number of areas.

ADMM is an algorithm that is intended to blend the decomposability of dual ascent with the superior convergence properties of the method of multipliers. The algorithm solves problems in the form:

$$\begin{aligned} &\min f(x) + g(z), \\ &\text{subject to } Ax + Bz = c, \end{aligned} \tag{9.22}$$

with variables $x \in R^n$ and $z \in R^m$, where $A \in R^{p\times n}$, $B \in R^{p\times m}$, and $c \in R^p$. It is assumed that f and g are convex; the only difference from the general linear equality-constrained problem is that the variable, called x there, has been split into two parts, called x and z here, with the objective function separable across this splitting. The optimal value of the problem (9.22) will be denoted by

$$p^* = inf\,\{f(x) + g(z)|Ax + Bz = c\}. \tag{9.23}$$

As in the method of multipliers, one form of the augmented Lagrangian is

$$L(x, z, y) = f(x) + g(z) + y^T(Ax + Bz - c) + (\rho/2)||Ax + Bz - c||_2^2. \tag{9.24}$$

ADMM consists of the following iterations:

$$x^{k+1} := arg \min_{x} L\left(x, z^k, y^k\right), \tag{9.25}$$

$$z^{k+1} := arg \min_{z} L\left(x^{k+1}, z, y^k\right),$$

$$y^{k+1} := y^k + \rho\left(Ax^{k+1} + Bz^{k+1} - c\right),$$

where $\rho > 0$. The algorithm is very similar to dual ascent and the method of multipliers: it consists of an x-minimization step, a z-minimization step, and a dual variable update. As in the method of multipliers, the dual variable update uses a step size equal to the augmented Lagrangian parameter ρ.

The method of multipliers for (9.24) has the form:

$$\left(x^{k+1}, z^{k+1}\right) := arg \min_{x,z} L\left(x, z, y^k\right), \tag{9.26}$$

$$y^{k+1} := y^k + \rho\left(Ax^{k+1} + Bz^{k+1} - c\right).$$

Here the augmented Lagrangian is minimized jointly with respect to the two primal variables. In ADMM, on the contrary, x and z are updated in an alternating or sequential fashion, which accounts for the term alternating direction. ADMM can be viewed as a version of the method of multipliers where a single Gauss–Seidel passing over x and z is used instead of the usual joint minimization. Separating the minimization over x and z into two steps is precisely what allows for decomposition when f or g is separable.

The algorithm state in ADMM consists of z^k and y^k. In other words, (z^{k+1}, y^{k+1}) is a function of (z^k, y^k). The variable x^k is not a part of the state; it is an intermediate result computed from the previous state (z^{k-1}, y^{k-1}).

If we switch (re-label) x and z, f and g, and A and B in the problem (9.24), we obtain a variation on ADMM with the order of the x-update and z-update steps (9.25) reversed. The roles of x and z are almost symmetric, however, the dual update is done after the z-update but before the x-update.

9.4.4 Model Selection Criterion

Many different approaches of penalization have been added in regression model to avoid over-fitting. To compare among different models, the Akaike information criterion (AIC) and the Bayesian information criterion (BIC) are frequently used.

The AIC is proposed by Akaike in the year 1974, and it is a criterion that can balance between model complexity and the goodness of data fitting, which is defined as

$$AIC = 2p - 2ln(L). \tag{9.27}$$

Here, p is the number of parameters and L is the likelihood function. The model with the smallest AIC should be chosen.

The BIC is similar to the AIC, and it was firstly proposed by Schwarz in the year 1978. The BIC is defined as

$$BIC = pln(n) - 2ln(L). \tag{9.28}$$

Here, p is the number of parameters, n is the number of samples, and L is the likelihood function. The penalization term $pln(n)$ is especially useful when $p >> n$, and it can effectively avoid the dimension disaster phenomenon.

Except the traditional AIC and BIC criterions, some novel criterions have also been introduced in existing references, including the extended BIC (EBIC), the AUC of ROC, and the generalized cross validation [64]. The EBIC is defined as

$$EBIC = -2ln(L) + log(n)d + 2\gamma ln(p)d. \tag{9.29}$$

Here, d denotes the number of nonzero features, and γ is a constant, usually taken as 1. A model with smaller $EBIC$ value is better.

9.4.5 Comparison Among Different Penalty Terms

As a summary of this section, we compare among different regularization terms, as shown in Tables 9.2 and 9.3. Here, we compare both the advantages and disadvantages, as well as the applications for each kind of regression and regularization methods.

9.5 Hidden Markov Random Field Model and Its Applications

The hidden Markov random field model (HMRFM)[88–91] has been frequently used to prioritize genes based on omics data. In this subsection, we mainly introduce the work by Hou et al. [88] in the year 2014.

Table 9.2 Comparison among different regression and regularization methods

Methods	Regularizations	Advantages	Disadvantages	Applications	Ref.
Linear regression	–	Linear relationships; fruitful theoretical results; simple and easy to be explained.	Covariates should be independent with each other or with low correlation; unable to deal with nonlinear correlation and discrete variables; Unsuitable for $p >> n$	Reveal linear correlations; variable selection; prediction.	[64]
Logistic regression	–	Binary classification, extendable to multi-classification, variables can be continuous or discrete.	Unsuitable for $p >> n$	Classification; probability prediction.	[64]
LASSO	$\lambda\|\|\beta\|\|_1$.	Sparsity, variable selection, both linear and nonlinear, wide applications.	Selected variables are no more than n for $p >> n$.	Classification; variable selection; prediction.	[68]
Ridge regression	$\lambda\|\|\beta\|\|_2^2$.	Enhance group effect; smoothness; suitable for $p >> n$; with analytical solution and high computational efficiency.	Not sparse; without oracle property; variables cannot be automatically selected.	Classification; prediction; feature selection.	[69]
Elastic net	$\lambda\alpha\|\|\beta\|\|_1 + \lambda(1-\alpha)\|\|\beta\|\|_2^2$.	Sparse; group effect; suitable for $p >> n$.	Without oracle property.	Classification; prediction; feature selection.	[70]
Bridge	$\lambda\|\|\beta\|\|^q, 0 < q < 1$.	Subset selection; sparse; oracle property; more efficient than LASSO and elastic net in identifying zero coefficients.	Nonconvex; unstable estimation around zero.	Classification; prediction; feature selection.	[65]
Adaptive LASSO	$\lambda\sum_{j=1}^{p}\|\hat{\beta}_j\|^{-\gamma}\|\beta_j\|$.	Unbiased; oracle property; sparse; avoiding over-fitting.	Affected by initial weights.	Classification; prediction; feature selection.	[79]
Adaptive ridge	$\lambda\sum_{j=1}^{p}\|\hat{\beta}_j\|^{-\gamma}\beta_j^2$.	Unbiased; oracle property; enhanced group effect.	Affected by initial weights.	Classification; prediction; feature selection.	[80]
Adaptive elastic net	$\lambda_1\sum_{j=1}^{p}\|\hat{\beta}_j\|^{-\gamma}\|\beta_j\| + \lambda_2\sum_{j=1}^{p}\|\hat{\beta}_j\|^{-\gamma}\beta_j^2$.	Unbiased; oracle property; sparse; enhanced group effect.	Affected by initial weights.	Classification; prediction; feature selection.	[81]
Adaptive bridge	$\lambda\sum_{j=1}^{p}\|\hat{\beta}_j\|^{-\gamma}\|\beta_j\|^q, 0, q < 1$.	Subset selection; sparse; oracle property.	Nonconvex; unstable estimation around zero; affected by initial weights.	Classification; prediction; feature selection.	[80]

Table 9.3 Comparing among different regression and regularization methods

Methods	Regularizations	Advantages	Disadvantages	Applications	Ref.
Fused LASSO	$\lambda_1\|\|\beta\|\|_1 + \lambda_2 \sum_{j=1}^{p} \|\beta_j - \beta_{j-1}\|$.	Appropriate for $p >> n$; sparse; Extensible to high-dimensional data.	Computationally costly; features should be ranked.	Classification; prediction; feature selection.	[71]
Group LASSO	$\lambda \sum_{g=1}^{\|G\|} \sqrt{d_g}\|\|\beta_{u_g}\|\|_2$.	Sparse; selected groups of features.	Unbounded penalization; difficult to be grouped, often needs a prior; Clustering methods affect the results.	Classification; prediction; feature selection.	[73]
Sparse-Group LASSO	$(1-\alpha)\lambda \sum_{j=1}^{\|G\|} \sqrt{pj}\|\|\beta^{(j)}\|\|_2 + \alpha\lambda\|\|\beta\|\|_1$.	Group; sparse; important feature selection.	Difficult to be grouped, often needs a prior; group affect the result.	Classification; prediction; feature selection.	[82]
Group bridge	$\lambda \sum_{j=1}^{\|G\|} C_j\|\|\beta^{(j)}\|\|^{\gamma}$.	Able to select important groups; sparse; oracle property.	Unbounded penalization; difficult to be grouped; often needs a prior; clustering methods affect the final results.	Classification; prediction; feature selection.	[83]
MCP	$\sum_{j=1}^{\|G\|} \rho_{\lambda,\gamma}(\beta_j)$; $\rho_{\lambda,\gamma}(\beta_j) = \lambda\beta - \beta^2/(2\gamma)$ if $\beta \le \gamma\lambda$, otherwise, $\rho_{\lambda,\gamma}(\beta_j) = \gamma\lambda^2/2$.	Asymptotically unbiased; with high probability selecting the right model.	Group information is not efficiently used; tends to select a few groups.	Classification; prediction; feature selection.	[84]
Group MCP	$\sum_{l=1}^{\|G\|} \rho_{\lambda,\gamma}(\sum_{j=1}^{pl} \rho_{\lambda,\gamma}(\|\beta_{lj}\|))$.	Able to select important covariates.	Difficult to be grouped; often needs a prior; clustering methods affect the results.	Classification; prediction; feature selection.	[85, 86]
Network constrained penalization	$\lambda_1\|\|\beta\|\|_1 + \lambda_2\beta^T L\beta$.	Sparse; similar coefficients for similar covariates; network information is used.	Need to construct network.	Classification; prediction; feature selection.	[72]
Pairwise structured penalization	$\lambda\alpha \sum_{j=1}^{p} \theta_j\|\beta_j\| + \lambda(1-\alpha) \sum_{i=1}^{p} \sum_{j=1}^{p} w_{ij}(\mu_i\beta_i - \mu_j\beta_j)^2$.	Sparse; unbiased; highly correlated genes can be selected simultaneously; sufficiently uses data.	Need to construct network.	Classification; prediction; feature selection.	[74]
SCAD	$\sum_{j=1}^{p} \rho_{\lambda,\gamma}(\beta_j)$.	Similar to the MCP.	Group strength is weaker than the MCP.	Classification; prediction; feature selection.	[87]

9.5.1 Measurement of Network Rewiring

The PCC was calculated for each pair of genes in treated samples and control samples, separately. Let r^t_{ij} denote the PCC of genes i and j in the treated samples, and r^c_{ij} that in the control samples. Previously, the difference between PCCs was used to measure differential rewiring [92]. Hu et al. [93] showed, by simulation, applying the following Fisher transformation:

$$F(r) = \frac{1}{2} ln \frac{1+r}{1-r}, \tag{9.30}$$

can improve the power to identify differentially rewired genes. Hou et al. [88] used the following Fisher's test of difference between two correlation coefficients, which considers both the change of PCC levels and effect of sample sizes:

$$rewire_{ij} = P\left\{ |X| \leq \left| \frac{F(r^t) - F(r^c)}{\sqrt{\frac{1}{n_t-3} + \frac{1}{n_c-3}}} \right| \right\}, X \sim N(0, 1). \tag{9.31}$$

Here, n_t, n_c denote the numbers of treated samples and control samples, respectively. The test statistic approximately follows the standard normal distribution under the null hypothesis of no difference in the PCC levels between treatments and controls. Thus, the rewiring information, $rewire_{ij}$, is defined as a value between 0 and 1, with larger value indicating more dramatic rewiring effect.

9.5.2 Network Dichotomization

The rewiring and co-expression networks are both weighted networks, with weights ranging between 0 and 1. However, the weights, rewiring information and absolute PCC values, are distinct concepts and not comparable by nature. To facilitate the comparison, Hou et al. [88] dichotomized the two networks in such a way that they had the same network density. In detail, the rewiring information of all gene pairs was ranked, and the 0.9, 0.95, and 0.99 quantiles were chosen as the hard thresholds. The resultant network densities were 0.1, 0.05, and 0.01, respectively. The static co-expression network is dichotomized likewise by hard thresholding on the absolute PCC values.

9.5.3 Markov Random Field Modeling

In omics data analysis, to prioritize disease-associated genes with network rewiring, Hou et al. [88] utilized the HMRFM to formulate the problem. In the network, each node is a gene, with an association label w_i, either $+1$ (associated) or -1 (not associated). A network configuration is the label vector of all nodes in the network, $(w_1, w_2, \ldots, w_p)^T$, where p is the number of genes considered. Two genes are connected ($e_{ij} = 1$) if they were co-expressed either in the disease state or healthy state. The threshold used to dichotomize the co-expression network was chosen by power law distribution. The degree of rewiring ($rewire_{ij}$) is described as in Eq. (9.31). The distribution of network configuration is defined as follows:

$$
\begin{aligned}
P(w_1, w_2, \ldots, w_p) = \frac{1}{Z} exp\left(-h \sum_{i=1}^{p} I(w_i = 1)\right) \qquad (9.32)\\
+\tau_1 \sum_{e_{ij}=1} rewire_{ij} \times I(w_i = 1, w_j = 1)\\
-\tau_2 \sum_{e_{ij}=1, rewire_{ij} > \delta} rewire_{ij} \times I(w_i = -1, w_j = -1),
\end{aligned}
$$

where (h, τ_1, τ_2) are hyper-parameters, $I(.)$ is an indicator function, and Z is the partition function.

In Eq. (9.32), τ_1, τ_2 are weighting parameters with positive values, which determine the influences of different types of edges. The impacts of these parameters can be better explained with Eq. (9.33):

$$
\begin{aligned}
log\frac{P(w_i = 1|w_{-i})}{P(w_i = -1|w_{-i})} = -h + \tau_1 \sum_{e_{ij}=1, w_j=1} rewire_{ij} \qquad (9.33)\\
+\tau_2 \sum_{e_{ij}=1, w_j=-1, rewire_{ij} > \delta} rewire_{ij}.
\end{aligned}
$$

Here, w_{-i} stands for all genes in the network except gene i. Eq. (9.33) shows the conditional probability for the association state of gene i. This probability is composed of three parts: (1) h is a constant defining the probability of being disease associated if the gene is isolated, thus no network information can be incorporated; (2) τ_1 indicates the contribution of rewiring degree of "associated" neighbors; while (3) τ_2 indicates those of "non-associated" neighbors. Suppose the association states for the neighbors of gene i are fixed, the larger τ_1 and τ_2 are, the more likely gene i is disease associated.

In Reference [88], δ was set to be 0.95, and a rewiring degree less than that indicates that the difference of PCC between treated and control samples is not significant. The odds of a gene to be associated with disease will increase with larger rewiring degree with its neighbors. The underlying biological assumption is that a group of erroneous gene interactions, which are present in one condition and absent in the other, probably reflect organizational changes of the cellular networks under different disease conditions.

Given the network structure and the association signals, the posterior probability of the network configuration can be inferred through Bayesian framework:

$$P(\Omega|Y) \propto P(Y|\Omega)P(\Omega). \tag{9.34}$$

The observed data $Y = (y_1, y_2, \ldots, y_n)^T$ are the normalized scores corresponding to the P-values in GWAS studies: $y_i = \phi^{-1}(1 - p_i)$, where ϕ is the cumulative distribution function of a standard normal variable. Under the null hypothesis that the gene is not associated with the disease, its P-value follows a $Uniform(0, 1)$ distribution. Thus, $P(y_i|w_i = -1) \sim N(0, 1)$. Under the alternative hypothesis, i.e., the association state is "+1," just the same as Chen et al. [89] by assuming $P(y_i|w_i = 1) \sim N(\mu_i, \sigma_i^2)$, and assign ϕ conjugate priors for μ_i and σ_i^2:

$$\mu_i|\sigma_i^2 \sim N(\bar{\mu}, \sigma_i^2/a), \sigma_i^2 \sim InverseGamma(v/2, vd/2). \tag{9.35}$$

The hidden states can be inferred by the iterated conditional modes algorithm.

Although this modelling framework is similar to that by Chen et al. [89], it is declared that the proposed approach by Hou et al. [88] is different in principle. First, the previous approach assumes that connected genes in a pathway tend to share association states, which is a "guilt by association" approach in the general sense. Second, the current approach incorporates the network structure at a systems level, and not restricted to connections defined within annotated pathways as proposed in the previous approach. Indeed, dynamic changes tend to involve genes between pathways, rather than within known pathways [88].

9.5.4 Choice of Hyper-Parameters

There are two sets of hyper-parameters in the above HMRFM, including network parameters (h, τ_1, τ_2) and GWAS parameters $(\bar{\mu}, a, v, d)$, respectively. The parameters (τ_1, τ_2) reflect the context-dependent contribution of network rewiring to the configuration distribution. The change of energy function caused by assigning node i to "+1" from "−1" (Eq. (9.36)) can be easily derived from Eq. (9.32). Hou et al. fixed both τ_1 and τ_2 as 1, since they assumed that the rewiring with both associated and non-associated genes increases the probability that this gene is associated.

$$-h + \tau_1 \sum_{e_{ij}=1, w_j=1} rewire_{ij} + \tau_2 \sum_{e_{ij}=1, w_j=-1, rewire_{ij}>\delta} rewire_{ij}. \tag{9.36}$$

The parameter h, a negative value, determines the distribution of network configuration when neither GWAS nor gene expression data is available. When (τ_1, τ_2) are fixed, a larger value of h favors network configurations with more nodes labeled as "not associated." h can be chosen empirically. The GWAS parameters have been previously discussed [89], where the authors noted that the results are not sensitive to these parameters based on simulation studies.

9.5.5 Applications of the HMRFM in Gene Prioritization

Hou et al. [88] considered the changes of co-expression networks in Crohn's disease patients and controls, and how network dynamics reveals information on disease associations. Their results demonstrate that network rewiring is abundant in the immune system, and disease-associated genes are more likely to be rewired in patients. To integrate this network rewiring feature and GWAS signals, they proposed to use the HMRFM framework to integrate network information to prioritize genes. Hou et al. [88] applied the method to the Crohn's disease and Parkinson's disease (Fig. 9.5), and they found that there is an increase of replication rate in the prioritized set of genes (Fig. 9.5a, c). It is worth noting that the prioritized genes at a moderate cut-off can achieve similar replication rate as genes at a more stringent cut-off without prioritization, so that genes with moderate P-values can be recovered by the prioritization method, without sacrificing the replication rate. By contrasting the replication rate from the real data to the permutation results (Fig. 9.5b, d), it revealed that the permutation results were inferior to those from the real data even though differential expression was kept intact between the permutated and real data. This demonstrates that the improvement of replication rate was largely attributed to the appropriate modeling and incorporation of rewiring information, not to the overlapping information between rewiring and differential expression. As a conclusion, applications of the HMRFM in Crohn's disease and Parkinson's disease show that this framework leads to more replicable results and implicate potentially disease-associated pathways. For more details of the applications of the HMRFM, one can refer to works [88–91] and others therein.

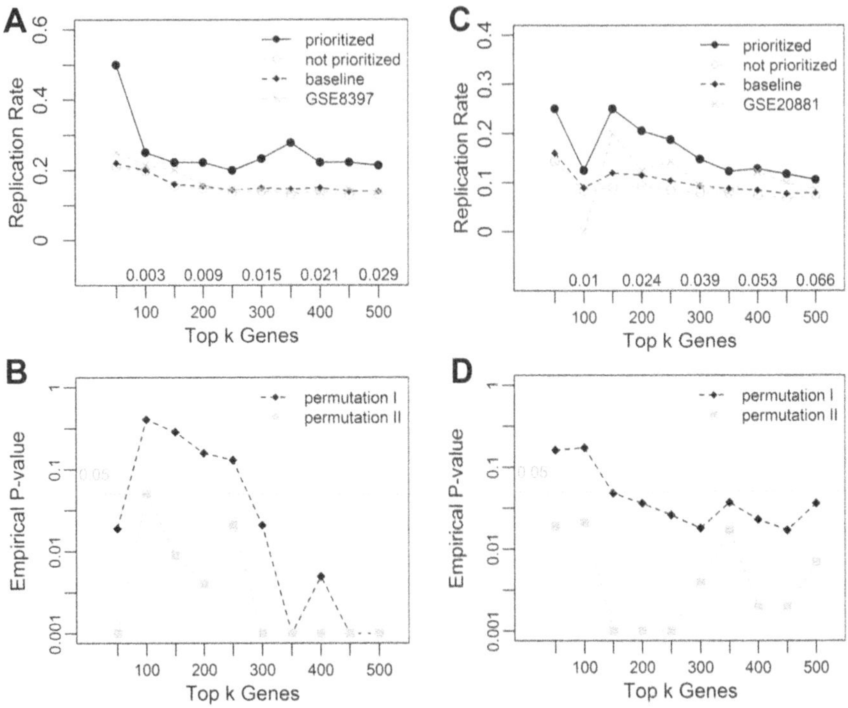

Fig. 9.5 Replication rates between independent cohorts in Crohn's disease (**a**), (**b**) and Parkinson's disease (**c**), (**d**) study. A gene is called replicable if its association P-value in the replication cohort is <0.05. (**a**) and (**c**). Black circles: prioritized genes. Gray circles: nonprioritized genes. Black diamonds: baseline replication rates for the top k genes. Gray crosses: prioritized genes with microarray dataset GSE20881 (a microarray dataset of Crohn's disease) and microarray dataset GSE8397 (a microarray data set of Parkinson's disease). The numbers on the top of x-axis are the association P-value of the kth genes in the discovery cohort. (**b**) and (**d**). Empirical P-values of the replication rate of the prioritized gene set derived from permutations. The dotted line indicates significance level at 0.05. The y-axis is drawn at the log_{10} scale but labeled with original P-values for ease of reading. Reproduced from Ref. [88], by permission of Oxford University Press

9.6 Discussions and Conclusions

With the accumulation of omics data, an arising challenge problem is how to effectively explore bioinformatics from these massive data [94–102]. In this chapter, we firstly introduce various HTS technologies and their applications, and then we introduce some popular methods to perform omics data analysis, including the WGCNA, GWAS, HMRF, and various penalized regression models. Among these methods, GCNs have wide applications.

Except the mentioned methods, one can also use some other variable selection methods from statistical theory. Variable selection plays an important role in high-dimensional data analysis. Marginal feature screening becomes indispensable for ultrahigh-dimensional data and has received much attention in the very recent literature. Various feature screening procedures have been proposed for linear models, generalized linear models, and robust linear models. Aforementioned model-based screening procedures perform well when the underlying models are correctly specified, but their performance may be quite poor in the presence of model mis-specification. Specifying a correct model for ultrahigh-dimensional data may be challenging. Thus, model-free sure screening procedures are appealing and have been developed by several authors [103–105].

References

1. Fan, L.: Bioinformatics, Hangzhou: Zhejiang University Press (2017)
2. Li, W., Li, J.: Modeling and analysis of RNA-seq data: a review from a statistical perspective. Quantitative Biol. **6**(3),195–209 (2018)
3. Mayer, B.: Bioinformatics for omics data. Humana Press (2011)
4. Kao, W.C.: Algorithms for next-generation high-throughput sequencing technologies. University of California at Berkeley (2011)
5. Roberts, G.G., Parrish, J.R., Mangiola, B.A., et al.: High-throughput *yeast* two-hybrid screening. Meth. Mol. Biol. (Clifton, N.J.) **812**, 39–61 (2012)
6. Trigg, S.A., Garza, R.M., Macwilliams, A., et al.: CrY2H-seq: a massively multiplexed assay for deep-coverage interactome mapping. Nat. Meth. **14**(8), 819–825 (2017)
7. Wang, P., Chen, Y., Lü, J., Wang, Q., Yu, X.: Graphical features of functional genes in human protein interaction network. IEEE Trans. Biomed. Circ. Syst. **10(3)**, 707–720 (2016)
8. Brown, K. R., Jurisica, I.: Online predicted human interaction database. Bioinformat. **21**, 2076–2082 (2005)
9. Peri, S., Navarro, J.D., Amanchy, R., et al.: Development of human protein reference database as an initial platform for approaching systems biology in humans. Genome Res. **13**, 2363–2371 (2003)
10. Stark, C., Breitkreutz, B.J., Reguly, T., et al.: BioGRID: a general repository for interaction datasets. Nucl. Acids Res. **34**, D535–D539 (2006)
11. Güldener, U., Münsterkötter, M., Oesterheld, M., et al.: MPact: the MIPS protein interaction resource on *yeast*. Nucl. Acids Res. **34**, D436–D441 (2006)
12. Bader, G.D., Hogue, C.W.: BIND–a data specification for storing and describing biomolecular interactions, molecular complexes and pathways. Bioinformat. **16**, 465–477 (2000)
13. Xenarios, I., Rice, D. W., Salwinski, L., et al.: DIP: the database of interacting proteins. Nucl. Acids Res. **28**, 289–291 (2000)
14. Zanzoni, A., Montecchi-Palazzi, L., Quondam, M., et al.: MINT: a molecular interaction database. FEBS Lett. **513**, 135–140 (2002)
15. Aranda, B., Achuthan, P., Alam-Faruque, Y., et al.: The IntAct molecular interaction database in 2010. Nucl. Acids Res. **38**, D525–D531 (2010)
16. Uetz, P., Giot, L., Cagney, G., et al.: A comprehensive analysis of protein-protein interactions in *Saccharomyces cerevisiae*. Nature **403**, 623–627 (2000)
17. Yu, H., Braun, P., Yıldırım, M.A., et al.: High-quality binary protein interaction map of the *Yeast* interactome network. Science **322**, 104–110 (2008)
18. Lander, E.S., Linton, L.M., Birren, B., et al., Int. Human Genome Seq. Consortium: Initial sequencing and analysis of the human genome. Nature **409**, 860–921 (2001)
19. Venter, J.C., Adams, M.D., Myers, E.W., et al.: The sequence of the human genome. Science **291**, 1304–1351 (2001)

20. Chinwalla, A., Cook, L., Delehaunty, K., et al.: Initial sequencing and comparative analysis of the *mouse* genome. Nature **420**, 520–562 (2002)
21. Gibbs, R.A., Weinstock, G.M., Metzker, M.L., et al., *Rat* genome sequencing project consortium: genome sequence of the Brown Norway *rat* yields insights into mammalian evolution. Nature **428**, 493–521 (2004)
22. Chimpanzee Sequencing and Analysis Consortium: Initial sequence of the *Chimpanzee* genome and comparison with the human genome. Nature **437**, 69–87 (2005)
23. Margulies, M., Egholm, M., Altman, W.E., et al.: Genome sequencing in microfabricated high-density picolitre reactors. Nature **437**, 376–380 (2005)
24. Reuter, J., Spacek, D.V., Snyder, M.: High-throughput sequencing technologies. Mol. Cell **58**(4), 586–597 (2015)
25. SEQC/MAQC-III Consortium: A comprehensive assessment of RNA-seq accuracy, reproducibility and information content by the sequencing quality control consortium. Nat. Biotechnol. **32**, 903–914 (2014)
26. Nagalakshmi, U., Wang, Z., Waern, K., et al.: The transcriptional landscape of the *yeast* genome defined by RNA sequencing. Science **320**, 1344–1349 (2008)
27. Core, L.J., Waterfall, J.J., Lis, J.T.: Nascent RNA sequencing reveals widespread pausing and divergent initiation at human promoters. Science **322**, 1845–1848 (2008)
28. Khodor, Y.L., Rodriguez, J., Abruzzi, K.C., et al.: Nascent-seq indicates widespread cotranscriptional premRNA splicing in *Drosophila*. Genes Dev. **25**, 2502–2512 (2011)
29. Churchman, L.S., Weissman, J.S.: Nascent transcript sequencing visualizes transcription at nucleotide resolution. Nature **469**, 368–373 (2011)
30. Ingolia, N.T., Ghaemmaghami, S., Newman, J.R.S., Weissman, J.S.: Genome-wide analysis in vivo of translation with nucleotide resolution using ribosome profiling. Science **324**, 218–223 (2009)
31. Hansen, R.S., Thomas, S., Sandstrom, R., et al.: Sequencing newly replicated DNA reveals widespread plasticity in human replication timing. Proc. Natl. Acad. Sci. USA. **107**, 139–144 (2010)
32. Lieberman-Aiden, E., van Berkum, N.L., Williams, L., et al.: Comprehensive mapping of long-range interactions reveals folding principles of the human genome. Science **326**, 289–293 (2009)
33. Fullwood, M.J., Liu, M.H., Pan, Y.F., et al.: An oestrogen-receptor-α-bound human chromatin interactome. Nature **461**(7269), 58–64 (2009)
34. Dostie, J., Richmond, T., Arnaout, R., et al.: Chromosome conformation capture carbon copy (5C): a massively parallel solution for mapping interactions between genomic elements. Genome Res. **16**, 1299–1309 (2006)
35. Chu, C., Qu, K., Zhong, F.L., Artandi, S.E., Chang, H.Y.: Genomic maps of long noncoding RNA occupancy reveal principles of RNA-chromatin interactions. Mol. Cell **44**, 667–678 (2011)
36. Meissner, A., Mikkelsen, T.S., Gu, H., et al.: Genome scale DNA methylation maps of pluripotent and differentiated cells. Nature **454**, 766–770 (2008)
37. Cokus, S.J., Feng, S., Zhang, X., et al.: Shotgun bisulphite sequencing of the *Arabidopsis* genome reveals DNA methylation patterning. Nature **452**, 215–219 (2008)
38. Crawford, G.E., Holt, I.E., Whittle, J., et al.: Genome-wide mapping of DNase hypersensitive sites using massively parallel signature sequencing (MPSS). Genome Res. 16, 123–131 (2006)
39. Buenrostro, J.D., Giresi, P.G., Zaba, L.C., Chang, H.Y., Greenleaf, W.J.: Transposition of native chromatin for fast and sensitive epigenomic profiling of open chromatin, DNA-binding proteins and nucleosome position. Nat. Meth. **10**, 1213–1218 (2013)
40. Kertesz, M., Wan, Y., Mazor, E., et al.: Genome-wide measurement of RNA secondary structure in *yeast*. Nature **467**, 103–107 (2010)
41. Ding, L., Ley, T.J., Larson, D.E., et al.: Clonal evolution in relapsed acute myeloid leukaemia revealed by whole-genome sequencing. Nature **481**, 506–510 (2012)

42. Buenrostro, J.D., Araya, C.L., Chircus, L.M., et al.: Quantitative analysis of RNA-protein interactions on a massively parallel array reveals biophysical and evolutionary landscapes. Nat. Biotechnol. **32**, 562–568 (2014)
43. Sephton, C.F., Cenik, C., Kucukural, A., et al.: Identification of neuronal RNA targets of TDP-43-containing ribonucleoprotein complexes. J. Biol. Chem. **286**, 1204–1215 (2011)
44. German, M.A., Pillay, M., Jeong, D.-H., et al.: Global identification of microRNA-target RNA pairs by parallel analysis of RNA ends. Nat. Biotechnol. **26**, 941–946 (2008)
45. Patwardhan, R.P., Hiatt, J.B., Witten, D.M., et al. Massively parallel functional dissection of mammalian enhancers in vivo. Nat. Biotechnol. **30**, 265–270 (2012)
46. Evans, C., Hardin, J., Stoebel, D.M.: Selecting between-sample RNA-seq normalization methods from the perspective of their assumptions. Brief. Bioinformat. **19**(5), 776–792 (2018)
47. Bullard, J.H., Purdom, E., Hansen, K.D., Dudoit, S.: Evaluation of statistical methods for normalization and differential expression in mRNA-seq experiments. BMC Bioinformat. **11**, 94 (2010)
48. Mortazavi, A., Williams, B.A., McCue, K., Schaeffer, L., Wold, B.: Mapping and quantifying mammalian transcriptomes by RNA-seq. Nat. Meth. **5**, 621–628 (2008)
49. Trapnell, C., Pachter, L., Salzberg, S.L.: Tophat: discovering splice junctions with RNA-seq. Bioinformat. **25**, 1105–1111 (2009)
50. Li, B., Dewey, C.N.: RSEM: accurate transcript quantification from RNA-seq data with or without a reference genome. BMC Bioinformat. **12**, 323 (2011)
51. Canzar, S., Andreotti, S., Weese, D., Reinert, K., Klau, G.W.: CIDANE: comprehensive isoform discovery and abundance estimation. Genome Biol. **17**, 16 (2016)
52. Stuart, J.M., Segal, E., Koller, D., Kim, S.K.: A gene coexpression network for global discovery of conserved genetic modules. Science **302**, 249–255 (2003)
53. Langfelder P., Horvath S.: WGCNA: an R package for weighted correlation network analysis. BMC Bioinformat. **9**, 559 (2008)
54. Song, L., Langfelder, P., Horvath, S.: Comparison of co-expression measures: mutual information, correlation, and model based indices. BMC Bioinformat. **13**, 328 (2012)
55. Zhang, B., Horvath, S.: A general framework for weighted gene co-expression network analysis. Stat. Appl. Genet. Mol. Biol. **4**, 17 (2005)
56. Ravasz, E., SomeraA. L., Mongru, D.A., Oltvai, Z.N., Barabàsi, A.L.: Hierarchical organization of modularity in metabolic networks. Science **297**, 1551–1555 (2002)
57. Oti, M., van Reeuwijk, J., Huynen, M.A. and Brunner, H.G.: Conserved co-expression for candidate disease gene prioritization. BMC Bioinformat. **9**, 208 (2008)
58. Segal, E., Shapira, M., Regev, A., Pe'er, D., Botstein, D., Koller, D., Friedman, N.: Module networks: identifying regulatory modules and their condition-specific regulators from gene expression data. Nat. Genet. **34**, 166–176 (2003)
59. The Wellcome Trust Case Control Consortium: Genome-wide association study of 14,000 cases of seven common diseases and 3000 shared controls. Nature **447**(7145), 661–678 (2007)
60. Bowcock, A.M.: Genomics: Guilt by association. Nature **447**(7145): 645–646 (2007)
61. Turnbull, C., Ahmed, S., Morrison, J., et al.: Genome-wide association study identifies five new breast cancer susceptibility loci. Nat. Genet. **42**(6), 504–507 (2010)
62. Corvin, A., Sanders, A.R., Maclean, A.W., et al.: Genome-wide association study identifies five new schizophrenia loci. Nat. Genet. **42**(1), 36–44 (2011)
63. Li, W.: Genome-wide association study. Encyclopedia Syst. Biol. 834–834 (2013)
64. Bühlmann P., van de Geer, S.: Statistics for high-dimensional data methods, theory and applications. Springer-Verlag Berlin Heidelberg (2011)
65. Liu, Z., Jiang, F., Tian, G., et al.: Sparse logistic regression with L_p penalty for biomarker identification. Stat. Appl. Genet. Mol. Biol. **6**(1),1–22 (2007)
66. Huang, J., Horowitz, J.L., Ma, S.: Asymptotic properties of bridge estimators in sparse high-dimensional regression models. Ann. Statist. **36**, 587–613 (2008)
67. Fu, W.J.: Penalized regressions: the bridge versus the lasso. J. Comput. Graph Statist. **7**, 397–416 (1998)

68. Tibshirani, R.: Regression shrinkage and selection via the lasso. J. Roy. Stat. Soc. Ser. B Stat. Methodol. **58**, 267–288 (1996)
69. Piepho, H.P.: Ridge regression and extensions for genomewide selection in *maize*. Crop Sci. **49**, 1165–1176 (2009)
70. Hastie, T.: Regularization and variable selection via the elastic net. J. Roy. Stat. Soc. Ser. B Stat. Methodol. **67**, 301–320 (2005)
71. Tibshirani, R.,Saunders, M., Rosset, S., et al.: Sparsity and smoothness via the fused lasso. J. Roy. Stat. Soc. Ser. B Stat. Methodol. **67**, 91–108 (2005)
72. Li, C., Li, H.: Network-constrained regularization and variable selection for analysis of genomic data. Bioinformat. **24**(21), 2566 (2008)
73. Yuan, M., Lin, Y.: Model selection and estimation in regression with grouped variables. J. Roy. Stat. Soc. Ser. B Stat. Methodol. **68**, 49–67 (2006)
74. Liu, C., Wong, H.S.: Structured penalized logistic regression for gene selection in gene expression data analysis. IEEE/ACM Trans. Comput. Biol. Bioinformat. **16**(1), 312–321 (2019)
75. Boyd, S., Parikh, N., Chu, E., et al.: Distributed optimization and statistical learning via the alternating direction method of multipliers. Found. Trend. Mach. Learn. **3**(1), 1–122 (2010)
76. Efron, B., Hastie, T., Tibshirani, J.R.: Least angle regression. Ann. Statist. **32**(2), 407–451 (2004)
77. Theodoridis, S.: Chapter 5—Stochastic gradient descent: the LMS algorithm and its family. Machine Learning—A Bayesian and Optimization Perspective, 161–231 (2015)
78. Milicic, N.: On a coordinate descent method. Math. Balkanica **7**, 257–263 (1977)
79. Zou, H.: The adaptive Lasso and its oracle properties. J. Indust. Manag. Optim. **101**(476), 1418–1429 (2006)
80. Chen, Z., Zhu, Y., Zhu, C.: Adaptive bridge estimation for high-dimensional regression models. J. Inequal. Appl. **2016**(1), 258 (2016)
81. Zou, H., Zhang, H.H.: On the adaptive elastic-net with a diverging number of parameters. Ann. Statist. **37**, 1733–1751 (2009)
82. Detmer, F.J., Slawski, M.: A note on coding and standardization of categorical variables in (sparse) group Lasso regression. arXiv, 1805.06915 (2018)
83. Huang, J., Shuange, M.A., Zhang, X.H.: A group bridge approach for variable selection. Biometrika **96**(2), 339–355 (2009)
84. Jiang, D., Huang, J., Zhang, Y.: The cross-validated AUC for MCP-logistic regression with high-dimensional data. Statist. Meth. Med. Res. **22**(5), 505–518 (2013)
85. Breheny, P., Huang, J.: Coordinate descent algorithms for nonconvex penalized regression, with applications to biological feature selection. Ann. Appl. Stat. **5**, 232–253 (2011)
86. Huang, J., Breheny, P., Ma, S.: A selective review of group selection in high dimensional models. Statist. Sci. **27**, 481–499 (2012)
87. Fan, J., Li. R.: Variable selection via nonconcave penalized likelihood and its oracle Properties. J. Amer. Statist. Assoc. **96**, 1348–1360 (2001)
88. Hou, L., Chen, M., Zhang, C.K., et al.: Guilt by rewiring: gene prioritization through network rewiring in genome wide association studies. Human Mol. Genet. **23**(10), 2780–2790 (2014)
89. Chen, M., Cho, J., Zhao, H.: Incorporating biological pathways via a Markov random field model in genome-wide association studies. PLoS Genet. **7**, e1001353 (2011)
90. Wei, Z., Li, H.: A Markov random field model for network-based analysis of genomic data. Bioinformat. **23**, 1537–1544 (2007)
91. Li, H., Wei, Z., Maris, J.M.: A hidden Markov random field model for genome-wide association studies. Biostat. **11**,139–150 (2010)
92. Hudson, N.J., Reverter, A., Dalrymple, B.P.: A differential wiring analysis of expression data correctly identifies the gene containing the causal mutation. PLoS Comput. Biol. **5**, e1000382 (2009)
93. Hu, R., Qiu, X., Glazko, G., Klebanov, L., Yakovlev, A.: Detecting intergene correlation changes in microarray analysis: a new approach to gene selection. BMC Bioinformat. **10**, 20 (2009)

94. Wang, Z, Yang, C, Chen, H, Wang, P, et al.: Multi-gene co-transformation can improve comprehensive resistance to abiotic stresses in *B. napus* L., *Plant Sci.*, **274**: 410–419 (2018)
95. Wang, P, Yang, C, Chen, H, et al.: Exploring transcriptional factors reveals crucial members and regulatory networks involved in different abiotic stresses in *Brassica napus* L., *BMC Plant Biol.*, **18**: 202 (2018)
96. Wang, P, Yang, C, Chen, H, et al.: Transcriptomic basis for drought-resistance in *Brassica napus* L., *Sci. Rep.*, **7**: 40532 (2017)
97. Shang, B, Zang, Y, Zhao, X, Zhu, J, et al.: Functional characterization of GhPHOT2 in chloroplast avoidance of *Gossypium hirsutum*, *Plant Physiol. Bioch.*, **135**: 51–60 (2019)
98. Qu, X, Cao, B, Kang, J, Wang, X, et al.: Fine-tuning stomatal movement through small signaling peptides, *Front Plant Sci.*, **10**: 69 (2019)
99. Wang, D, Yang, C, Dong, L, et al.: Comparative transcriptome analyses of drought-resistant and -susceptible *Brassica napus* L. and development of EST-SSR markers by RNA-Seq, *J. Plant Biol.*, **58**: 259–269 (2015)
100. Zhang, S, Li, X, Pan, J, et al.: Use of comparative transcriptome analysis to identify candidate genes related to albinism in *channel catfish* (*Ictalurus punctatus*). *Aquaculture*, **500**: 75–81 (2018)
101. Dong, W, Li, M., Li, Z., et al.: Transcriptome analysis of the molecular mechanism of *Chrysanthemum* flower color change under short-day photoperiods. *Plant Physiol. Bioch.*, **146**: 315–328 (2020)
102. Zhang, G., Yue, C., Lu, T., et al.: Genome-wide identification and expression analysis of NADPH oxidase genes in response to ABA and abiotic stresses, and in fibre formation in *Gossypium*, *PeerJ*, **8**: e8404 (2020)
103. Zhu, L.P., Li, L., Li, R., Zhu, L.X.: Model-free feature screening for ultrahigh dimensional data. J. Amer. Statist. Assoc. **106**, 1464–1475 (2011)
104. He, X., Wang, L., Hong, H.: Quantile-adaptive model-free variable screening for high-dimensional heterogeneous data. Ann. Statist. **41**, 342–369 (2013)
105. Cui, H., Li, R., Zhong, W.: Model-free feature screening for ultrahigh dimensional discriminant analysis. J. Amer. Stat. Assoc. **110**(510), 630–641 (2015)

Glossary

Bifurcation analysis: A dynamical analysis method that reveals how the states of a system rely on its parameters.

Bioinformatics: The terms bioinformatics and computational biology are often used interchangeably. However, bioinformatics more properly refers to the creation and advancement of algorithms, computational and statistical techniques, and theory to solve formal and practical problems inspired from the management and analysis of biological data. Computational biology, on the contrary, refers to hypothesis-driven investigation of a specific biological problem using computers, carried out with experimental or simulated data, with the primary goal of discovery and the advancement of biological knowledge. In other words, bioinformatics is concerned with the information, while computational biology is concerned with the hypotheses.

Central dogma: An explanation of the flow of genetic information within a biological system. It was first stated by Francis Crick in 1958. The central dogma of molecular biology deals with the detailed residue-by-residue transfer of sequential information. It states that such information cannot be transferred back from protein to either protein or nucleic acid. The central dogma has also been described as "DNA makes RNA and RNA makes protein."

Chemical master equation: It is used to describe the time evolution of a system that can be modeled as being in exactly one of the states at any given time, and where switching between states is treated probabilistically. The equations are usually a set of differential equations for the variation over time of the probabilities that the system occupies each of the different states.

Chemical Langevin equation: A stochastic differential equation describing the time evolution of a subset of the degrees of freedom. These degrees of freedom typically are collective (macroscopic) variables changing only slowly in comparison to the other (microscopic) variables of the system. The fast (microscopic) variables are responsible for the stochastic nature of the Langevin equation.

J. Lü, P. Wang, *Modeling and Analysis of Bio-molecular Networks*,
https://doi.org/10.1007/978-981-15-9144-0

Complex network: A graph consists of nodes and edges. Nodes correspond to individuals in a system, and edges represent the interaction among individuals. A complex network can be undirected or directed, weighted or unweighted, which depends on the property of the considered system.

Conserved genes: Genes that are evolutionary conserved in various species.

Disease genes: Genes that are related to disease phenotypes.

Essential genes: Genes that are essential or lethal for the survival of an organism. They can cause lethal phenotypes in gene knockout or mutation experiments.

Feed-forward loop: A typical network motif. Three-node FFLs consist of three nodes and three edges. Denote the three nodes as X, Y, Z, the three edges are $X \rightarrow Y$, $Y \rightarrow Z$, and $X \rightarrow Z$. In a genetic regulatory network, each edge can be repression or activation. Therefore, there are totally eight configurations. If the two routes from X to Z have the same sign, then such loops are called coherent FFLs; otherwise, they are called incoherent ones.

Gene ontology: A major bioinformatics initiative to unify the representation of gene and gene product attributes across all species. More specifically, the project aims to: (1) maintain and develop its controlled vocabulary of gene and gene product attributes; (2) annotate genes and gene products, and assimilate and disseminate annotation data; and (3) provide tools for easy access to all aspects of the data provided by the project, and to enable functional interpretation of experimental data using the GO, for example, via enrichment analysis.

Gene regulatory network: A typical directed biological network, where nodes represent genes and edges denote the interactions among genes. An edge can be repression or activation, which depends on whether the expression of a gene can repress or activate the targeted gene.

Hidden Markov random field model: HMRFM is derived from hidden Markov models (HMMs), which are defined as stochastic processes generated by a Markov chain whose state sequence cannot be observed directly, only through a sequence of observations. Each observation is assumed to be a stochastic function of the state sequence. The underlying Markov chain changes its state according to an $L \times L$ transition probability matrix, where L is the number of states.

Housekeeping genes: Genes that are required for maintaining basic cellular function; they are expressed in various tissues of an organism under normal and patho-physiological conditions.

Law of mass action: A mathematical model that explains and predicts behaviors of solutions in dynamic equilibrium. Simply put, it states that the rate of a chemical reaction is proportional to the product of the masses of the reactants. Necessarily, this implies that for a chemical reaction mixture that is in equilibrium, the ratio between the concentration of reactants and products is constant.

Network medicine: The application of network science toward identifying, preventing, and treating diseases. This field focuses on using network topology and network dynamics toward identifying diseases and developing medical drugs. Biological networks, such as PPIs and metabolic pathways, are utilized by network medicine. Disease networks, which map relationships between diseases and biological factors, also play an important role in the field. Epidemiology is extensively studied using network science as well; social networks and transportation networks are used to model the spreading of disease across populations.

Network motifs: Simple patterns that more frequently appeared in real-world networks than those in its randomized counterparts.

Omics: Omics include genomics, proteinomics, metabolomics, transcriptomics, lipidomics, immunomics, glycomics, RNomics, radiomics, ultrasomics, and so on. The root of the word "Omics" is "-ome." "-ome" refers to an ensemble of some kinds of individuals.

Personalized medicine: A medical model that separates patients into different groups—with medical decisions, practices, interventions, and/or products being tailored to the individual patient based on their predicted response or risk of disease. The terms personalized medicine, precision medicine, stratified medicine, and P4 medicine are used interchangeably to describe this concept though some authors and organizations use these expressions separately to indicate particular nuances.

Principal component analysis: A multivariate statistical method that transforms a set of dependent variables into independent integrative ones. It can be used in multivariate conditions for the purpose of simplifying data structure.

Protein–protein interaction network: A typical undirected bio-molecular network, where nodes represent proteins and edges denote the interactions among proteins.

Reaction rate equation: An equation that links the reaction rate with concentrations or pressures of reactants and constant parameters (normally rate coefficients and partial reaction orders).

ROC analysis: Receiver operating characteristic (ROC) is a graphical plot that illustrates the performance of a binary classifier system as its discrimination threshold is varied. The curve is created by plotting the true positive rate (TPR) against the false positive rate (FPR) at various threshold settings.

Scale-free network: A network with power-law degree distribution. A few nodes are with large degrees, while the degrees of the remaining large amount of nodes are very low.

Small-world network: A network encompasses high clustering coefficient and short average path length.

Systems biology: An emerging interdisciplinary scientific area that endeavors to clarify the components and the relationships of all the bio-molecules in life systems.

The largest connected component: A subnetwork that contains the most connected nodes of a complex network.

Tissue-enriched genes: Genes that only express in some specific tissues.

Transcription: The first step of gene expression, in which a particular segment of DNA is copied into RNA (mRNA) by the enzyme RNA polymerase.

Translation: The process by which a mature mRNA molecule is used as a template for synthesizing a new protein. Translation is carried out by ribosomes, large complexes of RNA and protein responsible for carrying out the chemical reactions to add new amino acids to a growing polypeptide chain by the formation of peptide bonds.

Viable genes: Genes that are not lethal for the survival of an organism. They have no lethal effect in gene knockout or mutation experiments.

The manufacturer's authorised representative in the EU is Springer Nature Customer Service Centre GmbH, Europaplatz 3, 69115 Heidelberg, Germany. If you have any concerns regarding our products, please contact ProductSafety@springernature.com

Printed and bound by CPI Group (UK) Ltd, Croydon, CR0 4YY
30/09/2026
02225291-0018